Statistical and Condensed Matter Physics

Serguei N. Burmistrov

Statistical and Condensed Matter Physics

Springer

Serguei N. Burmistrov
Kurchatov Institute
Moscow, Russia

ISBN 978-981-97-9074-6 ISBN 978-981-97-9072-2 (eBook)
https://doi.org/10.1007/978-981-97-9072-2

This Springer imprint is published by the registered company Springer Nature Singapore Pte Ltd.
The registered company address is: 152 Beach Road, #21-01/04 Gateway East, Singapore 189721, Singapore

For students studying the statistical and condensed matter physics

Preface to the English Edition

The Russian edition of this book came to light in 2023. The present English edition is augmented with a number of problems. The misprints revealed in the Russian edition are corrected as well.

Using this opportunity, the author would like to express his gratitude to all his colleagues at the Department of Theoretical Physics in Moscow Institute of Physics and Technology. In addition, I am also grateful to Prof. V. P. Krainov and Prof. I. Ya. Polishchuk for the fruitful discussions of the questions concerning the book.

Moscow, Russia
July 2024

Serguei N. Burmistrov

Preface

The subject of statistical physics is to study the physical properties and behavior of various macroscopic systems and condensed media under the thermodynamic (heat) equilibrium. The thermodynamic or heat equilibrium implies that the physical system can wholly be described by the canonical Gibbs distribution. The main difficulty that arises here is to calculate the partition function and corresponding thermodynamic potential.

At present, statistical physics is the basis of condensed matter physics and the tool for studying a variety of condensed media. The condensed medium or macroscopic system is usually understood as a physical system consisting of the macroscopic number of particles and having the macroscopic number of degrees of freedom. In essence, the term, macroscopic, implies implicitly the limiting transition to the infinite system having, for example, an infinite number of particles or volume. At the same time, the physical system in itself can also represent a single particle but interacting with the macroscopic thermal bath.

As a typical example of condensed media, we can mention gases, normal and superfluid liquids, crystalline and amorphous solids, superconductors, and various magnets.

The starting point in the study of condensed matter is the introduction and determination of the necessary thermodynamic quantities depending on the type of the physical system to be explored. For example, this may be temperature, volume, pressure, polarization, magnetization, and thermodynamic potentials as functions of these quantities, e.g. energy, free energy, and entropy.

The statistical physics is one of three main courses of theoretical physics which students of Moscow Institute of Physics and Technology study. Over the past 50 years, statistical physics and condensed matter physics have developed rapidly in the scientific context and have achieved outstanding success due, in particular, to the development and usage of specific models and mathematical methods. The latter often turns out to be very complicated, requires cumbersome calculations, and may simply be inaccessible without special mathematical training. However, these successes of statistical physics and condensed matter physics have not yet found an accessible and complete reflection in the textbooks proposed for the students of physical specialties. Apparently, this may be associated with the following reasons. Firstly, it is necessary to convey the material to students, using the simplest mathematical apparatus. Secondly, it is the limited time frame that is assigned to studying the subject. Currently, Moscow Institute of Physics and Technology, in contrast to the annual course of quantum mechanics, has a single semester for studying the statistical physics course. This is obviously insufficient due to the extent of the subject of statistical and condensed matter physics. Thus, it is necessary to enhance the requirements for selecting the instructional content that students of physical universities should study under single semester course.

The practice of teaching the statistical physics course has led me to the following. The content that relates to the principles, basic statements of statistical physics, and properties of classical and quantum ideal gases poses no difficulties for the MIPT students. The learning problems will concern a noticeable amount of students when they start to study the physical properties of non-ideal quantum systems and condensed media since this requires a comprehension of more complicated physical models and application of more sophisticated mathematical tools.

The purpose of the present textbook is to help the students, future physicists, to master the basic elements of statistical physics and learn how to apply its methods in practice. For understanding the content of the book, it is sufficient to be familiar with the basic concepts of statistical physics in the body of conventional general physics courses.

When writing the book, the author came across some difficulties associated with the reasonable text volume and, therefore, with the necessity to select the actual content from the numerous number of interesting physical phenomena in condensed media. Here the author's personal preferences and work experience in this field of physics are displayed as well as his viewpoint of statistical physics as a necessary part of the full course of theoretical physics.

The structure of the book proposed to the reader is easy to understand from its contents. In a few words, the author can say that the book is devoted both to the fundamental laws of statistical physics and to most interesting physical phenomena occurring in condensed media.

The book begins with explaining the basic principles of statistical physics based on the canonical Gibbs distribution and their connection with the laws of classical thermodynamics. Next, the thermodynamic properties are considered of non-interacting media representing classical (Boltzmann) ideal gases and quantum Fermi and Bose ideal gases. This section concludes with the thermal properties of black radiation and solid crystals.

The next section is devoted to the thermodynamic fluctuations, various phase transitions in the mean-field approximation, and basics of critical phenomena with the approximate calculation of critical indices.

The following sections deal with the properties and description of non-ideal quantum systems. First of all, these are normal Fermi liquid and non-ideal Fermi gas with the dipole interparticle interaction as an example of an anisotropic Fermi system. Thereafter the theory of superconductivity and its main manifestations are given in the framework of the BCS model by deriving the Ginzburg–Landau functional. The phenomenon of superconductivity is one of most striking examples of macroscopic quantum phenomena. One more fascinating example of macroscopic quantum phenomena is the Bose–Einstein condensation in a weakly ideal Bose gas, whose properties are interpreted in the framework of the Gross–Pitaevskii equation. Considerable attention is also paid to the theory of superfluidity in liquid helium by deriving the equations of two-fluid hydrodynamics.

Finally, the last section of the book deals both the fundamentals of magnetism and with the thermodynamic properties for the major types of magnetic structures. Main attention is paid to ferromagnets, antiferromagnets, and canted antiferromagnets with the Dzyaloshinskii–Moriya interaction.

As a rule, the introduction of some basic theoretical statement or question concludes with an illustrative example of one or more problems. Each of them is a useful addition for clarifying the material of the section or question considered. As for the readers, this is also an opportunity to master their practical skills and the comprehension of the section read. This book can be used as a supplement to the textbooks published on statistical physics and condensed matter physics.

Unfortunately, a complete bibliographic review of books on statistical and condensed matter physics would go far beyond our scope. For further details, a small list is given of the books and textbooks where the questions and problems raised in our book are discussed in more detail. Undoubtedly, this list is far from being exhaustive.

The author received a number of useful comments from Prof. V. P. Krainov, in particular, to the problem on the difference in the heat capacities of para- and orthohydrogen.

In conclusion, the author expresses his sincere gratitude to the entire team of teachers at the Theoretical Physics Department of Moscow Institute of Physics and Technology.

Moscow, Russia Serguei N. Burmistrov

About This Book

The book outlines the fundamentals of statistical and condensed matter physics. The selection of the material is mainly determined by the personal experience and preferences of the author, taking into account the program of Statistical Physics as a basic course of Theoretical Physics at Moscow Institute of Physics and Technology.

The book is intended for students and postgraduates studying statistical physics as one of the three main courses of theoretical physics: field theory, quantum mechanics, and statistical physics. To understand the subject matter, it is sufficient to know the traditional courses of General Physics at physical universities. The book can be used as a supplement to the textbooks published on statistical physics and condensed matter physics.

Contents

Notation

Operators are denoted by circumflex as ^
Mean values are denoted by angle brackets as $\langle \cdots \rangle$
Phase space
p, q the generalized momentum and coordinate
$d\mathbf{p}d\mathbf{q} = dp_1 \dots dp_d dq_1 \dots dq_d$ volume element in phase space of dimension d
$d\Gamma = d\boldsymbol{p}d\boldsymbol{q}/(2\pi\hbar)^d$
$\Gamma(\varepsilon)$ Number of states
$g(\varepsilon)$ Density of states
$\hat{w}$ Density matrix
Thermodynamic variables
T Temperature
$\beta = 1/T$ Inverse temperature
In all the formulas, temperature is expressed in energy units
Z Canonical partition function
E Energy
S Entropy
$F = E - TS$ Helmholtz free energy or free energy
P Pressure
V Volume
$C(T)$ Specific heat
$\widetilde{F}$ or $\Phi = E + PV$ Conjugate free energy or Gibbs free energy
$\mathcal{Z}$ Grand partition function
N Number of particles
μ Chemical potential
$\Omega = F - \mu N$ Grand thermodynamic potential
$\varphi(\boldsymbol{r})$ Order parameter
Gi The Ginzburg–Levanyuk number
$\boldsymbol{B}$ Magnetic induction
$\boldsymbol{h}$, $\boldsymbol{H}$ Magnetic field strength
$\boldsymbol{M}$ Magnetization
χ Magnetic susceptibility

ϕ_0 Magnetic flux quantum
$\hat{\boldsymbol{S}}_a$ Operator of spin at point $\boldsymbol{a}$
Θ_N Néel temperature
J Exchange coupling constant

Chapter 1
Main Principles of Statistical Physics

1.1 The Canonical Gibbs Distribution

According to the principles of quantum mechanics, the physical system, described by certain Hamiltonian $\hat{H}$, has a set of eigenstates denoted by vector $|k\rangle$. The vector $|k\rangle$ is an eigenfunction of the Hamiltonian

$$\hat{H}|k\rangle = \varepsilon_k|k\rangle$$

where ε_k is the corresponding eigenvalue of energy or the energy level of the system.

The statistical specification of the physical system implies its probabilistic description by introducing some distribution function $w_k = w(|k\rangle)$. The latter determines the probability to find the system in the state with vector $|k\rangle$. When we adhere to the mathematical positions alone, the choice of distribution functions w_k is very wide. From the physical point of view, the choice of the distribution function as the main principle of the theory should result in a set of physical consequences and conclusions fully consistent with the available experimental data and have predictive powers as well.

First of all, statistical physics studies the physical systems whose states can be specified by the *canonical Gibbs distribution*. The state when the physical system is described with the Gibbs distribution will be called the *thermodynamically equilibrium state*. In the thermodynamically equilibrium state, the probability to find the system with the state vector $|k\rangle$ depends on the state energy ε_k alone and is given by the *canonical Gibbs distribution*, as follows:

$$w_k = w(\varepsilon_k) = \frac{1}{Z}\exp\left(-\frac{\varepsilon_k}{T}\right) = \frac{1}{Z}\exp(-\beta\varepsilon_k).$$

S. N. Burmistrov, *Statistical and Condensed Matter Physics*,
https://doi.org/10.1007/978-981-97-9072-2_1

The parameter Z, determined by the normalization condition $\sum_k w_k = 1$, is equal to the statistical sum[1] over all the possible states weighed with the Gibbs exponential

$$Z = \sum_k \exp(-\varepsilon_k/T),$$

and is referred to as the *canonical partition function.*

According to this distribution, such a thermodynamically equilibrium system does not have the strictly fixed energy and thus is not the energetically closed system. We underline that each state in this sum is taken only *once*. The external non-negative parameter $T \geqslant 0$ is called the *absolute temperature* or briefly *temperature*[2] and, correspondingly, $\beta = 1/T$ is the inverse temperature.

The condition of non-negativity for temperature $T \geqslant 0$ means that in the thermodynamically equilibrium system, the states of higher energy become less probable at the fixed temperature. It is easy to see that the degenerate states of the same energy have equal probabilities in spite of various sets of quantum numbers. At zero $T = 0$ temperature, the physical system will be in the minimum energy or ground state which, in general, may be degenerate.

If the physical system under consideration consists of two independent and non-interacting subsystems, i.e. the total Hamiltonian $\hat{H} = \hat{H}_1 + \hat{H}_2$ is a sum of two Hamiltonians $\hat{H}_1$ and $\hat{H}_2$ and, respectively,

$$|k\rangle = |k_1\rangle \otimes |k_2\rangle \quad \text{and} \quad \varepsilon_k = \varepsilon_{k_1} + \varepsilon_{k_2} \,,$$

the Gibbs distribution represents the *simplest* and, in essence, single one satisfying the probability product condition for the statistically independent random variables, i.e.

$$w(\varepsilon_{k_1} + \varepsilon_{k_2}) = w(\varepsilon_{k_1})w(\varepsilon_{k_2}).$$

If two physical systems have the same temperature, i.e. $T_1 = T_2$, it is customary to say that these two systems are in the *thermodynamic* or *thermal equilibrium* with each other. Note the *transitivity property* of thermal equilibrium. In fact, when the first and second systems are separately in the thermal equilibrium with the third system, i.e. $T_1 = T_3$ and $T_2 = T_3$, then we find that the first and second systems are in thermal equilibrium with each other, i.e. $T_1 = T_2$.

The canonical partition function Z can be written in the operator form independent of the energetic representation chosen as

$$Z = \sum_k e^{-\varepsilon_k/T} = \sum_k \langle k|e^{-\varepsilon_k/T}|k\rangle = \sum_k \langle k|e^{-\hat{H}/T}|k\rangle = \operatorname{tr} e^{-\hat{H}/T}.$$

---

[1] The sign of the sum may also imply an integral which can be of the infinite dimensionality if the states of the system are specified, e.g. by the infinite set of continuous functions.

[2] The temperature is usually expressed in Kelvin degrees (K) instead of energy units. The conversion factor is called the *Boltzmann constant* k_B and equals $k_B = 1.38 \cdot 10^{-16}$ erg/K.

Accordingly, the *statistical operator* or *density matrix* is given by

$$\hat{w} = \frac{1}{Z} e^{-\hat{H}/T} = \frac{e^{-\hat{H}/T}}{\operatorname{tr} e^{-\hat{H}/T}}.$$

The *number of states* $\Gamma(\varepsilon)$ is introduced as the number of states with the energies not exceeding the given one

$$\Gamma(\varepsilon) = \sum_k \vartheta(\varepsilon - \varepsilon_k).$$

Here $\vartheta(\varepsilon)$ is the Heaviside step function.[3] The *density of states* $g(\varepsilon)$ is determined as a derivative of the number of states $\Gamma(\varepsilon)$

$$g(\varepsilon) = \frac{d\Gamma(\varepsilon)}{d\varepsilon} = \sum_k \delta(\varepsilon - \varepsilon_k).$$

The knowledge of the density of states is convenient since the partition function can be written as a single integral over energy

$$Z = \int g(\varepsilon) e^{-\varepsilon/T} d\varepsilon.$$

Problem

1. Express the elements of density matrix $\hat{w}$ for spin $S = 1/2$ in terms of the spin projections S_x, S_y, and S_z onto axes x, y, and z.

Solution. Let us write the density matrix in the general form as

$$\hat{w} = \begin{pmatrix} w_{11} & w_{12} \\ w_{21} & w_{22} \end{pmatrix}.$$

The magnitude of the average spin is determined by the trace

$$\boldsymbol{S} = \operatorname{tr}(\hat{w}\hat{\boldsymbol{S}}), \quad \hat{\boldsymbol{S}} = \hat{\boldsymbol{\sigma}}/2, \quad \hat{\boldsymbol{\sigma}} = (\hat{\sigma}_x, \hat{\sigma}_y, \hat{\sigma}_z)$$

where $\hat{\sigma}_x$, $\hat{\sigma}_y$, and $\hat{\sigma}_z$ are the Pauli matrices. Multiplying the matrices and calculating their traces, we arrive at the following relations:

$$S_x = (w_{12} + w_{21})/2,$$
$$S_y = i(w_{12} - w_{21})/2,$$
$$S_z = (w_{11} - w_{22}/2.$$

Then we augment them with the normalization condition $\operatorname{tr} \hat{w} = 1$ for the density matrix

$$w_{11} + w_{22} = 1$$

[3] $\vartheta(\varepsilon) = 1$ if $\varepsilon \geqslant 0$ and $\vartheta(\varepsilon) = 0$ if $\varepsilon < 0$.

and find the final answer from the above four equations

$$\hat{w} = \begin{pmatrix} \frac{1}{2} + S_z & S_x - iS_y \\ S_x + iS_y & \frac{1}{2} - S_z \end{pmatrix} = \frac{1}{2}\hat{1} + S_x\hat{\sigma}_x + S_y\hat{\sigma}_y + S_z\hat{\sigma}_z \, .$$

1.2 The Relation Between Statistical Physics and Classical Thermodynamics

To find the relation of statistical physics with classical phenomenological thermodynamics based on the longstanding experimental examinations, we define the *internal energy* E (or briefly *energy*) of the physical system as a usual average of statistical quantity governed by the Gibbs distribution with probability w_k

$$E = \langle \varepsilon \rangle = \sum_k \varepsilon_k w(\varepsilon_k) = \mathrm{tr}\,(\hat{H}\hat{w}).$$

Involving $\beta = 1/T$, we see the following relations: for the energy

$$E = -\frac{\partial \ln Z}{\partial \beta}$$

and for the mean square of energy fluctuations or dispersion

$$\langle (\Delta E)^2 \rangle = \langle (\varepsilon - \langle \varepsilon \rangle)^2 \rangle = \frac{\partial^2 \ln Z}{\partial \beta^2} \, .$$

The presence of energy fluctuations in the thermodynamic system results immediately from the fact that such a system is not energetically closed and its energy is not a fixed quantity. At zero temperature, the system stays only in the minimum energy state, and we must expect that the mean square of energy fluctuations vanishes at $T = 0$, i.e. $\langle (\Delta E)^2 \rangle_{T=0} = 0$.

The *specific heat* $C(T)$ of the system is determined as a derivative of energy with respect to the temperature

$$C(T) = \frac{\partial E}{\partial T} = \beta^2 \frac{\partial^2 \ln Z}{\partial \beta^2} = \frac{\langle (\Delta E)^2 \rangle}{T^2} \, .$$

We can see from the last equality that the thermodynamically equilibrium system should have the *positive* magnitude of specific heat $C(T) > 0$. The inequality $\partial E/\partial T > 0$ corresponds to the intuitive feeling that the energy of thermodynamically equilibrium system grows as its temperature increases.

The *Helmholtz free energy* $F(T)$ or for brevity, *free energy*, is introduced as follows:

$$F(T) = -T \ln Z.$$

Thus the canonical Gibbs distribution can be represented as

$$w_k = \exp\left(\frac{F - \varepsilon_k}{T}\right).$$

To obtain the well-known thermodynamic relations

$$F = E - TS \quad \text{and} \quad S = -\frac{\partial F}{\partial T}$$

established by the phenomenological thermodynamics between energy E, free energy F, and entropy S, we should determine the entropy[4] as an average value of logarithm of distribution function w_n with the minus sign

$$S = -\langle \ln w \rangle = -\sum_k w_k \ln w_k = -\mathrm{tr}\,(\hat{w} \ln \hat{w}).$$

In fact, we have

$$-\frac{F - E}{T} = \left(\ln \sum_k e^{\;\varepsilon_k/T} + \frac{1}{T} \frac{\sum_k \varepsilon_k e^{-\varepsilon_k/T}}{\sum_k e^{-\varepsilon_k/T}} \right) =$$

$$= \frac{\partial}{\partial T}\left(T \ln \sum_k e^{-\varepsilon_k/T} \right) = -\frac{\partial}{\partial T}(-T \ln Z) = -\frac{\partial F}{\partial T}.$$

On the other hand, using condition $\sum_k w_k = 1$, we obtain

$$-\frac{F - E}{T} = -\left(\sum_k w_k \ln \frac{1}{Z} + \sum_k w_k \ln e^{-\varepsilon_k/T} \right) =$$

$$= \sum_k w_k \ln \frac{e^{-\varepsilon_k/T}}{Z} = -\sum_k w_k \ln w_k\,.$$

Comparing these two equalities, we see the complete equivalence for the *thermodynamical* and *statistical* definitions of entropies in the thermodynamically equilibrium system.

[4] The so-defined entropy is often referred to as the Gibbs–Shannon entropy. Formula $S = -\mathrm{tr}\,(\hat{w} \ln \hat{w})$ is also called the definition of the von Neumann entropy. In the narrow sense, the Shannon or von Neumann entropy implies base 2 for logarithm $S = -\mathrm{tr}\,(\hat{w} \log_2 \hat{w})$.

The entropy is the *additive*[5] *quantity*. So, let the system with energy $E = E_1 + E_2$ consist of two non-interacting and thermodynamically equilibrium subsystems with energies E_1 and E_2, respectively. Then, $S(E_1 + E_2) = S(E_1) + S(E_2)$. In fact, for the non-interacting and statistically independent systems, we have that the distribution function of the total unified system represents a product of distribution functions of the separate systems: $w(\varepsilon_{k_1} + \varepsilon_{k_2}) = w(\varepsilon_{k_1})w(\varepsilon_{k_2})$. Then we find

$$
\begin{aligned}
S(E_1 + E_2) &= -\sum_{k_1,k_2} w(\varepsilon_{k_1})w(\varepsilon_{k_2}) \ln\big[w(\varepsilon_{k_1})w(\varepsilon_{k_2})\big] = \\
&= -\sum_{k_1,k_2} w(\varepsilon_{k_1})w(\varepsilon_{k_2}) \ln w(\varepsilon_{k_1}) - \sum_{k_1,k_2} w(\varepsilon_{k_1})w(\varepsilon_{k_2}) \ln w(\varepsilon_{k_2}) = \\
&= -\sum_{k_1} w(\varepsilon_{k_1}) \ln w(\varepsilon_{k_1}) - \sum_{k_2} w(\varepsilon_{k_2}) \ln w(\varepsilon_{k_2}) = S(E_1) + S(E_2).
\end{aligned}
$$

Provided that the physical system has only N states with the identical energy values $\varepsilon_1, \varepsilon_2, \ldots \varepsilon_N = \varepsilon$ and, therefore, has the equal probabilities $w_1, w_2, \ldots w_N = 1/N$ of filling the states, the entropy of the system will be equal to $S = \ln N$.

With the aid of the density of states $g(\varepsilon)$, the expression for entropy can be represented as follows:

$$
S = \int d\varepsilon\, g(\varepsilon) w(\varepsilon) \ln \frac{1}{w(\varepsilon)}\,.
$$

The relation $S = -\partial F/\partial T$ between free energy and entropy means that the total differential of free energy equals

$$
dF(T) = -S\,dT.
$$

To find the total differential of energy E, we employ relation $E = F + TS$, entailing the obvious answer

$$
dE = d(F + TS) = -S\,dT + T dS + SdT = T\,dS.
$$

The thermodynamic quantity $\delta Q = T\,dS$, arising here, represents the *amount of heat* transferred to the system as a result of the process called the *heat exchange*.

The energy $E = E(S)$ of the system is a function of entropy, and the following relation can be written: for the inverse temperature

[5] Note that there are non-additive generalizations of the Gibbs entropy. For example, the Tsallis entropy of index q reads $S_q = \big(1 - \mathrm{tr}\,\hat{w}^q\big)/(q-1)$ and, correspondingly, $S_q(E_1 + E_2) = S_q(E_1) + S_q(E_2) + (1-q)S_q(E_1)S_q(E_2)$. Another example is the Rényi information entropy of index $\alpha \geqslant 0$ according to $S_\alpha = \log\big(\mathrm{tr}\,\hat{w}^\alpha\big)/(1-\alpha)$. In the limit $q \to 1$ and $\alpha \to 1$, both generalizations go over to the Gibbs entropy.

$$\frac{1}{T} = \frac{\partial S}{\partial E}.$$

Inequality $1/T \geqslant 0$ means that the entropy is a monotonous function of energy. The amount of heat to be supplied to a body to produce a unit increase in its temperature is referred to as the *specific heat*. From the above equations, one can obtain the following formula for specific heat $C = \partial E/\partial T$:

$$C = T\frac{\partial S}{\partial T}.$$

Let us further consider two non-interacting subsystems together composing the closed system so that the total energy of the system conserves, i.e. $E = E_1 + E_2 =$ const, but the subsystems themselves can exchange energy with each other. Of interest is the behavior of the total entropy $S(E_1 + E_2)$ as a function of the energy, e.g. of the first subsystem, provided that both subsystems are in the thermal equilibrium with each other, i.e. $T_1 = T_2 = T$. So, for the derivative with respect to energy, we have involving $dE_1 = -dE_2$

$$\frac{\partial S(E_1 + E_2)}{\partial E_1} = \frac{\partial S(E_1)}{\partial E_1} + \frac{\partial S(E_2)}{\partial E_2}\frac{\partial E_2}{\partial E_1} = \left(\frac{1}{T_1} - \frac{1}{T_2}\right)_{T_1=T_2} = 0.$$

Hence we draw the first conclusion. The entropy of the closed system on the whole is *extremum* if its both subsystems relax to the thermal equilibrium state characterized by the same temperatures in both subsystems. Using the *transitivity property* for the thermodynamically equilibrium systems, this conclusion can be generalized to the case of any number of the subsystems in the thermal equilibrium with each other.

It is of interest to clarify whether the given extremum is maximum or minimum. Thus we need to calculate the second derivative and determine its sign. Then,

$$\begin{aligned}\frac{\partial^2 S(E_1 + E_2)}{\partial E_1^2} &= \frac{\partial}{\partial E_1}\left(\frac{1}{T_1}\right) - \frac{\partial}{\partial E_2}\left(\frac{1}{T_2}\right)\frac{\partial E_2}{\partial E_1} = \\ = -\frac{1}{T_1^2}\frac{\partial T_1}{\partial E_1} - \frac{1}{T_2^2}\frac{\partial T_2}{\partial E_2} &= -\left(\frac{1}{\langle(\Delta E_1)^2\rangle} + \frac{1}{\langle(\Delta E_2)^2\rangle}\right) = \\ = -\frac{2}{\langle(\Delta E_1)^2\rangle} &= -\frac{2}{T^2 C_1(T)} < 0.\end{aligned}$$

Here we have taken $\Delta E_1 = -\Delta E_2$ into account. The negative sign of the second derivative tells us that *the extremum is a maximum*. As a result, we can conclude that the achievement of complete thermal equilibrium state in the closed system is related with the maximum magnitude of its entropy.[6]

---

[6] Within this statement in the century before last, the hypothesis was put forward about the heat death of the Universe (Big Freeze), i.e. the Universe eventually has to evolve to a state of thermal equilibrium over time. In the modern cosmology, taking into account the forces of gravity, it is

In conclusion, we say a few words about the magnitudes of entropy and specific heat at zero temperature. At zero temperature, according to the canonical Gibbs distribution, the physical system is in the state with the minimum energy, which, in general, can be N times degenerate, i.e. has N different quantum states. Accordingly, the number of equally probable ways in which this state of the physical system can be realized is also equal to N. The entropy at $T = 0$ will be equal to $S(0) = \ln N$. If the state with the minimum energy is non-degenerated, i.e. $N = 1$, the entropy at zero temperature vanishes. Otherwise, when $S(0) \neq 0$, one talks about *residual entropy* which can be finite or infinite if the ground state of the physical system is infinitely degenerate. Usually, in the latter case, there is a sense to consider the difference $\Delta S(T) = S(T) - S(0)$ alone and analyze the change of entropy in the physical processes instead of its absolute magnitude.

For the finite entropy magnitude at $T = 0$, we can conclude that the specific heat vanishes in the zero temperature limit. From the expression for the specific heat, written as

$$C = T\frac{\partial S}{\partial T} = \frac{\partial S}{\partial \ln T},$$

we have the following limiting behavior:

$$C(T \to 0) \sim \big[S(T \to 0) - S(0)\big] \to 0.$$

In the case of entropy divergence in the limit $T \to 0$, say, as $S(T) \sim -\ln T$ typical for the non-quantum classical systems, e.g. classical ideal gas or classical magnetic moments, the specific heat at zero temperature will be characterized by the finite magnitude. Other examples of physical systems with the residual entropy are the following: frustrated antiferromagnets, spin glasses, quasi-crystals, and incommensurable systems.

Problems

1. Derive the above-obtained relations between free energy, energy, and entropy, using the operator representation for the density matrix $\hat{w}$.

Solution. Let us differentiate the following definition of free energy F with respect to the temperature:

$$e^{-F/T} = \operatorname{tr}\big(e^{-\hat{H}/T}\big).$$

As a result, we have the simple relation

$$\left(\frac{F}{T^2} - \frac{1}{T}\frac{\partial F}{\partial T}\right)e^{-F/T} = \frac{1}{T^2}\operatorname{tr}\big(\hat{H}e^{-\hat{H}/T}\big),$$

which reduces straightforwardly to the equation

$$F - T\partial F/\partial T = \operatorname{tr}\big(\hat{H}e^{(F-\hat{H})/T}\big) = \operatorname{tr}(\hat{H}\hat{w}) = E.$$

---

considered that a homogeneous isothermal state is not most probable and does not correspond to the entropy maximum. Other hypotheses have been put forward about the ultimate fate of the Universe, e.g. Big Rip, Big Bang, and Big Crunch.

Then the energy equals $E = F + TS$ where entropy is $S = -\partial F/\partial T$.

On the analogy we have for the entropy

$$S = -\mathrm{tr}\left(\hat{w} \ln \hat{w}\right) = -\mathrm{tr}\left(\hat{w}\frac{F - \hat{H}}{T}\right) = -\frac{F}{T}\,\mathrm{tr}\,(\hat{w}) + \frac{1}{T}\,\mathrm{tr}\,(\hat{w}\hat{H}) = -\frac{F - E}{T},$$

or the same usual expression $E = F + TS$.

2. The black hole entropy as a function of the energy is described as

$$S(E) = \frac{4\pi G}{\hbar c^5} E^2$$

where c is the light velocity and G is the gravitational constant. Show that the thermodynamically equilibrium state of such black holes is impossible.

Solution. Since in the thermodynamically stable state the energy should be minimum as a function of entropy, the second derivative $\partial^2 E/\partial S^2 = T/C(T)$ must be positive, $C(T)$ being the specific heat. If $S(E) = \left(E/E_0\right)^n$, we have

$$\frac{\partial^2 E}{\partial S^2} = E_0 \frac{1 - n}{n^2} S^{\frac{1-2n}{n}}.$$

To provide the positive inequality, it is necessary to have $n < 1$. Since $n = 2$, the black hole proves to be thermodynamically unstable and thus radiates.

3. Find the following fluctuations of energy: $\langle(\Delta E)^3\rangle$ and $\langle(\Delta E)^4\rangle$.

Solution. Let us express the cube of energy fluctuations in terms of the derivatives for partition function Z with respect to inverse temperature β

$$\langle(\Delta E)^3\rangle = -\frac{1}{Z}\frac{\partial^3 Z}{\partial \beta^3} + \frac{3}{Z^2}\frac{\partial Z}{\partial \beta}\frac{\partial^2 Z}{\partial \beta^2} - \frac{2}{Z^3}\left(\frac{\partial Z}{\partial \beta}\right)^3 = -\frac{\partial^3 \ln Z}{\partial \beta^3}.$$

Employing the relation for the free energy $\beta F = -\ln Z$, we find

$$\langle(\Delta E)^3\rangle = \frac{\partial^3}{\partial \beta^3}(\beta F).$$

The calculation for $\langle(\Delta E)^4\rangle$ is analogous to that considered above:

$$\frac{\partial^4 \ln Z}{\partial \beta^4} = \langle(\Delta E)^4\rangle - 3\langle(\Delta E)^2\rangle^2.$$

Hence we have readily

$$\langle(\Delta E)^4\rangle = 3\left(\frac{\partial^2}{\partial \beta^2}(\beta F)\right)^2 - \frac{\partial^4}{\partial \beta^4}(\beta F).$$

4. The heat engine operates in the Carnot cycle exchanging the heat between the hot and cold thermal reservoirs. The hot thermal reservoir has positive temperature $T_h > 0$, and the cold reservoir has negative temperature $T_c < 0$. Show that the efficiency η of such a Carnot heat engine equals unity.

Solution. Let us consider isothermal expansion at $T = T_h$ (Fig. 1.1). The amount of heat absorbed by the heat engine from Point 1 to Point 2 is equal to $Q_h = T_h(S_2 - S_1) = T_h \Delta S_h$. There is no further change in the entropy from Point 2 to Point 3. In the process of isothermal compression from Point 3 to 4, the amount of heat that the engine exchanges with the refrigerator at temperature

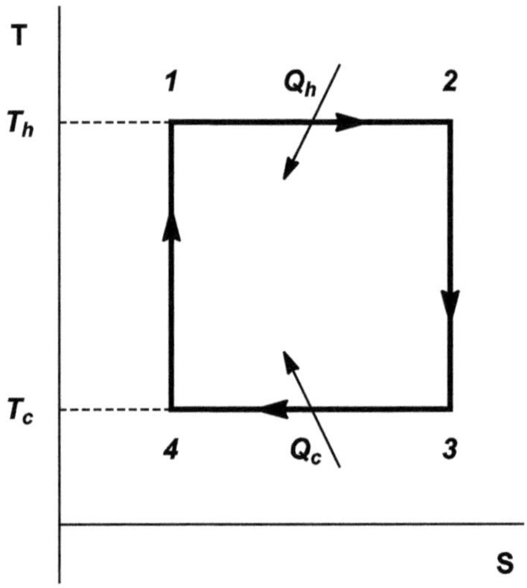

Fig. 1.1 The Carnot cycle in the temperature–entropy diagram

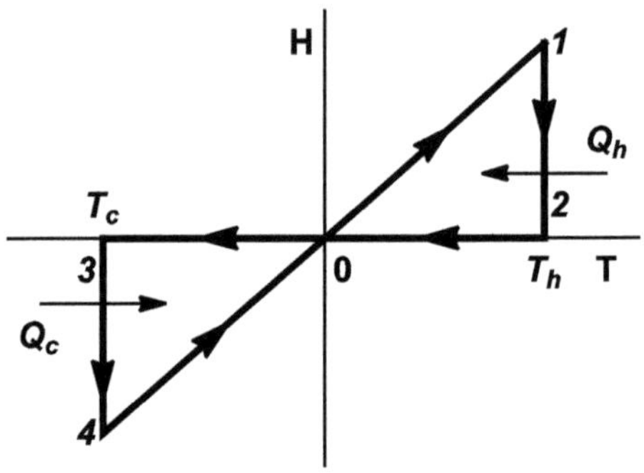

Fig. 1.2 The Carnot cycle of magnetic moment between the thermal reservoirs with the opposite signs of temperature

T_c equals $Q_c = T_c(S_4 - S_3) = T_c\Delta S_c$. The total entropy change $\Delta S = \Delta S_h + \Delta S_c$, as a result of the complete cycle, vanishes, i.e. $\Delta S_c = -\Delta S_h$ and, respectively,

$$Q_c = -Q_h \frac{T_c}{T_h}.$$

Since the heat engine returns to its initial state, the change in its internal energy vanishes. Accordingly, the total work A is given by changing the heat balance $A = Q_{in} + Q_{out}$ where Q_{in} and Q_{out} are the heat taken from and put into the system, respectively, at the isothermal stages of cycle. Let $Q_h > 0$. In the usual situation, when the temperatures T_h and T_c have the same signs, we have the heat leaving the engine to the cold reservoir due to $Q_c < 0$ and, thus, the heat input Q_{in} equals Q_h. The efficiency is determined by the ratio

$$\eta = A/Q_{in}$$

where $A = Q_h + Q_c$ is given by the classical expression $\eta = 1 - T_c/T_h < 1$.

If the temperatures of hot and cold reservoirs have the different signs ($T_c < 0$), the heat is absorbed with the heat engine at both isothermal stages, i.e. $Q_h > 0$ and $Q_c > 0$. Therefore, no amount of heat is removed from the engine, i.e. $Q_{out} = 0$, and the heat absorbed in the complete cycle will be equal to the work done by the system, entailing $Q_{in} = A$. As a result, we arrive at the efficiency equal to unity

$$\eta = 1 \quad (T_h > 0, \quad T_c < 0).$$

Since condition $\eta < 1$ does not hold for the different signs of temperatures T_h and T_c, the negative value of temperature cannot represent absolute temperature.[7]

In Fig. 1.2 the Carnot cycle is shown for the magnetic moment when the magnetic field plays a role of the working parameter. The cycle here can be represented as two Carnot cycles. The first cycle operates between zero temperature and temperature T_h with $\eta_1 = 1 - 0/T_h = 1$. The second does between T_c and zero temperature with $\eta_2 = 1 - 0/T_c = 1$. The total efficiency will be $\eta = 1$.

1.3 Thermodynamic Variables and Potentials

So far, we have paid no attention to a possible dependence of physical states in the system and its energy levels upon other allowable physical parameters determining the state of the system. Such parameters or *thermodynamic variables* can be, for example, number of particles N in the physical system, volume V in gases and liquids, surface area in droplets, electrical induction $\boldsymbol{D}$ in dielectrics, magnetic induction $\boldsymbol{B}$ in magnets, and strain tensor u_{ik} in solids. The choice of thermodynamic variables or parameters, which can be scalars, vectors, or tensors, as well as the choice of the number of variables, depend on the properties of the physical system, experimental conditions, types of physical processes, and specific purposes of describing the system under study.

Let λ be the very parameter which the Hamiltonian of the system $\hat{H}(\lambda)$ depends on and which, along with temperature, is chosen to characterize the thermodynamic state of the physical system. The dependence of the Hamiltonian on parameter λ results naturally in the dependence of the energy eigenvalues $\varepsilon_k(\lambda)$ of the system upon the same parameter λ. This dependence can be determined from the *Hellmann–Feynman theorem* according to the following relation:

$$\frac{\partial \varepsilon_k(\lambda)}{\partial \lambda} = \langle k_\lambda \left| \frac{\partial \hat{H}}{\partial \lambda} \right| k_\lambda \rangle .$$

Here $|k_\lambda\rangle$ is the eigenfunction of Hamiltonian $\hat{H}(\lambda)$.

Let us turn now to studying the free energy

$$F(\lambda) = -T \ln Z(\lambda) = -T \ln \sum_k e^{-\varepsilon_k(\lambda)/T}$$

[7] Efficiency $\eta = 1$ implies a perpetual motion machine of the second kind that can convert all the heat entirely into mechanical work, violating thus the subtle second law of thermodynamics for the cyclic Carnot process.

as a function of parameter λ. For this purpose, we differentiate this expression with respect to λ at the fixed temperature

$$\left(\frac{\partial F}{\partial \lambda}\right)_T = \sum_k \frac{\partial \varepsilon_k(\lambda)}{\partial \lambda} \frac{e^{-\varepsilon_k(\lambda)/T}}{Z(\lambda)} = \sum_k \frac{\partial \varepsilon_k(\lambda)}{\partial \lambda} w_k(\lambda) = \langle \frac{\partial \varepsilon(\lambda)}{\partial \lambda} \rangle .$$

Accordingly, taking the temperature dependence of free energy into account, we can represent its differential, as follows:

$$dF(T, \lambda) = \left(\frac{\partial F}{\partial T}\right)_\lambda dT + \left(\frac{\partial F}{\partial \lambda}\right)_T d\lambda = -S\,dT + \Lambda\,d\lambda.$$

Here we introduce a new thermodynamic variable as a partial derivative of free energy in accordance with the definition[8]

$$\Lambda = \left(\frac{\partial F}{\partial \lambda}\right)_T .$$

The pairs of thermodynamic quantities (T, S) and (Λ, λ) are referred to as the *conjugate thermodynamic variables*. If we wish to go over from potential $F(T, \lambda)$ to *conjugate thermodynamic potential* $\widetilde{F} = \widetilde{F}(T, \Lambda)$ depending already on the conjugate variable Λ, one should employ the *Legendre transformation*:

$$\widetilde{F}(T, \Lambda) = F - \Lambda\lambda.$$

It is easy to check that the following formula:

$$d\widetilde{F}(T, \Lambda) = (-S\,dT + \Lambda\,d\lambda) - \lambda\,d\Lambda - \Lambda\,d\lambda = -S\,dT - \lambda\,d\Lambda$$

is valid for the differential of the conjugate thermodynamic function $\widetilde{F}(T, \Lambda)$. The conjugate thermodynamic potential $\widetilde{F}$ proves to be the function of T and Λ and, correspondingly,

$$\lambda = -\left(\frac{\partial \widetilde{F}}{\partial \Lambda}\right)_T .$$

If necessary, one can simultaneously consider several parameters λ_i and, accordingly, introduce the same number of conjugate thermodynamic variables Λ_i.

Here are some examples of possible thermodynamic variables. For example, one of the possible characteristics of a condensed medium (body) may be its volume which is not necessarily constant and the energy of the condensed medium can be volume-dependent. In this case, the partial derivative of free energy

---

[8] For the clarity of physical interpretation, variable Λ can be defined with the opposite sign as well. For example, $\lambda = V$ is the volume and $P = -\Lambda$ is the pressure.

$$P = -\left(\frac{\partial F}{\partial V}\right)_T$$

with respect to volume V of the system (body) is called its *pressure*. The physical sense for the pressure is the density of force at which the system (body) acts on the boundary of its volume. The differential for the energy of the system (body) reads

$$dE = T\,dS - P\,dV = \delta Q + \delta R.$$

Accordingly, the change in the energy of the system in the thermodynamically equilibrium state can be represented as a sum of the *amount of heat* $\delta Q = T\,dS$ transferred to the system (body) due to heat exchange and the *work* $\delta R = -P\,dV$ done on the system (body)[9] by changing the volume.

The thermodynamic process that occurs without transferring the heat $\delta Q = 0$ is called the *adiabatic process*. The *isentropic process* implies no net transfer of heat and occurs at the constant entropy. The expression for the transferred amount of heat δQ has a universal form. On the contrary, the expression δR for the work done by the system depends on the physical properties of the system and the type of processes occurring when the state of the system changes.

The energy of the system, in general, may depend on the number of particles N in the system. If we take the number of particles in the system as an additional thermodynamic variable, the partial derivative of free energy

$$\mu = \left(\frac{\partial F}{\partial N}\right)_{T,V}$$

is called the *chemical potential* μ of the system (body) and characterizes the change of free energy by varying the particle number in the system by one particle. As a result, we have for the total differential of free energy $F(T, V, N)$

$$dF(T, V, N) = -S\,dT - P\,dV + \mu\,dN.$$

The partial derivatives of free energy F per unit volume with respect to electrical induction $\boldsymbol{D}$ and magnetic induction $\boldsymbol{B}$ determine the strengths of electric $\boldsymbol{E}$ and magnetic $\boldsymbol{H}$ fields as follows:

$$\boldsymbol{E} = 4\pi\left(\frac{\partial F}{\partial \boldsymbol{D}}\right)_{T,V} \quad \text{and} \quad \boldsymbol{H} = 4\pi\left(\frac{\partial F}{\partial \boldsymbol{B}}\right)_{T,V}$$

and, correspondingly, specify the following differential:

$$dF(T, \boldsymbol{D}, \boldsymbol{B}) = -s\,dT + \frac{\boldsymbol{E}\,d\boldsymbol{D} + \boldsymbol{H}\,d\boldsymbol{B}}{4\pi}$$

[9] Accordingly, $P\,dV$ is the work done by the system (body) when its volume changes.

where s is the amount of entropy per unit volume. The stress tensor σ_{ik} is the variable conjugated to the strain tensor u_{ik}

$$\sigma_{ik} = \left(\frac{\partial F}{\partial u_{ik}}\right)_T.$$

Here F is the free energy per unit volume, and the differential of free energy density equals $dF(T, u_{ik}) = -s\,dT + \sigma_{ik}du_{ik}$.

Depending on the experimental conditions and types of physical processes under study, it is convenient to introduce various thermodynamic potentials and variables. As we have seen above, the free energy $F(T, V)$ is the thermodynamic potential depending on the temperature and volume and its differential $dF = -SdT - PdV$ serves for determining the entropy S and pressure P.

The energy or internal energy $E(S, V) = F + TS$ is the thermodynamic potential expressed in terms of entropy and volume. The differential of energy $dE = TdS - PdV$ gives the following expressions for the temperature and pressure:

$$T = \left(\frac{\partial E}{\partial S}\right)_V \quad \text{and} \quad P = -\left(\frac{\partial E}{\partial V}\right)_S.$$

The variables of temperature T and pressure P are associated with the *thermodynamic Gibbs potential* Φ or *Gibbs free energy*[10] as well:

$$\Phi(T, P) = F + PV = E - TS + PV,$$
$$d\Phi(T, P) = -S\,dT + V\,dP.$$

Note that $F - \Phi = -PV$.

The *enthalpy* or *heat function* $W(S, P)$ is introduced as a thermodynamic potential depending on entropy and pressure

$$W(S, P) = E + PV = \Phi + TS,$$
$$dW(S, P) = T\,dS + V\,dP.$$

Provided that the number of particles in the system can vary, i.e. system with the variable number of particles under study, it is advisable to introduce the *grand thermodynamic potential* into consideration. We hereafter denote this potential as $\Omega(\mu)$, and its independent variable will be called the *chemical potential* μ conjugated to the number of particles N

$$\Omega(T, V, \mu) = F - \mu N.$$

Using $d(\mu N) = \mu\,dN + N\,d\mu$, we obtain for the differential $d\Omega$

[10] In this sense the Gibbs free energy Φ is conjugate to the Helmholtz free energy F.

$$d\Omega(T, V, \mu) = -S\,dT - P\,dV - N\,d\mu.$$

For the system (body) under the fixed volume, we find

$$d\Omega = -S\,dT - N\,d\mu.$$

The number of particles can be found by differentiating the grand potential Ω with respect to the chemical potential under constant temperature and volume

$$N = -\left(\frac{\partial\Omega}{\partial\mu}\right)_{T,V}.$$

For simple and spatially homogeneous systems, it is straightforward to obtain the relationship between thermodynamic Gibbs potential $\Phi(P, T, N)$ and chemical potential μ, using the property of extensivity of entropy $S = S(E, V, N)$. The extensivity of entropy implies that, when varying such parameters of the system as energy E, volume V, and number of particles N by a certain factor k, the entropy changes by the same factor

$$S(kE, kV, kN) = kS(E, V, N).$$

Differentiating this equality with respect to k and putting k equal to unity, we arrive at

$$S = \frac{\partial S}{\partial E}E + \frac{\partial S}{\partial V}V + \frac{\partial S}{\partial N}N = \frac{E}{T} + \frac{P}{T}V - \frac{\mu}{T}N\,.$$

Hence we have $E - TS + PV = N\mu$. On the other hand, the left-hand side of this equation represents the thermodynamic Gibbs free energy Φ and, therefore, $\Phi = N\mu$. Comparing two differentials

$$d\,\Phi = -S\,dT + V\,dP + \mu\,dN = d(\mu N) = N\,d\mu + \mu\,dN$$

yields the *Gibbs–Duhem equation*

$$d\mu = -\frac{S}{N}dT + \frac{V}{N}dP \quad \text{or} \quad d\left(\frac{\mu}{m}\right) = -\sigma dT + \frac{dP}{\rho}.$$

Here m is the mass of single particle, σ is the entropy per unit mass, and ρ is the density.

Problems

1. Let the Hamiltonian of system $\hat{H} = \hat{H}(\lambda)$ depend on parameter λ. Derive the formula for the free energy differential dF obtained above, using the matrix representation for density matrix $\hat{w}$.

Solution. Since $\operatorname{tr}\hat{w} = \operatorname{tr}[\exp(F - \hat{H})/T] = 1$, the differential $d\,(\operatorname{tr}\hat{w}) = \operatorname{tr}(d\hat{w})$ vanishes. So,

$$\operatorname{tr}\left[\left(\frac{dF - \frac{\partial\hat{H}}{\partial\lambda}d\lambda}{T} - \frac{F - \hat{H}(\lambda)}{T^2}dT\right)\hat{w}\right] = 0.$$

Hence we have

$$\left(dF-\frac{F}{T}\right)\mathrm{tr}\,\hat{w}+\frac{\mathrm{tr}\,(\hat{H}\hat{w})}{T}-\mathrm{tr}\left(\frac{\partial\hat{H}}{\partial\lambda}\hat{w}\right)d\lambda=dF-\frac{F-E}{T}-\langle\frac{\partial\hat{H}}{\partial\lambda}\rangle d\lambda=0$$

or

$$dF=-S\,dT+\langle\frac{\partial\hat{H}}{\partial\lambda}\rangle d\lambda=-S\,dT+\Lambda\,d\lambda.$$

2. Let the energy of the states in the thermodynamic system depend on parameter λ as $\tilde{\varepsilon}_k=\varepsilon_k+\lambda\Lambda_k$ for all states of the system with no exception.

(a) Find the mean value of the conjugate parameter Λ.

(b) Find the mean square fluctuation or dispersion of parameter Λ.

Solution. (a) Consider the partition function $\tilde{Z}=\sum_k\exp(-\tilde{\varepsilon}_k/T)$ and free energy $\tilde{F}=-T\ln\tilde{Z}$. Next, we calculate the derivative

$$\frac{\partial\tilde{F}}{\partial\lambda}=\frac{1}{\tilde{Z}}\sum_k e^{-\tilde{\varepsilon}_k/T}\frac{\partial\tilde{\varepsilon}_k}{\partial\lambda}=\sum_k\Lambda_k\frac{e^{-\tilde{\varepsilon}_k/T}}{\tilde{Z}}=\sum_k\Lambda_k\tilde{w}_k=\langle\Lambda_k\rangle=\Lambda$$

which delivers the mean value of parameter Λ_k. Note the normalization condition $\sum_k\tilde{w}_k=1$ for the distribution of thermodynamic probability of populating the states.

(b) Let us employ the following relation for derivatives:

$$\frac{\partial^2\ln\tilde{Z}}{\partial\lambda^2}=\frac{1}{\tilde{Z}}\frac{\partial^2\tilde{Z}}{\partial\lambda^2}-\left(\frac{\partial\ln\tilde{Z}}{\partial\lambda}\right)^2.$$

The calculation yields

$$\frac{\partial^2\ln\tilde{Z}}{\partial\lambda^2}=\frac{1}{T^2}\sum_k\Lambda_k^2\frac{e^{-\tilde{\varepsilon}_k/T}}{\tilde{Z}}-\frac{1}{T^2}\left(\sum_k\Lambda_k\frac{e^{-\tilde{\varepsilon}_k/T}}{\tilde{Z}}\right)^2=\frac{1}{T^2}\left(\sum_k\Lambda_k^2\tilde{w}_k-\big(\sum_k\Lambda_k\tilde{w}_k\big)^2\right).$$

Hence we find the simple formula for the mean square fluctuation

$$\langle(\Delta\Lambda)^2\rangle=\langle\Lambda_k^2\rangle-(\langle\Lambda_k\rangle)^2=T^2\frac{\partial^2\ln\tilde{Z}}{\partial\lambda^2}=-T\frac{\partial^2\tilde{F}}{\partial\lambda^2}.$$

Chapter 2
Ideal Boltzmann Gas

2.1 Partition Function of an Ideal Gas

An *ideal gas* is referred to as a gas in which the interaction is completely neglected between the particles composing the gas. Usually such an approximation is associated with a sufficiently low density or rarefaction of the condensed medium. On the neglect of the interparticle interaction the problem to calculate the partition function Z is greatly simplified and, due to the identity of the particles in the gas, can be reduced to determining the partition function of a single particle. From this point of view an ideal gas is the simplest physical system.

So, let the Hamiltonian for the system of N identical particles

$$\hat{H} = \hat{H}_1 + \hat{H}_2 + \ldots + \hat{H}_N$$

represent a sum of identical N Hamiltonians. On the whole, the state of the system is described by a set $\{|k_1\rangle, |k_2\rangle, \ldots |k_N\rangle\}$ of quantum numbers for each single Hamiltonian $\hat{H}_i$ and single ith particle specified with state vector $|k_i\rangle$. In the lack of interparticle interaction, the energy of the whole system is a sum of the energies of single particles

$$\varepsilon_{k_1,\ldots,k_N} = \varepsilon_{k_1} + \varepsilon_{k_2} + \ldots + \varepsilon_{k_N} .$$

As a result, the total partition function Z splits into a product of N identical partition functions taken over the states of each separate particle

$$\sum_{\{k_1,\ldots,k_N\}} e^{-(\varepsilon_{k_1}+\cdots+\varepsilon_{k_N})/T} = \sum_{\{k_1,\ldots,k_N\}} e^{-\varepsilon_{k_1}/T} \times \cdots \times e^{-\varepsilon_{k_N}/T} =$$

$$= \prod_{i=1}^{N} \sum_{k_i} e^{-\varepsilon_{k_i}/T} = \left(\sum_{k_i} e^{-\varepsilon_{k_i}/T} \right)^N .$$

S. N. Burmistrov, *Statistical and Condensed Matter Physics*,
https://doi.org/10.1007/978-981-97-9072-2_2

Finally the calculation of partition function for the total system reduces to the product of N identical one-particle partition functions z_0

$$Z = z_0^N \quad \text{where} \quad z_0 = \sum_k e^{-\varepsilon_k/T} = \int g(\varepsilon)\, e^{-\varepsilon/T}\, d\varepsilon.$$

The answer can be expressed in terms of one-particle density of states $g(\varepsilon) = \sum_k \delta(\varepsilon - \varepsilon_k)$.

2.2 Ideal Gas in the Classical Statistics

Let us treat motion of gas particles within the framework of classical mechanics. Then, calculating the partition function, the following specific features of the classical system must be taken into account. Firstly, the states of the system of N classical particles are classified by a set of momenta and coordinates

$$\{\mathcal{P}, \mathcal{Q}\} = \{(\boldsymbol{p}_1, \boldsymbol{q}_1), (\boldsymbol{p}_2, \boldsymbol{q}_2), \ldots, (\boldsymbol{p}_N, \boldsymbol{q}_N)\}$$

which represent a point in the phase space. The energy of the system $\varepsilon(\mathcal{P}, \mathcal{Q})$ is a simultaneous function of momenta and coordinates

$$\varepsilon(\mathcal{P}, \mathcal{Q}) = \varepsilon\big[(\boldsymbol{p}_1, \boldsymbol{q}_1), (\boldsymbol{p}_2, \boldsymbol{q}_2), \ldots, (\boldsymbol{p}_N, \boldsymbol{q}_N)\big].$$

The vector notations for momentum $\boldsymbol{p}_i = \{p_{\alpha i}\}$ and coordinate $\boldsymbol{q}_i = \{q_{\alpha i}\}$ implicate the spatial components $\alpha = 1, 2, \ldots, d$ as well. Here d is the dimensionality of geometrical space occupied by the particles of a gas or the number of degrees of freedom of a particle. In this case the sum over all possible states of a gas, i.e. sum over all possible momenta $\{\mathcal{P}\}$ and coordinates $\{\mathcal{Q}\}$, represents an integration over the whole region of accessible momenta $\{\mathcal{P}\}$ and coordinates $\{\mathcal{Q}\}$. The number of possible states $d\tau$ in the phase space element $d\mathcal{P}d\mathcal{Q}$ equals

$$d\tau = \prod_{i=1}^{N}\prod_{\alpha=1}^{d} \frac{dp_{\alpha i}\, dq_{\alpha i}}{2\pi\hbar} = \prod_{i=1}^{N} \frac{d\boldsymbol{p}_i\, d\boldsymbol{q}_i}{(2\pi\hbar)^d}\,.$$

Secondly, we should take the identity of like particles into account. In fact, if we transpose two identical particles, say i and j, the state of the whole system will be characterized by another point in the phase space, corresponding to permuting the ith particle coordinate and momentum with the jth particle coordinate and momentum, i.e. $(\boldsymbol{p}_i, \boldsymbol{q}_i \ldots \boldsymbol{p}_j, q_j) \to (\boldsymbol{p}_j, \boldsymbol{q}_j \ldots \boldsymbol{p}_i, \boldsymbol{q}_i)$.

Due to the principle of indistinguishability, such a permutation of particles does not vary the energy as well as the state of the whole system. In other words, both such points of phase space, though different, represent, in essence, the same state of the system in the case of identical particles. As a result, the total number of various states in the system is smaller than the integral $\int d\tau$ by a factor equal to the number of possible permutations between identical particles. For the system composed with N identical particles, the total number of possible permutations equals $N!$ and, therefore, the total number of states in the classical system is expressed by the following formula:

$$\frac{1}{N!}\int d\tau .$$

Thus, provided that the classical expressions for the energy of ideal gas composed of N identical particles are employed, the partition function of the gas is given by

$$Z_{\text{cl}} = \frac{z_0^N}{N!}$$

where one-particle partition function z_0 equals the integral

$$\begin{aligned} z_0 = \int \frac{dp_1 dp_2 \dots dp_d\, dq_1 dq_2 \dots dq_d}{(2\pi\hbar)^d}\, e^{-\varepsilon(\boldsymbol{p},\,\boldsymbol{q})/T} = \\ = \int \frac{d\boldsymbol{p}\, d\boldsymbol{q}}{(2\pi\hbar)^d}\, e^{-\varepsilon(\boldsymbol{p},\,\boldsymbol{q})/T} = \int g(\varepsilon)\, e^{-\varepsilon/T}\, d\varepsilon . \end{aligned}$$

The one-particle density of states $g(\varepsilon)$ is expressed via integral

$$g(\varepsilon) = \int \frac{d\boldsymbol{p}\, d\boldsymbol{q}}{(2\pi\hbar)^d}\, \delta\big(\varepsilon - \varepsilon(\boldsymbol{p},\, \boldsymbol{q})\big) .$$

The proper approximation used for $N!$ in the macroscopic limit $N \to \infty$ is given by Stirling's formula $N! \approx (N/e)^N$. Accordingly, we arrive at

$$Z_{\text{cl}} = \left(\frac{ez_0}{N}\right)^N \qquad (N \to \infty)$$

and obtain the following formula:

$$F = -NT \ln\big(ez_0/N\big)$$

for the free energy.

2.3 Equation of State for Ideal Gas

Analyzing the thermodynamic properties of an ideal gas in the lack of external field, we assume that the translational motion of gas particles obeys the laws of classical mechanics. Then, calculating the one-particle partition function z_0, we separate the total energy of a particle or gas molecule into the kinetic energy due to translational center-of-mass motion of a particle and into the internal energy related to the internal degrees of freedom of a particle or molecule

$$\varepsilon_k(\boldsymbol{p}) = \boldsymbol{p}^2/2m + \varepsilon_k^{(\mathrm{in})} .$$

The internal energy, depending on the type of a particle or molecule, can consist of the rotational, vibrational, and electronic modes of motion and excitations. In the simplest approximation, to describe the internal modes of motion, we assume their independence from each other. This allows us to represent the internal energy as a sum of rotational and vibrational energies and to take apart the contributions of electronic and nuclear degrees of freedom into account.

Such an approximation results in

$$z_0 = z_0^{\mathrm{kin}} z_0^{(\mathrm{in})} = z_0^{(\mathrm{in})} \int d^3 r \int \frac{d^3 p}{(2\pi\hbar)^3}\, e^{-p^2/2mT} = V \left(\frac{mT}{2\pi\hbar^2}\right)^{3/2} z_0^{(\mathrm{in})}$$

where V is the volume occupied with the gas, and the partition function for the internal states of a gas particle

$$z_0^{(\mathrm{in})} = \sum_k e^{-\varepsilon_k^{(\mathrm{in})}/T} = e^{-f(T)/T}$$

is a certain function depending on the temperature alone. Thus we obtain for the free energy of ideal gas

$$F(T,\, V) = -NT \ln \frac{eV}{N} \left(\frac{mT}{2\pi\hbar^2}\right)^{3/2} + Nf(T) .$$

The pressure of ideal gas is independent of the internal degrees of a gas particle and equals

$$P = -\partial F/\partial V = NT/V .$$

This yields the *equation of state of an ideal gas* or *ideal gas law*

$$PV = NT.$$

Having the formula for the free energy, we can find the other thermodynamic potentials and variables. So, the Gibbs free energy $\Phi = F + PV$ is proportional to the number of particles N and equals

$$\Phi(P,\, T) = NT \ln\left[\frac{P}{T}\left(\frac{2\pi\hbar^2}{mT}\right)^{3/2}\right] + Nf(T) = N\mu(P,\, T).$$

Here $\mu(P,\, T)$ is the chemical potential or Gibbs energy per particle.

The entropy, defined as $S = -\partial F/\partial T$, reads

$$S(T, V) = N \ln\left[\frac{V}{N}\left(\frac{mT}{2\pi\hbar^2}\right)^{3/2}\right] + \frac{5}{2}N - Nf'(T).$$

Finally, the energy is proportional to the number of particles N in the gas and equal to

$$E = N\left(\frac{3}{2}T + f(T) - Tf'(T)\right).$$

The energy of ideal gas per one particle depends on the temperature alone. The same statement refers to the enthalpy due to relation $W = E + PV = E + NT$.

The specific heats at constant pressure and volume

$$C_P = \left(\frac{\partial W}{\partial T}\right)_P \quad \text{and} \quad C_V = \left(\frac{\partial E}{\partial T}\right)_V$$

are also the functions of the temperature alone. Since $W - E = NT$, the difference

$$C_P - C_V = N$$

demonstrates the universal *Mayer's relation* for the ideal gas.

In conclusion, we note that the macroscopic description of the gas supposes the transition to the following limits: $N \to \infty$ and $V \to \infty$, the gas density $N/V =$ const being constant.

Problems

1. Show that the equation of state $PV = NT$ for an ideal gas is independent of the energy spectrum type of its particles $\varepsilon = \varepsilon(p)$ and spatial dimension d. Determine the energy and specific heat of the gas for the particle spectrum $\varepsilon_p = cp^k\,(c,\ k > 0)$.

Solution. The partition function equals $Z = z_0^N/N!$ where N is the number of particles and z_0 is the one-particle partition function

$$z_0 = \int d^d x \int \frac{d^d p}{(2\pi\hbar)^d} e^{-\varepsilon(p)/T} = Vf(T).$$

Here V is the d-dimensional volume occupied by the gas. Accordingly, the free energy equals $F = -T\ln[(Vf)^N/N!]$ and the gas pressure is $P = -\partial F/\partial V = NT/V$.

For spectrum $\varepsilon_p = cp^k$, we find $z_0 = AVT^{d/k}$ where A is the constant quantity independent of V and T. The free energy reads

$$F = -NT\ln\left(AVT^{d/k}\right) + T\ln N!\,.$$

The specific heat of a gas is temperature-independent

$$C = -T\left(\frac{\partial^2 F}{\partial V^2}\right) = \frac{d}{k}N$$

with the corresponding energy equal to $E = F + TS = (d/k)NT$. This can be interpreted as a classical *equipartition property* when each degree of freedom for the particle motion contributes an equal fraction $1/k$ to the energy and specific heat of an ideal gas.

2. Let us assume that entropy S is an additive quantity represented by the integral over the space and as a scalar is a relativistic invariant. In other words, the magnitude of entropy is independent of the choice of the coordinate frame.

Find how the temperature of a body could change from the viewpoint of an observer in the coordinate frame moving at velocity $\boldsymbol{v}$. Show that the equation of state $PV = NT$ for the ideal gas keeps the relativistic invariance.

Solution. The energy differential of a body moving at velocity $\boldsymbol{v}$ reads

$$dE = T\,dS + \boldsymbol{v}\,d\boldsymbol{P}$$

where T is the temperature and $\boldsymbol{P}$ is the momentum of a body. Then we have for the entropy differential

$$dS = \frac{1}{T}dE - \frac{\boldsymbol{v}}{T}d\boldsymbol{P}.$$

Let us give a relativistically invariant form to this expression, involving that 4-vector of momentum is presented by the covariant formula $dP_\alpha = \big(dE/c,\ -d\boldsymbol{P}\big)$.

So, we introduce the contravariant 4-vector of inverse temperature $\beta^\alpha = \big(c/T, \boldsymbol{v}/\boldsymbol{T}\big)$ and write the entropy differential according to the rule $x^\alpha y_\alpha = x^0 y_0 - \boldsymbol{x}\boldsymbol{y}$ of the 4-vector product $x^\alpha = (x^0, \boldsymbol{x})$ and $y_\alpha = (y^0, -\boldsymbol{y})$ in the identical form

$$dS = \beta^\alpha dP_\alpha = \frac{c}{T}\frac{dE}{c} - \frac{\boldsymbol{v}}{T}d\boldsymbol{P} = \frac{dE}{T} - \frac{\boldsymbol{v}}{T}d\boldsymbol{P}.$$

Such a record $dS = \beta^\alpha dP_\alpha = \beta_\alpha dP^\alpha$ has a form of relativistic invariant. For the square of inverse temperature 4-vector, according to $x^\alpha x_\alpha = (x^0)^2 - \boldsymbol{x}^2$, we have

$$\beta^\alpha\beta_\alpha = \frac{c^2 - \boldsymbol{v}^2}{T^2} = \left(\frac{c}{\gamma T}\right)^2, \quad \gamma = \frac{1}{\sqrt{1 - v^2/c^2}},$$

γ being the Lorentz factor. From invariance for the square of inverse temperature 4-vector, we can conclude the invariance of product γT under the Lorentz transformation. Therefore, body temperatures T_1 and T_2, observed in the reference frames moving at velocities v_1 and v_2, should be connected with the relation

$$\Theta(T_1, v_1) = \frac{T_1}{\sqrt{1 - v_1^2/c^2}} = \frac{T_2}{\sqrt{1 - v_2^2/c^2}} = \Theta(T_2, v_2).$$

The quantity $\Theta(T, v)$ is Lorentz-invariant. The thermodynamic equilibrium between two bodies should be disturbed under $\Theta(T_1, v_1) \neq \Theta(T_2, v_2)$.

If the entropy is invariant $dS' = dS$ and the energy and volume should transform as $E' = \gamma E$ and $V' = V/\gamma$, we arrive at

$$\frac{1}{T'} = \frac{\partial S'}{\partial E'} = \frac{\partial S}{\gamma \partial E} = \frac{1}{\gamma T}, \quad \text{i.e.}\ \ T' = \gamma T.$$

For the pressure, one has

$$P' = T'\frac{\partial S'}{\partial V'} = \gamma T\frac{\partial S}{\partial(V/\gamma)} = \gamma^2 T\frac{\partial S}{\partial V} = \gamma^2 P.$$

On the analogy we find for the chemical potential: $\mu' = \gamma\mu$.

Involving that $N = N'$, the equation of state for ideal gas proves to be Lorentz-invariant

$$\frac{P'V'}{N'T'} = \frac{PV}{NT} = 1$$

and the ideal gas remains ideal for any inertial observer.

2.4 The Boltzmann Distribution

When describing the properties of a gas, it is of interest to know the average number of gas particles $n(\varepsilon)$ in the state with energy ε. In the general case we imply that the particle energy $\varepsilon = \varepsilon(\boldsymbol{p}, \boldsymbol{q})$ depends on the momentum $\boldsymbol{p}$ and coordinate $\boldsymbol{q}$. According to the Gibbs distribution, the probability to find a particle in the state of energy ε is proportional to the exponential $\exp(-\varepsilon/T)$. Thus the average number of particles in this state will also be proportional to the same exponential, i.e. $n(\varepsilon) = a\exp(-\varepsilon/T)$. The constant a is determined with the condition of normalizing the function $n(\varepsilon)$ by the total number of gas particles

$$N = \int d\tau n(\varepsilon) = a \int d\tau \exp(-\varepsilon/T).$$

Taking into account that the partition function z_0 equals

$$z_0 = \int d\tau \exp(-\varepsilon/T),$$

we readily find that $a = N/z_0$ and

$$n(\varepsilon) = \frac{N}{z_0} e^{-\varepsilon/T}.$$

On the other hand, we have using the relation $ez_0/N = \exp(-F/NT)$ for free energy F and the equation of state $PV = NT$ for an ideal gas

$$z_0/N = e^{-(F+NT)/NT} = e^{-(F+PV)/NT} = e^{-\Phi/NT} = e^{-\mu/T}.$$

Here μ is the chemical potential of a gas. As a result, in the ideal classical gas the average number of particles in the state with energy ε or population of the states with energy ε is determined by the following formula:

$$n(\varepsilon) = e^{(\mu-\varepsilon)/T}.$$

This formula is called the *Boltzmann distribution*.

The ideal gas entropy, expressed via the Boltzmann distribution function $n(\varepsilon)$, can be written as

$$S = \int d\tau\, n \ln \frac{e}{n} .$$

If no external potential is in the space region occupied with the gas and the gas particle energy $\varepsilon = \boldsymbol{p}^2/2m$ depends on the momentum alone, we arrive at the distribution called the *Maxwell distribution*

$$n(\varepsilon) = \frac{N}{V} \frac{e^{-\frac{p^2}{2mT}}}{\int e^{-\frac{p^2}{2mT}} \frac{d^3p}{(2\pi\hbar)^3}} = \frac{N}{V}\left(\frac{2\pi\hbar^2}{mT}\right)^{3/2} e^{-\frac{p^2}{2mT}} ,$$

V being the volume occupied with the gas.

Consider next the gas of particles in the external potential field, the potential energy $u(\boldsymbol{r})$ of a particle being coordinate-dependent. Writing the energy of a particle as

$$\varepsilon(\boldsymbol{p}, \boldsymbol{r}) = \frac{\boldsymbol{p}^2}{2m} + u(\boldsymbol{r}),$$

we see that the particle number distribution starts to depend on the spatial position of particles with the probability proportional to the following exponential:

$$n(\varepsilon) \propto e^{-u(\boldsymbol{r})/T} .$$

As an example for the behavior of a gas in the external potential field, we consider an ideal gas of $N \gg 1$ particles in the harmonic potential $u(x, y, z)$ producing a trap for the particles of mass m:

$$u(x, y, z) = \frac{m}{2}\left(\omega_x^2 x^2 + \omega_y^2 y^2 + \omega_z^2 z^2\right).$$

Let us start to calculate the free energy of a gas in the trap from determining the one-particle density of states

$$g(\varepsilon) = \int d^3r \int \frac{d^3p}{(2\pi\hbar)^3} \delta\big(\varepsilon - \varepsilon(\boldsymbol{p}, \boldsymbol{r})\big)$$

where $\varepsilon(\boldsymbol{p}, \boldsymbol{r}) = \boldsymbol{p}^2/(2m) + u(\boldsymbol{r})$ is the energy of a trapped particle. Integrating over the particle momentum yields

$$g(\varepsilon) = \frac{4\pi\sqrt{2}m^{3/2}}{(2\pi\hbar)^3} \int\limits_{u(\boldsymbol{r})<\varepsilon} d^3r \sqrt{\varepsilon - u(\boldsymbol{r})} .$$

The last integral is calculated with the aid of substituting the variables $x \to x\sqrt{2/(m\omega_x^2)}$, etc. Hence

$$g(\varepsilon) = \frac{16}{8\pi^2\hbar^3\omega_x\omega_y\omega_z} \int\limits_{x^2+y^2+z^2<\varepsilon} dx\,dy\,dz\sqrt{\varepsilon - (x^2+y^2+z^2)} =$$

$$= \frac{8}{\pi\hbar^3\omega_x\omega_y\omega_z} \int\limits_{r^2<\varepsilon} dr\,r^2\sqrt{\varepsilon - r^2} = \frac{\varepsilon^2}{2\hbar^3\omega_x\omega_y\omega_z}\,.$$

Knowing the one-particle density of states, we find readily the free energy for the gas of N particles from the relation

$$F = -NT\ln\frac{e}{N}\int d\varepsilon\,g(\varepsilon)e^{-\varepsilon/T} =$$

$$= -NT\ln\left[\frac{e}{2\hbar^3\omega_x\omega_y\omega_z N}\int\limits_0^\infty d\varepsilon\,\varepsilon^2 e^{-\varepsilon/T}\right] = -NT\ln\frac{eT^3}{\hbar^3\omega_x\omega_y\omega_z N}\,.$$

Differentiating the free energy with respect to the particle number results in the chemical potential reading as

$$\mu = \frac{\partial F}{\partial N} = -3T\ln\frac{T}{\hbar(\omega_x\omega_y\omega_z N)^{1/3}}\,.$$

The specific heat of ideal gas in the harmonic trap equals

$$C(T) = -T\frac{\partial^2 F}{\partial T^2} = 3N$$

and proves to be constant and temperature-independent.

Note the specific feature of the transition to the macroscopic limit for an ideal gas in the harmonic trap. The following limits are meant here: $N \to \infty$ and $\omega_x,\ \omega_y,\ \omega_z \to 0$, the quantity $(\omega_x\omega_y\omega_z N)^{1/3} = \text{const}$ being kept constant.

Problem

1. The ideal gas of particles is in the centrally symmetric trap with the impenetrable core of radius R and external logarithmic potential

$$U(r) = \begin{cases} \infty\,, & r = R\,, \\ U_0\ln(r/R)\,, & r > R \quad (U_0 > 0)\,. \end{cases}$$

Find the specific heat of a gas and the limiting temperature above which it is impossible to heat the gas.

Solution. The free energy of a gas is calculated from the expression

$$F = -NT\ln\frac{ez_0}{N} = -NT\ln\frac{e}{N}\int\limits_{r>R} d^3r\int\frac{d^3p}{(2\pi\hbar)^3}\exp\left(-\frac{p^2}{2mT}-\frac{U(r)}{T}\right) =$$

$$= -NT\ln\left[\frac{e}{N}\frac{(2\pi mT)^{3/2}}{(2\pi\hbar)^3}\,4\pi\int\limits_R^\infty dr\,r^2\exp\left(-\frac{U_0}{T}\ln\frac{r}{R}\right)\right].$$

The last integral converges only if the temperature $T < U_0/3$. This condition determines the limiting temperature $T_{\rm lim} = U_0/3$ below which the thermodynamically equilibrium state of the gas can still exist in the trap. The physical reason is connected with the exponential growth of the one-particle density of states as the energy of particles increases in the logarithmic potential. The point is that, due to slow growth of the trap potential, the particles depart away from the trap center by increasing the energy of particles and their kinetic energy turns out to be small compared with the potential energy contribution.

Calculating the integral gives the magnitude of free energy

$$F = -NT\ln\left[\frac{e}{N}\left(\frac{mT}{2\pi\hbar^2}\right)^{3/2} 4\pi R^3\,\frac{T}{U_0-3T}\right].$$

Next, the specific heat of gas reads

$$C(T) = -T\frac{\partial^2 F}{\partial T^2} = N\left[\frac{3}{2}+\left(\frac{U_0}{U_0-3T}\right)^2\right]$$

with the corresponding energy $E = F + TS$ of the trapped gas

$$E = NT\left(\frac{3}{2}+\frac{U_0}{U_0-3T}\right).$$

As we see, when the temperature tends to the limiting value $U_0/3$, both the specific heat and the energy of the trapped gas diverge together with the amount of heat necessary for heating the gas in the logarithmic trap.

2.5 Monatomic Ideal Gas

Above, describing the thermodynamic properties of an ideal gas, we have separated the total energy of a particle or gas molecule into the kinetic energy due to translational center-of-mass motion and the internal energy associated with the internal degrees of freedom of a particle or molecule. The internal energy, depending on the type of a particle or molecule, can consist of the rotational, vibrational, and electronic modes of motion and excitations. In general, it is necessary to perform the numerical calculations of quantum states and energy levels of a single molecule.

In the simplest approximation for describing the internal modes of motion, their independence from each other is assumed below. The reason for such an approximation can be a noticeable difference in the frequencies that characterize these modes of motion. Usually, for the frequencies of rotational, vibrational, and electronic modes, one can approximately assume that $\omega_r \ll \omega_v \ll \omega_{el}$. This makes it possible to approximate the internal energy as a sum of rotational and vibrational energies and to involve separately the contributions of the electronic and nuclear degrees of freedom of a molecule. In this case the one-particle partition function $z^{(in)}$ reduces to the product of the partition functions for the corresponding modes of motion: $z^{(in)} = z_r z_v z_{el}$. Below we start by treating a monatomic gas.

Turning to the case of a monatomic gas, we will immediately imply the sufficicntly low temperatures as compared with the ionization energy of the atom and neglect completely any possibility of its ionization. Denoting the atomic energy levels that we count from the lowest one as ε_i and the corresponding degree of degeneracy as g_i, we arrive at the following one-particle partition function related to the electronic degrees of freedom

$$z_{in} = z_0^{(el)} = \sum_i e^{-\varepsilon_i/T}.$$

For free energy F, entropy S, and specific heat C, we obtain

$$F(T, V) = -NT \ln\left[\frac{eV}{N}\left(\frac{mT}{2\pi\hbar^2}\right)^{3/2}\right] + Nf_{el}(T),$$

$$S(T, V) = N \ln\left[\frac{eV}{N}\left(\frac{mT}{2\pi\hbar^2}\right)^{3/2}\right] + \frac{3}{2}N - Nf'_{el}(T),$$

$$C(T) = \frac{3}{2}N - NTf''_{el}(T),$$

where

$$f_{el}(T) = -T \ln z_{el} = -T \ln \sum_i g_i e^{-\varepsilon_i/T}.$$

This expression simplifies in two limiting cases: $T \ll \varepsilon_i$ and $T \gg \varepsilon_i$. For sufficiently low temperatures $T \ll \varepsilon_i$, we have $f'_{el}(T) = \ln g_0$ and g_0 is the multiplicity of atom ground state. In the opposite limit $T \gg \varepsilon_i$ we put approximately $f'_{el}(T) = \ln \sum_i g_i$.

Let us analyze specific features of this contribution to the specific heat. We consider a simple example when there are two energy levels, say, ground level with $\varepsilon_0 = 0$ and the excited one with energy ε_1, the difference $\Delta = \varepsilon_1 - \varepsilon_0$ being the excitation energy. Then we readily find

$$f_{el}(T) = -T \ln\left(g_0 + g_1 e^{-\Delta/T}\right)$$

and the contribution of these levels to the specific heat equals

$$C_{el}(T) = N\frac{\Delta^2}{T^2}\frac{g_0 g_1 e^{-\Delta/T}}{\left(g_0 + g_1 e^{-\Delta/T}\right)^2} = N\frac{\Delta^2}{T^2}\begin{cases} \frac{g_1}{g_0}e^{-\Delta/T}, & T \ll \Delta, \\ \frac{g_0 g_1}{(g_0+g_1)^2}, & T \gg \Delta. \end{cases}$$

So, if there are energy level splitting and corresponding fine (hyperfine) atomic structure, in the usual monotonous temperature behavior of specific heat we should expect a narrow maximum or peak with the width about Δ and at the temperatures about Δ. Such an anomalous behavior of the specific heat with a drastic maximum against the background of almost constant temperature behavior due to splitting the atomic level structure is referred to as the *Schottky anomaly*.[1]

2.6 Diatomic Gas: The Vibrational Degree of Freedom

Let us turn to thermodynamic properties of diatomic gases. Of course we suppose that the dissociation energy of gas molecules is large as compared with the temperatures analyzed. In comparison with the monatomic case, the molecule of diatomic gas has additional degrees of freedom, namely two rotational and one vibrational. The vibrational–rotational energy of a molecule, in first approximation, can be represented as an independent sum of the vibrational–rotational energies of the nuclei in a molecule. We start from analyzing the vibrational contribution.

According to the concepts of quantum mechanics, the system displaying the harmonic oscillations near the potential energy minimum at frequency Ω has an equidistant spectrum

$$\varepsilon_n = \hbar\omega\left(n + \frac{1}{2}\right), \quad n = 0,\ 1,\ 2, \ldots$$

Calculating the corresponding partition function $z_0^{(\mathrm{v})}$ is straightforward

$$z_0^{(\mathrm{v})} = \sum_{n=0}^{\infty} e^{-\frac{\hbar\omega}{T}\left(n+\frac{1}{2}\right)} = \frac{e^{-\hbar\omega/2T}}{1 - e^{-\hbar\omega/T}}\,.$$

Though the oscillation anharmonicity and relationship with the rotational degrees of freedom start to interplay for the moderate vibrational quantum numbers n, we have formally extended the summation to $n = \infty$. For our justification, we refer to the rapid convergence of an exponential series and conservation of qualitative specific features in the behavior of thermodynamic potentials.

As a result, we arrive at the vibrational contribution to the free energy for the gas of N diatomic molecules

$$F_{\mathrm{v}} = N\frac{\hbar\omega}{2} + NT\ln\left(1 - e^{-\hbar\omega/T}\right).$$

[1] This notion is widely used in solid state physics when the specific heat of a solid has a peak at low temperatures.

So, we have for the vibrational contribution to the gas energy

$$E_\mathrm{v} = N\frac{\hbar\omega}{2} + N\frac{\hbar\omega}{e^{\hbar\omega/T}-1} \approx N\begin{cases} N\hbar\omega\left(\frac{1}{2}+e^{-\hbar\omega/T}\right), & T \ll \hbar\omega, \\ NT\left(1+\frac{1}{12}\left(\frac{\hbar\omega}{T}\right)^2\right), & T \gg \hbar\omega, \end{cases}$$

and to the specific heat

$$C_\mathrm{v}(T) = N\left(\frac{\hbar\omega}{T}\right)^2\frac{e^{\hbar\omega/T}}{\left(e^{\hbar\omega/T}-1\right)^2} \approx N\begin{cases} \left(\frac{\hbar\omega}{T}\right)^2 e^{-\hbar\omega/T}, & T \ll \hbar\omega, \\ 1-\frac{1}{12}\left(\frac{\hbar\omega}{T}\right)^2, & T \gg \hbar\omega. \end{cases}$$

As one sees, in the high temperature $T \gg \hbar\omega$ limit we arrive at the classical values for energy $E_\mathrm{v} = NT$ and specific heat $C_\mathrm{v} = N$. In the low temperature $T \ll \hbar\omega$ range, the vibrational contribution to the specific heat freezes exponentially due to presence of energy gap $\hbar\omega$ in the excitation spectrum of the vibrational degree of freedom.

2.7 Diatomic Gas of Heteronuclear Molecules: The Rotational Degrees of Freedom

Below we proceed to analyzing a possible effect of diatomic molecule rotation relative to the center of inertia on the thermodynamic properties of an ideal gas of diatomic molecules. As a first approximation to describe the rotation of a diatomic molecule, we choose the model of a rigid rotator with the given moment of inertia J, which corresponds to the Hamiltonian $\hat{H} = \hbar^2\hat{\boldsymbol{K}}^2/2J$ and $\hbar\hat{\boldsymbol{K}}$ is the rotational moment operator of the molecule.[2] The rotational energy levels are written as follows:

$$E_{K,K_z} = \frac{\hbar^2}{2J}K(K+1).$$

Here $K = 0,\ 1,\ 2, \dots$ is the angular momentum quantum number, and each energy is $(2K+1)$-fold degenerate in the projection $\hat{K}_z$ of angular momentum operator $\hat{\boldsymbol{K}}$. The corresponding one-particle partition function is given by the expression

$$z_0^{(\mathrm{r})} = \sum_{K=0}^{\infty}(2K+1)\exp\left[-\frac{T_0}{T}K(K+1)\right].$$

[2] The product of the reduced mass for both atoms by the square of the internuclear spacing gives us an estimate for the inertia moment J.

Here $T_0 = \hbar^2/2J$ is the typical rotational energy of a molecule or its rotational constant. The rotational fraction of free energy F_r for the gas of N molecules is determined by formula $F_r = -NT \ln z_0^{(r)}$.

In general, the analytical calculation of partition function $z_0^{(r)}$ is rather complicated and we consider two limiting cases of low and high temperatures. We find in the low $T \ll T_0$ temperature limit

$$z_0^{(r)} = 1 + 3e^{-2T_0/T} + \cdots$$

To determine the asymptotics at high $T \gg T_0$ temperatures, we employ the *Euler–Maclaurin formula* to evaluate a finite sum for slowly varying functions

$$\sum_{n=0}^{N} f(n) = \int_{-1/2}^{N+1/2} f(n)\, dn - \frac{\Delta f'}{24} + \frac{7}{5760}\Delta f''' - \frac{31}{967680}\Delta f^{(V)} + \cdots$$

where $\Delta f^{(k)} = f^{(k)}(N + 1/2) - f^{(k)}(-1/2)$ is the difference in the derivatives of kth order at the boundary points $N + 1/2$ and $-1/2$.

The asymptotic high-temperature expansion of partition function $z_0^{(r)}$ reads

$$z_0^{(r)} = \frac{T}{T_0} + \frac{1}{3} + \frac{1}{15}\frac{T_0}{T} + \frac{4}{315}\frac{T_0^2}{T^2} + \cdots$$

Let us write first terms of the limiting behavior for the rotational contribution to the specific heat of a gas

$$C_r(T) = N \begin{cases} 12\left(\dfrac{T_0}{T}\right)^2 e^{-2T_0/T} + \cdots, & T \ll T_0\,, \\ 1 + \dfrac{1}{45}\left(\dfrac{T_0}{T}\right)^2 + \dfrac{16}{945}\left(\dfrac{T_0}{T}\right)^3 + \cdots, & T \gg T_0\,. \end{cases}$$

As a result, from classical magnitude $C_r = N$ in the high-temperature limit the rotational specific heat passes through the maximum at $T \sim 0.8T_0$ by lowering the temperature. The further decrease of temperature to absolute zero results in the exponential freezing of specific heat.

2.8 Diatomic Gas of Homonuclear Molecules: The Rotational Degrees of Freedom

If a diatomic molecule is composed of homonuclear atoms, there appear quantum effects due to permutation of identical particles and necessary corresponding symmetry for the wave functions of the nuclei. This quantum effect in the rotational specific heat is most clearly manifested for the hydrogen molecules where rotational energy $T_0 \approx 85$ K is one of largest values due to lighter hydrogen atom mass compared with other chemical elements. First of all, let us pay attention to the nucleus spin of hydrogen atom which is a proton and spin $i = 1/2$ fermion. Therefore the complete wave function of two nuclei, including the product of the spin and spatial (orbital) parts of the wave function, must be antisymmetric with respect to permuting the nuclei.[3] When we permute the nuclei, i.e. changing the direction of the radius vector equal to the difference for the radius vectors of nucleus positions, the spatial or orbital part of the wave function multiplies by factor $(-1)^K$ where K is the angular momentum quantum number. In its turn, changing the sign for the spin part of the total wave function depends on the total nuclear spin I. The latter for hydrogen molecule can be either zero for the antiparallel nuclear spins or one for the parallel nuclear spins. The number of states with the total nuclear spin $I = 0$ is only one, and this state is antisymmetric with respect to permuting the nuclei. The number of states with the total nuclear spin $I = 1$ equals three, and all these states are symmetric with respect to permuting the nuclei. As a result, the total wave function of the nuclei multiplies by factor $(-1)^{K+I-1}$. The antisymmetry requirement for the total wave function of two fermions means that the sum $K + I$ must be even.

Molecular hydrogen with two proton nucleus spins aligned parallel or with total nuclear spin $I = 1$ is called *orthohydrogen*, and that in the antiparallel nuclear spin state with $I = 0$ is referred to as *parahydrogen*. Thus, in orthohydrogen there are three nuclear states and angular momentum quantum number K takes the odd integer values alone. Accordingly, parahydrogen has a single nuclear state and angular momentum quantum number K takes the even integer values alone.

Note that the total nuclear spin of a hydrogen molecule can change as a result of the collisions between molecules (e.g. due to magnetic dipolar coupling). In this case one says about ortho–para conversion. With conversion there occurs a change of rotational number by unity $\Delta K = 1$ with the simultaneous transitions between the triplet $I = 1$ and singlet $I = 0$ nuclear states in a molecule. The ortho–para conversion grows by increasing the magnetic field gradient at the nuclei. In the low density gas, the probability of molecular collisions is small and the ortho–para conversion process is sufficiently slow.[4] The gas mixture in which all possible nuclear states are represented in the equal probabilities is called the *statistical mixture*. In the statistical mixture, the number of orthohydrogen molecules is three times larger than that of parahydrogen molecules.

In pure parahydrogen, the one-particle partition function can be written as follows:

$$z_p = \sum_{K=0,\,2,\,\ldots} (2K+1) \exp\left[-\frac{T_0}{T} K(K+1)\right]$$

[3] The nucleus spin of deuterium atom, i.e. spin of deuteron, is equal to $i = 1$ and the total wave function for two nuclei of deuterium molecule D_2 as bosons should be symmetric to the nucleus permutation.

[4] The conversion rate in the hydrogen gas under normal conditions is about several percents per week.

and in pure orthohydrogen as

$$z_o = \sum_{K=1,3,\ldots} 3(2K+1)\exp\left[-\frac{T_0}{T}K(K+1)\right].$$

Hence we find readily in the low temperature $T \ll T_0$ limit

$$z_p = 1 + 5\,e^{-6T_0/T} + \cdots,$$
$$z_o = 9\,e^{-2T_0/T} + 21\,e^{-12T_0/T} + \cdots.$$

For calculating the power asymptotics in the high temperature $T \gg T_0$ limit, we again employ the Euler–Maclaurin formula used in the previous paragraph for estimating a sum of slowly varying functions. The asymptotic expansions for the one-particle partition functions z_p and z_o into a power series have the same structure in both cases

$$z_p = \frac{1}{2}\frac{T}{T_0} + \frac{1}{6} + \frac{1}{30}\frac{T_0}{T} + \frac{2}{315}\frac{T_0^2}{T^2} + \cdots,$$
$$z_o = 3\left(\frac{1}{2}\frac{T}{T_0} + \frac{1}{6} + \frac{1}{30}\frac{T_0}{T} + \frac{2}{315}\frac{T_0^2}{T^2} + \cdots\right).$$

Let us write below the corresponding expansions for the rotational contribution to the specific heat of ideal para- and orthohydrogen gas per molecule in the low- and high-temperature range

$$C_p(T) = \begin{cases} 180\left(\dfrac{T_0}{T}\right)^2 e^{-6T_0/T} + \cdots, & T \ll T_0, \\ 1 + \dfrac{1}{45}\left(\dfrac{T_0}{T}\right)^2 + \dfrac{16}{945}\left(\dfrac{T_0}{T}\right)^3 + \cdots, & T \gg T_0; \end{cases}$$

$$C_o(T) = \begin{cases} \dfrac{700}{3}\left(\dfrac{T_0}{T}\right)^2 e^{-10T_0/T} + \cdots, & T \ll T_0, \\ 1 + \dfrac{1}{45}\left(\dfrac{T_0}{T}\right)^2 + \dfrac{16}{945}\left(\dfrac{T_0}{T}\right)^3 + \cdots, & T \gg T_0. \end{cases}$$

The coincidence of power expansion for the specific heat of para- and orthohydrogen means that the specific heat difference will be exponentially small in the high temperature $T \gg T_0$ range

$$C_p(T) - C_o(T) \sim \exp\left(-\frac{\pi^2}{4}\frac{T}{T_0}\right).$$

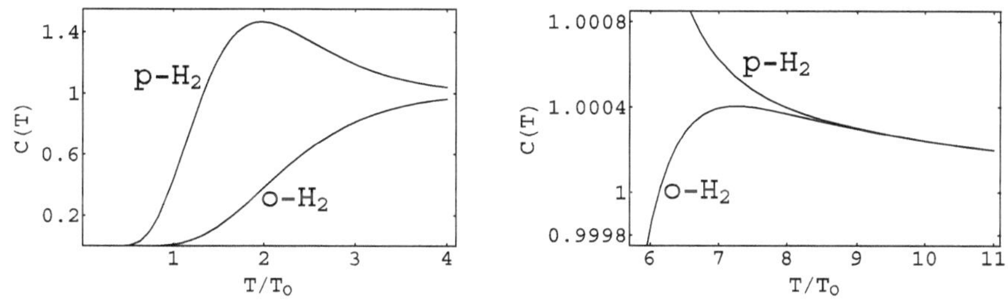

Fig. 2.1 The temperature behavior of rotational contribution to the specific heat of parahydrogen (p-H_2) and orthohydrogen (o-H_2) at low and intermediate temperatures

This high-temperature asymptotics behaves satisfactorily in the $T > 10\,T_0$ region. Both functions $C_p(T)$ and $C_o(T)$ pass through the maxima by lowering the temperature and then freeze exponentially out (Fig. 2.1). For orthohydrogen, the maximum of rotational specific heat is small numerically. The noticeable quantitative difference in the rotational specific heat of para- and orthohydrogen begins in the moderate temperature region at $T \lesssim 5\,T_0$. The magnitudes $6\,T_0$ and $10\,T_0$ in the exponents represent the energy gap between the first excited and ground levels, respectively.

Under complete thermal equilibrium between para- and ortho-molecules of a gas, each nuclear state, para or ortho, should once be involved into the partition function with weight $(2K + 1)\exp[-T_0\,K(K + 1)/T]$. Therefore, one-particle partition function will be a sum of the partition functions of para- and ortho-states

$$z_{eq} = z_p + z_o = \sum_{K=0}^{\infty}\bigl(2 - (-1)^K\bigr)(2K + 1)\exp\left[-\frac{T_0}{T}K(K + 1)\right].$$

In the gas mixture under thermal equilibrium, the probability to find a molecule in the state with angular momentum quantum number K equals

$$w_K = \frac{1}{z_{eq}}\bigl(2 - (-1)^K\bigr)(2K + 1)\exp\left[-\frac{T_0}{T}K(K + 1)\right].$$

The probability to find a molecule in the para-state with the even K numbers will equal a sum of probabilities w_K over all even rotational numbers. Accordingly, the probability to find in the ortho-state is a sum over all odd numbers. Hence, we have for these probabilities, respectively,

$$z_p/z_{eq} \quad \text{and} \quad z_o/z_{eq}\,.$$

Thus, under complete thermal equilibrium the ratio $x(T)$ of the number N_o of orthohydrogen molecules to the number N_p of parahydrogen molecules is temperature-dependent according to

$$x(T)=\frac{N_o}{N_p}=\frac{z_o(T)}{z_p(T)},$$

and varies from 0 to 3 as the temperature grows from $T=0$ to $T=\infty$.

The thermodynamic equilibrium at the ortho–para conversion can also be analyzed from the viewpoint of equilibrium condition for the chemical reaction like $o \leftrightarrows p$. The chemical equilibrium with respect to mutual conversion reaches under equality of chemical potentials for para- and ortho-molecules

$$\mu_p=-T\ln(z_p/N_p)=-T\ln(z_o/N_o)=\mu_o\,.$$

This means the same relation for $x(T)$. Using the equalities $N_p=Nz_p/(z_p+z_p)$ and $N_o=Nz_o/(z_o+z_o)$ where N is the total number of all molecules, one can be convinced that the free energy of thermodynamically equilibrium gas F_{eq} represents a sum of free energies of para- and orthohydrogen F_p and F_o, i.e.

$$F_{eq}=-NT\ln\frac{e(z_p+z_o)}{N}=-T\ln\left(\frac{e(z_p+z_o)}{N}\right)^{N_p}\left(\frac{e(z_p+z_o)}{N}\right)^{N_o}=$$
$$=-T\ln\left(\frac{ez_p}{N_p}\right)^{N_p}\left(\frac{ez_o}{N_o}\right)^{N_o}=-N_pT\ln\frac{ez_p}{N_p}-N_oT\ln\frac{ez_o}{N_o}=F_p+F_o.$$

If the hydrogen atom nucleus would be a spinless particle, the partition function would include the even angular momentum quantum numbers alone.

Problem

1. Estimate the limiting behavior for the difference in the specific heat of para- and orthohydrogen per molecule at high $T\gg T_0$ temperatures.

Solution. Let us write the one-particle partition functions for parahydrogen

$$z_p=\sum_{K=0,2,\dots}(2K+1)\exp\left[-\frac{T_0}{T}K(K+1)\right]=\sum_{n=0}^{\infty}(4n+1)e^{-an(2n+1)}$$

and for orthohydrogen

$$z_o=3\sum_{K=1,3,\dots}(2K+1)\exp\left[-\frac{T_0}{T}K(K+1)\right]=3\sum_{n=0}^{\infty}(4n+3)e^{-a(n+1)(2n+1)}.$$

Here we have introduced notation $a=2T_0/T$ for brevity. In the difference of the expressions

$$\Delta z=z_p-\frac{z_o}{3}=\sum_{n=0}^{\infty}\left[(4n+1)e^{-an}-(4n+3)e^{-a(3n+1)}\right]e^{-2an^2}=\sum_{n=0}^{\infty}g(n)e^{-2an^2}$$

at $a\ll 1$, the terms with numbers $n\lesssim n_m\sim 1/\sqrt{2a}$ provide us the main contribution while the factor $\exp(-2an^2)$ differs little from unity.

To estimate the sum, the *Poisson formula* is employed

$$\sum_{n=0}^{\infty} f(n) = \int_{-1/2}^{\infty} f(n)dn + 2\mathrm{Re}\sum_{k=1}^{\infty}\int_{-1/2}^{\infty} f(n)e^{2\pi ikn}dn$$

for $f(n) = g(n)\exp(-2an^2)$. Integrating the first integral by parts, we see that this integral vanishes. So,

$$\Delta z = 2\mathrm{Re}\sum_{k=1}^{\infty}\int_{-1/2}^{\infty} f(n)e^{2\pi ikn}dn = 2\mathrm{Re}\sum_{k=1}^{\infty}\int_{-1/2}^{\infty} g(n)e^{-2an^2+2\pi ikn}dn.$$

Let us estimate the integral using the saddle-point method. For the exponent

$$S(n) = -2an^2 + 2\pi ikn,$$

we find the saddle point

$$S'(n_0) = -4an_0 + 2\pi ik = 0, \quad n_0 = \frac{i\pi k}{2a}, \quad S(n_0) = -\frac{\pi^2k^2}{4a}, \quad S''(n_0) = -4a.$$

In the saddle-point method the preexponential factor equals

$$\sqrt{\frac{2\pi}{|S''(n_0)|}} = \sqrt{\frac{\pi}{2a}}\,.$$

Further, in the sum of integrals we keep the first $k = 1$ harmonic alone. The other $k \geqslant 2$ harmonics are neglected due to exponentially smaller contribution. Thus, we arrive at the following approximate estimate:

$$\Delta z \approx 2\Big(\mathrm{Re}\,g(n_0)\Big)\sqrt{\frac{\pi}{2a}}\exp\Big(-\frac{\pi^2}{4a}\Big).$$

The approximate calculation $g(n_0)$ with $a \ll 1$ yields

$$g(n_0) = -i(4n_0 + 1) - i(4n_0 + 3)e^{-a} \approx -8in_0 = 4\pi/a.$$

As a result, we obtain

$$\Delta z = z_p - \frac{z_o}{3} \approx 4\left(\frac{2\pi}{a}\right)^{3/2} e^{-\frac{\pi^2}{2a}} = 4\left(\frac{\pi T}{T_0}\right)^{3/2}\exp\Big(-\frac{\pi^2}{4}\frac{T}{T_0}\Big).$$

Then we proceed to estimating the difference in the specific heats of para- and orthohydrogen in the high ($T \gtrsim T_0$) temperature limit. For the difference in specific heats per molecule, we have

$$C_p - C_o = T\frac{\partial^2}{\partial T^2}\Big[T\big(\ln z_p - \ln z_o\big)\Big] = T\frac{\partial^2}{\partial T^2}\Big[T\big(\ln z_p - \ln(z_o/3)\big)\Big] =$$
$$= T\frac{\partial^2}{\partial T^2}\Big(T\ln\frac{z_p}{z_p - \Delta z}\Big) \approx T\frac{\partial^2}{\partial T^2}\Big(T\frac{\Delta z}{z_p}\Big) = 2T_0 T\frac{\partial^2}{\partial T^2}\Delta z.$$

Here, for the partition function z_p, we can put its limiting $T \to \infty$ value equal to $z_p = 1/a = T/2T_0$. The main contribution in the last formula gives the differentiation of the exponential alone, resulting in the final answer

$$C_p(T) - C_o(T) \sim 4\sqrt{\pi}\left(\frac{\pi^2}{4}\frac{T}{T_0}\right)^{5/2}\exp\left(-\frac{\pi^2}{4}\frac{T}{T_0}\right), \quad T \gg T_0\,.$$

2.9 Gas of Polyatomic Molecules

Below we make a few remarks about the thermodynamic properties for the gas consisting of polyatomic molecules. As above, in first approximation it is possible to represent the total energy of a molecule as a sum of two terms. The first is the kinetic energy associated with the translational center-of-mass motion of a molecule and the second is the internal energy associated with the internal degrees of freedom of a molecule. The internal energy of the molecule is the rotational, vibrational, and electronic modes of motion and excitations. In general, it is necessary to analyze the quantum states and calculate the energy levels of a molecule.

The molecule composed of n_a atoms has $3n_a$ degrees of freedom. Among them, three degrees of freedom correspond to the translational motion of a molecule. The number n_r of rotational degrees of freedom equals three for the nonlinear molecule and equals two in the case of a linear molecule. The other degrees of freedom are vibrational and the number of vibrational degrees of freedom v in the n_a-atomic molecule reads

$$v = 3n_a - 3 - n_r\,.$$

As a first approximation, the rotation of a molecule can be treated as a rigid body rotation described by the Hamiltonian

$$\hat{H} = \frac{\hbar^2}{2}\left(\frac{\hat{K}_\xi^2}{J_1} + \frac{\hat{K}_\eta^2}{J_2} + \frac{\hat{K}_\varsigma^2}{J_3}\right).$$

Here, the operators $\hat{K}_{\xi,\eta,\varsigma}$ are the angular momentum components in the rotating coordinate frame in which the axes are directed along three principal axes of inertia. Accordingly, the principal momenta of inertia of a molecule are denoted with J_1, J_2, and J_3. In the general case of *asymmetrical spinning top* when $J_1 \neq J_2 \neq J_3$ the calculation of the energy levels, in general, is a complicated problem. In this case, the degeneracy in the angular momentum directions relative to the spinning top is completely removed, i.e. for the given angular momentum quantum number K there are $2K + 1$ various non-degenerated levels.

In the case *symmetrical spinning top* $J_1 = J_2 > J_3$, the molecule has one symmetry axis higher than the second order and the expression for the rotational energy levels simplifies as

$$E = \frac{\hbar^2 K(K+1)}{2J_1} + \frac{\hbar^2 k^2}{2}\left(\frac{1}{J_3} - \frac{1}{J_1}\right).$$

Here we denote k as eigenvalues of operator $\hat{K}_\varsigma$, which run the values $k = -K,\ -K+1, \ldots, +K$. The level $k = 0$ is degenerated by $(2K + 1)$ times. The levels with $k \neq 0$ are degenerated twice as larger, i.e. $2(2K + 1)$ times.

The most simple case is the *ball top* when all three principal momenta of inertia are identical $J_1 = J_2 = J_3 = J$. In this case the energy of the levels is given by the simple formula

$$E = \frac{\hbar^2}{2J} K(K+1).$$

The corresponding degeneracy of the level[5] equals $(2K+1)^2$.

From the above expressions for the rotational energy levels, one can see that the analytical calculation of both corresponding one-particle partition function and rotational contribution to the specific heat is a complicated task. From the general consideration, the following conclusions can be drawn about the behavior of rotational specific heat. For sufficiently low $T \lesssim \hbar^2/2J$ temperatures, the rotational specific heat freezes exponentially out. As it concerns the high $T \gtrsim \hbar^2/2J$ temperature limit, the rotational specific heat tends to its classical magnitude.

The crossover to the classical high-temperature limit is most readily traced in the case of a ball top. Setting the typical energy interval between the rotational levels as $T_0 = \hbar^2/2J$, we have for the one-particle partition function

$$z_0^{(\mathrm{r})} = \sum_{K=0}^{\infty} (2K+1)^2 e^{-\frac{T_0}{T} K(K+1)}.$$

Replacing the summation with integrating over K is justified in the $T_0/T \ll 1$ limit. Then we find

$$z_0^{(\mathrm{r})} \approx \int_{-1/2}^{\infty} dK\, (2K+1)^2 e^{-\frac{T_0}{T} K(K+1)} = \sqrt{\pi} \left(\frac{T}{T_0} \right)^{3/2} e^{\frac{T_0}{4T}}.$$

Hence we obtain the classical magnitude for the rotational contribution to the specific heat per molecule

$$C_{\mathrm{r}}(T \gg T_0) = \frac{3}{2}.$$

For $T \to 0$, the rotational contribution to the specific heat freezes out according to the exponential law

$$C_{\mathrm{r}}(T) \approx 72 \left(\frac{T_0}{T} \right)^3 e^{-\frac{2T_0}{T}}.$$

As for the linear molecules, in which all atoms are located along one straight line, such molecules similar to diatomic molecules have only two rotational modes (degrees of freedom) and one moment of inertia J. At high $T \gg \hbar^2/2J$ temperatures, the specific heat tends to the constant classical limit $C_{\mathrm{cl}} = 1$ and freezes exponentially when the temperature lowers down to zero.

Here we should make the following remark. If we employ the classical description for the rotational modes of a molecule and calculate the corresponding rotational

[5] We do not consider the effects associated with the spins of homonuclear atoms in a molecule.

partition function $z_0^{(\mathrm{r})}$, we should keep in mind that a molecule may have several symmetric axes, when a rotation around these axes at a certain angle translates the molecule into itself and thus reduces to a permutation of atoms. Since these states are physically completely equivalent, then the statistical sum $z_{0,\mathrm{cl}}^{(\mathrm{r})}$ must be divided by the number of different rotations around the axes of symmetry.

The calculation of vibrational contribution to the thermodynamic properties of ideal gas can be performed in the analogous manner as for diatomic molecules. Unlike diatomic molecules, the polyatomic molecule has several vibrational modes as $v = 3n_a - 3 - n_r$. In first harmonic approximation, we can assume that the vibrational modes (or normal oscillations) are independent of each other and are characterized by frequencies ω_α, the index α numerating all v types of vibrational modes. Since in the harmonic approximation the vibrational modes are completely independent, the vibrational fraction of free energy reduces to a sum of contributions from each vibrational mode

$$F_{\mathrm{v}} = NT\sum_{\alpha=1}^{v}\ln\bigl(1 - e^{-\hbar\omega_\alpha/T}\bigr).$$

Correspondingly, the vibrational fraction of specific heat $C_{\mathrm{v}}(T)$ will be a sum of partial contributions delivered with each vibrational mode to the specific heat. It is clear that the vibrational contribution to specific heat freezes exponentially at the temperatures $T \lesssim \hbar\omega_{min}$, ω_{min} being the minimum frequency of vibrational modes. As the temperature grows and reaches the next magnitude about $\hbar\omega_\alpha$, the specific heat of a gas per molecule will increase by unity. Finally, at high temperatures exceeding the maximum magnitude among $\hbar\omega_\alpha$, the vibrational fraction of specific heat approaches its classical magnitude equal to $C_{\mathrm{v}}(T) = Nv$. Here it is implicitly assumed that the molecule does not decay at such temperatures yet.

Problem

1. Estimate the limiting behavior for the rotational specific heat of polyatomic molecule described by the model of ball spinning top with the moment of inertia J at high temperatures $T \gg T_0 = \hbar^2/2J$.

Solution. Let us write the one-particle partition function for the rotational mode

$$z_0^{(\mathrm{r})} = \sum_{K=0}^{\infty}(2K+1)^2 e^{-\frac{T_0}{T}K(K+1)} = 4e^{a/4}\sum_{n=0}^{\infty}(n+1/2)^2 e^{-a(n+1/2)^2}$$

where we denote $a = T_0/T$ for brevity. For $a \ll 1$, the main contribution results from the terms with large numbers while $n \lesssim 1/\sqrt{a}$.

To estimate, the *Poisson formula* is employed

$$\sum_{n=0}^{\infty} f(n) = \int_{-1/2}^{\infty} f(n)dn + 2\mathrm{Re}\sum_{k=1}^{\infty}\int_{-1/2}^{\infty} f(n)e^{2\pi ikn}dn$$

for $f(n) = (n+1/2)^2\exp\bigl[-a(n+1/2)^2\bigr]$. Integrating the first integral yields $\pi^{1/2}/(4a^{3/2})$. The integrals under the summation sign are estimated with the aid of the saddle-point method

$$\Delta z = 2\mathrm{Re}\sum_{k=1}^{\infty}\int_{-1/2}^{\infty} f(n)e^{2\pi ikn}dn = 2\mathrm{Re}\sum_{k=1}^{\infty}\int_{-1/2}^{\infty} g(n)e^{-2an^2+2\pi ikn}dn.$$

For the exponent

$$S(n) = -a(n+1/2)^2 + 2\pi ikn,$$

we determine the stationary point n_0

$$S'(n_0) = -2a\left(n_0+\frac{1}{2}\right)+2\pi ik = 0, \quad n_0+\frac{1}{2} = \frac{i\pi k}{a},$$

$$S(n_0) = -\frac{\pi^2k^2}{a} - \pi ik, \quad S''(n_0) = -2a.$$

In the saddle-point method the preexponential factor equals

$$\sqrt{\frac{2\pi}{|S''(n_0)|}} = \sqrt{\frac{\pi}{a}}.$$

Then we arrive at the following estimate of the integral:

$$-\sqrt{\frac{\pi}{a}}\frac{\pi^2k^2}{a^2}(-1)^k e^{-\pi^2k^2/a}.$$

Further, we keep only the first harmonic with $k=1$ in the sum. The other harmonics with $k \geqslant 2$ are neglected due to their exponentially smaller contribution. So, we obtain the following approximate estimate for the partition function:

$$z_0^{(\mathrm{r})} \approx \frac{\sqrt{\pi}}{a^{3/2}}e^{\frac{a}{4}}\left(1+\frac{2\pi^2}{a}e^{-\frac{\pi^2}{a}}\right)$$

and free energy

$$F_{\mathrm{r}} = -T\ln z_0^{(\mathrm{r})} \approx -T\ln\left(\sqrt{\pi}\frac{T_0^{3/2}}{T^{3/2}}e^{\frac{T_0}{4T}}\right) - \frac{2\pi^2T^2}{T_0}\exp\left(-\pi^2\frac{T}{T_0}\right).$$

Calculating the entropy, according to $S = -T\partial^2F/\partial T^2$, and differentiating only the large exponent in the correction term, we find

$$C_{\mathrm{r}}(T) \approx \frac{3}{2} + 2\left(\frac{\pi^2T}{T_0}\right)^3\exp\left(-\pi^2\frac{T}{T_0}\right).$$

Thus, in the temperature range $T \gtrsim T_0$ the rotational specific heat exceeds slightly the classical value 3/2. The latter numerically turns out to be a good approximation.

2.10 Thermal Ionization of Monatomic Gas

At sufficiently high temperatures in a neutral gas, the process of *ionization* becomes noticeable. In this process an electrically neutral atom or a molecule acquires a negative or positive charge by gaining or losing electrons. The *thermal ionization* results mainly from the collisions between the high-energy particles of a gas. The *collisional ionization* is the most important mechanism of ionization in gases and rarefied plasma. Below we consider the main specific features of thermal ionization on a simple example of a monatomic ideal gas, neglecting for simplicity the effects of multiple ionization and excited states of atoms.

So, let ionization of atoms occur in the closed system under constant volume at temperature T. We treat the process of singly thermal ionization of an atom A

$$A \leftrightarrows A^{+} + e^{-},$$

assuming the thermodynamic equilibrium in the system. The latter is composed of three ideal subsystems, namely subsystem of atoms A, subsystem of ions A^{+}, and subsystem of electrons e^{-}. The complete thermodynamic equilibrium in the system means the thermodynamic equilibrium between all three subsystems, setting in via the exchange of particles between these three subsystems.

Under thermodynamic equilibrium, the entropy of total system S due to property of additivity is a sum of the entropies of the subsystems with the number of atoms N_a, ions N_i, and electrons N_e. Each of the subsystems occupies the same volume V. Under thermodynamic equilibrium, the entropy of the whole system $S = S_a + S_i + S_e$ should reach its maximum magnitude via redistributing the number of particles in every subsystem. As a result, we should have for the differential of the total entropy in equilibrium

$$dS = dS_a + dS_i + dS_e = \left(\frac{dE_a}{T} + \frac{P_a}{T}dV - \frac{\mu_a}{T}dN_a\right) +$$
$$+\left(\frac{dE_i}{T} + \frac{P_i}{T}dV - \frac{\mu_i}{T}dN_i\right) + \left(\frac{dE_e}{T} + \frac{P_e}{T}dV - \frac{\mu_e}{T}dN_e\right) = 0.$$

Here P_a, P_i, P_e are the partial pressure and μ_a, μ_i, μ_e are the chemical potentials, respectively, for each of the subsystems. Above, we have taken the condition for an equality of the subsystem temperatures $T_a = T_i = T_e = T$ into account. This condition is necessary for the thermal equilibrium between all the subsystems. Involving the conservation of total energy $d(E_a + E_i + E_e) = 0$ and invariance of total volume $dV = 0$, we arrive at the following equation for the chemical potentials:

$$\mu_a dN_a + \mu_i dN_i + \mu_e dN_e = 0.$$

Involving the stoichiometry for ionization of an atom, i.e. relationship between the variations in the number of particles in each subsystem as $dN_i = -dN_a$ and $dN_e = dN_i$, we can write down the final equation determining the thermal equilibrium under the particle exchange between subsystems

$$\mu_a(P_a, T) - \mu_i(P_i, T) - \mu_e(P_e, T) = 0.$$

The chemical potential of a monatomic gas with the particle energy $\varepsilon_p = \varepsilon_0 + p^2/2m$ reads

$$\mu(P, T) = -T \ln \left[\frac{T}{P} \left(\frac{mT}{2\pi\hbar^2} \right)^{3/2} g\, e^{-\varepsilon_0/T} \right]$$

where g is the multiplicity of degeneracy or statistical weight for the lower level of a particle. Hence, in equilibrium this entails the equality

$$\varepsilon_a + T \ln \left[\frac{P_a}{g_a T} \left(\frac{2\pi\hbar^2}{M_a T} \right)^{3/2} \right] =$$

$$= \varepsilon_i + T \ln \left[\frac{P_i}{g_i T} \left(\frac{2\pi\hbar^2}{M_i T} \right)^{3/2} \right] + T \ln \left[\frac{P_e}{g_e T} \left(\frac{2\pi\hbar^2}{mT} \right)^{3/2} \right].$$

Here the electron energy is measured from zero $\varepsilon_e = 0$. The pressures P_a, P_i, and P_e are the partial ones in the subsystems of atoms, ions, and electrons. The corresponding masses are denoted as M_a, M_i, and m. Next, we have

$$T \ln \left[T \frac{P_a}{P_i P_e} \frac{g_e g_i}{g_a} \left(\frac{m M_i}{M_a} \frac{T}{2\pi\hbar^2} \right)^{3/2} \right] = \varepsilon_i - \varepsilon_a = I_i$$

where the ionization potential of atom is defined according to equality $I_i = \varepsilon_i - \varepsilon_a$.

Let us rewrite this equation using the ideal gas equation of state $P = nT$ for each of subsystems where n is the number density of the particles in each subsystem. Then we arrive at the *Saha ionization equation*[6]

$$\frac{n_i n_e}{n_a} = \frac{g_i g_e}{g_a} \left(\frac{m M_i}{M_a} \frac{T}{2\pi\hbar^2} \right)^{3/2} e^{-I_i/T}.$$

The region of applicability of the Saha ionization equation is limited. The equation is fulfilled only for the weakly ionized plasma when the Coulomb energy of the interparticle interaction is small as compared with the kinetic energy of particles which is of the order of the temperature.

We define the *degree of ionization* $\alpha = N_i/N$ of the gas as a ratio of the number of ions to the total number of atoms. Then the number of ions and electrons reads $N_i = N_e = \alpha N$ and the number of non-ionized atoms equals $N_a = (1 - \alpha)N$. Expressing the partial pressures of the subsystems in terms of the total pressure $P = P_a + P_i + P_e$, we have

$$\begin{cases} P_a \sim N_a \sim (1-\alpha)N \\ P_i \sim N_i \sim \quad \alpha N \\ P_e \sim N_e \sim \quad \alpha N \end{cases} \Longrightarrow P \sim (1+\alpha)N \Longrightarrow \begin{cases} P_a \sim \frac{1-\alpha}{1+\alpha} P \\ P_i \sim \frac{\alpha}{1+\alpha} P \\ P_e \sim \frac{\alpha}{1+\alpha} P \end{cases}.$$

[6] Also the Saha–Langmuir equation.

Taking into account that the atom and ion masses are practically the same $M_i \approx M_a$ due to $m \ll M_a$ and the multiplicity of degeneracy for the electrons of the same energy equals $g_e = 2$, we obtain the equation

$$\frac{1-\alpha^2}{\alpha^2 P} = \frac{g_a}{2g_i}\left(\frac{2\pi\hbar^2}{m}\right)^{3/2}\frac{e^{I_i/T}}{T^{5/2}} \equiv K(T)\,.$$

The quantity $K(T)$ is the *equilibrium constant*, and the formula

$$\alpha(P,\,T) = 1/\sqrt{1+PK(T)}$$

determines completely the dependence of the ionization degree as a function of temperature and pressure.

In the low-temperature region the degree of ionization is an exponentially small quantity. Let us estimate the temperature when the degree of ionization equals 50%. For convenience, we express the degree of ionization via the non-ionized gas density $n = N/V$, using the relation for the total pressure $P = (1+\alpha)nT$:

$$\alpha = \frac{\sqrt{1+4nTK(T)}-1}{2nTK(T)}\,.$$

It is convenient to represent quantity $nTK(T)$ in the following dimensionless form:

$$nTK(T) = \left(\frac{T_e}{T}\right)^{3/2} e^{I_i/T}$$

where temperature T_e is defined according to

$$T_e = \left(\frac{g_a}{g_i}\right)^{2/3} 2^{1/3}\pi\,\frac{\hbar^2 n^{2/3}}{m} \sim \frac{\hbar^2 n^{2/3}}{m}\,.$$

This temperature has a sense of the degeneracy temperature for the electron gas of the same density n as that of the non-ionized atom gas. At $nTK(T) = 2$ the gas becomes half-ionized. The corresponding temperature satisfies the equation

$$\left(\frac{T_e}{T}\right)^{3/2} e^{I_i/T} = 2 \quad \text{or} \quad T = \frac{I_i}{(3/2)\ln(2^{2/3}T/T_e)}\,.$$

The most interesting situation appears in the limit of sufficiently rarified gas under $T_e \ll I_i$. Solving the equation by the successive approximation method and using $T \approx I_i$ as a first approximation, we find in second approximation

$$T \approx \frac{I_i}{(3/2)\ln(2^{2/3}I_i/T_e)} \ll I_i \quad \text{if} \quad \ln(I_i/T_e) \gg 1\,.$$

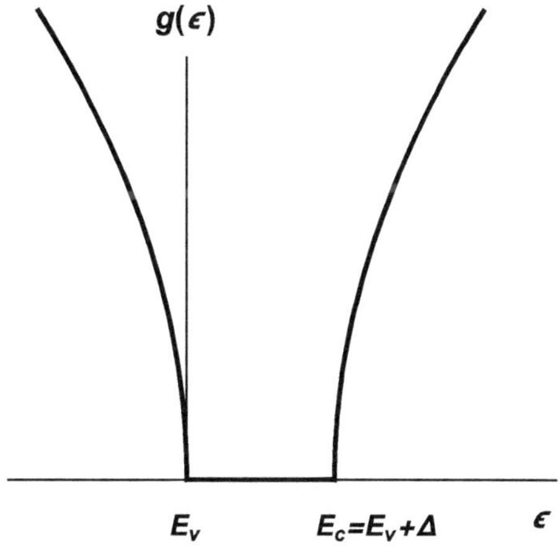

Fig. 2.2 The density of states in a semiconductor. The gap of forbidden band is Δ

Thus, the noticeable degree of ionization in the rarefied gas occurs starting from the temperature significantly lower than the ionization potential of an atom. In the dense gas for which the degeneracy temperature T_e is comparable with or exceeds the ionization potential I_i, the higher temperature comparable with the ionization potential is necessary to achieve the same degree of ionization.

Problems

1. Semiconductors are characterized by the presence of energy gap Δ or forbidden band in the density of electronic states (Fig. 2.2). The available states are separated into two bands, namely valence band whose states extend from $\varepsilon = -\infty$ to the band top $E_v = 0$ and the conduction band whose states extend from the band bottom $E_c = E_v + \Delta$ to $\varepsilon = +\infty$. At zero temperature, the states in the valence band are all occupied and in the conduction band the state are all free. As the temperature increases, there occurs thermal excitation of electrons into the conduction band and depletion of occupied electronic states in the valence band.

Find the electron concentration $n_e(T)$ in the conduction band as a function of temperature, assuming $T \ll \Delta$. The electron spectrum in the conduction band is described by the effective mass m_e and in the valence band by mass m_h.

Solution. Let us write the expression for the total number of electrons at an arbitrary temperature

$$N = \int_{-\infty}^{E_v} g_v(\varepsilon)d\varepsilon = \int_{-\infty}^{E_v} n(\varepsilon)g_v(\varepsilon)d\varepsilon + \int_{E_c}^{\infty} n(\varepsilon)g_c(\varepsilon)d\varepsilon$$

where $g_v(\varepsilon)$ and $g_c(\varepsilon)$ are the density of states in the valence and conduction bands, $n(\varepsilon) = \left[e^{(\varepsilon-\mu)/T} + 1\right]^{-1}$ being the Fermi–Dirac distribution. Then we rewrite this equality as follows:

$$\int_{-\infty}^{E_v} \left[1 - n(\varepsilon)\right]g_v(\varepsilon)d\varepsilon = \int_{E_c}^{\infty} n(\varepsilon)g_c(\varepsilon)d\varepsilon.$$

In this representation, this equality means that the number of electrons excited into the conduction band is equal to the number of free states or to the number of *holes* originated in the valence band. Then we denote $1 - n(\varepsilon)$ as $n_h(\varepsilon)$ where

$$n_h(\varepsilon) = 1 - n(\varepsilon) = \frac{1}{1 + e^{(\mu-\varepsilon)/T}} = \frac{1}{e^{(\varepsilon_h-\mu_h)/T} + 1}, \quad \varepsilon_h = -\varepsilon \text{ and } \mu_h = -\mu.$$

It is convenient to interpret the last expression as the Fermi–Dirac distribution for holes to which we can attribute the energy $\varepsilon_h(\boldsymbol{p}) = -\varepsilon(\boldsymbol{p})$ and the chemical potential $\mu_h = -\mu$ with the signs opposite to these quantities related to the valence band. It follows from this expression that the hole number N_h equals the number N_e of electrons occupying the conduction band. Since the total charge of all electrons in both valence and conduction bands remains unchanged, we can write equality $e_h N_h + eN_e = 0$ and interpret it as if the charge sign of holes is opposite to the electron one, i.e. positive $e_h = -e$.

Let us consider thermal excitation of electrons to the conduction band as a thermal ionization of the completely occupied valence band by producing the negatively charged electrons and positively charged holes. We have in the thermal equilibrium

$$\mu_h + \mu_e = 0.$$

Due to smallness $T \ll \Delta$, the concentrations of holes n_h and electrons n_e are not large and the Fermi–Dirac distribution can be replaced with the Boltzmann one. Then the chemical potentials are given for the ideal gases of electrons and holes with the particle masses m_e and m_h, respectively,

$$\mu_e = E_c + T \ln \frac{n_e}{2}\left(\frac{2\pi\hbar^2}{m_e T}\right)^{3/2},$$
$$\mu_h = -E_v + T \ln \frac{n_h}{2}\left(\frac{2\pi\hbar^2}{m_h T}\right)^{3/2}.$$

To derive these formulas, we have employed the following dispersion laws for electrons and holes:

$$\varepsilon_e(\boldsymbol{p}) = E_c + \boldsymbol{p}^2/2m_e \quad (E_c = \Delta),$$
$$\varepsilon_h(\boldsymbol{p}) = -E_v + \boldsymbol{p}^2/2m_h \quad (E_v = 0).$$

Substituting these chemical potentials into the above-derived equation of thermal equilibrium yields the analog of the Saha ionization equation

$$n_e n_h = 4\left(\frac{\sqrt{m_e m_h}\, T}{2\pi\hbar^2}\right)^3 e^{-\Delta/T}.$$

Taking the equality $n_h = n_e$ into account, we find

$$n_e(T) = n_h(T) = 2\left(\frac{\sqrt{m_e m_h}\, T}{2\pi\hbar^2}\right)^{3/2} e^{-\Delta/2T}.$$

The following chemical potential $\mu = \mu_e$ of a semiconductor corresponds to these concentrations:

$$\mu = \frac{\Delta}{2} + \frac{3}{4} T \ln \frac{m_h}{m_e} \approx \frac{\Delta}{2}.$$

The condition $\Delta \gg T$ justifies the use of the Boltzmann approximation. Within this accuracy, the chemical potential of an intrinsic (undoped) semiconductor is in the middle of the forbidden band.

2. A classical ideal gas of charged particles and antiparticles in volume V is in the thermodynamic equilibrium with the electromagnetic radiation (photon gas). Determine the energy $E = E(T)$ of the gas provided that the total charge of the gas is fixed.

Solution. Let us treat the particle–antiparticle production and annihilation as a thermodynamic equilibrium with respect to the process $a + \bar{a} \leftrightarrows \gamma$ by producing one or several photons. Involving that the chemical potential of photons is zero, we find the following relation between the chemical potentials of particles and antiparticles: $\mu_+ + \mu_- = 0$. Then we set $\mu_+ = \mu$ and $\mu_- = -\mu$. The condition of conserving the total electric charge means that difference $Q = N_+ - N_-$ between the numbers of particles and antiparticles is fixed.

Employing the Boltzmann distribution leads to

$$N = N_+ + N_- = V \int d\tau_p \, e^{\frac{\mu - \varepsilon_p}{T}} + V \int d\tau_p \, e^{\frac{-\mu - \varepsilon_p}{T}} = 2V \cosh \frac{\mu}{T} \int d\tau_p \, e^{-\frac{\varepsilon_p}{T}},$$

$$Q = N_+ - N_- = V \int d\tau_p \, e^{\frac{\mu - \varepsilon_p}{T}} - V \int d\tau_p \, e^{\frac{-\mu - \varepsilon_p}{T}} = 2V \sinh \frac{\mu}{T} \int d\tau_p \, e^{-\frac{\varepsilon_p}{T}}.$$

Here ε_p is the energy of a particle (antiparticle), and $d\tau_p = d^3p/(2\pi\hbar)^3$ is the phase volume element. Hence the total number $N(T)$ of all particles equals

$$N(T) = N_+ + N_- = \sqrt{Q^2 + V^2\chi^2(T)} \quad \text{and} \quad \chi(T) = 2 \int d\tau_p \, e^{-\frac{\varepsilon_p}{T}}.$$

The calculation of the gas energy is performed in a similar manner

$$E(T) = V \int d\tau_p \, \varepsilon_p e^{\frac{\mu - \varepsilon_p}{T}} + V \int d\tau_p \, \varepsilon_p \, e^{\frac{-\mu - \varepsilon_p}{T}} = 2V \cosh \frac{\mu}{T} \int d\tau_p \, \varepsilon_p e^{-\frac{\varepsilon_p}{T}}.$$

Introducing the average energy according to

$$\langle \varepsilon_p \rangle = \frac{\int d\tau_p \, \varepsilon_p \exp(-\varepsilon_p/T)}{\int d\tau_p \exp(-\varepsilon_p/T)},$$

we represent the result in the obvious form

$$E(T) = N(T)\langle \varepsilon_p \rangle.$$

For the spectrum $\varepsilon_p = p^2/2m$, we have $\langle c_p \rangle - 3T/2$ and $\langle \varepsilon_p \rangle = 3T$ if $\varepsilon_p = cp$.

The thermodynamic Gibbs free energy will be equal to

$$\Phi(T) = \mu_+ N_+ + \mu_- N_- = \mu(N_+ - N_-) = \mu(T)Q.$$

3. An ideal electron–positron plasma is in the thermodynamic equilibrium with the electromagnetic radiation (photon gas). Determine the chemical potential, and electron and positron densities as a function of temperature provided that the total electric charge of the plasma is fixed.

Solution. Let us treat the production and annihilation of electron–positron pairs as a thermodynamic equilibrium with respect to the process

$$e^+ + e^- \leftrightarrows \gamma$$

by producing one or several photons. Taking into account that the chemical potential of photons is zero, we have the following relation between the chemical potentials of electrons and positrons:

$$\mu_+ + \mu_- = 0.$$

Then we set $\mu_+ = \mu$ and $\mu_- = -\mu$. The condition of conserving the total electric charge means that the difference between the electron and positron densities is fixed as well

$$\Delta n = n_+ - n_- = 2 \int \frac{d^3p}{(2\pi\hbar)^3} \left(\frac{1}{e^{(\varepsilon_p - \mu)/T} + 1} - \frac{1}{e^{(\varepsilon_p + \mu)/T} + 1} \right) =$$

$$= 2 \int \frac{d^3p}{(2\pi\hbar)^3} \frac{\sinh \mu/T}{\cosh \varepsilon_p/T + \cosh \mu/T} = \text{const.}$$

Here $\varepsilon_p = c\sqrt{p^2 + m^2c^2}$ is the relativistic magnitude of energy and the equation determines implicitly the function $\mu = \mu(T)$. It is obvious that μ is an odd function Δn and $\mu = 0$ at $\Delta n = 0$. We set $\Delta n > 0$ below so that $\mu > 0$.

For $T = 0$, nonzero contribution arises from the electronic term alone and is limited by the Fermi energy in energy $\varepsilon_p \leqslant \mu(T = 0) = \varepsilon_F$ and by the Fermi momentum in momentum $p \leqslant p_F$. Then,

$$p_F = \hbar(3\pi^2\Delta n)^{1/3} \quad \text{and} \quad \varepsilon_F = c\sqrt{p_F^2 + m^2c^2}\,.$$

In the low-temperature limit $T \to 0$, the chemical potential behavior can be estimated with the aid of the expression for the density of states

$$\nu(\varepsilon) = \frac{\varepsilon}{\pi^2\hbar^3c^3}\sqrt{\varepsilon^2 - m^2c^4}\,,$$

according to

$$\mu(T) \approx \varepsilon_F - \frac{\pi^2T^2}{6}\frac{\nu'(\varepsilon_F)}{\nu(\varepsilon_F)} = \varepsilon_F\left[1 - \frac{\pi^2}{6}\frac{\varepsilon_F^2 + p_F^2c^2}{\varepsilon_F^2}\left(\frac{T}{p_Fc}\right)^2\right], \quad (T \ll cp_F).$$

The situation changes significantly in the high-temperature limit $T \to \infty$. Due to increasing the positron number as the temperature grows, the chemical potential continues to decrease and vanish, remaining always positive. In fact, assuming $\mu \ll T$ and $T \gg mc^2$, we find

$$\Delta n \approx 2\int \frac{d^3p}{(2\pi\hbar)^3}\frac{\mu/T}{\cosh\varepsilon_p/T + 1} \approx \frac{\mu T^2}{\pi^2\hbar^3c^3}\int_0^\infty \frac{x^2\,dx}{\cosh x + 1} = \frac{\mu T^2}{3\hbar^3c^3}\,.$$

We have set $x = pc/T$ and neglected the term $(mc^2/T)^2 \ll 1$ in the spectrum. Finally, we have for the high-temperature behavior $\mu(T)$ under the fixed value Δn

$$\mu(T) \approx \frac{3\hbar^3c^3\Delta n}{T^2} = \frac{(p_Fc)^3}{\pi^2T^2} \sim T^{-2}.$$

In the low-temperature limit the positron density can be estimated as

$$n_+(T) \approx 2\int \frac{d^3p}{(2\pi\hbar)^3}\, e^{-(\varepsilon_p+\mu)/T} \approx$$

$$\approx e^{-\varepsilon_F/T}\int_0^\infty \frac{p^2\,dp}{\pi^2\hbar^3}e^{-(mc^2+p^2/2m)/T} = 2\left(\frac{mT}{2\pi\hbar^2}\right)^{3/2} e^{-(\varepsilon_F+mc^2)/T},$$

and it is exponentially small due to presence of the gap in the energy spectrum of particles. In limit $T \gg mc^2$, the electron and positron densities become approximately equal and can be estimated with the integral at zero value of chemical potential with the ultrarelativistic energy spectrum $\varepsilon_p = pc$ of particles

$$n_+ \sim n_- \approx 2\int \frac{d^3p}{(2\pi\hbar)^3}\frac{1}{e^{pc/T} + 1} = \frac{T^3}{\pi^2\hbar^3c^3}\int_0^\infty \frac{x^2\,dx}{e^x + 1} = \frac{3\zeta(3)}{2\pi^2}\left(\frac{T}{\hbar c}\right)^3.$$

In the high temperature, the electron and positron densities in the plasma amount to three-quarters of the photon gas density, i.e. $n_- = n_+ = (3/4)n_\gamma$.

2.11 Chemical Equilibrium in the Ideal Gas Mixture

Let us apply the general statements of statistical physics and thermodynamics to describing the chemical equilibrium when the reactants and chemical reaction products are in the thermodynamic equilibrium. We consider below the chemical system as several subsystems numerated by index i, and each subsystem amounts to N_i particles (atoms or molecules) of ith component of the system. Each subsystem or component occupies the same overall volume V and has the identical temperature T as one of necessary conditions for the thermodynamic equilibrium between the subsystems. The physical process that makes it possible to establish the thermodynamic equilibrium between the subsystems consists in a possibility of transferring the particles from one subsystem to another.

According to additivity principle of entropy for the independent subsystems in the thermodynamic equilibrium, the entropy of the whole system $S = \sum_i S_i$ is a sum of entropies of each subsystems. The thermodynamic equilibrium of the closed system corresponds to the maximum magnitude of its entropy. Accordingly, we should have for the differential of the total entropy

$$dS = \sum_i dS_i = \frac{\sum_i dE_i}{T} + \frac{\sum_i P_i}{T} dV - \frac{\sum_i \mu_i dN_i}{T} = 0$$

where P_i is the partial pressure and μ_i is the chemical potential in ith subsystem. Taking into account that the entire system is closed, energy is conserved, i.e. $d(\sum_i E_i) = 0$, and the total volume remains unchanged, i.e. $dV = 0$, we arrive at the following relationship between the chemical potentials and the variations in the particle numbers of the subsystems:

$$\sum_i \mu_i dN_i = 0.$$

The chemical reaction or exchange with the reacting substances is represented as the following equation if all the terms are transferred to one side:

$$\sum_i \nu_i A_i \leftrightarrows 0.$$

Here the A_i is the symbol for denoting the reacting substances. The stoichiometric coefficients ν_i are the positive integers for the products of reaction and negative integers for the initial reactants of chemical conversion. The changes in the particle numbers N_i of components are related with the reaction equation. If, say, particle number N_1 changes by factor ν_1, each of the other numbers N_i will change by factor ν_i. Correspondingly,

$$\frac{dN_1}{\nu_1} = \frac{dN_2}{\nu_2} = \cdots \frac{dN_i}{\nu_i} = \cdots = dn.$$

Substituting this relation into the equation for the chemical potentials and canceling by dn, we derive the condition of thermodynamic equilibrium between the components of the system

$$\sum_i \nu_i \mu_i(P_i\,, T) = 0$$

where P_i is the partial pressure of ith components.

Let us apply the above-derived condition for chemical equilibrium to reactions in the ideal gas media. The general dependence for the chemical potential of ideal gas upon the pressure and temperature is given by the relation[7]

$$\mu_i(P_i,\ T) = T \ln P_i + \chi_i(T).$$

Then we have

$$\sum_i \nu_i \mu_i = T \sum_i \nu_i \ln P_i + \sum_i \chi_i(T) = 0.$$

Hence,

$$\prod_i P_i^{\nu_i} = e^{-\sum\limits_i \nu_i \chi_i / T} = K_p(T).$$

This relation is called the *law of mass action*. The right-hand side of equation, temperature-dependent alone, is referred to as the *equilibrium constant*.

This equation can be represented in the other way if we use the equation of ideal gas state $P_i = n_i T$ for each ith component, n_i being the particle density. Introducing the total pressure $P = \sum_i P_i$ and total particle density $n = \sum_i n_i$, we can represent the law of mass action in the following equivalent form:

$$\prod_i \left(\frac{n_i}{n}\right)^{\nu_i} = \frac{e^{-\sum\limits_i \nu_i \chi_i / T}}{P^{\sum\limits_i \nu_i}} = K_p(T) P^{-\sum\limits_i \nu_i} = K_c(P, T).$$

Determining the specific type of temperature behavior for the equilibrium constant $K_p(T)$ requires more knowledge about the properties of gases undergoing in the reaction.

Problem

1. Find the degree of dissociation for the ideal gas of diatomic molecules AB placed into the symmetrical harmonic trap of frequency ω_{ab}. The molecule consists of various atoms A and B. Consider the case of moderately low temperatures when the thermal excitation of the rotational and vibrational modes in the molecule can be neglected.

---

[7] The quantity $\varphi(T) = e^{\chi(T)/T}$ is often referred to as the *fugacity*.

Solution. The reaction looks like $AB \leftrightarrows A + B$. The thermodynamic equilibrium condition in the system consisting of three ideal gas subsystems implies the following equality for the chemical potentials:

$$\mu_{ab}(N_{ab}, T) = \mu_a(N_a, T) + \mu_b(N_b, T)$$

where N_{ab}, N_a, and N_b are the numbers of molecules and corresponding atoms. Substituting the values of chemical potentials yields

$$\varepsilon_{ab} - 3T \ln \frac{T}{\hbar\omega_{ab}N_{ab}^{1/3}} = \varepsilon_a - 3T \ln \frac{T}{\hbar\omega_a N_a^{1/3}} + \varepsilon_b - 3T \ln \frac{T}{\hbar\omega_b N_b^{1/3}}\,.$$

Here ε_{ab}, ε_a, and ε_b are the energies of molecule and atoms, respectively. The frequencies ω_{ab}, ω_a, and ω_b for the molecule and atoms under the trapping potential $U(r) = kr^2/2$ are related as $m_{ab}\omega_{ab}^2 = m_a\omega_a^2 = m_b\omega_b^2 = k$ via the corresponding masses of particles. This results in the equation

$$\frac{N_a N_b}{N_{ab}} = \left(\frac{T\omega_{ab}}{\hbar\omega_a\omega_b}\right)^3 e^{-\Delta/T} = N_0\left(\frac{T}{\hbar\omega_{ab}N_0^{1/3}}\,\frac{\sqrt{m_a m_b}}{m_{ab}}\right)^3 e^{-\Delta/T} = N_0 K(T, N_0),$$

where $\Delta = \varepsilon_a + \varepsilon_b - \varepsilon_{ab}$ is the dissociation energy for the molecule and N_0 is the initial number of molecules.

Let us introduce degree of dissociation as a ratio $\alpha = (N_0 - N_{ab})/N_0$ of the number of dissociated molecules to the initial total number of molecules. Accordingly, $N_a = N_b = \alpha N_0$. Then,

$$\frac{\alpha^2}{1-\alpha} = K(T, N_0) \quad \text{or} \quad \alpha = \frac{2K}{K + \sqrt{K^2 + 4K}}\,.$$

At low $T \ll \Delta$ temperatures it is expected $\alpha(T) \sim e^{-\Delta/2T}$.

Chapter 3
Quantum Ideal Gases

3.1 The Gibbs Distribution for the Systems with a Variable Number of Particles. The Grand Thermodynamic Potential

When treating the quantum thermodynamic systems with the macroscopically large number of interacting particles $N \to \infty$, in order to simplify the calculations, it is more convenient to assume that the number of particles in the system under study is not fixed and can vary. Thus, we should consider the thermodynamic potentials depending on the variable conjugated to the number of particles N, namely *chemical potential* μ. In other words, we introduce the *Grand thermodynamic potential* $\Omega(T, \mu)$ instead of the Helmholtz free energy $F(T, N)$ according to

$$\Omega(T, \mu) = F - \mu N.$$

Here the chemical potential μ plays a role of an independent thermodynamic variable instead of the number of particles N. Accordingly, the number of particles $N = N(\mu)$ in the system becomes a function of chemical potential and is determined with the formula relating the number of particles with the chemical potential in a thermodynamically equilibrium system

$$N = -\left(\frac{\partial \Omega}{\partial \mu}\right)_T.$$

On the analogy with the definition of partial function Z which allows us to calculate the free energy F, we can introduce the definition of the *grand partition function* in order to find the *grand thermodynamic potential* Ω and then to determine the

S. N. Burmistrov, *Statistical and Condensed Matter Physics*,
https://doi.org/10.1007/978-981-97-9072-2_3

number of particles in the system. In accordance with the content presented in the first chapter, for this purpose it is sufficient to consider the following Hamiltonian:

$$\hat{\mathcal{H}} = \hat{H} - \mu\hat{N}$$

where $\hat{N}$ is the particle number operator. The grand partition function $\mathcal{Z}$ corresponding to Hamiltonian $\hat{\mathcal{H}}$ is written as

$$\mathcal{Z} = \operatorname{tr} e^{-\hat{\mathcal{H}}/T}.$$

Here we should take into account that the state of the system under variable number of particles is already characterized by an additional parameter N_k or number of particles in state $|k\rangle$ over and above the set of quantum states $|k\rangle$ with the appropriate energies ε_k. Thus, the state of the system is classified with the enlarged state vector $|\mathcal{N}\rangle = |k, N_k\rangle$. The equivalent expression for the grand partition function reads

$$\mathcal{Z} = \sum_{k,N_k} \langle k, N_k| \exp\left(-\frac{\hat{H} - \mu\hat{N}}{T}\right)|k, N_k\rangle = \sum_{k,N_k} \exp\left(-\frac{\varepsilon_{k,N_k} - \mu N_k}{T}\right).$$

The probability to find the system in the state with vector $|\mathcal{N}\rangle = |k, N_k\rangle$ is given by the grand canonical distribution

$$\tilde{w}_{k,N_k} = \exp\left(\frac{\Omega + \mu N_k - \varepsilon_{k,N}}{T}\right).$$

The normalization condition $\sum_{\mathcal{N}} \tilde{w}_{\mathcal{N}} = 1$ agrees with the definition of grand thermodynamic potential $\Omega(\mu)$

$$\Omega(\mu) = -T \ln \sum_{k,N_k} e^{(\mu N_k - \varepsilon_{k,N_k})/T}.$$

It is easy to check the correspondence between the thermodynamic and statistical definitions of the average number of particles in the system. In fact,

$$N = -\left(\frac{\partial\Omega}{\partial\mu}\right)_T = T\frac{\sum_{\mathcal{N}} N_k e^{(\mu N_k - \varepsilon_{k,N_k})/T}}{\sum_{\mathcal{N}} e^{(\mu N_k - \varepsilon_{k,N_k})/T}} = \sum_{k,N_k} N_k \tilde{w}_{k,N_k}.$$

The thermodynamic properties of the system and its physical behavior are greatly affected by such a factor as the maximum number of particles M, which can simultaneously occupy the same quantum state and have the same quantum numbers. The possible sets of population numbers with vector $|k\rangle$ and energy ε_k, in general, run through the integers $N_k = 0, 1, 2, \ldots, M$. For the Fermi particles or *fermions* in accordance with the Pauli principle, this is $N_k = 0,\ 1\ (M = 1)$ alone or no more than one particle can be in the same quantum state. On the contrary, for the Bose particles or *bosons* an arbitrary number of particles with the identical quantum numbers can occupy the same quantum state. In the limiting case all the particles, at least, can populate the same quantum state. In formal words: $N_k = 0, 1, 2, \ldots \infty$ $(M = \infty)$.[1]

When we analyze the systems with the variable number of particles, the quantity $n(\varepsilon_k)$, i.e. average number of particles in the quantum state $|k\rangle$ with energy ε_k, plays a key role in the description of their physical properties. In the case of fermions, the average number of particles in state $|k\rangle$ is referred to as the *Fermi–Dirac distribution* and is found from the following sum over the occupation numbers:

$$n_F(\varepsilon_k) = \sum_{N_k=0}^{1} N_k \tilde{w}_{kN_k} = \frac{0 \cdot 1 + 1 \cdot e^{(\mu-\varepsilon_k)/T}}{1 + e^{(\mu-\varepsilon_k)/T}} = \frac{1}{e^{(\varepsilon_k-\mu)/T} + 1}\,.$$

The total number of particles N in the system can be found by summarizing over all possible states $|k\rangle$ of the system

$$N = \sum_k n_F(\varepsilon_k) = \sum_k \frac{1}{e^{(\varepsilon_k-\mu)/T} + 1}\,.$$

This equality determines the relation between chemical potential μ and the number of particles N in the Fermi system.

The grand potential Ω_F is determined from the relation

$$e^{-\Omega_F/T} = \sum_{k,\,N_k} e^{(\mu N_k - \varepsilon_k N_k)/T} = \sum_{N_k=0}^{1} \sum_k \left(e^{(\mu-\varepsilon_k)/T}\right)^{N_k} =$$
$$= \prod_k \sum_{N_k=0}^{1} \left(e^{(\mu-\varepsilon_k)/T}\right)^{N_k} = \prod_k \left(1 + e^{(\mu-\varepsilon_k)/T}\right)$$

and equals the following sum over the states $|k\rangle$ of the system:

$$\Omega_F = -T \sum_k \ln\left(1 + e^{(\mu-\varepsilon_k)/T}\right).$$

For the system of the Bose particles, the speculations are completely analogous. For the Bose system, the average number of particles in quantum state $|k\rangle$ with energy ε_k is referred to as the *Bose–Einstein distribution* and can be determined from the following sum over the occupation numbers:

[1] The intermediate case $1 < M < \infty$ is called *parastatistics* and is not examined in the book.

$$n_B(\varepsilon_k) = \sum_{N_k=0}^{\infty} N_k \tilde{w}_{kN_k} = \frac{0 \cdot 1 + 1 \cdot e^{(\mu-\varepsilon_k)/T} + 2 \cdot e^{2(\mu-\varepsilon_k)/T} + \cdots}{1 + e^{(\mu-\varepsilon_k)/T} + e^{2(\mu-\varepsilon_k)/T} + \cdots} =$$
$$= \frac{1}{e^{(\varepsilon_k-\mu)/T} - 1}\,.$$

The total number of particles N in the system is given by summarizing over all possible states $|k\rangle$ and equals

$$N = \sum_k n_B(\varepsilon_k) = \sum_k \frac{1}{e^{(\varepsilon_k-\mu)/T} - 1}\,.$$

This equality determines the relation between chemical potential μ and particle number N in the Bose system. For convergency and positive definiteness of the Bose–Einstein distribution in the thermodynamically equilibrium Bose system, the chemical potential cannot exceed the minimally possible magnitude of energy for the states in the system, i.e. $\mu \leqslant \varepsilon_{\min}$.

The grand thermodynamic potential Ω_B is given by the relation

$$e^{-\Omega_B/T} = \sum_{k,\,N_k} e^{(\mu N_k - \varepsilon_k N_k)/T} = \sum_{N_k=0}^{\infty} \sum_k \left(e^{(\mu-\varepsilon_k)/T}\right)^{N_k} =$$
$$= \prod_k \sum_{N_k=0}^{\infty} \left(e^{(\mu-\varepsilon_k)/T}\right)^{N_k} = \prod_k \left(1 - e^{(\mu-\varepsilon_k)/T}\right)^{-1}.$$

Hence the grand potential Ω in the Bose system equals the following sum over all the states $|k\rangle$:

$$\Omega_B = T \sum_k \ln\left(1 - e^{(\mu-\varepsilon_k)/T}\right).$$

The relation $N = -\partial\Omega_B/\partial\mu$ is obvious and remains valid as well.

In conclusion, we will show how the general scheme of speculations outlined above can be applied to the case of classical Boltzmann statistics. In the latter case, the unlimited number of particles can occupy the same state ($M = \infty$). However, before calculating the partition function

$$e^{-\Omega_{\rm cl}/T} = \mathcal{Z} = \sum_{k,N_k} \frac{e^{-(N_k\varepsilon_k - \mu N_k)/T}}{N_k!} = \sum_{N_k=0}^{\infty} \sum_k \frac{\left(e^{(\mu-\varepsilon_k)/T}\right)^{N_k}}{N_k!} =$$
$$= \prod_k \sum_{N_k=0}^{\infty} \frac{\left(e^{\mu-\varepsilon_k)/T}\right)^{N_k}}{N_k!} = \prod_k \exp\left(e^{(\mu-\varepsilon_k)/T}\right),$$

it is important to determine correctly the number of different physical states in the system of classical particles. Thus we have divided the above sum by the Gibbs factor $N_k!$, taking into account the number of the possible particle permutations which keep the state of the system unchanged. As a result, we arrive at the grand thermodynamic potential

$$\Omega_{\rm cl} = -T\sum_k e^{(\mu-\varepsilon_k)/T}.$$

The total number of particles in the system equals

$$N = \sum_k e^{(\mu-\varepsilon_k)/T}.$$

Obviously, this corresponds completely to the Boltzmann distribution

$$n_{\rm cl}(\varepsilon_k) = \exp(\mu-\varepsilon_k)/T.$$

Let us note here, in the limit $\exp(\varepsilon_k - \mu/T) \gg 1$ both Fermi–Dirac and Bose–Einstein distributions go over to the classical Boltzmann distribution. At the same time the occupation numbers of states $n(\varepsilon_k) \ll 1$ are small for all three statistics.

Problems

1. Find the mean square of the number of fermions $\langle n^2(\varepsilon_k)\rangle$ in the quantum state $|k\rangle$ with energy ε_k.

Solution. Using the definition for the mean value, we find

$$\langle n^2(\varepsilon_k)\rangle = \sum_{N_k=0}^{1} N_k^2\tilde{w}_{kN_k} = \frac{0^2\cdot 1 + 1^2\cdot e^{(\mu-\varepsilon_k)/T}}{1+e^{(\mu-\varepsilon_k)/T}} = \frac{1}{e^{(\varepsilon_k-\mu)/T}+1} = n_F(\varepsilon_k).$$

The mean square fluctuation of the number of fermions $\langle\big(n(\varepsilon_k) - n_F(\varepsilon_k)\big)^2\rangle$ in the quantum state $|k\rangle$ is equal to

$$\langle(\Delta n_k)^2\rangle = \langle n_k^2\rangle - \langle n_k\rangle^2 = n_F(\varepsilon_k)\big(1-n_F(\varepsilon_k)\big).$$

2. Find the mean square of the number of bosons $\langle n^2(\varepsilon_k)\rangle$ in the quantum state $|k\rangle$ with energy ε_k.

Solution. We find using the analogy with the previous problem

$$\langle n^2(\varepsilon_k)\rangle = \sum_{N_k=0}^{\infty} N_k^2\tilde{w}_{kN_k} = \frac{0^2\cdot 1 + 1^2\cdot e^{(\mu-\varepsilon_k)/T} + 2^2\cdot e^{2(\mu-\varepsilon_k)/T}+\cdots}{1+e^{(\mu-\varepsilon_k)/T}+e^{2(\mu-\varepsilon_k)/T}+\cdots} =$$

$$= \frac{e^{(\varepsilon_k-\mu)/T}+1}{\big(e^{(\varepsilon_k-\mu)/T}-1\big)^2} = 2n_B^2(\varepsilon_k) + n_B(\varepsilon_k).$$

Calculating the sum in the numerator, we employ the formula

$$\sum_{n=0}^{\infty} n^2 q^n = \frac{q(1+q)}{(1-q)^3} \quad (q<1).$$

The mean square fluctuation of the number of bosons $\langle (n(\varepsilon_k) - n_F(\varepsilon_k))^2 \rangle$ in the quantum state $|k\rangle$ is equal to

$$\langle (\Delta n_k)^2 \rangle = \langle n_k^2 \rangle - \langle n_k \rangle^2 = n_B(\varepsilon_k)\big(1 + n_B(\varepsilon_k)\big).$$

For the Boltzmann distribution, we have from the condition $n_F(\varepsilon_k) \ll 1$ or $n_B(\varepsilon_k) \ll 1$

$$\langle (\Delta n_k)^2 \rangle = \langle n_k \rangle.$$

3.2 Ideal Fermi Gas

Let us proceed to analyzing the physical properties of ideal Fermi systems. We assume below that the system consists of non-interacting fermions with spin $\sigma = 1/2$. So, the Fermi–Dirac distribution, representing the mean number of fermions $n_{k,\sigma}$ in quantum state $|k, \sigma\rangle$ with energy $\varepsilon_{k,\sigma}$, is given by the formula

$$n(\varepsilon_{k,\sigma}) = \sum_{k,\sigma} \frac{1}{e^{(\varepsilon_{k,\sigma}-\mu)/T} + 1}$$

and μ is the chemical potential of fermions. Summing over all possible states yields the total number of fermions N and grand potential $\Omega(T, \mu)$ in the system

$$N(\mu) = \sum_{k,\sigma} n(\varepsilon_{k,\sigma}) = \sum_{k,\sigma} \frac{1}{e^{(\varepsilon_{k,\sigma}-\mu)/T} + 1},$$

$$\Omega(T, \mu) = -T \sum_{k,\sigma} \ln\big(1 + e^{(\mu-\varepsilon_{k,\sigma})/T}\big).$$

These equations are the main ones describing the properties of an ideal Fermi gas.

Let us focus first on the low-temperature properties of a Fermi gas. The high-temperature properties can be described by the Boltzmann distribution and, therefore, are similar to those of a classical gas. Below we mean the Fermi gas of neutral particles or non-interacting spin-1/2 electrons with energy spectrum $\varepsilon_p = p^2/2m$.

At zero temperature the Fermi–Dirac distribution takes the step-like behavior $n(\varepsilon_{p,\sigma}) = \vartheta(\mu - \varepsilon_{p,\sigma})$. From the physical point of view, this means that the fermions occupy progressively all the energy states to the chemical potential magnitude in order to realize the state with the least possible total energy, i.e. ground state of the Fermi system. The boundary energy or chemical potential μ at $T = 0$ is called the *Fermi energy* $\varepsilon_F = \mu(T = 0)$. The corresponding boundary momentum $p_F = \sqrt{2m\varepsilon_F}$ is referred to as the *Fermi momentum*.

In the momentum space the particles occupy all the states inside the sphere of radius $p = p_F$, called the Fermi sphere or Fermi surface. The Fermi momentum depends on the gas density $n = N/V$ alone

$$p_F = \hbar(3\pi^2 n)^{1/3}$$

and is calculated from the equality

$$N = \sum_\sigma \int \frac{Vd^3p}{(2\pi\hbar)^3}\vartheta(\mu - \varepsilon_p) = 2\frac{V}{(2\pi\hbar)^3}\frac{4\pi}{3}p_F^3.$$

Here we have involved that the fermions with the opposite spin directions are two different states, though they have the same energies $\varepsilon_p = p^2/2m$. The Fermi energy equals

$$\varepsilon_F = \frac{p_F^2}{2m} = (3\pi^2)^{2/3}\frac{\hbar^2 n^{2/3}}{2m}.$$

The total energy of the gas reads

$$E(T=0) = 2\int \frac{Vd^3p}{(2\pi\hbar)^3}\,\varepsilon_p n(\varepsilon_p) = \frac{3}{5}\,\varepsilon_F N.$$

Hence we obtain the corresponding pressure or equation of states at $T = 0$

$$P(T=0) = \frac{(3\pi^2)^{2/3}}{5}\frac{\hbar^2}{m}\left(\frac{N}{V}\right)^{5/3} = \frac{2}{3}\frac{E}{V} \sim V^{-5/3}$$

and then we find the sound velocity c in an ideal Fermi gas at zero temperature

$$c = \sqrt{\partial P/\partial\rho} = v_F/\sqrt{3}.$$

Here the mass density is $\rho = mN/V$. The *Fermi velocity* $v_F = p_F/m$ is defined as the velocity of particles at the Fermi surface.

The Fermi gas at temperatures $T \ll \varepsilon_F$ is referred to as *degenerate* and as *non-degenerate* at $T \gg \varepsilon_F$. The temperature defined by relation $T \sim \varepsilon_F$ is often called the *degeneracy temperature*.

Problems

1. Let electron energy spectrum $\varepsilon(\boldsymbol{p})$ be given by the model of *effective mass tensor* where the constant-energy surface represents the ellipsoid with the following dispersion in the principal axes:

$$\varepsilon(\boldsymbol{p}) = \frac{p_x^2}{2m_x} + \frac{p_y^2}{2m_y} + \frac{p_z^2}{2m_z},$$

m_x, m_y, and m_z being the effective electron masses along the principal axes of the ellipsoid.

Find the Fermi energy and one-particle density of states if the electron density equals n.

Solution. The number of states with energy ε in the volume V is determined with the integral

$$\Gamma(\varepsilon) = 2V \int\limits_{\varepsilon(\boldsymbol{p})\leqslant\varepsilon} \frac{d^3p}{(2\pi\hbar)^3} = V\frac{\sqrt{m_x m_y m_z}\,(2\varepsilon)^{3/2}}{3\pi^2\hbar^3}.$$

At zero temperature all the states with the energy smaller than the Fermi energy are occupied with the electrons. In other words, the number of electrons $N = nV$ equals the number of states $\Gamma(\varepsilon_F)$.

Hence we have

$$\varepsilon_F = \frac{1}{2}\frac{(3\pi^2\hbar^3 n)^{2/3}}{(m_x m_y m_z)^{1/3}}.$$

The density of states $g(\varepsilon)$ is given by the routine differentiation

$$g(\varepsilon) = \frac{d\Gamma(\varepsilon)}{d\varepsilon} = V\frac{\sqrt{m_x m_y m_z}\,(2\varepsilon)^{1/2}}{\pi^2\hbar^3}.$$

2. The electrical field of strength E is applied in the normal direction to the surface of low doped n-type semiconductor. Find the electron density profile near the semiconductor surface, using the semiclassical *Thomas–Fermi approximation*. The permittivity equals κ and electron mass is m. The electron concentration n is very small, Fermi level ε_F lies near the conduction band bottom, and $na_B^3 \ll 1$ where $a_B = \kappa\hbar^2/me^2$ is the effective Bohr radius. The temperature of semiconductor is $T \ll \varepsilon_F$, and the gas of electrons is implied to be completely degenerate.

Solution. The electric field strength E can be related with the electron surface density N as follows:

$$E = \frac{4\pi eN}{\kappa}.$$

To find the electron density profile, we use the condition of the electrochemical potential constancy over the conductor volume

$$e\varphi(x) + \mu(x) = \varepsilon_F \approx 0.$$

Here $\varphi(x)$ is the electrostatic potential at distance x from the surface. The local chemical potential is related to the electron density $n(x)$ as usual

$$\mu(x) = \frac{\hbar^2\left[3\pi^2 n(x)\right]^{2/3}}{2m}.$$

The electrostatic potential and charge density satisfy the Poisson equation

$$\kappa\varphi''(x) = -4\pi e n(x).$$

Employing $\mu(x) = -e\varphi(x)$ and last two equations, we arrive at

$$a_B\left(-\frac{2me}{\hbar^2}\varphi''\right) = \frac{8}{3\pi}\left(-\frac{2me}{\hbar^2}\varphi\right)^{3/2}.$$

Solving this equation under condition $\varphi(x\to\infty) = 0$ discloses a power-like behavior

$$\varphi(x) = -\frac{225\pi^2}{8}\frac{e}{\kappa}\frac{a_B^3}{(x+d)^4}.$$

The integration constant d means the width of the surface charged layer. The electron density behavior is subjected to the power-like behavior as well

$$n(x) = \frac{1125\pi}{8}\frac{a_B^3}{(x+d)^6}.$$

The layer width d can be related to the surface electron density N as

$$N = \int_0^\infty n(x)dx = \frac{1125\pi}{8}\int_0^\infty \frac{a_B^3 dx}{(x+d)^6} = \frac{225\pi}{8}\frac{a_B^3}{d^5}.$$

Hence we find

$$d = a_B\left(\frac{225\pi}{8}\frac{1}{Na_B^2}\right)^{1/5} = a_B\left(\frac{225\pi^2}{2}\frac{e}{\kappa a_B^2}\frac{1}{E}\right)^{1/5}.$$

Let us discuss condition of small electron concentration n_0 corresponding to the Fermi energy $\varepsilon_F = \hbar^2(3\pi^2 n_0)^{2/3}/2m$. The approximation chosen supposes inequality $\varepsilon_F \ll |e\varphi(x \sim d)|$. Accordingly,

$$n_0 a_B^3 \ll \left(\frac{a_B}{d}\right)^6 \quad \text{or} \quad E \gg E_c \sim \frac{e}{\kappa a_B^2}\left(n_0 a_B^3\right)^{5/6}.$$

Thus, the strength E of the necessary electric field proves to be small as compared with the typical electric field strength $e/\kappa a_B^2$ due to inequality $n_0 a_B^3 \ll 1$.

3. An ideal gas of the Fermi particles with the fixed spin projection and energy spectrum $\varepsilon_p = p^2/2m$ is in the one-dimensional potential box of length L. If the particles do not penetrate through the box walls, the energy spectrum becomes discrete

$$\varepsilon_n = \frac{\hbar^2\pi^2}{2mL^2}n^2, \quad n = 1, 2, 3, \ldots$$

At zero temperature the particles occupy progressively the energy levels to the Fermi energy. The total energy for the gas of N particles equals

$$E(N) = \sum_{n=1}^{N}\varepsilon_n = \frac{\hbar^2\pi^2}{2mL^2}\,\frac{N(N+1)(2N+1)}{6}.$$

The box is split with the membrane into two sections of sizes L_1 and L_2 $(L = L_1 + L_2)$. In the sections there are N_1 and N_2 particles $(N = N_1 + N_2)$, respectively. The small holes in the membrane do not disturb the energy levels in the separate sections of the box but permit the particles to penetrate from the left-hand side of the box to the right-hand one and vice versa.

Find the *mesoscopic* oscillations of the force that acts on the membrane as a function of its position. The mesoscopic oscillations are due to finiteness of the particle numbers.

Solution. The total energy of the system is a sum of the energies of the left- and right-hand box sections $E = E_1 + E_2$, i.e.

$$E(N_1, N_2) = \frac{\hbar^2\pi^2}{12m}\left(\frac{N_1(N_1+1)(2N_1+1)}{L_1^2} + \frac{N_2(N_2+1)(2N_2+1)}{L_2^2}\right).$$

Since the left- and right-hand sections of the box can exchange the particles, the ground state of the system can be found by minimizing its total energy in N_1 under the fixed total number of particles $N = N_1 + N_2$. In the macroscopic limit when $N_{1,2}$ and $L_{1,2} \to \infty$, we have an equality of chemical potentials

$$\mu_1 = \partial E_1/\partial N_1 = \partial E_2/\partial N_2 = \mu_2,$$

resulting in the equalities of the particle densities and pressures in both sections

$$n = \frac{N_1}{L_1} = \frac{N_2}{L_2}, \quad P_1 = P_2 = \frac{\hbar^2\pi^2}{3m}n^3 \quad \text{and} \quad n = \frac{N_1 + N_2}{L_1 + L_2} = \frac{N}{L}.$$

Thus the net force acting on the membrane vanishes at any position of the membrane.

The finiteness of particle numbers in the system leads to the *mesoscopic effects*. Let us consider the membrane position characterized by the lengths L_1 and L_2 and compare the energy minima corresponding to the neighboring set of particle numbers (N_1, N_2) and $(N_1 + 1, N_2 - 1)$. The transition from one set to another becomes energetically favorable provided that

$$E(N_1+1,\, N_2-1) - E(N_1,\, N_2) = \frac{\hbar^2\pi^2}{2m}\left(\frac{(N_1+1)^2}{L_1^2} - \frac{N_2^2}{L_2^2}\right) = 0.$$

This yields the following values for the membrane positions:

$$\frac{N_1+1}{L_1} = \frac{N_2}{L_2} \quad \text{or} \quad \frac{L_1}{L} = \frac{N_1+1}{N+1} \quad \text{and} \quad \frac{L_2}{L} = \frac{N_2}{N+1}.$$

The force acting on the membrane in the state (N_1, N_2) is given by the difference of derivatives

$$F(N_1, L_1;\ N_2, L_2) = -\frac{\partial E_1(N_1,\, L_1)}{\partial L_1} + \frac{\partial E_2(N_2,\, L_2)}{\partial L_2} =$$
$$= \frac{\hbar^2\pi^2}{6m}\left(\frac{N_1(N_1+1)(2N_1+1)}{L_1^3} - \frac{N_2(N_2+1)(2N_2+1)}{L_2^3}\right).$$

When the numbers of the particles vary by unity on the left-hand and right-hand sides of the membrane, i.e. at $L_1/L = (N_1+1)/(N+1)$ and $L_2/L = N_2/(N+1)$, the magnitude of the force depends on the particle numbers in the left-hand and right-hand sides of the membrane

$$F(N_1,\, N_2) = -\frac{\hbar^2}{6m}\left(\frac{N+1}{L}\right)^3\left(\frac{3N_1+2}{(N_1+1)^2} + \frac{3N_2+1}{N_2^2}\right) < 0,$$
$$F(N_1+1,\, N_2-1) = \frac{\hbar^2\pi^2}{6m}\left(\frac{N+1}{L}\right)^3\left(\frac{3N_1+4}{(N_1+1)^2} + \frac{3N_2-1}{N_2^2}\right) > 0.$$

Every time when a particle transmits from one side of the membrane to another, a jump in the force, acting on the membrane, occurs from the negative value to the positive one. The magnitude of the jump depends on N_1 and N_2 and equals

$$F(N_1+1,\, N_2-1) - F(N_1,\, N_2) = \frac{\hbar^2\pi^2}{m}\left(\frac{N+1}{L}\right)^3\left(\frac{1}{N_1+1} + \frac{1}{N_2}\right).$$

This results in the force oscillations around zero macroscopic value as a function of the membrane position. The oscillation period is $L/(N+1)$ and the relative amplitude of oscillations is about

$$\frac{\Delta F}{\hbar^2\pi^2 n^3/m} \sim \frac{1}{\min(N_1,\, N_2)}.$$

The example of oscillations is shown in Fig. 3.1.

4. The size quantization of electron spectrum in a metal film influences the physical properties of the film.

Find the surface energy of the film and its surface tension, using the model of free non-interacting electrons. Assume that the film thickness L is much larger as compared with the mean spacing between electrons. The temperature of the film is zero and electron concentration is n.

Solution. In order to calculate the surface energy of an electron gas limited with size L in one of the directions, we should compare the energy of the system for the uniform electron density distribution over the film thickness with that when the density distribution becomes inhomogeneous near the film walls. It is more convenient to perform such a comparison for the energy as a function of chemical potential μ but as a function of the total electron number N. In other words, we will compare two grand thermodynamic potentials $\Omega(\mu)$ determined as

$$\Omega(\mu) = E(N) - \mu N.$$

Let us write energy for electron of mass m, which quantizes in the normal direction to the film walls and is continuous in the two longitudinal directions

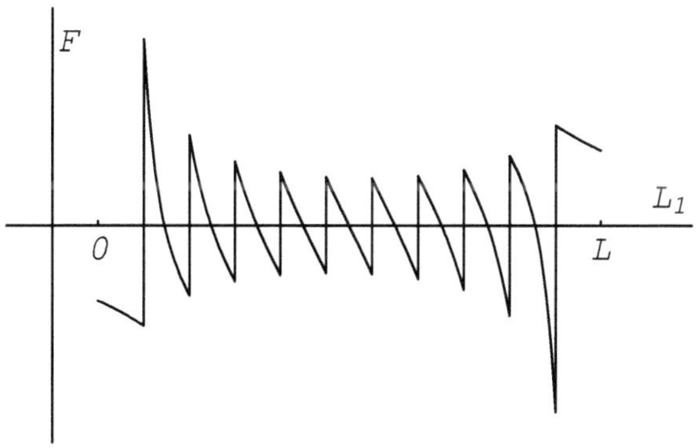

Fig. 3.1 The example of mesoscopic force oscillations depending on the membrane position L_1

$$\epsilon_k(p_x, p_y) = \frac{p_x^2 + p_y^2}{2m} + \varepsilon_k\,, \quad \varepsilon_k = \frac{\pi^2\hbar^2 k^2}{2mL^2} \quad k = 1,\, 2,\, 3\ldots$$

The total electron number N in the film of the wall area S is given by the following integral with the step-like Heaviside function ϑ:

$$N = S\sum_k \int \frac{2dp_x\, dp_y}{(2\pi\hbar)^2}\vartheta\big(\mu - \epsilon_k(p_x, p_y)\big) = S\frac{m}{\pi\hbar^2}\sum_k \int_0^\infty \vartheta(\mu - \varkappa - \varepsilon_k)d\varkappa =$$

$$= S\frac{m}{\pi\hbar^2}\sum_{k=1}^{K}\left(\mu - \frac{\pi^2\hbar^2 k^2}{2mL^2}\right) = S\frac{m}{\pi\hbar^2}\left(\mu K - \frac{\pi^2\hbar^2}{2mL^2}\frac{K(K+1)(2K+1)}{6}\right).$$

Here μ is the chemical potential and K is the number of the upper occupied energy level for the transverse motion. Accordingly, we obtain the relation

$$\mu = \frac{\pi^2\hbar^2 K^2}{2mL^2}\,.$$

Calculating the energy of a gas is performed in a similar way

$$E = S\sum_k \int \frac{2dp_x\, dp_y}{(2\pi\hbar)^2}\epsilon_k(p_x, p_y)\vartheta\big(\mu - \epsilon_k(p_x, p_y)\big) =$$

$$= S\frac{m}{\pi\hbar^2}\int_0^\infty (\varkappa + \varepsilon_k)\vartheta(\mu - \varkappa - \varepsilon_k)d\varkappa = S\frac{m}{2\pi\hbar^2}\sum_{k=1}^{K}\left[\mu^2 - \left(\frac{\pi^2\hbar^2}{2mL^2}\right)^2 k^4\right] =$$

$$= S\frac{m}{2\pi\hbar^2}\left[\mu^2 K - \left(\frac{\pi^2\hbar^2}{2mL^2}\right)^2\frac{K(K+1)(2K+1)(3K^2+3K-1)}{30}\right].$$

The grand potential $\Omega(\mu)$ reads

$$\Omega(\mu) = E - \mu N = -\frac{4}{15}S\frac{m}{\pi\hbar^2}\left(\frac{\pi^2\hbar^2}{2mL^2}\right)^2 K^5\left(1 - \frac{15}{16}\frac{1}{K} - \frac{1}{16}\frac{1}{K^4}\right).$$

In the expansion of grand potential $\Omega(\mu)$ in powers $K^{-1} \ll 1$ the term proportional to the film thickness L yields the magnitude of potential Ω_V for the homogeneous electron density distribution, and its magnitude is proportional to the film volume $V = SL$. The next expansion term proportional to the film area S will determine the surface contribution Ω_S to the total energy of the film. For $L \gg \hbar/p_F$ where p_F is the Fermi momentum, it is sufficient to restrict ourselves with the approximation

$$\Omega(\mu) = -\frac{4}{15}S\frac{m}{\pi\hbar^2}\left(\frac{\pi^2\hbar^2}{2mL^2}\right)^2 K^5\left(1-\frac{15}{16}\frac{1}{K}\right)+\cdots, \quad \mu = \frac{\pi^2\hbar^2}{2mL^2}K^2.$$

Hence we readily arrive at

$$\Omega(\mu) = -\frac{2}{5}SLn\mu + \frac{3}{8}Sn\mu\frac{\pi\hbar}{p_F} = \Omega_V + \Omega_S \quad \text{where} \quad n = \frac{p_F^3}{3\pi^2\hbar^3} \quad \text{and} \quad \frac{p_F^2}{2m} = \mu.$$

The first term in the sum corresponds to the bulk contribution and the second one does to the surface contribution. Since the film has two sides, we determine the surface tension coefficient σ from the relation $\Omega_S = 2\sigma S$. As a result,

$$\sigma = \frac{3\pi}{16}n\mu\frac{\hbar}{p_F} = \frac{(9\pi)^{1/3}}{16}\mu n^{2/3} = \frac{1}{32\pi}\frac{1}{m}\left(\frac{\hbar}{p_F}\right)^4.$$

The magnitude of surface tension proves to be about μ/a^2, $a \sim n^{-1/3}$ being the mean distance between electrons.

3.3 Specific Heat and the Pauli Paramagnetism of a Degenerate Fermi Gas

The thermodynamic potentials of a Fermi gas can readily be found using the one-particle density of states $g(\varepsilon)$. So, the total energy of a gas with N fermions in volume V can be calculated as follows:

$$E = \int_0^\infty d\varepsilon\, \varepsilon g(\varepsilon) f(\varepsilon) = \int_0^\infty d\varepsilon\, \frac{\varepsilon g(\varepsilon)}{e^{(\varepsilon-\mu)/T}+1},$$

$$N = \int_0^\infty d\varepsilon\, g(\varepsilon) f(\varepsilon) = \int_0^\infty d\varepsilon\, \frac{g(\varepsilon)}{e^{(\varepsilon-\mu)/T}+1},$$

$$g(\varepsilon) = 2V\frac{4\pi p^2(\varepsilon)}{(2\pi\hbar)^3}\frac{dp}{d\varepsilon} = V\frac{mp(\varepsilon)}{\pi^2\hbar^3}, \quad p(\varepsilon) = \sqrt{2m\varepsilon}.$$

The chemical potential $\mu = \mu(N)$ is implicitly determined by the second equation. For the low temperature $T \ll \epsilon_F$ calculations, the following *Sommerfeld expansion* is useful:

$$\int_0^\infty F(\varepsilon)\left(-\frac{\partial n(\varepsilon)}{\partial\varepsilon}\right) = F(\mu) + \frac{\pi^2T^2}{6}F''(\mu) + \cdots$$

for the smooth functions within the vicinity $\varepsilon = \mu$.

Integrating the expressions for the energy and particle number by parts and using the *Sommerfeld expansion*, we obtain

$$E(T) = \int_0^{\mu} \varepsilon g(\varepsilon) d\varepsilon + \frac{\pi^2 T^2}{6}\big(\mu g(\mu)\big)' + \dots = E(\mu) + \frac{\pi^2 T^2}{6}\big(\mu g(\mu)\big)' + \dots ,$$

$$N = \frac{V}{3}\frac{(2m\varepsilon_F)^{3/2}}{\pi^2\hbar^3} = \frac{V}{3}\frac{(2m\mu)^{3/2}}{\pi^2\hbar^3} + \frac{\pi^2 T^2}{6} g'(\mu) + \dots$$

Employing these two equations, we find first terms of the low-temperature expansion for the total energy and chemical potential

$$E(T) = E(0) + \frac{\pi^2}{6} T^2 g(\varepsilon_F) + \dots \quad \text{and} \quad \mu(T) = \varepsilon_F - \frac{\pi^2 T^2}{6}\frac{g'(\varepsilon_F)}{g(\varepsilon_F)} + \dots$$

This gives straightforwardly the linear law for the low-temperature behavior of specific heat and entropy in the degenerate Fermi gas

$$C(T) = \frac{\pi^2}{3} T g(\varepsilon_F) \quad \text{and} \quad S(T) = \frac{\pi^2}{3} T g(\varepsilon_F).$$

The qualitative nature of this behavior is related to the effective number of fermions involved in the thermal excitation near the Fermi surface. The number of such fermions is about the product of temperature T by the density of states $g(\varepsilon_F)$. Each fermion near the Fermi surface gives an increase of the order of temperature to the gas energy. As a result, the gas energy will increase by the order of $T \times T g(\varepsilon_F) \sim T^2 g(\varepsilon_F)$.

The linear law for the temperature growth of specific heat and entropy in an ideal Fermi gas is independent of the spatial dimensionality d. However, this property does not hold for the chemical potential. Since for the d-dimensional gas the density of states reads $g(\varepsilon) \sim \varepsilon^{(d-2)/2}$, the derivative of chemical potential with respect to temperature $d\mu/dT^2 \sim (2-d)$ changes the sign from positive at $d = 1$ to negative at $d = 3$ and larger.

The origination of magnetization in a neutral fermionic gas[2] in the magnetic field results from the presence of the spin and corresponding magnetic moment. The latter interacts already with the magnetic field. Let us write the energy of a spin-1/2 fermion in the magnetic field H as

$$\varepsilon(p, H) = \varepsilon(p) \pm \frac{1}{2}\mu_{\text{eff}} H$$

where μ_{eff} is the effective fermionic magneton. Two signs in the energy correspond to two values ($\pm 1/2$) in the fermion spin projections onto the magnetic field directions.

[2] These can be the ^{3}He atoms whose magnetic moments are due to nuclear spins.

The number of fermions with the spin projections parallel or antiparallel to the magnetic field is equal to

$$N_+ = \frac{1}{2}\int_0^\infty n\big(\varepsilon - \frac{1}{2}\mu_{\text{eff}}H\big)g(\varepsilon)d\varepsilon, \quad N_- = \frac{1}{2}\int_0^\infty n\big(\varepsilon + \frac{1}{2}\mu_{\text{eff}}H\big)g(\varepsilon)d\varepsilon.$$

Factor 1/2 reflects the point that both possible spin projections are incorporated into the definition of the density of states $g(\varepsilon)$. The total magnetic moment $M = (1/2)\mu_{\text{eff}}(N_+ - N_-)$ then reads

$$M = \frac{\mu_{\text{eff}}}{4}\int_0^\infty \big[n\big(\varepsilon - \frac{1}{2}\mu_{\text{eff}}H\big) - n\big(\varepsilon + \frac{1}{2}\mu_{\text{eff}}H\big)\big]g(\varepsilon)d\varepsilon.$$

In the weak magnetic field, we can decompose the difference in the distribution functions under the integral sign and find the *paramagnetic Pauli susceptibility* of fermionic gas according to $M = \chi_P H$

$$\chi_P = \frac{\mu_{\text{eff}}^2}{4}\int_0^\infty \Big(-\frac{\partial n(\varepsilon)}{\partial \varepsilon}\Big)g(\varepsilon)d\varepsilon.$$

At low temperatures $T \ll \varepsilon_F$, the spin (paramagnetic) susceptibility or *Pauli susceptibility* of degenerate Fermi gas is given by the expansion

$$\chi_P = \frac{\mu_{\text{eff}}^2}{4}g(\varepsilon_F)\Big(1 + \frac{\pi^2 T^2}{6}\frac{\partial^2}{\partial \varepsilon_F^2}\ln g(\varepsilon_F)\Big).$$

Remaining finite at zero temperature, the Pauli spin susceptibility behaves differently by increasing the temperature, depending on the dimensionality d of the system d. The behavior is similar to that of chemical potential. This conclusion results from the relation $\partial^2 \ln g(\varepsilon)/\partial\varepsilon^2 = -(d/2 - 1)/\varepsilon^2$.

The high temperature $T \gg \varepsilon_F$ behavior of susceptibility can be obtained if we take into account that, for the high temperatures, the Fermi–Dirac distribution goes over into the Boltzmann one for which $-\partial n/\partial\varepsilon = n/T$. Then

$$\chi_P = \frac{\mu_{\text{eff}}^2}{4T}\int_0^\infty n(\varepsilon)g(\varepsilon)d\varepsilon = \frac{\mu_{\text{eff}}^2}{4T}N,$$

and the susceptibility follows the classical *Curie law*.

Problems

1. How do the Fermi momentum and energy vary under complete spin polarization of an ideal ^{3}He atom gas in the magnetic field?

Solution. Under complete spin polarization, a single fermion alone can occupy the state with the same momentum instead of two fermions as in the case of zero magnetic field. Then the number of atoms is determined by the relation

$$N = \int\limits_{p\leqslant p_F} \frac{V d^3 p}{(2\pi\hbar)^3} = V\frac{p_F^3}{6\pi^2\hbar^3}\,.$$

Therefore, the Fermi sphere radius increases and equals $p_F = 2^{1/3}p_0$, p_0 being the Fermi momentum in zero field. The Fermi energy equals $\varepsilon_F = 2^{2/3}\varepsilon_0$ where ε_0 is the Fermi energy in zero field.

2. Graphene is a single layer of graphite. Near the conical points the dependence of electron energy as a function of momentum $\boldsymbol{p} = (p_x,\ p_y)$ is described by the *massless Dirac spectrum* $\varepsilon_{\boldsymbol{p}} = v\boldsymbol{\sigma}\,\boldsymbol{p}$ where v is the electron velocity and $\boldsymbol{\sigma} = (\sigma_x,\ \sigma_y)$ are the Pauli matrices. The corresponding dispersion law $\varepsilon_{\boldsymbol{p}} = \pm v|\boldsymbol{p}|$ represents two subbands. In pure graphene, the chemical potential $\mu = 0$ is temperature-independent and crosses always the conical Dirac point $\varepsilon = 0$ relating to charge neutrality.

Determine the one-particle density of states $g(\varepsilon)$. Find the specific heat $C(T)$ of graphene. (Take into account the double electron spin degeneracy and the presence of two conical points.)

Solution. As the chemical potential crosses the conical point of the electronic spectrum, the Fermi surface degenerates into a point and the single energy parameter is temperature. Let us write one-particle density of states per unit area

$$g(\varepsilon) = \nu\int\frac{d^2p}{(2\pi\hbar)^2}\big[\delta(\varepsilon - vp) + \delta(\varepsilon + vp)\big]= \nu\frac{|\varepsilon|}{2\pi\hbar^2 v^2}$$

where $\nu = 2\times 2 = 4$ is the degree of degeneracy resulted from the spin and the number of conical points. Differentiating the expression for energy at $\mu = 0$

$$E(T) = \int_{-\infty}^{\infty}\frac{d\varepsilon\ \varepsilon g(\varepsilon)}{e^{\varepsilon/T}+1}$$

with respect to temperature, we find the specific heat

$$C(T) = \frac{\nu}{2\pi\hbar^2v^2}\int\limits_{-\infty}^{\infty}\frac{d\varepsilon\ |\varepsilon|\varepsilon}{4T\cosh^2\varepsilon/2T}\,\frac{\varepsilon}{T} = \frac{\nu T^2}{2\pi\hbar^2v^2}2\int\limits_{0}^{\infty}\frac{x^3\,dx}{4\cosh^2(x/2)} = \nu\frac{9\zeta(3)}{2\pi}\,\frac{T^2}{\hbar^2v^2}\,.$$

Such quadratic dependence results directly from the conical shape of isoenergetic surface and the density of states $g(\varepsilon)\sim|\varepsilon|$. In fact, the electrons of energy $|\varepsilon|\lesssim T$ near the conical point contribute mostly to the thermal excitation of the system. The number of such electrons is $n(T)\sim Tg(\varepsilon\sim T)$ and, therefore, the thermal excitation energy reads $E(T)\sim Tn(T)\sim T^3$. This entails $C(T)\sim T^2$.

3. Find the paramagnetic susceptibility of graphene under conditions of the previous problem.

Solution. The energy of an electron in the magnetic field equals $\varepsilon_p\pm\beta H$, β being the effective magnetic moment. The total magnetic moment, expressed with the aid of distribution function $n(\varepsilon)$ can be written as

$$M = \frac{\beta}{2}\int_{-\infty}^{\infty} d\varepsilon\ g(\varepsilon)\big[n(\varepsilon-\beta H) - n(\varepsilon+\beta H)\big].$$

Differentiating the magnetization with respect to magnetic field yields the paramagnetic susceptibility $\chi_P = \partial M/\partial H$

$$\chi_P = -\frac{\beta^2}{2}\int_{-\infty}^{\infty} d\varepsilon\, g(\varepsilon)\left(\frac{\partial n(\varepsilon-\beta H)}{\partial\varepsilon} - \frac{\partial n(\varepsilon+\beta H)}{\partial\varepsilon}\right) =$$

$$= \frac{\beta^2}{2}\int_{-\infty}^{\infty} d\varepsilon\left[g(\varepsilon-\beta H)+g(\varepsilon+\beta H)\right]\left(-\frac{\partial n(\varepsilon)}{\partial\varepsilon}\right) =$$

$$= \frac{\nu}{2\pi\hbar^2 v^2}\,\frac{\beta^2}{2}\int_{-\infty}^{\infty}\frac{d\varepsilon}{4T}\,\frac{|\varepsilon-\beta H|+|\varepsilon+\beta H|}{\cosh^2(\varepsilon/2T)}.$$

Using the value of the integral

$$\int_{-\infty}^{\infty}\frac{|x+a|}{\cosh^2 x}\,dx = 2\ln(2\cosh a)\,,$$

we obtain the following answer:

$$\chi_P(T,H) = 2\beta^2 g(T)\ln\left(2\cosh\frac{\beta H}{T}\right).$$

Let us give the limiting expressions for the low- and high magnetic fields

$$\chi_P(H) = \begin{cases} (2\ln 2)\,\beta^2 g(T)\sim T, & \beta H \ll T \ \text{(low field)},\\ \beta^2 g(\beta H)\sim H\,, & \beta H \gg T \ \text{(high field)}. \end{cases}$$

4. A *tachyon* is a hypothetical particle which is introduced to describe the faster-than-light motion. Its spectrum is given by the relativistic formula

$$\varepsilon_p = \sqrt{p^2c^2 - m^2c^4} \quad (p \geqslant mc)$$

where c is the light velocity, p is the momentum, and m is the tachyonic mass. The tachyons are assumed to be fermions with the number g of internal degree of freedom.

Find the Fermi momentum p_F, energy density E, pressure P, and sound velocity u in an ideal tachyonic gas with the particle density n at zero temperature. The sound velocity in relativistic gas is determined by the relation $u = c(\partial P/\partial E)^{1/2}$.

Solution. The one-particle density of states per unit volume equals

$$\nu(\varepsilon) = \frac{g}{2\pi^2\hbar^3c^3}\,\varepsilon\sqrt{\varepsilon^2+m^2c^4}\,.$$

We find the Fermi momentum p_F from the condition

$$n = \int_{mc}^{p_F} g\frac{4\pi p^2\,dp}{(2\pi\hbar)^3} = g\frac{p_F^3 - m^3c^3}{6\pi^2\hbar^3} \quad\text{and}\quad p_F = mc\left(1+\frac{6\pi^2\hbar^3 n}{gm^3c^3}\right)^{1/3}.$$

Accordingly, the Fermi energy reads

$$\varepsilon_F = c\left[\left(6\pi^2\hbar^3 n/g + m^3c^3\right)^{2/3} - m^2c^2\right]^{1/2} = mc^2\sqrt{\beta^2-1} \quad\text{where}\quad \beta = p_F/mc.$$

The energy density of the gas at temperature $T = 0$ equals the integral

$$E = \frac{g}{(2\pi\hbar)^3}\int\limits_{mc}^{p_F} \varepsilon_p\, 4\pi p^2\, dp = \frac{gm^4c^5}{16\pi^2\hbar^3}\int\limits_1^{\beta} y^2\sqrt{y^2-1}\, dy =$$

$$= \frac{gm^4c^5}{16\pi^2\hbar^3}\big[(2\beta^3-\beta)\sqrt{\beta^2-1} - \ln(\beta+\sqrt{\beta^2-1}\,)\big].$$

We determine the pressure at $T = 0$ from the relation $P = -\partial(EV)/\partial V$ where V is the volume and EV is the total energy of the gas. This entails $P = n\partial E/\partial n - E$. We can use the definition of potential $\Omega = -PV$ as well. Then,

$$P = T\int g\frac{d^3p}{(2\pi\hbar)^3}\ \ln\big(1+e^{(\mu-\varepsilon_p)/T}\big)$$

and μ is the chemical potential. For $T = 0$, this expression goes over to the simple relation

$$P = g\int\frac{d^3p}{(2\pi\hbar)^3}\,(\varepsilon_F-\varepsilon_p) = n\varepsilon_F - E.$$

As a result, the pressure of ideal tachyonic gas equals

$$P = \frac{gm^4c^5}{16\pi^2\hbar^3}\big[\big(\frac{2}{3}\beta^3+\beta-\frac{8}{3}\big)\sqrt{\beta^2-1} + \ln(\beta+\sqrt{\beta^2-1}\,)\big].$$

The square of sound velocity reads

$$u^2 = c^2\,\partial P/\partial E = c^2\,\frac{\partial P/\partial\beta}{\partial E/\partial\beta} = \frac{c^2}{3}\,\frac{\beta^2+\beta+1}{\beta^2+\beta}\,.$$

Since for tachyons we have $p_F \geqslant mc$ ($\beta \geqslant 1$), the sound velocity varies from $c/\sqrt{2}$ for the small density gas ($\beta - 1 \ll 1$) to $c/\sqrt{3}$ in the large density gas ($\beta \gg 1$), remaining always smaller than the speed of light c.

3.4 Degenerate Fermi Gas in the Harmonic Trap

Let us turn to considering the properties of a degenerate ideal Fermi gas in the harmonic trap with the confining potential

$$U(x, y, z) = m\big(\omega_x^2x^2 + \omega_y^2y^2 + \omega_z^2z^2\big)/2.$$

The energy levels of a particle in the harmonic potential are given by the simple formula

$$\varepsilon(n_1, n_2, n_3) = \hbar\omega_x\big(n_1 + 1/2\big) + \hbar\omega_y\big(n_2 + 1/2\big) + \hbar\omega_z\big(n_3 + 1/2\big)$$

where n_1, n_2, and n_3 are the integer non-negative numbers. Below we assume the macroscopic limit when the number of fermions $N \gg 1$ is large and the energy levels are arranged sufficiently close to each other. In this case for calculating the thermodynamic functions of the gas, we can employ the semiclassical approximation and replace the summation over the neighboring discrete levels with the corresponding integration approximated with the continuous level spectrum.

The simplest method to introduce the semiclassical description is to apply the local density approximation for the Fermi–Dirac distribution

$$n(\boldsymbol{r}, \boldsymbol{p}) = \left(e^{\left(\varepsilon(\boldsymbol{p},\boldsymbol{r})-\mu\right)/T} + 1\right)^{-1}.$$

Here $\varepsilon(\boldsymbol{r}, \boldsymbol{p}) = \boldsymbol{p}^2/2m + U(\boldsymbol{r})$ is the classical energy of a fermion, and μ is the chemical potential determined with the normalization condition for the total number of fermions

$$N = 2\int \frac{d^3r\, d^3p}{(2\pi\hbar)^3} n(\boldsymbol{r}, \boldsymbol{p}) = \int_0^\infty \frac{g(\varepsilon)d\varepsilon}{e^{(\varepsilon-\mu)}/T + 1}.$$

Here we have introduced the one-particle density of states according to

$$g(\varepsilon) = 2\int \frac{d^3r\, d^3p}{(2\pi\hbar)^3} \delta\left(\varepsilon - \varepsilon(\boldsymbol{p}, \boldsymbol{r})\right) = \frac{\varepsilon^2}{\hbar^3\omega_x\omega_y\omega_z}.$$

The use of the one-particle density of states[3] allows us to apply the results from the previous sections for calculating the thermodynamic functions. So, for $T = 0$ from the condition

$$N = \int_0^{\varepsilon_F} g(\varepsilon)d\varepsilon = \int_0^{\varepsilon_F} \frac{\varepsilon^2}{\hbar^3\omega_x\omega_y\omega_z} d\varepsilon = \frac{\varepsilon_F^3}{3\hbar^3\omega_x\omega_y\omega_z},$$

we obtain the Fermi energy $\varepsilon_F = \mu(T = 0)$

$$\varepsilon_F = \hbar(3N\omega_x\omega_y\omega_z)^{1/3}.$$

The total energy of the gas E_0 at $T = 0$ equals

$$E_0 = \int_0^{\varepsilon_F} \varepsilon g(\varepsilon)\, d\varepsilon = \frac{\varepsilon_F^4}{4\hbar^3\omega_x\omega_y\omega_z} = \frac{3}{4}N\varepsilon_F\,.$$

The low temperature $T \ll \varepsilon_F$ corrections to the energy, chemical potential, entropy, and specific heat are obvious

[3] Note that from the formal point of view, such a behavior of the density of states as a function of energy would be the same for the spatially homogeneous gas of particles with the conventional energy dispersion $\varepsilon_{\boldsymbol{p}} = \boldsymbol{p}^2/2m$ in the six-dimensional space $d = 6$.

$$E(T) = E_0 + \frac{\pi^2}{6} T^2 g(\varepsilon_F), \quad \mu(T) = \varepsilon_F - \frac{\pi^2}{3} \frac{T^2}{\varepsilon_F},$$
$$S(T) = \frac{\pi^2}{3} T g(\varepsilon_F), \qquad C(T) = \frac{\pi^2}{3} T g(\varepsilon_F).$$

Here $g(\varepsilon_F) = \varepsilon_F^2/(\hbar^3 \omega_x \omega_y \omega_z) = 3N/\varepsilon_F$ is the density of one-particle states at the Fermi surface.

Of interest is the spatial distribution of the particle density $n(\boldsymbol{r})$ in the trap. It can be found by integrating the distribution function $n(\boldsymbol{r}, \boldsymbol{p})$ with respect to momentum. For $T = 0$, we have $n(\boldsymbol{r}, \boldsymbol{p}) = \vartheta\big(\varepsilon_F - \varepsilon(\boldsymbol{r}, \boldsymbol{p})\big)$ and then

$$n(\mathbf{r}) = 2\int \frac{d^3p}{(2\pi\hbar)^3}\, \vartheta\big(\varepsilon_F - \varepsilon(\boldsymbol{r}, \mathbf{p})\big) = \frac{(2m)^{3/2}}{3\pi^2\hbar^3}\big(\varepsilon_F - U(\boldsymbol{r})\big)^{3/2}$$
$$= \frac{8}{\pi^2} \frac{N}{R_x R_y R_z}\left(1 - \frac{x^2}{R_x^2} - \frac{y^2}{R_y^2} - \frac{z^2}{R_z^2}\right)^{3/2}.$$

The above density distribution of fermions in the trap is called the *Thomas–Fermi distribution*. In the harmonic trap, it has the ellipsoid shape with semi-axes R_x, R_y, and R_z whose magnitudes are determined by the following relations:

$$\frac{1}{2} m\omega_i^2 R_i^2 = \varepsilon_F = \frac{p_F^2}{2m} \quad (i = x,\, y,\, z).$$

The particle distribution over momenta $n(\boldsymbol{p})$ can be found by integrating the total distribution function $n(\boldsymbol{r}, \boldsymbol{p})$ over the coordinate. At temperature $T = 0$ we find

$$n(\boldsymbol{p}) = \frac{2}{(2\pi\hbar)^3}\int d^3r\, \vartheta\big(\varepsilon_F - \varepsilon(\boldsymbol{r}, \boldsymbol{p})\big) = \frac{8}{\pi^2}\frac{N}{p_F^3}\left(1 - \frac{p^2}{p_F^2}\right)^{3/2}$$

if $p < p_F$ and $n(\boldsymbol{p}) = 0$ on the contrary. This equation is a complete analog of the momentum distribution of particles similar to the step-like behavior $n(p) = \vartheta(1 - p^2/p_F^2)$ in the spatially homogeneous Fermi gas. Note that the magnitudes of the boundary Fermi momentum $p_F = \hbar[3\pi^2 n(r = 0)]^{1/3}$ are proved to be the same in both cases.

Problem

1. The classical string with the fixed end-points has an infinite set of oscillation frequencies $\omega_l = l\omega$, which are the multiples of the fundamental frequency ω. The fermionic quantum non-relativistic string represents a set of the infinite number of harmonic oscillators with frequencies ω, 2ω, 3ω, ... and energies

$$E_l = \hbar\omega_l(n_l + 1/2), \quad \omega_l = l\omega, \quad l = 1, 2, 3, \ldots$$

According to the Pauli principle, the occupation numbers n_l for each oscillator takes the values 0 and 1 alone.

For the fermionic string, find the free energy $F(T)$, entropy $S(T)$, specific heat $C(T)$, and number of states $\Gamma(E)$ in the high energy limit $E \gg \hbar\omega$.

Solution. Since the oscillators do not interact with each other, the total partition function reduces to the product of the partition functions of independent oscillators

$$Z = \prod_{l=1}^{\infty} z_l, \quad z_l = \sum_{n_l=0}^{1} e^{-\frac{\hbar\omega_l}{T}(n_l+1/2)} = \frac{e^{-\hbar\omega_l/2T}}{1+e^{-\hbar\omega_l/T}}.$$

Hence the free energy equals

$$F = E_0 - T\sum_{l=1}^{\infty}\ln\bigl(1+e^{-\hbar\omega_l/T}\bigr), \quad E_0 = \sum_{l=1}^{\infty}\hbar\omega_l/2,$$

E_0 being the energy of zero oscillations. Then we find the entropy and specific heat

$$S(T) = \sum_{l=1}^{\infty}\left[\frac{\hbar\omega_l/T}{e^{\hbar\omega_l/T}+1} + \ln\bigl(1+e^{-\hbar\omega_l/T}\bigr)\right],$$

$$C(T) = \sum_{l=1}^{\infty}\left[\frac{\hbar\omega_l/2T}{\cosh(\hbar\omega_l/2T)}\right]^2, \quad \omega_l = l\omega.$$

For low temperatures ($T \ll \hbar\omega$), both these quantities vanish exponentially

$$S(T) = (\hbar\omega/T)e^{-\hbar\omega/T}, \quad C(T) = (\hbar\omega/T)^2 e^{-\hbar\omega/T}.$$

The behavior at high temperatures ($T \gg \hbar\omega$) is of more interest. In this limit the main contribution to the sums over l results from the terms with the large numbers $l \gg 1$. So, we can convert the sum over l into an integration over l and arrive at

$$F - E_0 = -T\int_0^{\infty} dl\, \ln\bigl(1+e^{-\hbar\omega l/T}\bigr) = -\frac{T^2}{\hbar\omega}\int_0^{\infty} dx\, \ln\bigl(1+e^{-x}\bigr) = -\frac{\pi^2}{12}\frac{T^2}{\hbar\omega},$$

$$C(T) = \int_0^{\infty} dl\left(\frac{\hbar\omega l/2T}{\cosh(\hbar\omega l/2T)}\right)^2 = \frac{2T}{\hbar\omega}\int_0^{\infty} dx\,\frac{x^2}{\cosh^2 x} = \frac{\pi^2}{6}\frac{T}{\hbar\omega}.$$

To find the number of the string states $\Gamma(E)$ at high energies, we employ the relation $S(E) = \ln\Gamma(E)$. From the equations

$$E - E_0 = \pi^2 T^2/12\hbar\omega \quad \text{and} \quad S = \pi^2 T/6\hbar\omega$$

valid at $T \gg \hbar\omega$, we find

$$\ln\Gamma(E) = 2\pi\sqrt{\frac{E-E_0}{12\hbar\omega}} = 2\pi\sqrt{\frac{N}{12}} \quad \text{where} \quad \Gamma(E) \sim e^{2\pi\sqrt{N/12}}, \quad N = \frac{E-E_0}{\hbar\omega} \gg 1.$$

Here the number $N = n_1 + 2n_2 + 3n_3 + \ldots$ plays a role of the particle number expressed in terms of the occupation numbers.

3.5 Diamagnetism of Ideal Electron Gas

Diamagnetism of an electron gas[4] is due to electric charge of an electron as well as to the interaction between the electric charge and magnetic field. We start analyzing the diamagnetic properties of the electron gas with the calculation of the thermodynamic grand potential Ω. First of all, it is necessary to determine the electron energy levels quantized in the magnetic field. The electron energy levels in the magnetic field H parallel to the z-axis or the *Landau levels* are given by the formula

$$\varepsilon_n(p_z, x_0) = \hbar\omega_H\left(n + \frac{1}{2}\right) + \frac{p_z^2}{2m^*}, \quad \omega_H = \frac{eH}{m^*c}, \quad n = 0,\ 1,\ 2, \ldots$$

where $\omega_H = eH/m^*c$ is the *cyclotronic frequency*.To generalize our speculations, we introduce the effective electron mass m^*, supposing the dispersion law $\varepsilon_p = p^2/2m^*$. The electron spectrum in the magnetic field is degenerate in the orbit center $x_0 = cp_y/eH$ which is proportional to momentum p_y normal to the magnetic field direction. In view of double degeneracy in the electron spin, we write the number of states at each Landau level within the momentum interval $dp_z/(2\pi\hbar)$ as

$$2\frac{eH}{2\pi\hbar c}V\frac{dp_z}{2\pi\hbar}$$

where V is the volume occupied with electrons.

Let us calculate the grand potential

$$\Omega_H(\mu) = -VT\sum_{n=0}^{\infty}\int_{-\infty}^{\infty}\frac{dp_z}{2\pi\hbar}\frac{2eH}{2\pi\hbar c}\ln\left(1 + e^{\frac{\mu-\varepsilon_n(p_z)}{T}}\right) =$$

$$= -\frac{VT}{(\pi\hbar)^2}\frac{eH}{c}\sum_{n=0}^{\infty}\int_{\hbar\omega_H(n+\frac{1}{2})}^{\infty}d\varepsilon\,\frac{dp_z(\varepsilon)}{d\varepsilon}\ln\left(1 + e^{\frac{\mu-\varepsilon}{T}}\right),$$

putting $p_z = \sqrt{2m^*\big(\varepsilon - \hbar\omega_H(n + 1/2)\big)}$ and μ as the chemical potential. Integrating by parts results in

$$\Omega_H(\mu) = -\frac{V}{(\pi\hbar)^2}\frac{eH}{c}\sum_{n=0}^{\infty}\int_{\hbar\omega_H(n+1/2)}^{\infty}d\varepsilon\,\frac{\sqrt{2m^*\big(\varepsilon - \hbar\omega_H(n + \frac{1}{2})\big)}}{e^{(\varepsilon-\mu)/T} + 1}.$$

This expression is valid at any temperature. First, we restrict ourselves with the case $T \ll \mu$ and put $T = 0$ as a first approximation. Then the Fermi–Dirac distribution

---

[4] The Landau diamagnetism as well.

can be replaced with the step-like function. So,

$$\Omega_H(\mu) = -\frac{V}{(\pi\hbar)^2}\frac{eH}{c}\sum_n \int\limits_{\hbar\omega_H(n+1/2)}^{\infty} d\varepsilon\,\theta(\mu-\varepsilon)\sqrt{2m^*\big(\varepsilon - \hbar\omega_H(n+1/2)\big)} =$$

$$= -\frac{2\sqrt{2m^*}}{3}\frac{V}{(\pi\hbar)^2}\frac{eH}{c}\sum_{n=0}^{N}\big(\mu - \hbar\omega_H(n+1/2)\big)^{3/2}$$

where the integer N equal to[5]

$$N = \left[\frac{\mu}{\hbar\omega_H} - \frac{1}{2}\right]$$

has the sense of the last Landau level occupied completely.

For low magnetic fields H and large numbers N, the sum can be estimated with aid of the formula

$$\sum_{n=0}^{N} f(n) = \int\limits_{-1/2}^{N+1/2} f(n)\,dn - \frac{f'(N+1/2) - f'(-1/2)}{24} + \cdots$$

As a result, we find the thermodynamic grand potential equal to

$$\Omega_H(\mu) = -\frac{2\sqrt{2m^*}}{3}\frac{V}{(\pi\hbar)^2}\frac{eH}{c}\left(\frac{5}{2}\frac{\mu^{5/2}}{\hbar\omega_H} - \frac{3}{2}\frac{\hbar\omega_H\mu^{1/2}}{24} + \cdots\right) =$$

$$= -\frac{5}{3}\frac{\sqrt{2m^*}V}{\pi^2\hbar^3}m^*\mu^{5/2} + \frac{V}{24\pi^2\hbar}\left(\frac{eH}{c}\right)^2\sqrt{\frac{2\mu}{m*}} + \cdots$$

The total magnetic moment of electron gas is determined by the following derivative:

$$\boldsymbol{M} = -\frac{\partial\Omega_H}{\partial\boldsymbol{H}} = -\frac{V}{12\pi^2\hbar}\frac{e^2}{c^2}\sqrt{\frac{2\mu}{m^*}}\boldsymbol{H} =$$

$$= -\frac{V}{3}\left(\frac{e\hbar}{2m^*c}\right)^2\frac{m^* p_F}{\pi^2\hbar^3}\boldsymbol{H} = -\frac{1}{3}\left(\frac{e\hbar}{2m^*c}\right)^2 g(\mu)\boldsymbol{H}$$

where $g(\mu)$ is the one-particle density of states at the Fermi level and $\mu(T=0) = \varepsilon_F$. Thus, the magnetic susceptibility of electron gas, associated with the Landau levels of a charged particle in the magnetic field, results in the diamagnetic character of the response to the magnetic field. The zero temperature susceptibility reads

[5] Here we mean the domain of integers n for which the square root argument is non-negative.

$$\chi_{\text{dia}} = -\frac{1}{3}\left(\frac{m_e}{m^*}\right)^2\left(\frac{e\hbar}{2m_e c}\right)^2 g(\varepsilon_F) = -\frac{1}{3}\left(\frac{m_e}{m^*}\right)^2\frac{\mu_{\text{eff}}^2}{4}g(\varepsilon_F).$$

Here we have introduced the notation m_e for the genuine electron mass and $\mu_{\text{eff}} = 2\mu_B$ for the magnetic moment equal to the double Bohr magneton $\mu_B = e\hbar/2m_e c$.

If we compare the diamagnetic susceptibility χ_{dia} with the paramagnetic susceptibility χ_P for the free electron gas when $m^* = m_e$, we see that the diamagnetic response compensates the paramagnetic response by one third alone. Therefore, the free electron gas remains paramagnetic on the whole.

On the other hand, an existence of diamagnetic metals is related to the point that the energy spectrum of electron excitations in a metal, as a result of the interactions between the particles composing the metal, demonstrates much more complicated dependence upon the momentum as compared with the free electron spectrum. As we have seen above on the example of the simplest model for the energy spectrum with effective mass $m^* \neq m_e$, the ratio of susceptibilities reads

$$\chi_{\text{dia}}/\chi_P = (1/3)(m_e/m^*)^2.$$

This magnitude can exceed unity at $m^*/m_e < 1/\sqrt{3}$.

3.6 The de Haas–van Alphen Effect

We have considered above the diamagnetic susceptibility of an ideal degenerate electron gas in first approximation in the magnetic field strength. In first approximation, the correction to the thermodynamic grand potential is a quadratic one of the order $(\mu_{\text{eff}}H/\mu)^2$. Below we analyze the next corrections to the thermodynamic grand potential. The corrections will prove to be oscillating as a function of the magnetic field and result in the magnetization oscillations in the electron gas, the oscillation period being proportional to the inverse magnetic field magnitude $1/H$. The phenomenon of magnetization oscillations in the magnetic field is called the *de Haas–van Alphen effect*.

So, we write down the thermodynamic grand potential found above

$$\Omega_H(\mu) = -VT\sum_{n=0}^{\infty}\int_{-\infty}^{\infty}\frac{dp_z}{(2\pi\hbar)^2}\frac{2eH}{c}\ln\left(1 + e^{\frac{\mu-\varepsilon_n(p_z)}{T}}\right).$$

For the further analysis of the this expression, we employ the *Poisson summation formula*:

$$\sum_{n=0}^{\infty} f(n) = \int_{-1/2}^{\infty} f(n)\,dn + 2\text{Re}\sum_{k=1}^{\infty}\int_{-1/2}^{\infty} f(n)\,e^{2\pi ikn}\,dn\,.$$

Putting

$$f(n) = -2VT \int_{-\infty}^{\infty} \frac{2\pi m^* dp_z}{(2\pi\hbar)^3} \hbar\omega_H \ln\left(1 + e^{\frac{\mu - \varepsilon_n(p_z)}{T}}\right),$$

we represent the grand potential $\Omega_H(\mu)$ as

$$\Omega_H(\mu) = \Omega_0(\mu) + \Delta\Omega_H(\mu) = \Omega_0(\mu) - 2\frac{2VT}{(2\pi\hbar)^3}\sum_{k=1}^{\infty}(-1)^k \mathrm{Re}\, I_k$$

where $\Omega_0(\mu)$ is the potential in zero magnetic field

$$\Omega_0(\mu) = -2VT \int_{-\infty}^{\infty} \frac{2\pi m^* dp_z}{(2\pi\hbar)^3} \int_0^{\infty} d\varepsilon_\perp \ln\left(1 + e^{\frac{\mu - \varepsilon_\perp - p_z^2/2m^*)}{T}}\right) =$$

$$= -2VT \int \frac{d^3 p}{(2\pi\hbar)^3} \ln\left(1 + e^{\frac{\mu - p^2/2m^*)}{T}}\right),$$

and the integral I_k is determined by the following expression:

$$I_k = 2\pi m^* \int_{-\infty}^{\infty} dp_z \int_{-1/2}^{\infty} dn\, \hbar\omega_H \ln\left(1 + e^{\frac{\mu - \varepsilon_n(p_z)}{T}}\right) e^{2\pi i k(n+1/2)} =$$

$$= 4\pi m^* \int_0^{\infty} dp_z \int_0^{\infty} d\eta \ln\left(1 + e^{\frac{\mu - \eta - p_z^2/2m^*}{T}}\right) e^{2\pi i k\eta/\hbar\omega_H}.$$

Integrating the last integral twice by parts yields

$$\frac{\mathrm{Re}\, I_k}{4\pi m^*} = -\mathrm{Re} \int_0^{\infty} dp_z \frac{\hbar\omega_H}{2\pi i k T} \left[T \ln\left(1 + e^{\frac{\mu - p_z^2/2m^*}{T}}\right) + \right.$$

$$\left. + \frac{\hbar\omega_H}{2\pi i k} \left(\frac{1}{1 + e^{-\frac{\mu - p_z^2/2m^*}{T}}} - \int_0^{\infty} \frac{d\eta}{T} \frac{e^{2\pi k\eta/\hbar\omega_H}}{4\cosh^2 \frac{\eta + p_z^2/2m^* - \mu}{2T}} \right) \right] =$$

$$= \mathrm{Re} \frac{(\hbar\omega_H)^2}{(2\pi k)^2 T} \int_0^{\infty} dp_z \left(\int_0^{\infty} \frac{e^{2\pi i k\eta/\hbar\omega_H}}{4T\cosh^2 \frac{\eta + p_z^2/2m^* - \mu}{2T}} d\eta - \frac{1}{1 + e^{-\frac{\mu - p_z^2/2m^*}{T}}} \right).$$

In the remaining integrals we replace the variables $(\eta - \mu)/T = \xi$ and $p_z = q\sqrt{2m^*T}$. Then,

$$\frac{\mathrm{Re}\, I_k}{4\pi m^*} = \frac{(\hbar\omega_H)^2\sqrt{2m^*}}{(2\pi k)^2 T^{1/2}} \times$$

$$\times\, \mathrm{Re} \int\limits_0^\infty dq \left(\int\limits_{-\mu/T}^{\infty} e^{\frac{2\pi i k\mu}{\hbar\omega_H}} \frac{e^{2\pi i k\xi T/\hbar\omega_H}}{4\cosh^2(\xi + q^2)/2} d\xi - \frac{1}{1 + e^{(q^2 - \mu/T)}} \right).$$

In the integral over ξ the lower limit equal to $-\mu/T$ can be substituted for $-\infty$ due to condition of gas degeneracy $\mu \gg T$. Then, after shifting the variable ξ with q^2, we integrate with respect to ξ, using the magnitude of integral

$$\int\limits_{-\infty}^{\infty} \frac{e^{iax}}{4\cosh^2 x/2} = \frac{\pi a}{\sinh \pi a}.$$

As a result, we have

$$\frac{\mathrm{Re}\, I_k}{4\pi m^*} = \frac{(\hbar\omega_H)^2\sqrt{2m^*}}{(2\pi k)^2 T^{1/2}} \times$$

$$\times\, \mathrm{Re} \int\limits_0^\infty dq \left(e^{-\frac{2\pi i k q^2 T}{\hbar\omega_H}} \frac{2\pi^2 kT/\hbar\omega_H}{\sinh(2\pi^2 kT/\hbar\omega_H)} e^{\frac{2\pi i k\mu}{\hbar\omega_H}} - \frac{1}{1 + e^{(q^2 - \mu/T)}} \right).$$

The further integration with respect to q is performed with the aid of integrals:

$$\int\limits_0^\infty e^{-iax} dx = \frac{e^{-i\pi/4}}{2}\sqrt{\frac{\pi}{a}} \quad (a > 0) \quad \text{and} \quad \int\limits_0^\infty \frac{dx}{1 + e^{x^2 - a}} \approx \sqrt{a}, \quad a \gg 1.$$

This already delivers the clear expression

$$\frac{\mathrm{Re}\, I_k}{4\pi m^*} = \frac{(\hbar\omega_H)^2\sqrt{2m^*\mu}}{(2\pi k)^2 T} \left[\sqrt{\frac{\hbar\omega_H}{8\mu k}} \frac{2\pi^2 kT/\hbar\omega_H}{\sinh(2\pi^2 kT/\hbar\omega_H)} \cos\left(\frac{2\pi k\mu}{\hbar\omega_H} - \frac{\pi}{4} \right) - 1 \right].$$

Finally, we find for the magnetic field-dependent part of grand potential $\Delta\Omega_H(\mu)$

$$\Delta\Omega_H(\mu) = -\frac{4V}{(2\pi\hbar)^3} \sum_{k=1}^{\infty} (-1)^k \frac{(m^*)^{3/2}(\hbar\omega_H)^{5/2}}{2\pi k^{5/2}} \times$$

$$\times \left(\frac{2\pi^2 kT/\hbar\omega_H}{\sinh(2\pi^2 kT/\hbar\omega_H)} \cos\left(\frac{2\pi k\mu}{\hbar\omega_H} - \frac{\pi}{4} \right) - \sqrt{\frac{8\mu k}{\hbar\omega_H}} \right).$$

Let us now analyze the result obtained. First of all, we underline that the thermodynamic grand potential contains the oscillating terms, the oscillation period $\Delta(1/H)$ being determined by condition $\mu/\hbar\omega_H = 1$. From the physical point of view, the oscillations occur when the chemical potential passes through the next Landau energy level by varying the magnetic field and the Landau level becomes either filled or emptied. This condition is satisfied if $(n + 1/2)\hbar\omega_H = \mu$ where n is the integer. Accordingly, the oscillation period is determined by varying the number n by unity, i.e. if variation $\Delta(\mu/\hbar\omega_H) = 1$ or

$$\Delta\left(\frac{1}{H}\right) = \frac{e\hbar}{m^* c\mu} = \frac{2e\hbar}{cp_F^2}\,.$$

The measurement of oscillation period allows one to determine the boundary Fermi momentum in the direction normal to the magnetic field.

At the finite temperature, the oscillation amplitude even for the first harmonic demonstrates the exponentially drastic reduction proportional to $\exp[-2\pi^2 T/\hbar\omega_H]$ as the temperature grows. Thus, starting from $T > 2\pi^2\hbar\omega_H$, the de Haas–van Alphen oscillations become insignificant. The physical reason lies in the temperature broadening of the energy levels and in the smearing of the discrete structure of Landau levels.

As we have seen, the relative magnitude of the oscillating term in the thermodynamic potential is of the order $(\hbar\omega_H/\mu)^{5/2} \ll 1$. The non-oscillating term in the thermodynamic potential has the larger relative magnitude about $(\hbar\omega_H/\mu)^2$. In this sense the oscillating term as compared with the monotonous one is small. However, in such quantities as magnetic moment and susceptibility, the role of the oscillating terms increases by many times and becomes essential and basic. In fact, we have for the magnetic moment

$$M = -\frac{\partial \Delta\Omega_H}{\partial H} =$$
$$= \frac{4V}{(2\pi\hbar)^3}\frac{e\hbar}{c}\sum_{k=1}^{\infty}(-1)^k \frac{\mu\sqrt{m^*\hbar\omega_H}}{k^{3/2}}\,\frac{2\pi^2 kT/\hbar\omega_H}{\sinh(2\pi^2 kT/\hbar\omega_H)}\,\sin\left(\frac{2\pi k\mu}{\hbar\omega_H} - \frac{\pi}{4}\right).$$

Here we have differentiated the most rapidly changing multipliers alone, i.e. cosines, with respect to magnetic field H. The relative magnitude of the magnetic moment oscillations becomes already predominant:

$$M_{\text{osc}}/M_0 \sim \left(\mu/\hbar\omega_H\right)^{1/2} \gg 1.$$

Thus, at low temperatures the oscillating part of magnetic moment and susceptibility gives significantly larger contribution as compared with the monotonous one.

3.7 Ideal Fermi Gas with the Spin–Orbit Spectrum of Dispersion

Here we discuss the degenerate ideal gas of electrons whose energy spectrum has the spin–orbit effect. As an instructive model, which is often applied to describing the thin metal plates or films with the strong spin–orbit interaction, we will consider the two-dimensional ideal gas in which the electron spectrum is given by the following *Bychkov–Rashba Hamiltonian*:

$$H = \frac{\boldsymbol{p}^2}{2m} + \alpha(\boldsymbol{\sigma} \times \boldsymbol{p})_z .$$

Here the z-axis is chosen perpendicular to the film plane (x, y). Vector $\boldsymbol{\sigma} = (\sigma_x, \sigma_y)$ and σ_z are the Pauli matrices, $\boldsymbol{p} = (p_x, p_y)$ is the momentum of an electron, and m is its mass. The parameter $\alpha > 0$ denotes the spin–orbit Rashba coupling constant.

Due to spin–orbit effect, there occurs a splitting of the electron spectrum into two subbands with respect to the spin direction, respectively, corresponding to the electron spin orientations *up* (+) and *down* (-):

$$\varepsilon_\pm(p) = \frac{p^2}{2m} \mp \alpha p = \xi \mp \sqrt{2\Delta\xi}$$

where $\xi = p^2/2m$ and $\Delta = m\alpha^2$.

The density of states per unit volume equals

$$N_\pm(\varepsilon) = \int \frac{dp_x\, dp_y}{(2\pi\hbar)^2}\delta\big(\varepsilon - \varepsilon_\pm(\boldsymbol{p})\big) = \frac{m}{2\pi\hbar^2}\int\limits_0^\infty d\xi\, \delta\big[\varepsilon - (\xi \mp \sqrt{2\Delta\xi}\,)\big]$$

for the spins directed *up* and *down*. Calculating the integral results in the expression

$$N_\pm(\varepsilon) = \frac{m}{2\pi\hbar^2}\Sigma\left|\frac{\partial\varepsilon}{\partial\xi}\right|^{-1}_{\xi=\xi_0}$$

where the sign Σ implies the sum over all the roots of equation $\varepsilon_\pm = \xi \mp \sqrt{2\Delta\xi}$ in the region $\xi \geqslant 0$.

For the spin *down* direction, there is a single root alone

$$\xi_1^{(-)} = \varepsilon + \Delta - \sqrt{2\varepsilon\Delta + \Delta^2} \quad \text{at} \quad \varepsilon \geqslant 0.$$

Then we obtain for the density of states with the spin *down* projection

$$N_-(\varepsilon) = \frac{m}{2\pi\hbar^2}\left(1 - \frac{\Delta}{\sqrt{2\varepsilon\Delta + \Delta^2}}\right)\vartheta(\varepsilon).$$

For the spin *up* direction, there may exist one or two roots of equation $\varepsilon_+(\xi) = 0$, depending on the sign ε. If $\varepsilon \geqslant 0$, there is a single root

$$\xi_1^{(+)} = \varepsilon + \Delta + \sqrt{2\varepsilon\Delta + \Delta^2} \quad \text{at} \quad \varepsilon \geqslant 0.$$

As a result, we find for the density of states $N_+(\varepsilon)$ with the spin projection *up*

$$N_+(\varepsilon) = \frac{m}{2\pi\hbar^2}\left(1 + \frac{\Delta}{\sqrt{2\varepsilon\Delta + \Delta^2}}\right)\vartheta(\varepsilon) \quad \text{at} \quad \varepsilon \geqslant 0.$$

In the energy region $\varepsilon_{min} \leqslant \varepsilon \leqslant 0$, there are two roots $\xi_1^{(+)}$ and $\xi_2^{(+)}$

$$\xi_1^{(+)} = \varepsilon + \Delta + \sqrt{2\varepsilon\Delta + \Delta^2} \quad \text{and} \quad \xi_2^{(+)} = \varepsilon + \Delta - \sqrt{2\varepsilon\Delta + \Delta^2}.$$

The minimum energy $\varepsilon_{min} = -\Delta/2$ as possible in the electron spectrum is achieved at $\xi = \xi_m = \Delta/2$.

Finally, taking into account that $\partial\varepsilon/\partial\xi < 0$ at $\xi = \xi_2^{(+)}$, we obtain for the density of states $N_+(\varepsilon)$ at $\varepsilon \leqslant 0$

$$N_+(\varepsilon) = \frac{m}{2\pi\hbar^2}\left[\left(\frac{\Delta}{\sqrt{2\varepsilon\Delta + \Delta^2}} - 1\right) + \left(1 + \frac{\Delta}{\sqrt{2\varepsilon\Delta + \Delta^2}}\right)\right] = \\ = \frac{m}{\pi\hbar^2}\frac{\Delta}{\sqrt{2\varepsilon\Delta + \Delta^2}} \quad \text{for} \quad \varepsilon_{min} \leqslant \varepsilon \leqslant 0.$$

The total density of states, involving both spin projections, will equal

$$N(\varepsilon) = N_-(\varepsilon) + N_+(\varepsilon) = \frac{m}{\pi\hbar^2}\begin{cases} \Delta/\sqrt{2\varepsilon\Delta + \Delta^2}, & \varepsilon_{min} \leqslant \varepsilon \leqslant 0, \\ 1, & \varepsilon \geqslant 0, \end{cases}$$

and $N(\varepsilon < \varepsilon_{min}) = 0$.

At zero temperature the electrons occupy only those states whose energy does not exceed the chemical potential or Fermi energy. In this case the electron density n is related with the density of states $N(\varepsilon)$

$$n = \int_{\varepsilon_{min}}^{\varepsilon_F} N(\varepsilon)\, d\varepsilon = \frac{m}{\pi\hbar^2}\begin{cases} \sqrt{2\varepsilon_F\Delta + \Delta^2}\,, & \varepsilon_F \leqslant 0, \\ \varepsilon_F + \Delta, & \varepsilon_F \geqslant 0. \end{cases}$$

Hence we arrive at the Fermi energy as a function of electron concentration

$$\varepsilon_F = \Delta\begin{cases} -\frac{1}{2}\left[1 - \left(\frac{n}{n_c}\right)^2\right], & n \leqslant n_c\,, \\ \left(\frac{n}{n_c} - 1\right), & n \geqslant n_c \quad \left(n_c = \frac{m\Delta}{\pi\hbar^2}\right) \end{cases}$$

where n_c is the critical electron density meaning the start of the upper subband occupation for $n > n_c$.

At zero temperature, the following integral represents the total energy of the gas in the two-dimensional film of area A:

$$E = \int_{\varepsilon_{min}}^{\varepsilon_F} \varepsilon N(\varepsilon)\, d\varepsilon = \frac{\pi \hbar^2}{6m} A \begin{cases} -n(3n_c^2 - n^2)/n_c, & n \leqslant n_c\,, \\ (3n^2 - 6nn_c + n_c^2), & n \geqslant n_c\,. \end{cases}$$

On the analogy with the bulk case, we determine the pressure of two-dimensional gas as a derivative of the total energy with respect to area

$$P = -\left(\frac{\partial E}{\partial A}\right)_N$$

where $N = nA$ is the total particle number in the gas. The simple calculation yields

$$P = \frac{\pi \hbar^2}{6m} \begin{cases} 2n^3/n_c, & n \leqslant n_c\,, \\ (3n^2 - n_c^2), & n \geqslant n_c\,. \end{cases}$$

We can now find the square of sound velocity in such two-dimensional gas at zero temperature. In fact, introducing the mass density of gas as $\rho = mn$, we have

$$c^2 = \frac{\partial P}{\partial \rho} = \frac{v_F^2}{2} \begin{cases} n/n_c, & n \leqslant n_c\,, \\ 1, & n \geqslant n_c\,, \end{cases}$$

where $v_F = \hbar\sqrt{2\pi n}/m$ is the Fermi velocity in the ideal Fermi gas of density n.

Let us turn to determining the spin magnetization M of electrons at zero temperature when the electrons occupy the states with the energies smaller than the Fermi energy. So, we have

$$M = \beta \int_{\varepsilon_{min}}^{\varepsilon_F} \left[N_+(\varepsilon) - N_-(\varepsilon)\right] d\varepsilon,$$

and β is the effective magnetic moment of an electron. It is obvious that, if the Fermi energy ε_F does not exceed zero magnitude or the electron concentration $n < n_c$, the upper subband, corresponding to the spin *down* direction, remains unoccupied and $N(\varepsilon) = N_+(\varepsilon)$. Accordingly, all the electrons have the spin and the magnetization will be maximum

$$M = \beta n, \quad n \leqslant n_c\,.$$

In the regions of larger density magnitudes $n > n_c$, the simple calculation leads to the following result:

$$M = \beta n_c \left(\frac{2n}{n_c} - 1 \right)^{1/2}, \quad n \geqslant n_c .$$

An existence of spontaneous magnetization due to spin–orbital effects leads to appearing of the current states in the electron gas. This results from the relation $\boldsymbol{j} = c \operatorname{rot} \boldsymbol{M}$ between the magnetization current $\boldsymbol{j}$ and magnetization $\boldsymbol{M}$. Since the electron magnetic moment is $\beta \sim e\hbar/(2mc)$, the magnetization current magnitude is independent of the luminal speed c. Such states appear in the regions of spin magnetization inhomogeneity, for example, at the film edges or at the boundaries of domains with the different electron polarization of electrons. In this case such states are referred to as *edge current states*.

3.8 Ideal Bose Gas

In the region of temperatures small as compared with the degeneracy one, the properties of the Bose systems are drastically different from those of fermionic systems. First of all, the point is that the Bose gas at zero temperature must occupy the ground state with the lowest energy. To satisfy the ground state of a gas with the particle spectrum $\varepsilon_p = p^2/2m$ at $T = 0$, all the Bose particles should be placed at the same state with the lowest possible energy, i.e. with zero momentum $p = 0$. According to the Bose statistics, the unlimited number of particles can be in the same quantum state. The wave function of the ground state in an ideal Bose gas will be the product of the one-particle wave functions of all particles. The physical phenomenon of macroscopically large population of the same quantum state is called the *Bose–Einstein condensation*. The accumulation of particles in the same state is referred to as the condensation of particles in this state.

The particles in the macroscopically populated state are called the *condensate particles* or *condensate*. The particles beyond the condensate and occupying the excited states are the *overcondensate particles* or *overcondensate*. As the temperature grows, the number of particles in the excited $p > 0$ states increases and the number of condensate particles decreases. The transition temperature T_c, at which the macroscopically large population of the quantum state disappears, i.e. density of condensate particles vanishes, is called the *Bose–Einstein condensation temperature*. The transition of a Bose gas to the condensed state from the non-condensed one and, *vice versa*, is an example of phase transition in the condensed medium from one thermodynamic state to another.

The maximum possible number of particles that can be in all excited states will be achieved at the maximum possible value of the chemical potential. In an ideal gas of bosons this takes place at $\mu = 0$. For the gas of N spinless particle in volume V, we obtain

$$\max[N_{\text{exc}}(T)] = \int\limits_{p>0} \frac{V d^3 p}{(2\pi\hbar)^3} \frac{1}{e^{\varepsilon_p/T} - 1} = V \frac{m^{3/2}}{\sqrt{2}\pi^2\hbar^3} \int\limits_{\varepsilon=+0}^{\infty} \frac{\sqrt{\varepsilon}\, d\varepsilon}{e^{\varepsilon/T} - 1} =$$

$$= V \left(\frac{mT}{2\pi\hbar^2}\right)^{3/2} \frac{2}{\sqrt{\pi}} \int\limits_0^{\infty} \frac{\sqrt{x}\, dx}{e^x - 1} = V \left(\frac{mT}{2\pi\hbar^2}\right)^{3/2} \zeta(3/2).$$

The value of the Riemann zeta-function equals approximately $\zeta(3/2) \approx 2.612$.

At sufficiently low temperatures, the particle number not smaller than $N - \max[N_{\text{exc}}(T)]$ should be in the ground state with zero momentum $p = 0$, i.e. in the condensate. The Bose–Einstein condensation temperature T_c is determined from condition $N = \max[N_{\text{exc}}(T)]$ and depends on the Bose gas density $n = N/V$ as

$$T_c = \frac{2\pi}{\zeta^{2/3}(3/2)} \frac{\hbar^2 n^{2/3}}{m} \approx 3.31 \frac{\hbar^2 n^{2/3}}{m}.$$

If one lowers the temperature of a Bose gas under its fixed density, the chemical potential of gas takes zero value at $T = T_c$ and this zero value conserves down to absolute zero, i.e. $\mu(T \leqslant T_c) = 0$. The overcondensate particle density at $T \leqslant T_c$ reads

$$n_{\text{exc}}(T) = \int\limits_{p>0} \frac{d^3 p}{(2\pi\hbar)^3} \frac{1}{e^{\varepsilon_p/T} - 1} = \left(\frac{mT}{2\pi\hbar^2}\right)^{3/2} \zeta(3/2) = n\left(\frac{T}{T_c}\right)^{3/2}.$$

Accordingly, the condensate particle density will be finite and equal to

$$n_0(T) = n - n_{\text{exc}}(T) = n\left[1 - \left(\frac{T}{T_c}\right)^{3/2}\right].$$

The energy of the condensed gas at $T < T_c$ is governed with the particles whose energy $\varepsilon_p > 0$. Hence the energy is given by the following integral:

$$E = \int \frac{V d^3 p}{(2\pi\hbar)^3} \frac{\varepsilon_p}{e^{\varepsilon_p/T} - 1} = VT \left(\frac{mT}{2\pi\hbar^2}\right)^{3/2} \frac{2}{\sqrt{\pi}} \int\limits_0^{\infty} \frac{dx\, x^{3/2}}{e^x - 1} =$$

$$= \frac{3\zeta(5/2)}{2} VT \left(\frac{mT}{2\pi\hbar^2}\right)^{3/2} = \frac{3\zeta(5/2)}{2\zeta(3/2)} T n_{\text{exc}}(T) V \sim VT^{5/2}.$$

The specific heat C_v at constant volume equals

$$C_v = \left(\frac{\partial E}{\partial T}\right)_V = \frac{5}{2}\frac{E}{T} = \frac{15}{4}\frac{\zeta(5/2)}{\zeta(3/2)} \left(\frac{T}{T_c}\right)^{3/2} N \sim VT^{3/2}$$

and remains finite at the Bose–Einstein condensation point as $C_v(T_c) \approx 1.93N$. This magnitude of specific heat is slightly higher than $1.5N$ in the limit of high $T \gg T_c$ temperatures.

The condensed gas pressure P can be found, for example, from the grand thermodynamic potential $\Omega = -PV$

$$\Omega = T \int \frac{V d^3 p}{(2\pi\hbar)^3} \ln\left(1 - e^{-\varepsilon_p/T}\right) =$$
$$= VT \left(\frac{mT}{2\pi\hbar^2}\right)^{3/2} \frac{2}{\sqrt{\pi}} \int_0^\infty dx\, x^{1/2} \ln\left(1 - e^{-x}\right) =$$
$$= -VT \left(\frac{mT}{2\pi\hbar^2}\right)^{3/2} \zeta(5/2) = -\frac{2}{3} E\,.$$

Then we obtain the pressure of ideal Bose–Einstein condensed gas

$$P(T) = T \left(\frac{mT}{2\pi\hbar^2}\right)^{3/2} \zeta(5/2) \sim T^{5/2}.$$

The pressure is independent of the gas volume since the condensate particles with momentum $p = 0$ produce no contribution to the pressure. The volume-independent pressure results in the divergent magnitudes of isothermal expansion and compressibility as well as specific heat at constant pressure.

Since $\mu = 0$, the free energy $F(T, V)$ of condensed gas can readily be found from relation $F = \mu N + \Omega$

$$F = \Omega = -\frac{2}{3} E \sim -VT^{5/2}.$$

Hence we obtain the entropy S according to the formula $S = -\partial F/\partial T$

$$S = \frac{5}{3}\frac{E}{T} = \frac{2}{3} C_v \sim VT^{3/2}.$$

Since the entropy of a gas depends on its volume, the adiabatic coefficients of expansion and compressibility conserve their finite magnitudes in contrast to the isothermal counterparts. While varying the temperature and volume along the adiabat, the ratio T/T_c remains unchanged and, therefore, the relation between the overcondensate and condensate fractions holds for a constant in the adiabatic ($S = \text{const}$) process. In the ideal condensed Bose gas, we have $PV^{5/3} = \text{const}$ along the adiabatic curve. This entails the following magnitude for the adiabatic compressibility:

$$-\frac{1}{V}\left(\frac{\partial V}{\partial P}\right)_S = \frac{3}{5}\frac{1}{P(T)} \sim T^{-5/2}$$

which remains finite at $T \neq 0$ unlike the infinite magnitude of isothermal compressibility. Using the equality $\rho(\partial P/\partial \rho) = -V(\partial P/\partial V)$ where $\rho \sim 1/V$ is the gas density, we find the square of adiabatic sound velocity from the formula

$$c_s^2 = \left(\frac{\partial P}{\partial \rho}\right)_S = \frac{5}{3}\frac{P(T)}{\rho} \sim T^{5/2}.$$

The magnitude of adiabatic sound velocity $T \neq 0$ is finite as well. This is in contrast to zero magnitude of isothermal sound velocity determined by equation $c_T^2 = (\partial P/\partial \rho)_T$.

In conclusion, we would like to discuss a possibility of the Bose–Einstein condensation in an ideal gas as a function of the spatial dimensionality d and one-particle density of states $g(\varepsilon)$. Let us write down the maximum possible number of particles in all excited states $\varepsilon > 0$

$$\max[N_{\text{exc}}(T)] = \int\limits_0^\infty \frac{g(\varepsilon)\, d\varepsilon}{e^{\varepsilon/T} - 1} = T\int\limits_0^\infty \frac{g(xT)\, dx}{e^x - 1}.$$

The answer depends on the behavior $g(\varepsilon)$ for the small values of energy ε. If the above integral is convergent and unlimitedly decreases by lowering the temperature, the particle condensation to the lowest energy level is necessary at some temperature since all the particles cannot be put into the excited states.

Provided that the integral is divergent and the value $\max[N_{\text{exc}}(T \neq 0)]$ exceeds the total particle number N, the normalization with the total particle number satisfies always at the chemical potential value smaller than its maximum possible one, i.e. at $\mu < 0$. The population of the ground level $\varepsilon = 0$ will be proportional to the factor $\left[\exp(-\mu/T) - 1\right]^{-1}$ limited in magnitude and thus no macroscopic population of the ground level appears.

In the space of dimensionality d and for usual dispersion law $\varepsilon_p = p^2/2m$, the one-particle density of states $g(\varepsilon)$ is proportional to $\varepsilon^{(d-2)/2}$. Since the integral above converges if only $d > 2$, we draw a conclusion about the absence of the Bose–Einstein condensation phenomenon for the ideal gas of bosons in the space of dimensionality $d \leqslant 2$ at any finite temperature.

For bosons with the relativistic energy spectrum $\varepsilon = cp$, the one-particle density of states reads $g(\varepsilon) \sim \varepsilon^{(d-1)}$ and the Bose–Einstein condensation in the ideal gas of bosons will be possible in the spatial dimensionality $d > 1$.

Problems

1. Find the behavior for specific heat $C_p(T)$ in the ideal Bose gas at the constant pressure in the vicinity of the Bose–Einstein condensation temperature.

Solution. We employ the relation between the specific heats at constant pressure C_p and constant volume C_v

$$C_p = C_v\,\frac{(\partial P/\partial V)_S}{(\partial P/\partial V)_T}.$$

The quantities C_v and $(\partial P/\partial V)_S$ are continuous at the phase transition point T_c and, therefore, divergency C_p is associated with zero value of $(\delta P/\delta V)_T$. We have at the transition point

$$C_v\left(\frac{\partial P}{\partial V}\right)_S = -\frac{25}{4}\frac{P^2(T_c)}{T_c} = -\left(\frac{5\zeta(5/2)}{2\zeta(3/2)}\right)^2\left(\frac{N}{V}\right)^2 T_c\,.$$

Then we determine pressure P, using the relation $(\partial P/\partial\mu)_{T,V} = N/V$. So,

$$P = P_0(T) + \mu\,\frac{N}{V}$$

where $P_0(T)$ is the pressure at $\mu = 0$. Next, we have for the derivative with respect to volume

$$\left(\frac{\partial P}{\partial V}\right)_T = \frac{\partial}{\partial V}\left(\frac{N}{V}\,\mu(T,\,V)\right).$$

In the vicinity of phase transition temperature T_c, the chemical potential μ equals approximately

$$\mu = -\frac{9\,\zeta^2(3/2)}{16\pi}\frac{(T-T_c)^2}{T_c}\,\theta(T-T_c).$$

Taking $\partial T_c/\partial V = -(2/3)\,T_c/V$ into account, we find the leading contribution to the inverse compressibility at $T \to T_c$

$$\left(\frac{\partial P}{\partial V}\right)_T = -\frac{3\,\zeta^2(3/2)}{4\pi}\frac{N}{V^2}\,(T-T_c)\,\theta(T-T_c).$$

Finally, unlike the specific heat C_v at the constant volume, we get the divergent behavior near $T = T_c$

$$C_p = \frac{25\pi\,\zeta^2(5/2)}{3\,\zeta^4(3/2)}\,N\,\frac{T_c}{T-T_c} \sim \frac{1}{T-T_c}\,.$$

The jump in the second derivative of chemical potential with respect to temperature

$$\frac{\partial^2\mu}{\partial T^2} = -\frac{9\zeta^2(3/2)}{8\pi T_c}\theta(T-T_c)$$

at the transition point leads to the corresponding discontinuity in derivative $\partial C_v/\partial T$. In fact, using the relation for the energy of a gas

$$E = \frac{3}{2}\Big(P_0(T)V + \mu N\Big) = E_0(T) + \Delta E,$$

we have for the difference in the specific heats of the non-condensed and condensed states near the transition point

$$\Delta\left(\frac{\partial C_v}{\partial T}\right) = \frac{\partial}{\partial T}\left(\frac{\partial\Delta E}{\partial T}\right) = \frac{\partial}{\partial T}\left(\frac{3}{2}\frac{\partial\mu}{\partial T}N\right) = -\frac{27\zeta^2(3/2)}{16\pi}\frac{N}{T_c} \approx -3.66\frac{N}{T_c}\,.$$

2. There is an ideal gas of charged bosons on the neutralizing homogeneous background of opposite-sign charges.

Consider the screening of a trial charge Q, using *the self-consistent Debye–Hückel approximation*. The charge of spinless bosons equals q.

Solution. The trial charge produces an electric field with certain potential $\varphi(r)$. We have for the density $n(r)$ of the boson distribution

$$n(r) = \int \frac{d^3p}{(2\pi\hbar)^3} n_B(\varepsilon_p + q\varphi(r) - \mu)$$

where n_B is the Bose–Einstein distribution for the particles with dispersion ε_p and chemical potential μ. Far from the trial charge, where $\varphi \to 0$, the density of induced charges should cross over into the average density n_0 of bosons.

The potential φ is connected with the charge density by means of the Poisson equation

$$\nabla^2\varphi = -4\pi q\big(n(r) - n_0\big).$$

Next, we apply the Debye–Hückel approximation, assuming the relative weakness of interaction $q\varphi$

$$n(r) = n_0 + q\varphi(r)\left(-\frac{\partial n_0}{\partial \mu}\right)\bigg|_{\varphi=0}.$$

As a result, we obtain equation $\nabla^2\varphi = \varkappa^2\varphi$ where

$$\varkappa^2 = 4\pi q^2\frac{\partial n_0}{\partial \mu} = 4\pi q^2 \int \frac{d^3p}{(2\pi\hbar)^3}\frac{\partial n_B(\varepsilon_p - \mu)}{\partial \mu} = -4\pi q^2 \int g(\varepsilon) n'_B(\varepsilon - \mu)d\varepsilon$$

and $g(\varepsilon)$ is the density of states. The centrally symmetric solution, satisfying $\varphi \to Q/r$ as $r \to 0$, has the form of Debye screening

$$\varphi(r) = \frac{Q}{r}e^{-\varkappa r}.$$

The length $1/\varkappa$ has a sense of the screening radius (*Debye radius*). For the temperatures higher as compared with the degeneracy temperature of ideal boson gas, the Bose–Einstein distribution can be replaced with the Boltzmann one $n_B \approx \exp[(\mu - \varepsilon_p)/T]$. Hence we obtain the answer for the inverse Debye radius in the classical ideal plasma

$$\varkappa = \sqrt{\frac{4\pi q^2 n_0}{T}}.$$

For the gas of bosons with dispersion $\varepsilon_p = p^2/2m$, it is interesting to consider the behavior of the Debye radius at the temperatures close to the Bose–Einstein condensation temperature $T = T_c$. As $T \to T_c$, the estimate of the above integral yields

$$\varkappa^2 = 4\pi q^2\left(\frac{mT}{2\pi\hbar^2}\right)^{3/2}\frac{\sqrt{\pi}}{T}\sqrt{\frac{T}{|\mu|}} \to \frac{4\pi q^2 n_0}{T_c}\frac{4\pi}{3\zeta^2(3/2)}\frac{T_c}{T - T_c}.$$

Thus, in the ideal gas of charged bosons as $T \to T_c$ the Debye screening radius tends to zero according to $1/\varkappa \sim \sqrt{T - T_c}$.

Note that, for bosons with the relativistic spectrum $\varepsilon_p = pc$, the Debye radius remains finite at the temperatures near the condensation temperature T_c

$$\varkappa(T \to T_c) = \frac{\pi}{\sqrt{6\zeta(3)}}\sqrt{\frac{4\pi q^2 n_0}{T_c}}.$$

The example considered demonstrates that the behavior of a Bose gas at temperatures below the degeneracy temperature is very sensitive to the type of the particle energy dispersion and, correspondingly, density of states.

3. Find the isothermal sound velocity c_{T} in the two-dimensional ideal gas of spinless bosons of mass m and particle density n.

Solution. The square of the isothermal sound velocity is determined by the derivative of pressure P with respect to the mass density $\rho = mn$:

$$c_{\mathrm{T}}^2=\left(\frac{\partial P}{\partial \rho}\right)_T=\frac{1}{m}\left(\frac{\partial P}{\partial n}\right)_T=\frac{n}{m}\left(\frac{\partial \mu}{\partial n}\right)_T$$

where μ is the chemical potential which can be found from the equation

$$n=\int\frac{d^2p}{(2\pi\hbar)^2}\frac{1}{e^{(\varepsilon_p-\mu)/T}-1}=-\frac{mT}{2\pi\hbar^2}\ln\left(1-e^{\mu/T}\right),\quad \varepsilon_p=\frac{p^2}{2m}.$$

Hence chemical potential equals

$$\mu=T\ln\left(1-e^{-2\pi\hbar^2n/mT}\right).$$

Differentiating the chemical potential yields the final answer

$$c_{\mathrm{T}}=\frac{\sqrt{2\pi\hbar^2n}}{m}\left(e^{2\pi\hbar^2n/mT}-1\right)^{-1/2}\approx\begin{cases}\frac{\sqrt{2\pi\hbar^2n}}{m}\exp\left(-\frac{\pi\hbar^2n}{mT}\right), & T\ll\frac{2\pi\hbar^2n}{m}\\ \sqrt{\frac{T}{m}}, & T\gg\frac{2\pi\hbar^2n}{m}\end{cases}$$

according to $c_{\mathrm{T}}=\sqrt{n\mu'(n)/m}$. The isothermal sound velocity vanishes at $T=0$.

4. Find the isothermal sound velocity c_{T} in the two-dimensional ideal gas of spin-1/2 fermions of mass m and particle density n.

Solution. The square of the isothermal sound is determined by the derivative of pressure P with respect to the mass density $\rho=mn$:

$$c_{\mathrm{T}}^2=\left(\frac{\partial P}{\partial \rho}\right)_T=\frac{1}{m}\left(\frac{\partial P}{\partial n}\right)_T=\frac{n}{m}\left(\frac{\partial \mu}{\partial n}\right)_T$$

where μ is the chemical potential which can be found from the equation

$$n=2\int\frac{d^2p}{(2\pi\hbar)^2}\frac{1}{e^{(\varepsilon_p-\mu)/T}+1}=\frac{mT}{\pi\hbar^2}\ln\left(1+e^{\mu/T}\right),\quad \varepsilon_p=\frac{p^2}{2m}.$$

Hence the chemical potential equals

$$\mu=T\ln\left(e^{\pi\hbar^2n/mT}-1\right).$$

Differentiating the chemical potential yields the final answer

$$c_{\mathrm{T}}=\frac{\sqrt{\pi\hbar^2n}}{m}\left(1-e^{-\pi\hbar^2n/mT}\right)^{-1/2}\approx\begin{cases}\frac{v_F}{\sqrt{2}}, & T\ll\frac{mv_F^2}{2}\\ \sqrt{\frac{T}{m}}, & T\gg\frac{mv_F^2}{2}\end{cases}$$

according to $c_{\mathrm{T}}=\sqrt{n\mu'(n)/m}$ and $v_F=\sqrt{2\pi\hbar^2n}/n$ is the Fermi velocity.

Note the different low-temperature behavior of isothermal sound velocity in the fermion and boson ideal gases at the temperatures below the degeneracy one. As is expected, the sound velocity in both systems has the same classical magnitude in the high-temperature limit.

3.9 The Degenerate Bose Gas in the Harmonic Trap

Let us consider the behavior of ideal degenerate Bose gas in the trap with the harmonic confining potential

$$U(x, y, z) = m\left(\omega_x^2 x^2 + \omega_y^2 y^2 + \omega_z^2 z^2\right)/2.$$

The energy levels of a particle in the harmonic potential are given by the simple formula

$$\varepsilon(n_1, n_2, n_3) = \hbar\omega_x\left(n_1 + 1/2\right) + \hbar\omega_y\left(n_2 + 1/2\right) + \hbar\omega_z\left(n_3 + 1/2\right)$$

where n_1, n_2, and n_3 are the non-negative integers. Below we assume the macroscopic limit when the number of bosons $N \gg 1$ is large and the energy levels in the trap are located sufficiently close to each other. In this case, instead of summing over the energy levels, we can use the semiclassical approximation for calculating the thermodynamic functions of a gas. So, we replace the sum over the close discrete levels with the integral, thus assuming an approximation with the continuous spectrum of levels.

First of all, we determine the one-particle density of states in the trap

$$g(\varepsilon) = \int \frac{d^3p\, d^3r}{(2\pi\hbar)^3}\, \delta\left(\varepsilon - \frac{p_x^2 + p_y^2 + p_z^2}{2m} - \frac{m(\omega_x^2 x^2 + \omega_y^2 y^2 + \omega_z^2 z^2)}{2}\right) =$$
$$= \int \frac{d^3p\, d^3r}{\pi^3\hbar^3\omega_x\omega_y\omega_z}\, \delta\left(\varepsilon - \boldsymbol{p}^2 - \boldsymbol{r}^2\right).$$

The last integral can be interpreted as an integral in the six-dimensional space $\boldsymbol{R} = (\boldsymbol{p}, \boldsymbol{r})$. Then,

$$g(\varepsilon) = \int \frac{d^6R}{(\pi\hbar)^3\omega_x\omega_y\omega_z}\, \delta\left(\varepsilon - \mathbf{R}^2\right) = \int \frac{\pi^3 R^5\, dR}{(\pi\hbar)^3\omega_x\omega_y\omega_z}\, \delta\left(\varepsilon - R^2\right).$$

Here we have taken into account that the area for the six-dimensional sphere of unit radius equals π^3. Calculating the last integral with the substitution $R^2 = \rho$, we find the one-particle density of states in the harmonic trap

$$g(\varepsilon) = \frac{\varepsilon^2}{2\hbar^3\omega_x\omega_y\omega_z}.$$

The Bose–Einstein condensation T_c is determined with the aid of condition

$$N = \int_0^\infty \frac{d\varepsilon\, \varepsilon^2}{2\hbar^3\omega_x\omega_y\omega_z}\, \frac{1}{e^{\varepsilon/T_c} - 1} = \frac{T_c^3}{2\hbar^3\omega_x\omega_y\omega_z}\, 2\zeta(3).$$

Hence the condensation temperature equals

$$T_c = \hbar\big(\omega_x\omega_y\omega_z N/\zeta(3)\big)^{1/3}.$$

Here we emphasize the following point. Unlike the conventional trap as a box of volume $V = L_x L_y L_z$ when the macroscopic limit means $N \to \infty$, $V \to \infty$, and the gas density N/V to be held constant, the macroscopic description of the gas condensation in the trap assumes the following limiting conditions: $N \to \infty$, $(\omega_x\omega_y\omega_z)^{1/3} \to 0$ under $N(\omega_x\omega_y\omega_z)^{1/3}$ =const.

The number of overcondensate particles is given by the integral

$$N_{\text{exc}}(T) = \int_0^\infty \frac{d\varepsilon\,\varepsilon^2}{2\hbar^3\omega_x\omega_y\omega_z}\,\frac{1}{e^{\varepsilon/T}-1} = \frac{T^3}{\hbar^3\omega_x\omega_y\omega_z}\,\zeta(3) = N\left(\frac{T}{T_c}\right)^3.$$

The energy of condensed gas is determined with the expression

$$E(T) = \int_0^\infty \frac{d\varepsilon\,\varepsilon^2}{2\hbar^3\omega_x\omega_y\omega_z}\,\frac{\varepsilon}{e^{\varepsilon/T}-1} = \frac{T^4}{2\hbar^3\omega_x\omega_y\omega_z}\,\frac{\pi^4}{15} = \frac{\pi^4}{30\zeta(3)}\,T N_{\text{exc}}(T).$$

The specific heat of condensed gas is found by differentiating with respect to temperature

$$C(T) = 4\,\frac{E(T)}{T} = \frac{2\pi^4}{15\zeta(3)}\,N_{\text{exc}}(T) \sim T^3.$$

Let us analyze the behavior of chemical potential $\mu(T)$ near the condensation temperature T_c. For this aim, we write

$$\begin{aligned} N &= \int_0^\infty \frac{d\varepsilon\,\varepsilon^2}{2\hbar^3\omega_x\omega_y\omega_z}\,\frac{1}{e^{(\varepsilon-\mu/T)}-1} = \frac{T^3}{2\hbar^3\omega_x\omega_y\omega_z}\int_0^\infty \frac{dx\,x^2}{e^{-\mu/T}e^x-1} \\ &\approx \frac{T^3}{2\hbar^3\omega_x\omega_y\omega_z}\left(2\zeta(3) + \frac{\pi^2}{3}\,\frac{\mu}{T}\right) = \frac{T_c^3}{\hbar^3\omega_x\omega_y\omega_z}\,\zeta(3)\,. \end{aligned}$$

Estimating the integral, we have taken into account that the chemical potential $|\mu| \ll T \sim T_c$ near T_c. Then we have

$$\mu(T) = -\frac{18\zeta(3)}{\pi^2}\,(T-T_c)\theta(T-T_c) \quad \text{at} \quad |T-T_c| \ll T_c\,.$$

Unlike the usual case of three-dimensional trap as a cubic box, we see that the first derivative $\partial\mu/\partial T$ experiences already the finite jump.

Let us turn to variable $x = \varepsilon/T$ and differentiate the energy of a gas with respect to temperature. As a result we represent the specific heat as

$$C(T) = \frac{\partial}{\partial T}\int_0^\infty \frac{dx\; x^2}{2\hbar^3\omega_x\omega_y\omega_z}\,\frac{x}{e^{x-\mu/T}-1}$$

$$= \frac{4T^3}{2\hbar^3\omega_x\omega_y\omega_z}\int_0^\infty \frac{dx\; x^3}{e^{x-\mu/T}-1} + \frac{T^4}{2\hbar^3\omega_x\omega_y\omega_z}\int_0^\infty \frac{dx\; x^3 e^{x-\mu/T}}{(e^{x-\mu/T}-1)^2}\,\frac{\partial(\mu/T)}{\partial T}\,.$$

Involving the continuity of chemical potential μ and the jump of derivative $\partial\mu/\partial T$, we obtain near the transition temperature T_c

$$C(T_c+0) = C(T_c-0) + \frac{T_c^4}{2\hbar^3\omega_x\omega_y\omega_z}\int_0^\infty \frac{dx\; x^3 e^x}{(e^x-1)^2}\left(-\frac{18\,\zeta(3)}{\pi^2 T_c}\right).$$

Substituting the value of integral equal to $6\zeta(3)$, we arrive at the magnitude of the specific heat jump at the transition point $T = T_c$

$$\Delta C(T_c) = C(T_c-0) - C(T_c+0) = \frac{54\,\zeta(3)}{\pi^2}\,N\,.$$

The Bose–Einstein condensation in the harmonic trap is an example of the second-order phase transition.

Problems

1. The classical string with the fixed end-points has an infinite set of oscillation frequencies $\omega_l = l\omega$, which are the multiplies of the fundamental frequency ω. Its quantum mechanical generalization or bosonic quantum non-relativistic string represents a set of the infinite number of harmonic oscillators with frequencies ω, 2ω, 3ω, ... and energies

$$E_l = \hbar\omega_l(n_l + 1/2), \quad \omega_l = l\omega, \quad l = 1,\ 2,\ 3, \ldots$$

The occupation numbers $n_l = 0,\ 1,\ 2, \ldots$ for each oscillator are unlimited and vary from zero to infinity.

Find the free energy $F(T)$, entropy $S(T)$, specific heat $C(T)$, and the number of states $\Gamma(E)$ for the quantum Bose string in the high energy $E \gg \hbar\omega$ limit (*Hardy–Ramanujan formula*).

Solution. Since the oscillators do not interact with each other, the total partition function reduces to a product of partition functions for the separate oscillators

$$Z = \prod_{l=1}^\infty z_l, \quad z_l = \sum_{n_l=0}^\infty e^{-\frac{\hbar\omega_l}{T}(n_l+1/2)} = \frac{e^{-\hbar\omega_l/2T}}{1-e^{-\hbar\omega_l/T}}\,.$$

Hence the free energy reads

$$F = E_0 + T\sum_{l=1}^\infty \ln\left(1-e^{-\hbar\omega_l/T}\right), \quad E_0 = \sum_{l=1}^\infty \hbar\omega_l/2,$$

E_0 being the energy of zero oscillations. Then we find the entropy and specific heat

$$S(T) = \sum_{l=1}^{\infty}\left[\frac{\hbar\omega_l/T}{e^{\hbar\omega_l/T}-1} - \ln\left(1-e^{-\hbar\omega_l/T}\right)\right],$$

$$C(T) = \sum_{l=1}^{\infty}\left[\frac{\hbar\omega_l/2T}{\sinh(\hbar\omega_l/2T)}\right]^2, \quad \omega_l = l\omega.$$

For the low ($T \ll \hbar\omega$) temperatures, all these quantities vanish exponentially as

$$S(T) = (\hbar\omega/T)e^{-\hbar\omega/T}, \quad C(T) = (\hbar\omega/T)^2 e^{-\hbar\omega/T}.$$

The behavior at high ($T \gg \hbar\omega$) temperature is more interesting. In this limit the main contribution to the sum over l is given by the terms with large $l \gg 1$ numbers, and we can approximately convert the sum into the integral. Then,

$$F - E_0 = T\int_0^{\infty} dl\, \ln\left(1-e^{-\hbar\omega l/T}\right) = \frac{T^2}{\hbar\omega}\int_0^{\infty} dx\, \ln\left(1-e^{-x}\right) = -\frac{\pi^2}{6}\frac{T^2}{\hbar\omega},$$

$$C(T) = \int_0^{\infty} dl \left(\frac{\hbar\omega l/2T}{\sinh(\hbar\omega l/2T)}\right)^2 = \frac{2T}{\hbar\omega}\int_0^{\infty} dx\,\frac{x^2}{\sinh^2 x} = \frac{\pi^2}{3}\frac{T}{\hbar\omega}.$$

Note that the specific heat $C(T)$ of quantum string does not cross over into the usual classical behavior $C(T) = \text{const}$ even in the high-temperature limit.

To determine the number of states $\Gamma(E)$ of the string, we employ the relation $S(E) = \ln\Gamma(E)$. From the relations

$$E - E_0 = \pi^2T^2/6\hbar\omega \quad \text{and} \quad S = \pi^2 T/3\hbar\omega$$

valid at $T \gg \hbar\omega$, we find

$$\ln\Gamma(E) = S(E) = 2\pi\sqrt{\frac{E-E_0}{6\hbar\omega}} = 2\pi\sqrt{\frac{N}{6}}, \quad N = \frac{E-E_0}{\hbar\omega} \gg 1.$$

Here $N = n_1 + 2n_2 + 3n_3 + \ldots$ plays a role of particle number expressed in terms of occupation numbers.

In fact, function $\Gamma(N)$ coincides completely with that $p(N)$ called the *partition function* and is determined as the number of possible partitions of a non-negative integer N onto the integer terms regardless of their order. For instance, for $N = 4$ we have five partitions: 4, 3+1, 2+2, 2+1+1, 1+1+1+1, and $p(4) = 5$. We here present more precise *Hardy–Ramanujan asymptotic* for the number of string states at $N \gg 1$, containing one more expansion term

$$\ln\Gamma(N) \approx 2\pi\sqrt{N/6} - \ln(4\sqrt{3}\,N) \quad \text{and} \quad \Gamma(N) \approx (4\sqrt{3}\,N)^{-1}e^{2\pi\sqrt{N/6}}.$$

Emphasize that the thermodynamic properties for the fermionic and bosonic strings are isomorphic to each other in the high $T \gg \hbar\omega$ limit.

2. Determine the Bose–Einstein condensation temperature for an ideal gas consisting of $N \gg 1$ spinless bosons having mass M and restricted in motion with the spherical surface of radius R. Find the number $N_0(T)$ of condensate particles.

Solution. The energy $\varepsilon_{l,m}$ of a boson equals

$$\varepsilon_l = \frac{\hbar^2}{2MR^2}l(l+1), \quad l = 0,\ 1,\ 2, \ldots$$

where l is the orbital quantum number and each energy level has the $2l+1$ states degenerated with respect to magnetic quantum number $m=-l,\ -l+1,\ldots,\ l-1,\ l$. The maximum possible number of particles which can be in all excited states $l>0$ will be achieved at the maximum possible value of the chemical potential μ. In our case this is $\mu=0$ and we have

$$N=\sum_{l=1}^{\infty}\frac{2l+1}{e^{\varepsilon_l/T}-1}=\sum_{l=1}^{\infty}\frac{2l+1}{e^{al(l+1)}-1}, \quad a=\frac{\hbar^2}{2MR^2T}.$$

This equation determines the Bose–Einstein condensation temperature T_c.

Assuming that the condensation temperature is large on the scale of energy level discreteness, i.e. $T_c\gg\hbar^2/2MR^2$, we estimate the sum by the following integral within the logarithmic accuracy:

$$N\approx\int\limits_{l=1}^{\infty}\frac{(2l+1)\,dl}{e^{al(l+1)}-1}=-\frac{\ln(1-e^{-2a})}{a}\approx\frac{1}{a}\ln\frac{1}{2a}\quad(a\ll 1).$$

Solving this equation, we find the condensation temperature within the logarithmic accuracy

$$T_c\approx\frac{\hbar^2}{2MR^2}\frac{N}{\ln(cN)}$$

where $c\sim 1$ is the number of about unity. In the crossover limit $R\to\infty$ to the plane surface at the fixed gas density $n=N/4\pi R^2$, as is expected, we have $T_c\to 0$ for the finite magnitude R.

The number of condensate particles $N_0(T)$ at $T<T_c$ is given by the following difference:

$$N_0(T)=N-\sum_{l=1}^{\infty}\frac{2l+1}{e^{\varepsilon_l/T}-1}\approx\frac{1}{a_c}\ln\frac{1}{2a_c}-\frac{1}{a}\ln\frac{1}{2a}$$

or

$$\frac{N_0(T)}{N}\approx 1-\frac{a_c}{a}\frac{\ln(1/2a)}{\ln(1/2a_c)}=1-\frac{T}{T_c}\frac{\ln(MR^2T/\hbar^2)}{\ln(MR^2T_c/\hbar^2)}.$$

3.10 Ideal Gas of Elementary Bose Excitations

The Bose statistics can be applied to studying the thermal equilibrium of electromagnetic radiation and to calculating the thermodynamic properties for a series of condensed media. The latter can be described as a set of Bose *elementary excitations*. The radiation or electromagnetic waves can be represented as a set of electromagnetic field quanta, i.e. *photons*, whose energy is given by the momentum and polarization.

Another object of such a approach is solids or elastic condensed media in which excitations are vibrations of atoms near their equilibrium position. On the whole, such collective vibrations of a deformed medium correspond to the sound waves described as a set of quanta of elastic deformation or *phonons* whose energy is also determined by the momentum and type (branch) of the corresponding sound wave. The sound waves are simplest in an isotropic liquid which has a single *acoustic* branch. In the crystalline medium the sound spectrum is more complicated. There are always one *longitudinal* and two *transverse* sound waves. In the media with the complex crystalline lattice containing $\nu > 1$ atoms in a *unit cell*, there are extra $3(\nu - 1)$ sound branches called the *optical ones*. As concerns the acoustic branches, the sound wave frequency vanishes at zero wave vector. As for the optical branches, the sound frequency at zero wave vector remains finite.

Another example of condensed media, where elementary excitations of the Bose type exist, are magnetically ordered media. The rotation of magnetic moments or spins is accompanied with the propagation of waves through the magnetic system. The waves are referred to as *spin waves*, and the corresponding quanta are called *magnons*.

The physical process that provides us the thermal equilibrium of excitations is the generation and absorption of elementary excitations with the condensed medium. This circumstance leads to the following feature of elementary excitations. The total number of elementary excitations N in the condensed medium is not fixed and does not conserve. Accordingly, the number of excitations is wholly determined by the conditions of thermodynamic equilibrium alone, i.e. temperature, and $N = N(T)$. Thus, the total number of elementary excitations cannot be treated as an independent thermodynamic variable that can be set at its ambiguity. It must be determined from the condition of thermodynamic equilibrium, i.e. from the minimum of thermodynamic potentials, e.g. free energy F or Gibbs potential Φ. Since the derivative of potentials with respect to the number of particles yields the chemical potential

$$\mu = \left(\frac{\partial F}{\partial N}\right)_{T,V} = \left(\frac{\partial \Phi}{\partial N}\right)_{P,T},$$

the requirement for the minimum coincides with the condition $\mu = 0$.

Zero value of chemical potential is a general property for elementary excitations when their total number is not fixed and does not conserve. This property can also be examined from the viewpoint of the chemical equilibrium condition for the reaction of producing and absorbing an elementary excitation in a medium according to

$$\text{medium} + \text{excitation} \leftrightarrows \text{medium}.$$

Since under thermodynamic (thermal) equilibrium $\mu_{\mathrm{m}} + \mu_{\mathrm{ex}} = \mu_{\mathrm{m}}$, we expect $\mu_{\mathrm{ex}} = 0$.

In first approximation and at low temperatures when the number of elementary excitations is small, the excitations can be treated as an ideal gas of particles with zero chemical potential $\mu = 0$. The energy of such a particle is $\varepsilon_s(\boldsymbol{k}) = \hbar\omega_s(\boldsymbol{k})$ where $\omega_s(\boldsymbol{k})$ is the frequency of sth branch or excitation type, $\boldsymbol{k}$ is the wave vector, and $\boldsymbol{p} = \hbar\boldsymbol{k}$ is the momentum of excitation. The distribution of the Bose-type excitations or their average number in the quantum states with wave vector $\boldsymbol{k}$ and frequency $\omega_s(\boldsymbol{k})$ is given by the formula

$$n\big(\omega_s(\boldsymbol{k})\big) = \frac{1}{e^{\hbar\omega_s(\boldsymbol{k})/T} - 1}$$

which represents the *Planck distribution*. Below, we consider the application of this approach for studying some physical systems.

3.11 Black-Body Radiation

Let us apply the Planck distribution for describing the *black-body radiation* or thermodynamically equilibrium radiation representing a gas of photons. The photons are governed by the Bose statistics since they carry the integer spin angular momentum, which is related to photon polarization. Bearing in mind the two possible polarizations and dispersion of photons $\omega(k) = ck$ where c is the speed of light, we find the density of equilibrium photon gas at temperature T

$$n(T) = \frac{N(T)}{V} = 2\int \frac{d^3k}{(2\pi)^3}\frac{1}{e^{\hbar ck/T}-1} = \frac{T^3}{\pi^2\hbar^3}\int_0^\infty \frac{x^2\,dx}{e^x-1} = \frac{2\zeta(3)}{\pi^2}\left(\frac{T}{\hbar c}\right)^3.$$

With the aid of grand potential Ω and condition $\mu = 0$, we calculate the free energy F of photon gas

$$F = \Omega = 2TV\int \frac{d^3k}{(2\pi)^3}\ln\left(1 - e^{-\hbar ck/T}\right) = \frac{TV}{\pi^2}\int_0^\infty k^2\ln\left(1 - e^{-\hbar ck/T}\right)dk.$$

Introducing variable $x = \hbar ck/T$ and integrating by parts, we arrive at

$$F = -V\frac{T^4}{3\pi^2\hbar^3c^3}\int_0^\infty \frac{x^2\,dx}{e^x-1} = -\frac{\pi^2}{45}\frac{VT^4}{(\hbar c)^3}.$$

Hence the expression for entropy reads

$$S = -\frac{\partial F}{\partial T} = \frac{4\pi^2}{45}\left(\frac{T}{\hbar c}\right)^3 V$$

and, accordingly, the specific heat at constant volume is given by

$$C_V = T\left(\frac{\partial S}{\partial T}\right)_V = \frac{4\pi^2}{15}\left(\frac{T}{\hbar c}\right)^3 V.$$

As concerns the specific heat C_p at constant pressure, it diverges, i.e. $C_p = \infty$, since the pressure of equilibrium photon gas

$$P = -\left(\frac{\partial F}{\partial V}\right)_T = \frac{\pi^2}{45}\frac{T^4}{(\hbar c)^3}$$

is volume-independent.

The energy density $u(T)$ of equilibrium photon gas or bulk density of black-body radiation will equal

$$u(T) = \frac{E(T)}{V} = 2\int \frac{d^3k}{(2\pi)^3}\frac{\hbar ck}{e^{\hbar ck/T} - 1} = \frac{4\sigma T^4}{c}, \quad \sigma = \frac{\pi^2}{60\hbar^3 c^2}.$$

The factor σ is referred to as the *Stefan–Boltzmann constant*. The energy density of black-body radiation $u(T)$ can be associated with the energy density of electromagnetic field and with the time-averaged squares $\langle \boldsymbol{E}^2\rangle$ and $\langle \boldsymbol{H}^2\rangle$ of the electric and magnetic field strengths, as follows:

$$u = \frac{\langle \boldsymbol{E}^2\rangle + \langle \boldsymbol{H}^2\rangle}{8\pi} = \frac{\langle \boldsymbol{E}^2\rangle}{4\pi}, \quad \langle \boldsymbol{E}^2\rangle = \langle \boldsymbol{H}^2\rangle = \frac{16\pi\sigma}{c}T^4.$$

In these relations we have involved that for the electromagnetic field in the space devoid of a matter, the electric and magnetic field components have the same magnitudes.

Since all photons have the same energy-independent velocity, the radiant energy flux density Φ or the energy transfer per unit area and unit time can straightforwardly be evaluated as a product of photon velocity by the radiation energy density $u(T)$, i.e.

$$\Phi = cu(T).$$

In order to characterize the radiant energy flux in one direction or another, one introduces the concept of *radiant intensity* $I(\Omega)$ as a density of the radiated energy flux $d\Phi$ coming into the element of solid angle $d\Omega$, according to equation $d\Phi = I(\Omega)d\Omega$. Due to complete spatial isotropy of black-body radiation,[6] the identical radiation flux emits equally in all directions from each volume element, i.e. $I(\Omega) =$ const. Therefore, the same energy flux fraction equal to $\Phi/4\pi$ gets per unit solid angle since the full solid angle is 4π.

The radiant capacity of a body is described by another quantity referred to as *luminosity* $L = dF/dA$. The latter equals a ratio of the total radiant energy flux dF emitted with the surface element of area dA to the particular side of this surface element, say, outward in all directions and accordingly within the solid angle equal to 2π.

To determine the luminosity of the radiation emitted with the surface element dA, we need to integrate the vector of the radiant energy flux $d\boldsymbol{F} = d\boldsymbol{F}(\Omega)$ over the whole solid angle corresponding to the hemisphere and equal to 2π. In general, the energy flux vector could depend on the direction of radiation. Due to full independence

[6] The black-body radiation is a non-polarized one.

of black-body radiation from its direction, i.e. from solid angle Ω, the resultant energy flux $d\boldsymbol{F}$ will be directed along the normal $\boldsymbol{n}$ to the surface element dA and its magnitude will be equal to the integral taken over a half of the total solid angle from the projection of vector $d\boldsymbol{F}$ onto the surface normal $\boldsymbol{n}$. Writing the projection of radiant energy flux onto the normal as $dF_n = \boldsymbol{n} \cdot d\boldsymbol{F} = dF \cos\vartheta$ where ϑ is the angle between the normal to the surface and the direction of radiation[7] and expressing the flux $dF = I dA$ in terms of radiant intensity I, we arrive at the luminosity of the surface, as follows:

$$L = \frac{dF}{dA} = \int \frac{dF_n}{dA} d\Omega = \int_0^{\pi/2} \sin\vartheta d\theta \int_0^{2\pi} d\varphi\, I \cos\vartheta = \pi I.$$

The relation $L = \pi I$ expresses *Lambert's law* for the sources of completely isotropic radiation.

Using the relation between the black-body radiant intensity I and the radiant energy density $u(T)$, we obtain the following formula for luminosity:

$$L = \pi \frac{cu(T)}{4\pi} = \frac{c}{4} u(T) = \sigma T^4,$$

$\sigma = 5.67{\cdot}10^{-5}$ erg·s^{-1}·cm^{-2}·K^{-4} being the Stefan–Boltzmann constant. The relationship $L = \sigma T^4$ between the luminosity of a black body and its temperature represents the *Stefan–Boltzmann law*. The total integral power of radiation over the entire frequency spectrum of black-body radiation or the total luminosity of the sphere surface area $4\pi R^2$ at temperature of T will be equal to $W = 4\pi R^2 \sigma T^4$.

Problems

1. Find the attractive force of a hydrogen atom in the ground state against the black-body sphere of radius R at temperature T. The temperature T is small as compared with the excitation energy in the atom. Neglect all the possible transitions of the atom to the excited state.

Solution. The thermal radiation of a black body produces the temperature-dependent electrical field in its ambient space. Neglecting the possible transitions of the atom into an excited state due to inequality $T \ll U_{\text{exc}} \sim 10^4$ K, we can estimate the energy shift of the atom as a result of the Stark effect with the aid of the static polarizability $\varkappa$ for the hydrogen atom as

$$\Delta U = -\frac{\varkappa}{2} \langle E^2(\boldsymbol{r}) \rangle.$$

Here $\langle E^2(\boldsymbol{r}) \rangle$ is the time-averaged square of electric field at point $\boldsymbol{r}$, $\varkappa = (9/2)a_B^3$ is the polarizability of hydrogen atom in the ground state, and $a_B = \hbar^2/me^2$ is the Bohr radius. Using the relation

$$\langle E^2 \rangle = \frac{4\pi^3}{15} \frac{T^4}{(\hbar c)^3}$$

[7] The black body is a source of radiation obeying *Lambert's emission law*. The energy flux emitted with the surface element in the direction at angle ϑ to the normal is proportional to a product of energy flux in the normal direction by the solid angle magnitude $d\Omega = \sin\vartheta d\varphi$ and cosine of angle ϑ between the normal and radiant direction.

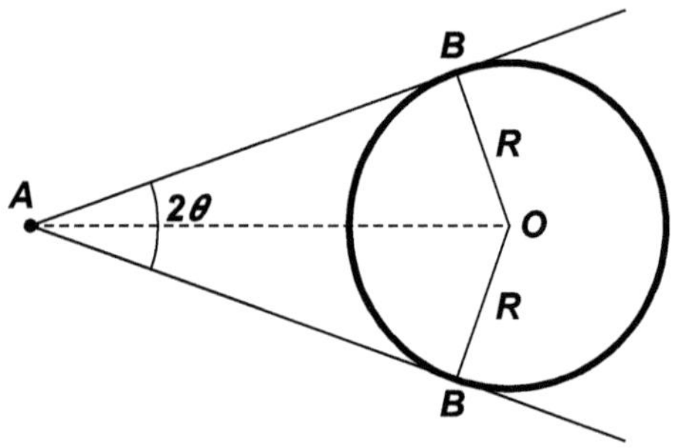

Fig. 3.2 The diagram explaining the geometry of the incoming radiation

for the shift of the hydrogen atom ground state, we find

$$\Delta U_0 = -\frac{2}{15}\varkappa\frac{\pi^3 T^4}{(\hbar c)^3} = -\frac{3\pi^3}{5}\left(\frac{a_B}{\hbar c}\right)^3 T^4 = -\frac{3\pi^3}{5\alpha^3}\frac{T^4}{(mc^2)^3}$$

where $\alpha = e^2/\hbar c$ is the *fine-structure constant* and m is the electron mass.

The thermal radiation of a heated sphere produces a spatial variation of the electromagnetic field. The radiation of the sphere is isotropic and identical in all directions. As is seen from Fig. 3.2, only a fraction of the total radiation, proportional to solid angle Ω under which the atom sees the sphere, can reach the hydrogen atom. The solid angle Ω equals

$$\Omega = 2\pi(1-\cos\vartheta) = 2\pi\left(1 - \frac{\sqrt{r^2-R^2}}{r}\right).$$

Involving the property of the whole isotropy of black-body radiation, we have the following expression for the mean square of the electric field strength at distance r from the center of the sphere:

$$\langle E^2(r)\rangle = \frac{\Omega}{4\pi}\langle E^2(R)\rangle = \frac{1}{2}\left(1 - \frac{\sqrt{r^2-R^2}}{r}\right)\langle E^2(R)\rangle.$$

Then we obtain the energy shift in the hydrogen atom

$$\Delta U(r) = \frac{1}{2}\left(1 - \frac{\sqrt{r^2-R^2}}{r}\right)\Delta U_0 = -\frac{3\pi^3}{10}\left(1 - \frac{\sqrt{r^2-R^2}}{r}\right)\left(\frac{a_B}{\hbar c}\right)^3 T^4.$$

The magnitude of the shift decays as $1/r^2$ at distances $r \gg R$.

Due to negative sign $\Delta U(r)$ for the atom in the ground state, the force exerted to the atom from the black-body radiation will be directed to the sphere center. In other words, the atom tends to be in the region of larger electromagnetic radiation. The magnitude of the force, attracting the atom to the heated body, is given by the relation

$$F(r) = -\frac{\partial\,\Delta U(r)}{\partial r} = \frac{R^2}{2r^2\sqrt{r^2-R^2}}\Delta U_0 = -\frac{3\pi^3}{20}\left(\frac{a_B}{\hbar c}\right)^3\frac{R^2}{r^2\sqrt{r^2-R^2}}T^4.$$

At large $r \gg R$ distances the attraction force decays as R^2/r^3.

In spite of small magnitudes of the atom energy shift and attraction force, such a black-body radiation effect is important as it concerns the problems of accuracy for the atomic frequency standards.

2. The black hole is known to absorb the particles. Due to quantum effects, the black hole can also evaporate the particles. The composition of evaporated particles depends on the black hole mass. These can be photons, gravitons, neutrinos, electrons, positrons, and other particles. The thermodynamic equilibrium of the black hole can be described as a state of a black body heated to the temperature as

$$T = \frac{\hbar\varkappa}{2\pi\hbar c} \quad \text{and} \quad \varkappa = \frac{GM}{R_g^2}.$$

Here $\varkappa$ is the free fall acceleration at the black hole surface, G is the gravitational constant, M is the black hole mass, and $R_g = 2GM/c^2$ is the black hole size or gravitational Schwarzschild radius.

Estimate the lifetime for the black hole of mass M, assuming that the black hole evaporation is mainly associated with emitting the photons.

Solution. The total power of photon radiation from the black body of area $4\pi R_g^2$ equals $4\pi R_g^2\sigma T^4$ where σ is the Stefan–Boltzmann constant. The loss of black hole mass in the evaporation process is determined with the energy balance equation

$$-\frac{d}{dt}\left(Mc^2\right) = 4\pi R_g^2\sigma T^4 = \frac{1}{2^{10}}\frac{1}{15\pi}\frac{\hbar c^6}{G^2M^2}.$$

Hence the lifetime of black hole proves to be finite and equal to

$$t = 15\cdot 2^{10}\pi\frac{G^2}{\hbar c^4}\int\limits_0^M M^2 dM = 5\cdot 2^{10}\pi\frac{G^2M^3}{\hbar c^4} = \frac{320}{\sqrt{2\pi}}\left(\frac{M}{M_p}\right)^3 t_p.$$

The latter equality is expressed in the terms of the *Planck units* of mass $M_p = \left(\hbar c/8\pi G\right)^{1/2}$ and time $t_p = \left(\hbar G/c^5\right)^{1/2} = l_p/c$. The length l_p is the *Planck length* or the size of photon when its energy $mc^2 = \hbar\omega \sim \hbar c/l_p$ would be comparable with its own gravitational energy Gm^2/l_p.

3.12 Thermal Properties of Solid Bodies: Phonons in a Solid Body

Excitations in the elastic condensed medium correspond to the vibrations of atoms near the equilibrium positions. In first approximation we assume that the vibrations of atoms are small and, therefore, we will describe them as *harmonic vibrations*. Such small elastic excitations in the medium are consistent with the sound waves. The latter ones are treated as a set of quanta of elastic deformation field or *phonons* whose energy is determined by the momentum and the type or branch of sound wave. The concept about gas of phonons can be applied for describing the thermal properties of a medium, resulting from the thermal vibrations around the equilibrium positions of atoms or lattice sites.

The structure of sound waves is simplest in an isotropic liquid which has a single longitudinal acoustic branch. In solids or crystals, the sound vibration spectrum is more complicated. If there are 3ν types of normal vibrations in the crystal lattice with dispersion $\omega = \omega_s(\boldsymbol{k})$, $\boldsymbol{k}$ being the wave vector and $s = 1, 2, \ldots, 3\nu$, the phonon part of free energy in a solid is given by the usual expression

$$F_{ph} = VT\sum_{s=1}^{3\nu}\frac{d^3k}{(2\pi)^3}\ln\left(1 - e^{-\hbar\omega_s(\mathbf{k})/T}\right) = VT\int d\omega\, g(\omega)\ln\left(1 - e^{-\hbar\omega/T}\right).$$

Here we have introduced the spectral density of phonon vibrations $g(\omega)$

$$g(\omega) = \sum_{s=1}^{3\nu}\int\frac{d^3k}{(2\pi)^3}\delta\left(\omega - \omega_s(\mathbf{k})\right),$$

i.e. the density of phonon states per unit volume of a body. The spectral density of phonon vibrations is the essential characteristic that determines the thermal properties of a solid body.

The general number of various normal vibrations or total number of phonon states is determined with the number of vibrational degrees of freedom of a body after subtracting all three degrees of translational motion and three degrees of rotational motion of the body as a whole. So, the total number of phonon states equals $3N - 6$ where N is the number of atoms or particles in the body. On the neglect of number 6 as compared with the number of atoms $N \gg 1$, we arrive at the following condition:

$$V\int g(\omega)\, d\omega = 3N.$$

The phonon spectrum is limited in frequencies $\omega \leqslant \omega_{max}$ from above. Since the lattice of atoms is discrete, the wavelength of sound cannot be smaller than the minimal spacing a between the atoms. In the low frequency $\omega \to 0$ limit, the sound excitations represent the long-wave acoustic phonons with frequencies $\omega \sim k$. The quadratic behavior of spectral density of states $g(\omega) \sim \omega^2$ corresponds to the acoustic phonon spectrum.

Using relation $E = F - T(\partial F/\partial T)$ between the energy and free energy, we can find the phonon contribution to the energy of a solid body

$$E_{ph}(T) = V\int d\omega\, g(\omega)\frac{\hbar\omega}{e^{\hbar\omega/T} - 1},$$

and then the phonon contribution to the specific heat at constant volume according to $C = (\partial E/\partial T)_V$

$$C_{ph}(T) = V\int d\omega\, g(\omega)\frac{(\hbar\omega/2T)^2}{\sinh^2 \hbar\omega/2T}.$$

In the high-temperature limit $T \gg \hbar\omega_{max}$, using approximations $e^x - 1 \approx x$ and $x/\sinh x \approx 1$ for small values $x \ll 1$, we arrive at the classical magnitudes of energy and specific heat

$$E_{ph} = V \int d\omega\, g(\omega) T = 3NT \quad \text{and} \quad C_{ph} = V \int d\omega\, g(\omega) = 3N.$$

Such a temperature-independent and universal behavior signify the *Dulong–Petit law* for the phonon specific heat at high temperatures.

3.13 Debye's Interpolation Model

Since it is impossible to perform an accurate analytical calculation of phonon spectral density $g(\omega)$, the simple *Debye model* is often used for approximating the spectral density. In this model the low-frequency quadratic behavior of spectral density $g(\omega) = \alpha\omega^2$ extends over the whole frequency range to the maximum frequency ω_D called the *Debye frequency*. The order-of-magnitude estimate for the Debye frequency can be given as $\omega_D \sim u/a$ where u is the typical sound velocity in a solid and a is the interatomic distance. The Debye frequency can be associated with the *Debye temperature* equal to $\Theta_D = \hbar\omega_D$. The proportionality coefficient α and Debye frequency ω_D must be consistent with each other through the normalization condition, resulting in the interpolation formula for the spectral density in the Debye model

$$g(\omega) = 9\frac{N}{V}\frac{\omega^2}{\omega_D^3}\vartheta(\omega_D - \omega).$$

As a result, we arrive at the following energy and specific heat in the Debye model:

$$E_{ph} = \frac{9N}{\omega_D^3}\int_0^{\omega_D} d\omega\, \omega^2 \frac{\hbar\omega}{e^{\hbar\omega/T} - 1} = 9N\frac{T^4}{\Theta_D^3}\int_0^{\Theta_D/T} dx\, \frac{x^3}{e^x - 1} = 3NT\, D_3\left(\frac{\Theta_D}{T}\right),$$

$$C_{ph} = \frac{9N}{\omega_D^3}\int_0^{\omega_D} d\omega\, \omega^2 \frac{\left(\hbar\omega/2T\right)^2}{\sinh^2 \hbar\omega/2T} = \frac{9N}{4}\frac{T^3}{\Theta_D^3}\int_0^{\Theta_D/T} dx\, \frac{x^4}{\sinh^2 x/2}$$

where $D_3(x)$ is the Debye function[8] of the third kind.

For the low $T \ll \Theta_D$ temperature region, the upper limit of integration can be extended to infinity. Then,

$$E_{ph} = \frac{3\pi^4}{5} N \frac{T^4}{\Theta_D^3} \quad \text{and} \quad C_{ph} = \frac{12\pi^4}{5} N \left(\frac{T}{\Theta_D}\right)^3.$$

In the high $T \gg \Theta_D$ temperature limit, the classical Dulong–Petit result reproduces.

---

[8] The Debye function of nth kind is defined by integral $D_n(x) = \frac{n}{x^n}\int_0^x \frac{t^n dt}{e^t - 1}$.

Solids with more than one atom in the unit cell have the optical phonon branches. As a rule, the optical frequencies depend weakly on the vector, i.e. $\omega(\boldsymbol{k}) \approx \text{const} = \omega_0$. In this case, the *Einstein model* may be applicable if the dispersion of the optical phonon branch is completely neglected. Accordingly, the optical mode contribution to the spectral density of states $g(\omega)$ is approximated with the δ-shaped peak near the frequency ω_0 as $g(\omega) \sim \delta(\omega - \omega_0)$. When combining the Debye and Einstein terms in the spectral density of states, the Debye temperature will be lowered since summing over all the phonon states would give the same number of vibrational degrees of freedom. For the temperatures $T \ll \hbar\omega_0$, the contribution of the optical phonon mode to the specific heat of a solid freezes exponentially and, thereby, does not affect the low-temperature power-like asymptotic behavior of the specific heat.

Problem

1. Calculate the *Grüneisen parameter* $\gamma = V(\partial P/\partial E)_V$ in the Debye model.

Solution. Let us employ the following relation between the derivative of free energy $F_{ph}(T)$ with respect to the Debye temperature Θ_D and the energy $E_{ph}(T)$, being valid in the Debye model:

$$\left(\frac{\partial F_{ph}(T)}{\partial \Theta_D}\right)_{N,T} = \frac{E_{ph}(T)}{\Theta_D}.$$

Determining the pressure P as usual, we obtain

$$P = -\left(\frac{\partial F_{ph}(T)}{\partial V}\right)_{N,T} = -\frac{\partial F_{ph}}{\partial \Theta_D}\frac{\partial \Theta_D}{\partial V} = -E_{ph}\frac{\partial \ln \Theta_D}{\partial V}$$

where V is the volume of a body. Then we find the Grüneisen parameter

$$\gamma = -\frac{\partial \ln \Theta_D}{\partial \ln V}.$$

In the Debye model, the Grüneisen parameter proves to be temperature-independent and characterizes the effect of varying the body volume as a function of the Debye temperature. The magnitude of parameter γ is usually about unity, i.e. $\gamma \sim 1$.

Let us express the Grüneisen parameter via the coefficients of thermal expansion $\alpha = V^{-1}\partial V/\partial T$ and isothermal compressibility $\beta = -V^{-1}\partial V/\partial P$ according to

$$\gamma = V\left(\frac{\partial P}{\partial T}\right)_V \frac{1}{C_V} = -\frac{(\partial V/\partial T)_P}{(\partial V/\partial P)_T}\frac{V}{C_V} = \frac{\alpha V}{\beta C_V}$$

where $C_V = C_{ph}(T)$ is the specific heat of a body at constant volume V. Since in the Debye model the coefficients γ and β are temperature-independent, we can draw a conclusion that the thermal expansion coefficient $\alpha(T)$ and specific heat $C_V(T)$ have the identical temperature behavior (*Grüneisen law*). As usual, in the solid dielectrics this law is fairly a good approximation.

3.14 The Phonon Spectrum of Crystalline Lattice

Let us consider in more detail the physical properties of phonon spectrum in a crystal. In order to determine the phonon spectrum in the crystalline lattice of atoms, it is necessary to set the forces acting between the atoms and describe the motion of atoms.

The spatial arrangement of atoms represents the *crystalline lattice sites* which can be given by a set of vectors

$$\boldsymbol{l} = l_1\boldsymbol{a}_1 + l_2\boldsymbol{a}_2 + l_3\boldsymbol{a}_3$$

where l_1, l_2, and l_3 are the integers. The arrangement of atoms in the whole crystal can be determined by specifying the single *unit cell*, e.g. by the parallelepiped constituted with the *primitive translational vectors* $\boldsymbol{a}_1$, $\boldsymbol{a}_2$, $\boldsymbol{a}_3$. Thus, the crystal is a repeating set of identical unit cells. In general, in each unit cell there may be a few of various atoms. Let index s numerate the atoms in the unit cell ($s = 1, 2, \ldots, \nu$), ν being the number of atoms in the unit cell.

Let us introduce notation $\boldsymbol{u}_{sl}$ for the displacement vector of sth atom with mass M_s in the $\boldsymbol{l} = (l_1, l_2, l_3)$ unit cell from the equilibrium position or crystal lattice site. Then the kinetic energy of a crystal reads

$$E_{\text{kin}} = \frac{1}{2}\sum_{sl} M_s |\dot{\boldsymbol{u}}_{sl}|^2.$$

In order to determine the potential energy, it is necessary to specify the interatomic forces which we will characterize with the aid of some function representing the potential energy of atom vibrations $U(\boldsymbol{u}_{sl})$ and depending on the coordinates of all atoms or on their displacements $\boldsymbol{u}_{sl}$ from the crystal lattice sites. Next, we assume that the potential energy $U(\boldsymbol{u}_{sl})$ reaches its absolute minimum when all the displacement vectors $\boldsymbol{u}_{sl}$ vanish. Thereby, the arrangement of atoms at the crystal lattice sites corresponds to the stable equilibrium of the crystal.

Applying the methods of small vibration theory, we expand the potential energy $U(\boldsymbol{u}_{sl})$ in the powers of displacement vectors $\boldsymbol{u}_{sl}$, restricting ourselves with the *harmonic approximation*, i.e. with the terms of expansion not higher than second order:

$$U(\boldsymbol{u}_{sl}) = U(0) + \frac{1}{2}\sum_{sl,\, s'l'} \boldsymbol{u}_{sl}\frac{\partial^2 U(0)}{\partial \boldsymbol{u}_{sl}\partial \boldsymbol{u}_{s'l'}}\boldsymbol{u}_{s'l'} + \ldots$$

The linear term in the expansion over displacement vectors $\boldsymbol{u}_{sl}$ vanishes due to assumption about the extremum of potential energy $U(\boldsymbol{u}_{sl})$. The constant $U(0)$ in the expansion is unessential from now on. Let us introduce the *force tensor*

$$\boldsymbol{\Lambda}_{sl,\, s'l'} = \left.\frac{\partial^2 U}{\partial \boldsymbol{u}_{sl}\partial \boldsymbol{u}_{s'l'}}\right|_{\boldsymbol{u}_{sl}=\boldsymbol{u}_{s'l'}=0}$$

for brevity and consider its simplest properties. Firstly, the force tensor $\boldsymbol{\Lambda}_{sl,\, s'l'}$ depends only on the difference between vectors $\boldsymbol{l}$ and $\boldsymbol{l}'$, i.e.

$$\boldsymbol{\Lambda}_{sl,\, s'l'} = \boldsymbol{\Lambda}_{s\,s'}(\boldsymbol{l} - \boldsymbol{l}'),$$

since the forces interacting between atoms are only related with their relative position in the lattice. Secondly, the force tensor has the following property of symmetry:

$$\mathbf{\Lambda}_{s\,s'}(\boldsymbol{l}) = \mathbf{\Lambda}_{s'\,s}(-\boldsymbol{l}).$$

And thirdly, the force tensor coefficients $\mathbf{\Lambda}_{s\,s'}(\boldsymbol{l})$ are also related with each other by the relation expressing the following. The uniform displacement of the crystal as a whole ($\boldsymbol{u}_{sl} = \text{const}$) does not vary its energy, the crystalline lattice stays in the equilibrium position, and there appear no additional forces acting on the atoms in the lattice. In fact, from the lack of any additional forces

$$-\frac{\partial U}{\partial \boldsymbol{u}_{sl}} = -\sum_{s'\,l'} \mathbf{\Lambda}(\boldsymbol{l}-\boldsymbol{l}')\boldsymbol{u}_{s'l'}\bigg|_{u_{s'\,l'}\equiv\text{const}} = 0,$$

we obtain the following relation:

$$\sum_{s'\,l} \mathbf{\Lambda}_{s\,s'}(\boldsymbol{l}) = 0.$$

From the Lagrange function for the crystal treated as a mechanical system

$$L = E_{\text{kin}} - U = \frac{1}{2}\sum_{sl} M_s|\dot{\boldsymbol{u}}_{sl}|^2 - U(0) - \frac{1}{2}\sum_{sl,\,s'l'} \boldsymbol{u}_{sl}\frac{\partial^2 U}{\partial \boldsymbol{u}_{sl}\partial \boldsymbol{u}_{s'l'}}\boldsymbol{u}_{s'l'}\,,$$

the following system for the equations of motion results in

$$M_s\ddot{\boldsymbol{u}}_{sl} = -\sum_{s'l'} \mathbf{\Lambda}_{s\,s'}(\boldsymbol{l}-\boldsymbol{l}')\boldsymbol{u}_{s'\,l'}\,.$$

It is convenient to examine the system of linear equations with the right-hand side representing a convolution by using the Fourier transformation, i.e. in the form of a set of monochromatic plane waves:

$$\boldsymbol{u}_{sl}(t) = \boldsymbol{e}_s(\boldsymbol{k})e^{i\boldsymbol{k}\boldsymbol{l}-i\omega_{\boldsymbol{k}}t}.$$

Here $\boldsymbol{e}_s(\boldsymbol{k})$ is the *polarization vector*, $\boldsymbol{k}$ is the *wave vector*, and $\omega_{\boldsymbol{k}}$ is the corresponding frequency of the wave. Then we have

$$M_s\omega_{\boldsymbol{k}}^2\boldsymbol{e}_s(\boldsymbol{k}) - \sum_{s'=1}^{\nu} \mathbf{\Lambda}_{s\,s'}(\boldsymbol{k})\boldsymbol{e}_{s'}(\boldsymbol{k}) = 0$$

where we have introduced the Fourier transform of the force tensor or *dynamic matrix* as follows:

$$\boldsymbol{\Lambda}_{s\,s'}(\boldsymbol{k}) = \sum_{l} \boldsymbol{\Lambda}_{s\,s'}(\boldsymbol{l})e^{-i\boldsymbol{k}\boldsymbol{l}}.$$

The dynamic matrix has the following properties:

$$\boldsymbol{\Lambda}_{s\,s'}(\boldsymbol{k}) = \boldsymbol{\Lambda}_{s'\,s}(-\boldsymbol{k}) = \boldsymbol{\Lambda}^{*}_{s'\,s}(\boldsymbol{k})$$

which characterizes it as a Hermitian matrix.

As a result, involving three projections x, y, and z for the displacement vectors $\boldsymbol{u}_{sl}$ and ν of polarization vectors $\boldsymbol{e}_s(\boldsymbol{k})$, we arrive at the uniform system of 3ν linear equations. The nontrivial solutions for this system of equations are determined by vanishing the characteristic polynomial $P(\omega_{\boldsymbol{k}}^2)$ equal to the $3\nu \times 3\nu$ determinant

$$P(\omega_{\boldsymbol{k}}^2) = \det \left\| \delta_{s\,s'} M_{s'} \omega_{\boldsymbol{k}}^2 - \boldsymbol{\Lambda}_{s\,s'}(\boldsymbol{k}) \right\| = 0.$$

Thus we get the algebraic equation of degree 3ν with respect to $\omega_{\boldsymbol{k}}^2$. In the general case, analyzing the dispersion equation $P(\omega_{\boldsymbol{k}}^2) = 0$, we should find 3ν independent solutions for frequencies $\omega_{\boldsymbol{k}}^2$ which determine the *dispersion* or frequency spectrum $\omega(\boldsymbol{k})$ of normal harmonic vibrations in the crystalline lattice as a function of wave vector $\boldsymbol{k}$. Each of the 3ν normal vibrations is referred to as the *branch of phonon spectrum*.

Let us note a number of general properties for the phonon spectrum. They follow from the properties for the Fourier transform of dynamic matrix $\boldsymbol{\Lambda}_{s\,s'}(\boldsymbol{k})$. Changing the direction of wave vector $\boldsymbol{k}$ to the opposite one, we obtain for the matrix transpose

$$\boldsymbol{\Lambda}_{s\,s'}(-\boldsymbol{k}) = \boldsymbol{\Lambda}^{T}_{s'\,s}(\boldsymbol{k}).$$

Since the transposition of the matrix keeps the magnitude of the determinant unchanged, we can assert that the dispersion equation $P(\omega_{-\boldsymbol{k}}^2) = 0$ remains the same and, therefore, the phonon frequencies are the even functions of the wave vector

$$\omega(\boldsymbol{k}) = \omega(-\boldsymbol{k}).$$

From the physical point of view, this property means the invariance of the Lagrangian and equations of motion with respect to the time reversal. If the wave vibration propagates in one direction, its propagation in the opposite direction is possible as well. The change of wave propagation direction corresponds to varying the sign of wave vector.

For describing the further properties of the phonon spectrum, we will need the concept of reciprocal lattice closely related to the periodicity of direct or real crystalline lattice. *The set of wave vectors* $\boldsymbol{b}$ *is called the reciprocal lattice if the plane wave* $e^{i\boldsymbol{b}\boldsymbol{r}}$ *has a periodicity of the direct crystalline lattice*. In other words, for an arbitrary radius vector $\boldsymbol{r}$ and an arbitrary vector of the direct lattice $\boldsymbol{l}$, the following equation is valid:

$$e^{i\boldsymbol{b}(\boldsymbol{r}+\boldsymbol{l})} = e^{i\boldsymbol{b}\boldsymbol{r}}.$$

Thus the reciprocal lattice is a set of wave vectors $\boldsymbol{k}$ satisfying the condition

$$e^{ibl} = 1.$$

For the set $\boldsymbol{a} = (\boldsymbol{a}_1, \boldsymbol{a}_2, \boldsymbol{a}_3)$ of primitive vectors, the reciprocal lattice is generated with the following set of three primitive vectors $\boldsymbol{b} = (\boldsymbol{b}_1, \boldsymbol{b}_2, \boldsymbol{b}_3)$ of reciprocal lattice:

$$\boldsymbol{b}_1 = 2\pi \frac{\boldsymbol{a}_2 \times \boldsymbol{a}_3}{\boldsymbol{a}_1 \cdot (\boldsymbol{a}_2 \times \boldsymbol{a}_3)}, \quad \boldsymbol{b}_2 = 2\pi \frac{\boldsymbol{a}_3 \times \boldsymbol{a}_3}{\boldsymbol{a}_1 \cdot (\boldsymbol{a}_2 \times \boldsymbol{a}_3)}, \quad \boldsymbol{b}_3 = 2\pi \frac{\boldsymbol{a}_1 \times \boldsymbol{a}_2}{\boldsymbol{a}_1 \cdot (\boldsymbol{a}_2 \times \boldsymbol{a}_3)}$$

such as $\boldsymbol{a}_i \cdot \boldsymbol{b}_j = 2\pi\delta_{ij}$. Setting an arbitrary vector of reciprocal lattice

$$\boldsymbol{b} = p_1\boldsymbol{b}_1 + p_2\boldsymbol{b}_2 + p_3\boldsymbol{b}_3$$

where p_1, p_2, p_3 are arbitrary integers and an arbitrary direct lattice vector $\boldsymbol{l} = l_1\boldsymbol{a}_1 + l_2\boldsymbol{a}_2 + l_3\boldsymbol{b}_3$, we can readily check the necessary relation

$$\boldsymbol{b} \cdot \boldsymbol{l} = 2\pi(p_1l_1 + p_2l_2 + p_3l_3).$$

The product $\boldsymbol{b} \cdot \boldsymbol{l}$ must always be a multiple of 2π for any choice of integers l_1, l_2, l_3. This can only be satisfied for integers p_1, p_2, p_3. The reciprocal lattice to the reciprocal one is the original direct lattice.[9]

According to the definition of the Fourier transform for dynamic matrix $\boldsymbol{\Lambda}_{s\,s'}(\boldsymbol{k})$, we have its periodicity with the reciprocal lattice vector $\boldsymbol{b}$:

$$\boldsymbol{\Lambda}_{s\,s'}(\boldsymbol{k} + \boldsymbol{b}) = \boldsymbol{\Lambda}_{s\,s'}(\boldsymbol{k}).$$

Such a translational property of dynamic matrix, as it follows from the dispersion equation, results in the phonon spectrum periodicity with the period of reciprocal lattice vector $\boldsymbol{b}$

$$\omega(\boldsymbol{k} + \boldsymbol{b}) = \omega(\boldsymbol{k}).$$

Due to periodicity of phonon spectrum, it is usual to consider a limited range of varying the vectors $\boldsymbol{k}$, corresponding only to one of the reciprocal lattice units. Usually, *the first Brillouin zone* is chosen as a locus of points in the reciprocal space that are closer to the origin $\boldsymbol{k} = 0$ of the reciprocal lattice than they are to any other reciprocal lattice point given by vector $\boldsymbol{b} = p_1\boldsymbol{b}_1 + p_2\boldsymbol{b}_2 + p_3\boldsymbol{b}_3$ in which at least one of the integer coefficients p_i is nonzero.[10] When wave vector $\boldsymbol{k}$ takes all the values from the first Brillouin zone, the phonon frequency $\omega(\boldsymbol{k})$ in each of 3ν spectrum branches runs the values filling some frequency interval or *band*. The various bands,

[9] The reciprocal lattice to the simple cubic one with the side a will be a simple cubic lattice with $2\pi/a$.

[10] For the simple cubic lattice with period a, the first Brillouin zone is the following region of reciprocal space: $-\pi/a \leqslant k_i \leqslant \pi/a$, where $i = x, y, z$.

in general, can overlap each other. This depends on the specific crystalline lattice symmetry to a great degree.

The condition of stability for the crystalline lattice at its small deformations together with the absolute minimum of potential energy $U(\boldsymbol{u}_{sl})$ at $\boldsymbol{u}_{sl} = 0$ reguire non-negativity of phonon frequencies $\omega^2(\boldsymbol{k}) \geqslant 0$ for all wave vectors $\boldsymbol{k}$. The value $\omega^2(\boldsymbol{k}_0) < 0$ would mean an instability of crystalline lattice with respect to distorting the lattice under wave vector $\boldsymbol{k}_0$.

For the crystalline lattice generated with the simple unit of one atom $\nu = 1$ (index S takes single value $s = 1$), it is possible to clarify the common features for the behavior of the phonon frequency in the long-wave limit of small wave vectors $k \to 0$. In this case, the expansion of the dynamic matrix starts from the terms quadratic in k:

$$\Lambda(\boldsymbol{k}) = \sum_{l} \boldsymbol{\Lambda}(\boldsymbol{l}) e^{-i\boldsymbol{k}\boldsymbol{l}} = -\frac{1}{2} \sum_{l} \boldsymbol{\Lambda}(\boldsymbol{l}) \big(\boldsymbol{k} \cdot \boldsymbol{l}\big)^2 + \ldots$$

Here we have taken the parity $\Lambda(\boldsymbol{l})$ and relation $\sum_l \Lambda(\boldsymbol{l}) = 0$ into account. Since the dynamic matrix is characterized by three spatial indices in addition, we write the detailed expression for it as

$$\Lambda_{ik}(\boldsymbol{k}) \approx \lambda_{iklm} k_l k_m \quad (i,\ k,\ l,\ m = x,\ y,\ z).$$

The dispersion equation reads

$$\det \left\| M\omega_{\boldsymbol{k}}^2 \delta_{ik} - \lambda_{iklm} k_l k_m \right\| = 0$$

and represents the equation of third degree with respect to ω^2 which, in general, has three various roots or three branches of phonon spectrum. The frequencies $\omega(\boldsymbol{k})$ are the homogeneous functions of the first kind from components k_x, k_y, k_z of wave vector $\boldsymbol{k}$ and vanish at $\boldsymbol{k} = 0$, i.e. at small k:

$$\omega(\boldsymbol{k}) = k f(\boldsymbol{k}/k).$$

The waves of such a type are called *acoustic*. They are characterized with the finite sound velocity $c = \omega/k$ depending on the wave propagation direction and type of their polarization.

In the anisotropic crystals, unlike isotropic media,[11] the wave propagation direction, in general, does not coincide with the direction of wave vector $\boldsymbol{k}$.

In the complex crystalline lattices with more than one atom in the unit cell ($\nu > 1$), there exist $3(\nu - 1)$ branches of phonon spectrum in addition to three acoustic ones.

---

[11] In elastic isotropic media, the acoustic (sound) branches are represented by *longitudinal wave* $\omega = c_l k$ in which the displacement $\boldsymbol{u}$ is directed along the wave propagation and by two coincident *transverse waves* $\omega = c_t k$ with the different polarizations when displacement $\boldsymbol{u}$ takes place in the plane normal to the propagation direction. As a rule, $c_l > c_t$. In the anisotropic crystals the acoustic waves, in general, are neither longitudinal nor transverse. The corresponding waves have three various sound velocities. The sound velocities are unambiguously related to the elastic constants.

In these spectrum branches called *optical*,[12] the frequency does not vanish at $\boldsymbol{k} = 0$ but tends to some finite limit ω_0 as $\boldsymbol{k} \to 0$. Due to parity of spectrum $\omega(\boldsymbol{k})$ near $\boldsymbol{k} = 0$, we can write the following expansion:

$$\omega(\boldsymbol{k}) = \omega_0 + \alpha_{ij} k_i k_j \quad (i,\ j = x,\ y,\ z).$$

As concerns the optical vibrations at the limiting case $\boldsymbol{k} = 0$, the atoms in the unit cell oscillate relative to each other, the center of gravity of the unit cell being immobile. In fact, summing the equation of motion over index s results in

$$\sum_{s=1}^{\nu} M_s \omega_{\boldsymbol{k}}^2 \boldsymbol{e}_s(\boldsymbol{k}) = \sum_{s=1}^{\nu} \sum_{s'=1}^{\nu} \boldsymbol{\Lambda}_{s\,s'}(\boldsymbol{k}) \boldsymbol{e}_{s'}(\boldsymbol{k}).$$

Employing the following property for zero Fourier transform of dynamic matrix:

$$\sum_{s=1}^{\nu} \boldsymbol{\Lambda}_{s\,s'}(\boldsymbol{k} = 0) = \sum_{s=1}^{\nu} \boldsymbol{\Lambda}_{s'\,s}(\boldsymbol{k} = 0) = 0,$$

we obtain for $\boldsymbol{k} = 0$

$$\omega_0^2 \sum_{s=1}^{\nu} M_s \boldsymbol{e}_s(\boldsymbol{k} = 0) = 0.$$

Accordingly, for $\omega_0 \neq 0$ and relation $\boldsymbol{e}_s(0) = \boldsymbol{u}_{sl}(t) e^{i\omega_0 t}$ between the polarization vector and the displacement vector, we arrive at the result

$$\sum_{s=1}^{\nu} M_s \boldsymbol{u}_{sl}(t) = 0$$

expected for the optical phonon branch at zero wave vector.

In conclusion, we emphasize once again that the consideration above has been performed in the *harmonic approximation* when the decomposition for the potential energy of crystalline lattice over the atom displacements is limited to the terms not exceeding second order. In the harmonic approximation, the phonons propagate freely over the crystalline lattice, do not interact with each other, and represent the undamped *elementary excitations* in the crystalline lattice.

Beyond the framework of harmonic approximation, the involvement of *anharmonic*, third and fourth-order terms in decomposing the potential energy over the atom displacements results in the phonon coupling and in such effects as scattering the phonons, their decay, damping, and formation of bound states. The number of phonons in these processes may not conserve. Here one discerns two types of phonon scattering. The first is referred to as the *normal process* or *N-scattering* when the total

[12] The acoustic and optical modes can be comprehended as if we generalize three translational and $3(\nu - 1)$ vibrational degrees of freedom of ν-atomic molecule to the case of a crystal.

values for the initial and the final wave vectors of the phonons participating in the collision are strictly equal to each other, thus conserving the total momentum. The second is referred to as the *Umklapp process* or *U-scattering* when the total values for the initial and final wave vectors of the phonons differ by *nonzero vector* of the reciprocal lattice. The total phonon momentum changes in the Umklapp processes.

The scattering and damping of phonon can originate from the breaking of strict periodicity in the crystalline lattice due to presence of impurity atoms, violation of isotope atom composition, and other lattice defects. The presence of defects can result in changing the phonon vibration spectrum, displaying them in the emergence of new frequencies related to quasi-local vibrations of lattice atoms near the defect as an impurity atom.

Problem

1. For a simple cubic crystal, determine the sound velocity of acoustic waves propagating in the cubic facet plane (crystallographic (001) plane).

Solution. Let us direct axes x, y, and z along the cube edges and compose the dispersion equation

$$\det \begin{Vmatrix} M\omega^2 - \lambda_1 k_x^2 - \lambda_3 k_y^2 & -2\lambda_2 k_x k_y & 0 \\ -2\lambda_2 k_x k_y & M\omega^2 - \lambda_1 k_y^2 - \lambda_3 k_x^2 & 0 \\ 0 & 0 & M\omega^2 - \lambda_3(k_x^2 + k_y^2) \end{Vmatrix} = 0.$$

In a cubic crystal the dynamic matrix elements $\lambda_{xxxx} = \lambda_1$, $\lambda_{xyxy} = \lambda_2$, $\lambda_{xxyy} = \lambda_3$ differ from zero as well as the dynamic matrix elements λ_{iklm} of the same magnitude by replacing the indices x, y with the others from the set x, y, z. (The matrix elements λ_1, λ_2, and λ_3, divided by the unit cell volume, represent all the three independent components of the elastic modulus tensor in the case of cubic crystal.)

Solving the dispersion equation yields the frequencies of all three sound branches

$$M\omega_{1,2}^2 = (k^2/2)\left\{\lambda_1 + \lambda_3 \pm \left[(\lambda_1 - \lambda_3)^2 - (\lambda_1 - \lambda_3 + 2\lambda_2)(\lambda_1 - \lambda_3 - 2\lambda_2)\sin^2 2\theta\right]^{1/2}\right\},$$

$$M\omega_3^2 = \lambda_3 k^2$$

where θ is the angle between axis x and wave vector $\boldsymbol{k}$ lying in the (x, y) plane. For $\theta = 0$ (wave vector $\boldsymbol{k}$ is along the x-axis), we have

$$M\omega_1^2 = \lambda_1 k^2 \quad \text{and} \quad M\omega_2^2 = \lambda_3 k^2.$$

At angle $\theta = \pi/4$ when vector $\boldsymbol{k}$ is directed along the diagonal of the cube facet, we find

$$M\omega_1^2 = (\lambda_1 + \lambda_3 + 2\lambda_2)k^2/2 \quad \text{and} \quad M\omega_2^2 = (\lambda_1 + \lambda_3 - 2\lambda_2)k^2/2.$$

In the theory of elasticity the modulus $\tilde{\lambda}_2$ is usually introduced according to $\tilde{\lambda}_2 = 2\lambda_2 - \lambda_3$.

3.15 Spectral Density of Lattice Vibrations

In the course of studying the physical properties of crystals, as a rule, it is necessary to know the number of oscillations $dN(\omega)$ for each oscillation branch in the frequency range from ω to $\omega + d\omega$ or the density of states $g(\omega)$ according to $dN(\omega) = g(\omega)d\omega$.

The number of states per interval $d^3k = dk_x dk_y dk_z$ of wave vector components is equal to $dk_x dk_y dk_z/(2\pi)^3$.

Let us consider the surfaces of constant frequency or isofrequency surfaces $\omega(\boldsymbol{k}) = \text{const}$ in the wave vector space. In each point of $\boldsymbol{k}$-space, the gradient $\nabla\omega(\boldsymbol{k}) = \partial\omega/\partial\boldsymbol{k}$ is aligned along the normal passing through this point of isofrequency surface. Since $d\omega = d\boldsymbol{k}\cdot\nabla\omega(\boldsymbol{k})$, the spacing along the normal between two close isofrequency surfaces ω and $\omega + d\omega$ can be represented as $d\omega/|\nabla\omega(\boldsymbol{k})|$. Multiplying this spacing by area df_k of isofrequency surface element and calculating the surface integral over all the surface $\omega(\boldsymbol{k}) = \omega$, we obtain the necessary portion d^3k of the $\boldsymbol{k}$-space volume. Dividing this volume by $(2\pi)^3$, we arrive at the following expression for the spectral density of lattice oscillations in the form of surface integral:

$$g(\omega) = \frac{1}{(2\pi)^3}\int\limits_{\omega(\boldsymbol{k})=\omega}\frac{df_k}{|\nabla\omega(\boldsymbol{k})|}.$$

Here the space of wave vectors $\boldsymbol{k}$ belongs to the first Brillouin zone, and the denominator is the group velocity of phonons $\partial\omega(\boldsymbol{k})/\partial\boldsymbol{k} = \nabla\omega(\boldsymbol{k})$.

This formula allows us to clarify the nature of singularities or *Van Hove singularities* for the function $g(\omega)$ in the crystals. The singularities in the spectral density should appear in the *critical points* of the Brillouin zone at which the group velocity $\partial\omega/\partial\boldsymbol{k}$ vanishes or at the frequency spectrum boundaries.

Let point $\boldsymbol{k}_0$ be critical and expand continuous function $\omega(\boldsymbol{k})$ into a series of second order in the vicinity of this point

$$\omega(\boldsymbol{k}) = \omega_0 + \alpha_{il}(k_i - k_{i0})(k_l - k_{l0}) + \ldots \quad (i, l = x, y, z).$$

The linear term of expansion is absent due to $\partial\omega(\boldsymbol{k})/\partial\boldsymbol{k} = 0$. By choosing the coordinate axes along the principal ones for this symmetrical quadratic form, we reduce it to the canonical form

$$\omega(\boldsymbol{k}) = \omega_0 + \alpha_1\xi_1^2 + \alpha_2\xi_2^2 + \alpha_3\xi_3^2 + \ldots$$

where variables ξ_i are the components of the vector departing from the critical point in the principal axes chosen. The coefficients $\alpha_1, \alpha_2, \alpha_3$ are the principal values of symmetrical tensor α_{il} determined with the second-order derivatives of frequency $\omega(\boldsymbol{k})$ with respect to the components of wave vector $\boldsymbol{k}$.

Let all the coefficients $\alpha_1, \alpha_2, \alpha_3$ have the same sign, i.e. critical point corresponds to the *maximum* or *minimum* of function $\omega(\boldsymbol{k})$, e.g. all $\alpha_1, \alpha_2, \alpha_3 < 0$ for definiteness. Thus, the surfaces of constant frequency are ellipsoids. The volume confined with the ellipsoid around the point $\boldsymbol{k}_0$ will equal

$$\frac{4\pi}{3}\frac{(\omega_0 - \omega)^{3/2}}{|\alpha|^{1/2}} \quad (\alpha = \alpha_1\alpha_2\alpha_3).$$

Differentiating over frequency ω and dividing by $(2\pi)^3$ yields a root singularity in the spectral density near the critical point provided that $\omega < \omega_0$. So, we have

$$\delta g(\omega) = g(\omega) - g(\omega_0) = \begin{cases} \frac{1}{2\pi^2}\frac{(\omega_0-\omega)^{1/2}}{|\alpha|^{1/2}}, & \omega < \omega_0, \\ 0, & \omega > \omega_0, \end{cases}$$

where we have also taken into account that for $\omega > \omega_0$, the vicinity of point $\boldsymbol{k}_0$ gives no contribution to $g(\omega)$.

The singular behavior for the spectral density in the case of the minimum of function $\omega(\boldsymbol{k})$ (all $\alpha_1, \alpha_2, \alpha_3 > 0$) is analogous and has the same root singularity

$$\delta g(\omega) = g(\omega) - g(\omega_0) = \begin{cases} \frac{1}{2\pi^2}\frac{(\omega-\omega_0)^{1/2}}{|\alpha|^{1/2}}, & \omega > \omega_0, \\ 0, & \omega < \omega_0. \end{cases}$$

The derivative $dg/d\omega$ at the critical point ω_0 becomes infinite.

There are other possibilities for the coefficients α_1, α_2, α_3 when one of them is positive and the other two are negative and *vice versa*. In this case the *saddle point* appears. A somewhat more complicated consideration results in a similar answer. So, for α_1, $\alpha_2 > 0$ and $\alpha_3 < 0$, there originates the same root singularity in the saddle-point vicinity:

$$\delta g(\omega) = g(\omega) - g(\omega_0) = \begin{cases} -\frac{1}{2\pi^2}\frac{|\omega_0-\omega|^{1/2}}{|\alpha|^{1/2}}, & \omega < \omega_0, \\ 0, & \omega > \omega_0. \end{cases}$$

For the saddle point with $\alpha_1, \alpha_2 < 0$ and $\alpha_3 > 0$, we have the same result by permuting the regions $\omega < \omega_0$ and $\omega > \omega_0$, i.e. the root singularity at $\omega > \omega_0$.

In total, we can note the following. At the minimum and maximum points of spectra $\omega(\boldsymbol{k})$, the variation reads $\delta g(\omega) > 0$ and we have always $\delta g(\omega) < 0$ for the saddle points of spectrum.

In conclusion, let us remark about the Brillouin zone center, i.e. point $\boldsymbol{k} = 0$. Though the absolute minimum of function $\omega(\boldsymbol{k})$ occurs for all three acoustic branches at this point, no singular behavior appears since the group velocity $\partial\omega/\partial\boldsymbol{k}$ of acoustic branch does not vanish. The singularities in the spectral density of states of crystalline lattice oscillations do affect on the thermodynamic and kinetic properties of crystals. The experimental investigation of the Van Hove singularities gives us information on the elementary excitation spectrum in a crystal.

Problem

Determine the type of the Van Hove singularities for the spectral density $g(\omega)$ of lattice oscillations in a two-dimensional crystal.

Solution. Let point $\boldsymbol{k}_0$ be critical one at which velocity $\partial\omega/\partial\boldsymbol{k}$ vanishes. Then we decompose $\omega(\boldsymbol{k})$ in the vicinity of point $\boldsymbol{k}_0$ as

$$\omega(\boldsymbol{k}) - \omega_0 = \alpha_{il}(k_i - k_{i0})(k_l - k_{l0}) + \dots \quad (i, l = x, y)$$

and reduce it to the canonical form by choosing the coordinate axes along the principal ones of symmetrical quadratic form

$$\omega(\boldsymbol{k}) - \omega_0 = \alpha_1 \xi_1^2 + \alpha_2 \xi_2^2 + \dots$$

Here quantities ξ_i are the vector components departing from the critical point in the principal axes chosen. The coefficients α_1, α_2 are the principal values of symmetrical tensor α_{il}. These coefficients may have the same sign ($\alpha_1\alpha_2 > 0$) in the case of maximum or minimum for function $\omega(\boldsymbol{k})$ or various signs in the saddle-point case ($\alpha_1\alpha_2 < 0$).

For definiteness, let this point be of minimum, i.e. both α_1 and $\alpha_2 > 0$. Then the line of constant frequency is the ellipse which can readily be parametrized as follows:

$$\xi_1 = \frac{(\omega - \omega_0)}{|\alpha_1|^{1/2}} \cos t \quad \text{and} \quad \xi_2 = \frac{(\omega - \omega_0)}{|\alpha_2|^{1/2}} \sin t,$$

the parameter t varying from $-\pi$ to π. The variation of spectral density $\delta g(\omega) = g(\omega) - g(\omega_0)$ is given with the integral estimated along the line of constant frequency

$$\delta g(\omega) = \frac{1}{(2\pi)^2} \int \frac{dl_k}{|\nabla\omega(\boldsymbol{k})|} = \frac{1}{(2\pi)^2} \int\limits_{-\pi}^{\pi} \frac{\sqrt{\xi_1'^2(t) + \xi_2'^2(t)}}{\sqrt{(2\alpha_1\xi_1)^2 + (2\alpha_2\xi_2)^2}} dt =$$

$$= \frac{1}{(2\pi)^2} \int\limits_{-\pi}^{\pi} \frac{dt}{2(\alpha_1\alpha_2)^{1/2}} = \frac{1}{4\pi(\alpha_1\alpha_2)^{1/2}}.$$

Thus, the spectral density $g(\omega)$ demonstrates a discontinuity of finite magnitude at the minimum of function $\omega(\boldsymbol{k})$. The same answer remains valid for the case when the critical point is a maximum of function $\omega(\boldsymbol{k})$.

In the saddle-point case ($\alpha_1\alpha_2 < 0$), the lines of constant frequency are hyperbolas. Putting $\alpha_1 > 0$, $\alpha_2 < 0$, and $\omega - \omega_0 > 0$ for definiteness, we employ the natural parameterization

$$\xi_1 = \frac{(\omega - \omega_0)}{|\alpha_1|^{1/2}} \cosh t \quad \text{and} \quad \xi_2 = \frac{(\omega - \omega_0)}{|\alpha_2|^{1/2}} \sinh t.$$

For the region of integration, we take the region for the deviation of wave vectors ξ_{1m} and ξ_{2m} lying sufficiently far from the saddle point but sufficiently close to be within the quadratic expansion, i.e. $k_{i0} \gg \xi_{im} \gg |k_i - k_{i0}|$. (The value ξ_{1m} will affect the type of singularity examined.) Denoting

$$\xi_{1m}^2 = \frac{\Omega}{|\alpha_1|} = \frac{(\omega - \omega_0)}{|\alpha_1|} \cosh^2 t_0\,, \qquad \cosh^2 t_0 = \frac{\Omega}{(\omega - \omega_0)} \gg 1,$$

and $\xi_{2m}^2 = (\Omega - (\omega - \omega_0))/|\alpha_2| \sim \Omega/|\alpha_2|$, we calculate the necessary integral

$$g(\omega) = \frac{1}{(2\pi)^2} \int \frac{dl_k}{|\nabla\omega(\boldsymbol{k})|} = \frac{1}{(2\pi)^2} \int\limits_{-t_0}^{t_0} \frac{\sqrt{\xi_1'^2(t) + \xi_2'^2(t)}}{\sqrt{(2\alpha_1\xi_1)^2 + (2\alpha_2\xi_2)^2}} dt =$$

$$= \frac{1}{(2\pi)^2} \int\limits_{-t_0}^{t_0} \frac{dt}{2(\alpha_1\alpha_2)^{1/2}} = \frac{t_0}{4\pi^2(|\alpha_1\alpha_2|)^{1/2}}.$$

Taking into account that $t_0 \gg 1$ and $\cosh^2 t_0 \approx e^{t_0}/4$, we arrive at the logarithmic singularity in the spectral density at the saddle point

$$g(\omega) \approx \frac{1}{4\pi^2(|\alpha_1\alpha_2|)^{1/2}} \ln \frac{4\Omega}{(\omega - \omega_0)}.$$

In the general form, this can be represented as a sum of some constant and logarithmically divergent singularity at the saddle point

$$g(\omega) = \text{const} + \frac{1}{4\pi^2(|\alpha_1\alpha_2|)^{1/2}} \ln \frac{\omega_0}{|\omega - \omega_0|} .$$

In the one-dimensional case (chain of atoms), the saddle points are absent and the Van Hove singularity is associated with the point of the minimum or maximum value in the frequency spectrum. In the vicinity of such a point, the spectral density $g(\omega) \sim 1/|\omega - \omega_0|^{1/2}$ diverges in a root-like manner.

Chapter 4
Phase Transitions and Critical Phenomena

4.1 Fluctuations of Thermodynamic Variable

In thermodynamically equilibrium systems as a result of their openness and interaction with the environment, the *thermal fluctuations* represent random fluctuations of physical variables from their average magnitudes. From the formal and mathematical point of view the fluctuations of any physical variables can be explained with the statistical nature of describing the system, when each possible state in the system realizes at certain probability governed by the distribution function. Usually, the thermal fluctuations grow as the temperature increases. In addition, an anomalous growth of fluctuations takes place for a number of physical variables near the critical points and second-order phase transitions. At zero temperature, the fluctuations of physical variables are primarily due to quantum mechanical phenomena.

In order to have the quantitative characteristic for the fluctuations of a random variable Λ, the *variance* $\langle(\Delta\Lambda)^2\rangle$ is introduced as an expected value of the squared deviation from the mean $\bar{\Lambda} = \langle\Lambda\rangle$ of random variable Λ, i.e. $\langle(\Delta\Lambda)^2\rangle = \langle(\Lambda - \langle\Lambda\rangle)^2\rangle$. The *standard deviation* is referred to as the square root of the variance. The mutual influence of several fluctuating variables Λ_i is characterized with their *correlation* $\langle\Delta\Lambda_i\Delta\Lambda_k\rangle$ where $\Delta\Lambda_{i,k} = \Lambda_{i,k} - \langle\Lambda_{i,k}\rangle$. For the statistically independent variables, we have $\langle\Lambda_i\Lambda_k\rangle = \langle\Lambda_i\rangle\langle\Lambda_k\rangle$ or their correlation vanishes, i.e. $\langle\Delta\Lambda_i\Delta\Lambda_k\rangle = 0$. For calculating the thermodynamic fluctuations, it is sufficient to write the corresponding value of fluctuation and then to average it with the aid of corresponding distribution function. In the thermodynamically equilibrium system, this function is the Gibbs distribution. Below we turn to calculating the fluctuations.

Let us treat the thermodynamic system in which the energy levels $\varepsilon_\Lambda = \varepsilon(\Lambda)$ as well as the corresponding state vectors $|\Lambda\rangle$ depend somehow on the fluctuating physical variable Λ of our interest. Next, we introduce the thermodynamic parameter λ conjugated to thermodynamic variable Λ. On the analogy with the Legendre transformation, we then determine an auxiliary variable of the dimension of energy

S. N. Burmistrov, *Statistical and Condensed Matter Physics*,
https://doi.org/10.1007/978-981-97-9072-2_4

$\tilde{\varepsilon}_\Lambda$ specified for all the states in the system and numerated with state vector $|\Lambda\rangle$, as follows:

$$\tilde{\varepsilon}_\Lambda = \varepsilon(\Lambda) + \lambda\Lambda\,.$$

And finally we write the corresponding expression[1] for the partition function

$$Z(\lambda) = \sum_\Lambda \exp(-\tilde{\varepsilon}_\Lambda/T)$$

which allows us to find the free energy according to definition $F(\lambda) = -T\ln Z(\lambda)$.

In order to find the mean value of thermodynamic variable Λ, we calculate the first derivative of free energy F with respect to parameter λ

$$\frac{\partial F}{\partial \lambda} = \frac{1}{Z}\sum_\Lambda e^{-\tilde{\varepsilon}_\Lambda/T}\frac{\partial \tilde{\varepsilon}_\Lambda}{\partial \lambda} = \sum_\Lambda \Lambda \frac{e^{-\tilde{\varepsilon}_\Lambda/T}}{Z} = \bar{\Lambda}.$$

For brevity, we have here denoted the mean value Λ as $\bar{\Lambda} = \langle\Lambda\rangle$. On the account for the temperature dependence of free energy, this means that the differential of free energy F can be represented as

$$dF = -S\,dT + \bar{\Lambda}\,d\lambda.$$

Accordingly, the differential for the conjugate thermodynamic potential $\tilde{F}(T,\bar{\Lambda}) = F - \lambda\bar{\Lambda}$, as a function of $\bar{\Lambda}$, will be equal to

$$d\tilde{F} = -S\,dT - \lambda\,d\bar{\Lambda}.$$

In the final expressions one should put $\lambda = 0$. This is equivalent to the fact that all the next derivatives must be calculated at the equilibrium values of the thermodynamic variable Λ. So, we find the average required

$$\bar{\Lambda} = \left(\frac{\partial F}{\partial \lambda}\right)_{\lambda=0},$$

being the derivative of free energy with respect to thermodynamic parameter λ.

In what follows, it is convenient to use the following formula:

$$\langle\Lambda^n\rangle = (-T)^n \frac{1}{Z}\frac{\partial^n Z}{\partial \lambda^n}$$

[1] As will be seen below, the partition function $Z(\lambda)$ plays a role of the *generating functional* for calculating various average quantities.

for calculating the mean values in power n. On the use of relation for second derivative with respect to parameter λ as

$$\frac{\partial^2 \ln Z}{\partial \lambda^2} = \frac{1}{Z}\frac{\partial^2 Z}{\partial \lambda^2} - \left(\frac{1}{Z}\frac{\partial Z}{\partial \lambda}\right)^2,$$

it is readily to see the following expression for the variance of variable Λ:

$$\langle(\Delta\Lambda)^2\rangle = T^2\frac{\partial^2 \ln Z}{\partial \lambda^2} = -T\frac{\partial^2 F}{\partial \lambda^2}.$$

This answer can be expressed via derivatives of the conjugate thermodynamic potential $\tilde{F} = \tilde{F}(\bar{\Lambda})$

$$\langle(\Delta\Lambda)^2\rangle = T\left(\frac{\partial^2 \tilde{F}}{\partial \bar{\Lambda}^2}\right)^{-1}$$

if one employs the following sequence of equalities:

$$\frac{\partial^2 F}{\partial \lambda^2} = \frac{\partial \bar{\Lambda}}{\partial \lambda} = \frac{1}{\partial \lambda/\partial \bar{\Lambda}} = -\frac{1}{\partial^2 \tilde{F}/\partial \bar{\Lambda}^2}.$$

Let us pay attention for the relation

$$\langle(\Delta\Lambda)^2 = -T\frac{\partial \bar{\Lambda}}{\partial \lambda}$$

between the variance and the derivative of mean value $\bar{\Lambda}$ with respect to external parameter λ. Within the accuracy to the sign, this derivative can be treated as a susceptibility $\partial\bar{\Lambda}/\partial\lambda$ or linear response to perturbation of the system with external impact given by quantity $-\lambda$. The positive definiteness of variance implies the fulfillment of the following *thermodynamic inequalities*: $\partial^2 F/\partial\lambda^2 < 0$ or $\partial^2 \tilde{F}/\partial\Lambda^2 > 0$. The state of thermodynamic system is absolutely unstable if this condition is violated.

Due to equivalence of conjugate variables λ and Λ, one can set the parameter λ as an independent thermodynamic fluctuating variable and determine similarly the variance of fluctuation $\langle(\Delta\lambda)^2\rangle$ for parameter λ. Of interest here is the correlation $\langle\Delta\Lambda\Delta\lambda\rangle$ of two conjugate variables. From the equality

$$\begin{aligned}\frac{\partial^2 \ln Z}{\partial\Lambda\partial\lambda} &= \frac{1}{Z}\frac{\partial^2 Z}{\partial\Lambda\partial\lambda} - \left(\frac{1}{Z}\frac{\partial Z}{\partial\Lambda}\right)\left(\frac{1}{Z}\frac{\partial Z}{\partial\lambda}\right) = \\ &= \sum_{\Lambda}\frac{\Lambda\lambda}{T^2}\frac{e^{-\tilde{\varepsilon}_\Lambda/T}}{Z} - \left(\sum_{\Lambda}\frac{\Lambda}{T}\frac{e^{-\tilde{\varepsilon}_\Lambda/T}}{Z}\right)\left(\sum_{\Lambda}\frac{\lambda}{T}\frac{e^{-\tilde{\varepsilon}_\Lambda/T}}{Z}\right),\end{aligned}$$

we get the following equality for the mean variables:

$$T^2 \frac{\partial^2 \ln Z}{\partial \Lambda \partial \lambda} = \langle \Lambda \lambda \rangle - \langle \Lambda \rangle \langle \lambda \rangle = \langle \Delta \Lambda \Delta \lambda \rangle.$$

Thus, we arrive at the expression

$$\langle \Delta \Lambda \Delta \lambda \rangle = -T \frac{\partial^2 F}{\partial \Lambda \partial \lambda}.$$

Taking into account that $\Lambda = \partial F / \partial \lambda$, we find the correlation for two conjugate thermodynamic variables

$$\langle \Delta \Lambda \Delta \lambda \rangle = -T.$$

This relation has a universal answer. The result obtained means that the fluctuations of conjugate thermodynamic variables are statistically dependent.

The method of the generating functional allows us to calculate higher powers of mean deviations as well. Let us start from calculating the mean $\langle (\Delta \Lambda)^3 \rangle$ and write the following equality:

$$\frac{\partial^3 \ln Z}{\partial \lambda^3} = \frac{1}{Z} \frac{\partial^3 Z}{\partial \lambda^3} - 3 \left(\frac{1}{Z} \frac{\partial Z}{\partial \lambda} \right) \left(\frac{1}{Z} \frac{\partial^2 Z}{\partial \lambda^2} \right) + 2 \left(\frac{1}{Z} \frac{\partial Z}{\partial \lambda} \right)^3.$$

Hence it is straightforwardly to see that

$$-T^3 \frac{\partial^3 \ln Z}{\partial \lambda^3} = \langle \Lambda^3 \rangle - 3 \langle \Lambda^2 \rangle \langle \Lambda \rangle + 2 \langle \Lambda \rangle^3 = \langle (\Lambda - \langle \Lambda \rangle)^3 \rangle = \langle (\Delta \Lambda)^3 \rangle.$$

As a result, we find the simple expression for the mean cube of fluctuations

$$\langle (\Delta \Lambda)^3 \rangle = T^2 \frac{\partial^3 F}{\partial \lambda^3} = T^2 \frac{\partial^2 \bar{\Lambda}}{\partial \lambda^2} = -T \frac{\partial}{\partial \lambda} \langle (\Delta \Lambda)^2 \rangle.$$

This answer can be expressed in terms of the derivatives of thermodynamic potential $\tilde{F} = \tilde{F}(\bar{\Lambda})$, depending on variable $\bar{\Lambda}$ conjugated to parameter λ:

$$\langle (\Delta \Lambda)^3 \rangle = -T^2 \frac{\partial^3 \tilde{F} / \partial \bar{\Lambda}^3}{\left(\partial^2 \tilde{F} / \partial \bar{\Lambda}^2 \right)^3} = \frac{T^2}{\partial^2 \tilde{F} / \partial \Lambda^2} \frac{\partial}{\partial \Lambda} \left(\frac{1}{\partial^2 \tilde{F} / \partial \Lambda^2} \right).$$

The next equality helps us to estimate the fluctuations of fourth power:

$$\frac{\partial^4 \ln Z}{\partial \lambda^4} = \frac{1}{Z} \frac{\partial^4 Z}{\partial \lambda^4} - 4 \left(\frac{1}{Z} \frac{\partial^3 Z}{\partial \lambda^3} \right) \left(\frac{1}{Z} \frac{\partial Z}{\partial \lambda} \right) + 6 \left(\frac{1}{Z} \frac{\partial^2 Z}{\partial \lambda^2} \right) \left(\frac{1}{Z} \frac{\partial Z}{\partial \lambda} \right)^2 - \\ -3 \left(\frac{1}{Z} \frac{\partial Z}{\partial \lambda} \right)^4 - 3 \left[\frac{1}{Z} \frac{\partial^2 Z}{\partial \lambda^2} - \left(\frac{1}{Z} \frac{\partial Z}{\partial \lambda} \right)^2 \right]^2.$$

In accordance with the definitions of the averages, we find that

$$T^4\frac{\partial^4 \ln Z}{\partial\lambda^4} = \langle\Lambda^4\rangle - 4\langle\Lambda^3\rangle\langle\Lambda\rangle + 6\langle\Lambda^2\rangle\langle\Lambda\rangle^2 - 3\langle\Lambda\rangle^4 - 3\big(\langle\Lambda^2\rangle - \langle\Lambda\rangle^2\big)^2 = \\ = \langle(\Delta\Lambda)^4\rangle - 3\langle(\Delta\Lambda)^2\rangle^2.$$

Finally we arrive at the comprehensible formula for the fluctuations of fourth power

$$\langle(\Delta\Lambda)^4\rangle = 3T^2\left(\frac{\partial^2 F}{\partial\lambda^2}\right)^2 - T^3\frac{\partial^4 F}{\partial\lambda^4} = \\ = 3T^2\frac{1}{\left(\partial^2\tilde{F}/\partial\bar{\Lambda}^2\right)^2} + 3T^3\frac{\left(\partial^3\tilde{F}/\partial\bar{\Lambda}^3\right)^2}{\left(\partial^2\tilde{F}/\partial\bar{\Lambda}^2\right)^5} - T^3\frac{\partial^4\tilde{F}/\partial\bar{\Lambda}^4}{\left(\partial^2\tilde{F}/\partial\bar{\Lambda}^2\right)^4} = \\ = \frac{3T^2}{f} + T^3\left(\frac{1}{f^2}\frac{\partial^2(1/f)}{\partial\Lambda^2} + \frac{1}{f}\left(\frac{\partial(1/f)}{\partial\Lambda}\right)^2\right) \quad \text{where} \quad f = \frac{\partial^2\tilde{F}}{\partial\Lambda^2}.$$

As an example, let us consider isothermal fluctuations of the total volume of the system. The volume of the system V should be chosen as a fluctuating parameter Λ and, correspondingly, we choose pressure P as a volume-conjugated variable λ. Then the free energy F represents in itself the Gibbs free energy $\Phi = \Phi(T, P)$ and its differential reads $d\Phi = V\,dP$. The potential $\tilde{F}$ conjugated to the latter one is the Helmholtz free energy $\tilde{F} = \tilde{F}(T, V)$ with the differential $d\tilde{F} = -P\,dV$. So, for the fluctuations of volume under constant temperature, we have

$$\langle(\Delta V)^2\rangle = -T\left(\frac{\partial V}{\partial P}\right)_T = -\frac{T}{\left(\partial P/\partial V\right)_T},$$

$$\langle(\Delta V)^3\rangle = T^2\left(\frac{\partial^2 V}{\partial P^2}\right)_T = -T^2\frac{\left(\partial^2 P/\partial V^2\right)_T}{\left(\partial P/\partial V\right)_T^3},$$

$$\langle(\Delta V)^4\rangle = 3T^2\left(\frac{\partial V}{\partial P}\right)_T^2 - T^3\left(\frac{\partial^3 V}{\partial P^3}\right)_T = \\ = \frac{3T^2}{\left(\partial P/\partial V\right)_T^2} - 3T^3\frac{\left(\partial^2 P/\partial V^2\right)_T^2}{\left(\partial P/\partial V\right)_T^5} + T^3\frac{\left(\partial^3 P/\partial V^3\right)_T}{\left(\partial P/\partial V\right)_T^4}.$$

For the second way of illustration, we treat the fluctuations of particle number N in the thermodynamically equilibrium system. Here we choose the following thermodynamic variables: chemical potential μ and the number of particles N. Let us put $\Lambda = N$ and $\lambda = -\mu$ so that $\tilde{\varepsilon}(\lambda, \mu) = \varepsilon(N) - \mu N$. Then the differential of free energy $F(\lambda)$ equals $dF = \bar{N}d(-\mu) = -\bar{N}d\mu$, and it can be identified with the thermodynamic grand potential $\Omega = \Omega(T, \mu)$. The conjugate potential $\tilde{F} = \tilde{F}(\bar{N})$ will coincide with the Helmholtz free energy having the differential $dF = \mu\,d\bar{N}$.

On completion, the mean number of particles is determined by the following derivative:

$$\langle N\rangle = \bar{N} = -\partial\Omega/\partial\mu\,,$$

and the variance equals

$$\langle(\Delta N)^2\rangle = -T\frac{\partial^2\Omega}{\partial\mu^2} = T\frac{\partial N}{\partial\mu} = \frac{T}{\partial\mu/\partial N}\,.$$

The fluctuations of third and fourth powers are, as follows:

$$\langle(\Delta N)^3\rangle = T^2\frac{\partial^2 N}{\partial\mu^2} = -T^2\frac{\partial^2\mu/\partial N^2}{(\partial\mu/\partial N)^3}\,,$$

$$\langle(\Delta N)^4\rangle = 3T^2\left(\frac{\partial N}{\partial\mu}\right)^2 + T^3\frac{\partial^3 N}{\partial\mu^3} =$$

$$= \frac{3T^2}{\left(\partial\mu/\partial N\right)^2} + 3T^3\frac{\left(\partial^2\mu/\partial N^2\right)^2}{\left(\partial\mu/\partial N\right)^5} - T^3\frac{\partial^3\mu/\partial N^3}{\left(\partial\mu/\partial N\right)^4}\,.$$

Problems

1. Find the isothermal fluctuations of volume V in an ideal gas and calculate the skewness S and kurtosis κ.

Solution. Differentiating the equation $V = NT/P$ yields

$$\langle(\Delta V)^2\rangle = \frac{V^2}{N}\,,\quad \langle(\Delta V)^3\rangle = \frac{2V^3}{N^2}\,,\quad \langle(\Delta V)^4\rangle = \frac{3(N+2)}{N^3}V^4.$$

The *skewness* S shows the measure of the distribution function asymmetry around the mean value and equals

$$S = \frac{\langle(\Delta V)^3\rangle}{\langle(\Delta V)^2\rangle^{3/2}} = \frac{2}{\sqrt{N}}\,.$$

The *kurtosis* κ characterizes the measure of the acuteness for the distribution function peak and equals

$$\kappa = \frac{\langle(\Delta V)^4\rangle}{\langle(\Delta V)^2\rangle^2} = 3 + \frac{6}{N}\,.$$

As the particle number N increases, the characteristics of distribution function approach the normal or Gaussian distribution for which $S = 0$ and $\kappa = 3$.

2. Find the fluctuations of the particle number in an ideal gas and calculate the skewness S and kurtosis κ.

Solution. Let us use relation for derivative $\partial N/\partial\mu = N/T$ valid in an ideal gas. Then we find

$$\langle(\Delta N)^2\rangle = N,\quad \langle(\Delta N)^3\rangle = N,\quad \langle(\Delta N)^4\rangle = 3N^2 + N.$$

The skewness S equals

$$S = \frac{\langle(\Delta N)^3\rangle}{\langle(\Delta N)^2\rangle^{3/2}} = \frac{1}{\sqrt{N}}\,.$$

The kurtosis κ reads

$$\kappa = \frac{\langle(\Delta N)^4\rangle}{\langle(\Delta N)^2\rangle^2} = 3 + \frac{1}{N}.$$

3. Let $P(N)$ be some discrete distribution function describing the fluctuations of particle number $N = 0, 1, 2, \ldots$

Find the characteristic or generating function for calculating the mean value for deviation $\langle(\Delta N)^l\rangle$ of particle number from its mean value $\langle N\rangle$ for an arbitrary power $l > 1$.

Solution. Characteristic (generating) function equals

$$K(\mu) = \ln \sum_{N=0}^{\infty} e^{i\mu N} P(N) = i\mu\kappa_1 + \frac{(i\mu)^2}{2!}\kappa_2 + \frac{(i\mu)^3}{3!}\kappa_3 + \cdots = \sum_{l=1}^{\infty} \frac{(i\mu)^l}{l!}\kappa_l$$

and the expansion coefficients (*cumulants*) κ_l in a Maclaurin series determine the averages required

$$\langle(\Delta N)^l\rangle = \kappa_l = (-i)^l \frac{d^l K(\mu)}{d\mu^l}\bigg|_{\mu=0} \quad (l > 1).$$

Coefficient κ_1 represents the mean value of distribution function, i.e. mean number of particles

$$\langle N\rangle = -i\frac{dK(\mu)}{d\mu}\bigg|_{\mu=0} = \sum_{N=0}^{\infty} NP(N).$$

4. Find the correlation between the fluctuations of chemical potential and particle number.

Solution. In the present case the free energy potential F is represented with thermodynamic potential $\Omega(\mu)$. The correlation of fluctuations is given by the formula

$$\langle\Delta\mu\Delta N\rangle = -T\frac{\partial^2\Omega}{\partial N\partial\mu} = T\frac{\partial N}{\partial N} = T.$$

5. Find the fluctuations of entropy and its correlation $\langle\Delta T\Delta S\rangle$ as a function of temperature.

Solution. We find using the general formulas $dF = -S\,dT$ and $d\tilde{F} = T\,dS$ in which $\lambda = T$ and $\Lambda = -S$ are to take

$$\begin{aligned}
\langle(\Delta S)^2\rangle &= -T\frac{\partial^2 F}{\partial T^2} = T\frac{\partial S}{\partial T} = C(T), \\
\langle(\Delta S)^3\rangle &= T^2\frac{\partial^3 F}{\partial T^3} = -T\frac{\partial^2 S}{\partial T^2} = -T^2\frac{\partial}{\partial T}\left(\frac{C(T)}{T}\right), \\
\langle(\Delta S)^4\rangle &= 3C^2(T) + T^3\frac{\partial^2}{\partial T^2}\left(\frac{C(T)}{T}\right)
\end{aligned}$$

where $C(T)$ is the specific heat of thermodynamic system. Since temperature and entropy are the conjugate variables, we have from the general formula: $\langle\Delta T\Delta S\rangle = T$.

4.2 Fluctuations of Several Thermodynamic Variables

We have considered above the fluctuations of a single thermodynamic variable from its mean value and have paid no attention to other variables. Let us turn now for calculating the simultaneous fluctuations of several thermodynamic quantities and their correlations.

Let energy levels $\varepsilon_{\mathbf{\Lambda}} = \varepsilon(\Lambda_1 \ldots \Lambda_n)$ of physical states in the thermodynamic system depend on a set of fluctuating variables $\mathbf{\Lambda} = \Lambda_1, \ldots, \Lambda_n$. Next, we introduce the corresponding conjugate variables $\lambda_1, \ldots, \lambda_n$ so that

$$\tilde{\varepsilon}_{\mathbf{\Lambda}} = \varepsilon(\Lambda_1 \ldots \Lambda_n) + \lambda_1\Lambda_1 + \ldots + \lambda_n\Lambda_n \,.$$

Then we consider the partition function or the generating functional corresponding to the above set of energy states numerated with the state vector $|\mathbf{\Lambda}\rangle$

$$Z(\lambda_1 \ldots \lambda_n) = \sum_{\mathbf{\Lambda}} \exp(-\tilde{\varepsilon}_{\mathbf{\Lambda}}/T)$$

and define the corresponding free energy $F(\lambda_1 \ldots \lambda_n) = -T \ln Z$.

Let us start first from calculating the mean deviations for two variables. For the second derivative of $\ln Z$, we find

$$\frac{\partial^2 \ln Z}{\partial\lambda_i\lambda_k} = \sum_{\mathbf{\Lambda}} \frac{\Lambda_i}{T}\frac{\Lambda_k}{T}\frac{e^{-\tilde{\varepsilon}_{\mathbf{\Lambda}}/T}}{Z} - \sum_{\mathbf{\Lambda},\mathbf{\Lambda}'} \left(\frac{\Lambda_i}{T}\frac{e^{-\tilde{\varepsilon}_{\mathbf{\Lambda}}/T}}{Z}\right)\left(\frac{\Lambda'_k}{T}\frac{e^{-\tilde{\varepsilon}_{\mathbf{\Lambda}'}/T}}{Z}\right).$$

One can readily see from this equality that

$$\langle\Lambda_i\Lambda_k\rangle - \langle\Lambda_i\rangle\langle\Lambda_k\rangle = T^2\frac{\partial^2 \ln Z}{\partial\lambda_i\lambda_k}.$$

Finally, we arrive at the simple answer for the correlation between the fluctuations of two thermodynamic variables

$$\langle\Delta\Lambda_i\Delta\Lambda_k\rangle = T^2\frac{\partial^2 \ln Z}{\partial\lambda_i\lambda_k} = -T\frac{\partial^2 F}{\partial\lambda_i\lambda_k} = -T\frac{\partial\Lambda_k}{\partial\lambda_i}\,.$$

This result can also be represented in the terms of conjugate thermodynamic potential $\tilde{F}$ expressed in variables Λ_i, Λ_k as

$$\langle\Delta\Lambda_i\Delta\Lambda_k\rangle = T\left(\frac{\partial^2\tilde{F}}{\partial\Lambda_i\partial\Lambda_k}\right)^{-1}.$$

For the non-conjugate mutually independent thermodynamic variables λ_i and Λ_k $(i \neq k)$ and from equality

$$\langle\Delta\lambda_i\Delta\Lambda_k\rangle = -T\frac{\partial^2 F}{\partial\Lambda_k\partial\lambda_i} = -T\frac{\partial\Lambda_i}{\partial\Lambda_k} = 0 \quad \text{if} \quad i \neq k,$$

one sees the mutual correlation of fluctuations is absent as it should be expectable.

For the correlation between three thermodynamic variables Λ_i, Λ_k, and Λ_l, expressing via the averages as follows:

$$\langle \Delta\Lambda_i \Delta\Lambda_k \Delta\Lambda_l \rangle = \langle \Lambda_i \Lambda_k \Lambda_l \rangle - \langle \Lambda_i \rangle \langle \Lambda_k \Lambda_l \rangle - \\ - \langle \Lambda_i \Lambda_k \rangle \langle \Lambda_l \rangle - \langle \Lambda_i \Lambda_l \rangle \langle \Lambda_k \rangle + 2 \langle \Lambda_i \rangle \langle \Lambda_k \rangle \langle \Lambda_l \rangle,$$

we take into account that

$$\langle \Lambda_1 \ldots \Lambda_n \rangle = (-T)^n \frac{1}{Z} \frac{\partial^n Z}{\partial \lambda_1 \ldots \partial \lambda_n},$$

and find after the simple calculation

$$\langle \Delta\Lambda_i \Delta\Lambda_k \Delta\Lambda_l \rangle = T^2 \frac{\partial^3 F}{\partial \lambda_i \partial \lambda_k \partial \lambda_l}.$$

Let us express the same answer in terms of the conjugate thermodynamic potential $\tilde{F}$ depending on the variables Λ. We denote $\tilde{F}_{ik}(\Lambda)$ as the following derivative of second order:

$$\tilde{F}_{ik} = \frac{\partial^2 \tilde{F}}{\partial \Lambda_i \partial \Lambda_k}.$$

Employing the following relations for derivatives:

$$\frac{\partial \lambda_k}{\partial \Lambda_l} = -\frac{\partial^2 \tilde{F}}{\partial \Lambda_k \partial \Lambda_l} = -\tilde{F}_{kl},$$
$$\frac{\partial^2 \lambda_k}{\partial \lambda_i \partial \Lambda_l} = \sum_{p=1}^{n} \frac{\partial^2 \lambda_k}{\partial \Lambda_p \partial \Lambda_l} \frac{\partial \Lambda_p}{\partial \lambda_i} = \sum_{p=1}^{n} \frac{\partial \tilde{F}_{kl}}{\partial \Lambda_p} \tilde{F}_{pi}^{-1},$$

we get a simple formula for the triple correlations

$$\langle \Delta\Lambda_i \Delta\Lambda_k \Delta\Lambda_l \rangle = -T^2 \tilde{F}_{kl}^{-2} \sum_{p=1}^{n} \frac{\partial \tilde{F}_{kl}}{\partial \Lambda_p} \tilde{F}_{pi}^{-1} = T^2 \sum_{p=1}^{n} \tilde{F}_{ip}^{-1} \frac{\partial \tilde{F}_{kl}^{-1}}{\partial \Lambda_p}.$$

For the correlation of four variables, one can specify in a similar way that

$$\langle \Delta\Lambda_i \Delta\Lambda_k \Delta\Lambda_l \Delta\Lambda_m \rangle = -T^3 \frac{\partial^4 F}{\partial \lambda_i \partial \lambda_k \partial \lambda_l \partial \lambda_m} + \\ + T^2 \left(\frac{\partial^2 F}{\partial \lambda_i \partial \lambda_k} \frac{\partial^2 F}{\partial \lambda_l \partial \lambda_m} + \frac{\partial^2 F}{\partial \lambda_i \partial \lambda_l} \frac{\partial^2 F}{\partial \lambda_k \partial \lambda_m} + \frac{\partial^2 F}{\partial \lambda_i \partial \lambda_m} \frac{\partial^2 F}{\partial \lambda_k \partial \lambda_l} \right) = \\ = T^3 \sum_{p,q=1}^{n} \left(\frac{\partial^2 \tilde{F}_{ik}^{-1}}{\partial \Lambda_p \partial \Lambda_q} \tilde{F}_{pl}^{-1} \tilde{F}_{qm}^{-1} + \frac{\partial \tilde{F}_{ik}^{-1}}{\partial \Lambda_p} \frac{\partial \tilde{F}_{pl}^{-1}}{\partial \Lambda_q} \tilde{F}_{qm}^{-1} \right) + \\ + T^2 \left(\tilde{F}_{ik}^{-1} \tilde{F}_{lm}^{-1} + \tilde{F}_{il}^{-1} \tilde{F}_{km}^{-1} + \tilde{F}_{im}^{-1} \tilde{F}_{kl}^{-1} \right).$$

4.3 The Gaussian Approximation for the Fluctuations of Thermodynamic Variables

As a rule with the exception of the immediate vicinity of critical points and lines of continuous phase transitions, the average fluctuations of third and fourth powers are small as compared with the mean square fluctuations. In this case, one can use the Gaussian approximation to describe the mean square fluctuations. We consider such an approximation on the example of energy fluctuations in the thermodynamically equilibrium system.

Let us write the Gauss distribution for the energy fluctuation with the variance equal to $\langle(\Delta E)^2\rangle$

$$W(E)dE = \frac{1}{\sqrt{2\pi\langle(\Delta E)^2\rangle}}\exp\left(-\frac{(E-\bar{E})^2}{2\langle(\Delta E)^2\rangle}\right)dE$$

so that

$$\int_{-\infty}^{\infty}(E-\bar{E})^2 W(E)\,dE = \langle(\Delta E)^2\rangle \quad \text{and} \quad \int_{-\infty}^{\infty} W(E)\,dE = 1.$$

The Gauss distribution[2] has a drastic maximum at the mean value of energy $E = \bar{E}$ and decays exponentially by increasing the energy E in the symmetrical way on both sides from the mean value $\bar{E}$. For this reason, we have extended the integration region to all values from $-\infty$ to $+\infty$ in spite of the fact that this region may not coincide with that for the possible variation of energy E.

Using the expression for the variance of energy fluctuation as a second derivative of entropy $S = S(E)$ of the system

$$\frac{1}{\langle(\Delta E)^2\rangle} = -\frac{\partial^2 S}{\partial E^2},$$

we represent the expression for the probability distribution as follows:

$$W(E) \propto \exp\left(\frac{1}{2}\frac{\partial^2 S}{\partial E^2}(E-\bar{E})^2\right).$$

Then we note that the entropy in the thermodynamically equilibrium system reaches the maximum value at the mean energy $E = \bar{E}$. Thus, we have

$$\left.\frac{\partial S}{\partial E}\right|_{E=\bar{E}} = 0.$$

[2] Any distribution with a drastic narrow maximum can be chosen as an approximating one.

Provided that the fluctuations are small, in the expansion of entropy $S(E)$ we can limit ourselves only with the terms not higher than second order and write

$$S(E) \approx S(\bar{E}) + \frac{1}{2}\frac{\partial^2 S}{\partial E^2}\bigg|_{E=\bar{E}} (E - \bar{E})^2.$$

As a result, we arrive at the following approximate formula for the probability distribution in the system:

$$W(E) \approx \text{const} \cdot e^{S(E)}.$$

This formula will be applicable for studying the small fluctuations in the thermodynamically equilibrium system. The formula can be rewritten as

$$W(\Delta E) = A\, e^{\Delta S(\Delta E)} \quad \text{where} \quad \Delta S = S(E) - S(\bar{E}) = \frac{1}{2}\frac{\partial^2 S}{\partial E^2}(\Delta E)^2.$$

The normalization constant A is determined with the condition $\int W(\Delta E)dE = 1$.

This formula can also be represented in the form of the thermal activation Gibbs distribution with some energy $U_{\min}$ of thermodynamic fluctuation

$$W \propto \exp\left(-\frac{U_{\min}}{T}\right).$$

Assuming the fluctuations in our approximation to be small as compared with the mean values of the variables themselves and limiting ourselves with linear approximation, we will find in accordance with the following chain of equalities:

$$\begin{aligned} \frac{\partial^2 S}{\partial E^2}(\Delta E)^2 &= \Delta\left(\frac{\partial S}{\partial E}\right)\Delta E = \Delta\left(\frac{1}{T}\right)T\Delta S = \\ &= -\frac{\Delta T\,\Delta S}{T} = -\frac{1}{T}\Delta\left(\frac{\partial E}{\partial S}\right)\Delta S = -\frac{1}{T}\frac{\partial^2 E}{\partial S^2}(\Delta S)^2 \end{aligned}$$

that the minimum energy[3] of thermodynamic fluctuation $U_{\min}$ will be equal to

$$U_{\min} = \frac{1}{2}\frac{\partial^2 E}{\partial S^2}(\Delta S)^2 = \frac{1}{2}\Delta T\,\Delta S.$$

The last expression for $U_{\min}$ is convenient since it is represented in the symmetrical form over the conjugate fluctuating variables as temperature and entropy. The preexponential factor in the Gibbs distribution can readily be found from the condition of

[3] In a number of textbooks, the energy $U_{\min}$ of fluctuation is defined as a minimum work $R_{\min}$ necessary to produce the assigned variations of thermodynamic variables in the reversible way.

normalizing the total probability to unity. The minimum energy of thermodynamic fluctuation can also be written as follows:

$$U_{\min} = \Delta E - T\,\Delta S.$$

This results immediately from the equality

$$\Delta E - T\,\Delta S = \frac{\partial E}{\partial S}\Delta S + \frac{1}{2}\frac{\partial^2 E}{\partial S^2}(\Delta S)^2 - T\,\Delta S = \frac{1}{2}\frac{\partial^2 E}{\partial S^2}(\Delta S)^2$$

which is valid in the linear approximation for sufficiently small fluctuations in the thermodynamic system.

Problem

Find the expression for the minimum energy $U_{\min}$ of fluctuations in the case of two independent thermodynamic variables as temperature T and variable Λ, using the Gauss approximation $W \propto \exp\big(\Delta S(E,\Lambda)\big)$ for the distribution function of fluctuations.

Solution. Let us define the thermodynamic potentials of energy E and free energy F according to the differentials: $dE(S,\Lambda) = T\,dS + \lambda\,d\Lambda$ and $dF(T,\Lambda) = -S\,dT + \lambda\,d\Lambda$. The change ΔS of entropy within accuracy to the square terms in deviations is given by

$$\Delta S(E,\Lambda) = \frac{1}{2}\frac{\partial^2 S}{\partial E^2}(\Delta E)^2 + \frac{\partial^2 S}{\partial E\partial\Lambda}\Delta E\,\Delta\Lambda + \frac{1}{2}\frac{\partial^2 S}{\partial\Lambda^2}(\Delta\Lambda)^2.$$

Here we have taken into account that the entropy in the state of thermodynamic equilibrium has the extremum value and, correspondingly, first derivatives $\partial S/\partial E = 0$ and $\partial S/\partial\Lambda = 0$ vanish at the equilibrium values E and Λ. Then,

$$\Delta S = \frac{1}{2}\Delta\left(\frac{\partial S}{\partial E}\right)\Delta E + \frac{1}{2}\Delta\left(\frac{\partial S}{\partial\Lambda}\right)\Delta\Lambda = \frac{1}{2}\Delta\left(\frac{1}{T}\right)(T\,\Delta S + \lambda\,\Delta\Lambda) - \frac{1}{2}\Delta\left(\frac{\lambda}{T}\right)\Delta\Lambda =$$
$$= -\frac{\Delta T\,\Delta S + \Delta\lambda\,\Delta\Lambda}{2T} = -\frac{U_{\min}}{T}.$$

The expression for the minimum energy $U_{\min}$ can also be rewritten in the following way:

$$U_{\min} = \frac{1}{2}\Delta\left(\frac{\partial E}{\partial S}\right)\Delta S + \frac{1}{2}\Delta\left(\frac{\partial E}{\partial\Lambda}\right)\Delta\Lambda = \Delta E - T\,\Delta S - \lambda\,\Delta\Lambda.$$

It is convenient to express the energy $U_{\min}$ with the aid of the derivatives of free energy $F = F(T,\Lambda)$. In fact, using

$$-(\Delta T\,\Delta S + \Delta\lambda\,\Delta\Lambda) = \Delta T\,\Delta\left(\frac{\partial F}{\partial T}\right) - \Delta\left(\frac{\partial F}{\partial\Lambda}\right)\Delta\Lambda =$$
$$= \Delta T\left(\frac{\partial^2 F}{\partial T^2}\Delta T + \frac{\partial^2 F}{\partial T\partial\Lambda}\Delta\Lambda\right) - \left(\frac{\partial^2 F}{\partial\Lambda\partial T}\Delta T + \frac{\partial^2 F}{\partial\Lambda^2}\Delta\Lambda\right)\Delta\Lambda,$$

we obtain that

$$-\frac{U_{\min}}{T} = \Delta S = \frac{1}{2T}\frac{\partial^2 F}{\partial T^2}(\Delta T)^2 - \frac{1}{2T}\frac{\partial^2 F}{\partial\Lambda^2}(\Delta\Lambda)^2.$$

In the Gauss approximation, the expression $W \propto \exp\big(-U_{\min}/T\big)$ for the distribution function splits into two multipliers depending on ΔT and $\Delta\Lambda$ alone. As is expected, the fluctuations of temperature T and variable Λ prove to be statistically independent, i.e. $\langle \Delta T\, \Delta\Lambda\rangle = 0$. Accordingly, the variances of the temperature T and variable Λ will be equal to

$$\langle(\Delta T)^2\rangle = -\frac{T}{\partial^2 F/\partial T^2} = \frac{T}{\partial S/\partial T} = \frac{T^2}{C_\Lambda},$$
$$\langle(\Delta\Lambda)^2\rangle = \frac{T}{\partial^2 F/\partial\Lambda^2} = \frac{T}{\partial\lambda/\partial\Lambda} = T\left(\frac{\partial\Lambda}{\partial\lambda}\right)_T$$

where C_Λ is the specific heat at constant parameter Λ.

4.4 Phase Equilibrium and Phase Transitions

The various states of substance that can simultaneously be in the thermodynamic equilibrium with each others represent the different *phases* (or phase states) of the same substance. Let us proceed to analyzing the necessary conditions for the coexistence of two phases. First of all, for the thermal equilibrium, it is necessary to have an equality of temperatures of Phases 1 and 2

$$T_1 = T_2\,.$$

The equality of pressures in both phases results from the mechanical equilibrium at the interface between Phases 1 and 2

$$P_1 = P_2\,.$$

This implies that the forces which the phases act on each other are equal and opposite in direction.[4]

The condition of equilibrium with respect to the transfer of particles from one phase to the other and *vice versa* is expressed by the equality in the chemical potentials of Phases 1 and 2

$$\mu_1(P_1,\, T_1) = \mu_2(P_2,\, T_2)\,.$$

Denoting temperature T and pressure P common for both phases, we arrive at the following equation:

$$\mu_1\big(P(T),\, T\big) = \mu_2\big(P(T),\, T\big).$$

This equation determines implicitly the *phase equilibrium line* or *binodal* $P = P(T)$ at which both phases can coexist. Let us differentiate this equation with respect to temperature

$$\frac{\partial\mu_1}{\partial P}\frac{dP}{dT} + \frac{\partial\mu_1}{\partial T} = \frac{\partial\mu_2}{\partial P}\frac{dP}{dT} + \frac{\partial\mu_2}{\partial T} \quad \text{or} \quad v_1\frac{dP}{dT} - s_1 = v_2\frac{dP}{dT} - s_2.$$

[4] Strictly speaking, here we imply the flat interface and the phases at rest.

Here v_1, s_1 and v_2, s_2 are the specific volumes and entropies, i.e. per one particle of the phase. Introducing the notion of *latent heat* in the phase transition according to $L = T(s_2 - s_1)$, we arrive at the *Clausius–Clapeyron equation*

$$\frac{dP(T)}{dT} = \frac{L}{T(v_2 - v_1)}\,.$$

This equation specifies the pressure variation along the phase equilibrium line. The latent heat L is positive if the heat is absorbed in the course of the transition from phase 1 to phase 2, and $L < 0$ if the heat releases during this transition. Since the entropy of thermodynamically equilibrium system at $T \to 0$ should tend to a constant limit, the latent heat of the phase transition is expected to vanish at zero temperature.

If equation $\mu_1 = \mu_2$ has the solutions not for all temperatures, the phase equilibrium line $P = P(T)$ has an end-point called the *critical point*. There exist no two various phases beyond the critical point. The critical point is a singular point, at least, for one of chemical potentials μ_1 or μ_2. The temperature and pressure at the critical point are called, respectively, *critical temperature* T_c and *critical pressure* $P_c = P(T_c)$.

For the phase equilibrium of three various phases of the same substance, it is necessary to satisfy the equalities of temperatures $T_1 = T_2 = T_3$, pressures $P_1 = P_2 = P_3$, and chemical potentials $\mu_1 = \mu_2 = \mu_3$.[5] This results in the following equations:

$$\mu_1(P, T) = \mu_2(P, T) = \mu_3(P, T)$$

with the two unknowns P and T having the fixed pair of values P_t and T_t as solutions of the equation above. The point $(P_t\,, T_t)$, in which the three phases of the same substance are in the thermodynamic equilibrium, is referred to as the *triple point*. The triple point is the intersection one of phase equilibrium lines $P = P(T)$ for each two from three possible phases. (See, for example, Fig. 4.1). If the first-order phase transition line crosses over continuously to the two second-order phase transition lines at the triple point, such a point is referred to as the *tricritical* one. The phase equilibrium of four or more phases in the one-component system is impossible. In the multi-component systems, for example in the solutions or alloys, coexistence of four or more different phases becomes possible.

The phase transition is classified as *first-order phase transition* when the first-order derivatives of chemical potential $(\partial\mu/\partial T)_P = -s$ or $(\partial\mu/\partial P)_T = v$ demonstrate the jump-like behavior at the phase transition point, i.e. $s_1 \neq s_2$ or $v_1 \neq v_2$. As a rule, first-order phase transition means simply the transition at nonzero latent heat $L \neq 0$.

When first derivatives of chemical potentials are identical for two phases at the transition point, such a transition is usually referred to as *second-order phase*

[5] In the complex and multi-component systems, the chemical potential may depend on other thermodynamic variables as well.

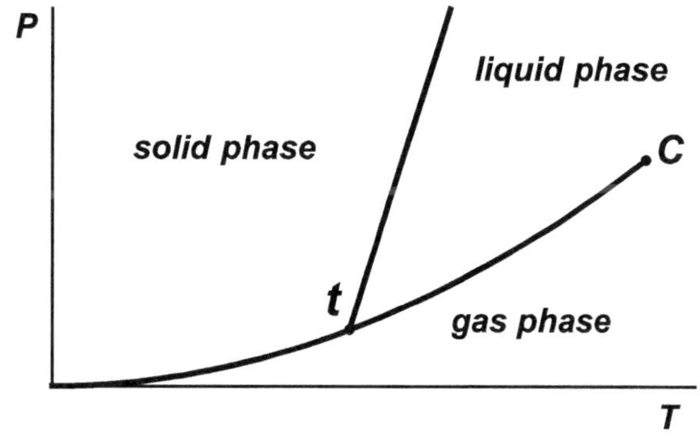

Fig. 4.1 The example of phase diagram for one-component substance. Here t is the triple point and C is the critical one

transition or *continuous phase transition*.[6] The specific anomalies, observed in the second-order phase transitions and associated with the divergence of second derivatives for chemical potentials at the phase transition point, are the subject of *critical phenomena*. These phenomena are connected, first of all, with anomalous growth of the thermodynamic fluctuations for a number of physical quantities in the phase transition region.

The phase transition existing at absolute zero is called the *quantum phase transition*. Such transition results from varying some physical parameter in the system or substance, which is different from temperature, e.g. pressure. The quantum phase transition represents the transition from one ground state of the system to another ground state. The value of parameter at which the phase transition occurs is called the *quantum critical point*.

Beyond the phase equilibrium line or spinodal $P = P(T)$ the choice of one or another phase of substance is governed with the energetic reasons alone. The absolutely steady or *stable phase* will be the phase with the minimum value of chemical potential. The phases with the larger values of chemical potential become unsteady or *unstable*. For specifying the type of instability and kind of phase transition, an important role plays the region for possible phase existence beyond the phase equilibrium line $P = P(T)$ and the position of the *boundary for the thermodynamic phase stability* or *spinodal* $P = P_s(T)$ associated also with violating one of thermodynamic inequalities $(\partial P/\partial v)_T < 0$ or $C_P = T(\partial s/\partial T)_P > 0$. In the region where these inequalities do not hold for, the corresponding phase becomes *absolutely unstable* and ceases to exist.

Two variants are possible here. The first, when the spinodal line is separated from the binodal, is typical for the first-order phase transitions. Then in the region of the parameters lying between the binodal and spinodal, the phase with the larger value of chemical potential may exist in the metastable state. The *metastable phase* represents the incomplete thermodynamic equilibrium and, therefore, has a finite lifetime by decaying into the stable phase via phase transition. Though the metastable phase is

[6] There are examples of continuous phase transitions of infinite order, e.g. Berezinskii–Kosterlitz–Thouless transition in the two-dimensional XY model.

unstable on the whole, it can remain stable against sufficiently small *heterogeneous fluctuations*. The existence of metastable region makes it possible for the *phenomena of overheating* or *overcooling* when the metastable phase may be in the *homogeneous* state for a sufficiently long time on the scale of experimental realization. The maximum possible limits of overheating or overcooling are confined with the spinodal.

The second option is when the spinodal line coincides with the binodal. In this case, the region of any existence for the phase with the smaller value of chemical potential is completely absent on the other side of phase transition line. There will be no phenomena of overheating or overcooling as well. Such a behavior is typical for second-order phase transitions.

If the line of first-order phase transition $P = P(T)$ ends at the critical point, the spinodal terminates at the same point as well.

4.5 Law of Corresponding States

The empirical Redlich–Kwong equation of state

$$P = \frac{NT}{V - Nb} - \frac{N^2 a}{\sqrt{T}\, V(V + Nb)},$$

as a rule, interpolates the equation of state for the real gas more precisely as compared with the traditional van der Waals equation of state at temperatures above the critical temperature. The Redlich–Kwong equation of state can also be employed for the qualitative comprehension of the gas–liquid phase transition. Like the van der Waals equation, there are fitting parameters in the Redlich–Kwong one. The constant a takes the attraction between gas particles into account. The constant b describes the repulsion resulting in effective limitation of the gas volume accessible to the particles.

Let us determine free energy of a gas $F(T, V)$ by integrating the relation $P = -\left(\partial F/\partial V\right)_T$. While integrating, we take into account that our result in the limit of large volume V should go over into the answer for an ideal gas. Then,

$$F = F_{\text{id}} - NT \ln \frac{V - Nb}{V} - \frac{Na}{b\sqrt{T}} \ln \frac{V}{V + Nb}$$

where F_{id} is the free energy of an ideal gas. Hence we get the entropy of the gas

$$S = S_{\text{id}} + N \ln \frac{V - Nb}{V} - \frac{Na}{2bT^{3/2}} \ln \frac{V}{V + Nb}$$

and then its energy $E = F + TS$ as

$$E = E_{\text{id}} - \frac{Na}{2b\sqrt{T}} \ln \frac{V}{V + Nb}.$$

Unlike the van der Waals gas, the specific heat $C_V = \left(\partial E/\partial T\right)_V$ in this case does not equal that of ideal gas and depends on the gas density as well:

$$C_V = C_{\text{id}}(T) + \frac{1}{4}\frac{Na}{bT^{3/2}} \ln \frac{V}{V+Nb}\,.$$

The critical temperature, pressure, and volume can be expressed via parameters a and b. So, it is necessary to write the following equations:

$$\left(\frac{\partial P}{\partial V}\right)_T = 0 \quad \text{and} \quad \left(\frac{\partial^2 P}{\partial V^2}\right)_T = 0.$$

Together with the equation of state $P = P(T, V)$, the above equations determine the infection point at the isotherm. As a result, we arrive at the equation determining the critical volume

$$x^3 - 3x^2 - 3x - 1 = 0 \quad \text{where} \quad x = V/(Nb).$$

We find the root $x = 1/(\sqrt[3]{2} - 1)$ with the aid of substitution $x = 1/(y - 1)$. Finally, the critical values of volume, temperature, and pressure are equal to

$$V_c = \frac{Nb}{\sqrt[3]{2}-1}, \quad T_c = \left(3(\sqrt[3]{2}-1)^2\frac{a}{b}\right)^{2/3},$$
$$P_c = \frac{(\sqrt[3]{2}-1)T_c}{3b} = \frac{(\sqrt[3]{2}-1)^{7/3}}{3^{1/3}}\frac{a^{2/3}}{b^{5/3}}\,.$$

The *Boyle temperature* T_b is defined as the temperature for which the second *virial coefficient* becomes zero in the decomposition of the equation of state in the powers of gas density

$$P = \frac{NT}{V}\left(1 + \frac{NB_2(T)}{V} + \frac{N^2B_3(T)}{V^2} + \cdots\right).$$

Decomposing the equation of state in N/V, we find the Boyle temperature

$$T_b = \left(\frac{a}{b}\right)^{2/3} = \left(3(\sqrt[3]{2}-1)^2\right)^{-2/3} T_c \approx 2.90\,T_c\,.$$

Approximating the equation of state with the Redlich–Kwong one is satisfactory in the region of moderate pressures and high temperatures: $P/P_c \lesssim T/2T_c$.

Let us introduce the reduced temperature, pressure, and volume according to the relations

$$t = \frac{T}{T_c} \quad p = \frac{P}{P_c}, \quad v = \frac{V}{V_c}\,.$$

The Redlich–Kwong equation, expressed in terms of these reduced quantities, takes the form

$$p = \frac{3t}{v-c} - \frac{1}{c\sqrt{t}}\frac{1}{v(v+c)} \quad \text{where} \quad c = \sqrt[3]{2} - 1.$$

This equation involves the reduced quantities alone and no other ones that characterize the given substance. The states of substances having the same reduced quantities are called the *corresponding states. The law of corresponding states* implies that, if two substances have the same two of three reduced quantities, these two substances have the same third quantity as well. The isotherms in the reduced quantities will be the same for all such substances. Note that the law of corresponding states results from the equations of states containing the only two parameters (a and b). In the reduced units the free energy $f = F/NT_c$ reads

$$f(t, v) = f_{\rm id}(t, v) - t\ln\frac{v-c}{v} + \frac{1}{3c^2}\frac{1}{\sqrt{t}}\ln\frac{v}{v+c}.$$

The isotherms determined with the Redlich–Kwong equation of state are in the qualitative correspondence with the isotherms of the interpolating van der Waals equation and are represented in Fig. 4.2. At high $T > T_c$ temperatures, the isotherms pass over the critical point C and represent the monotonically decreasing functions $P = P(V)$ as the gas volume increases. For the temperatures below the crit-

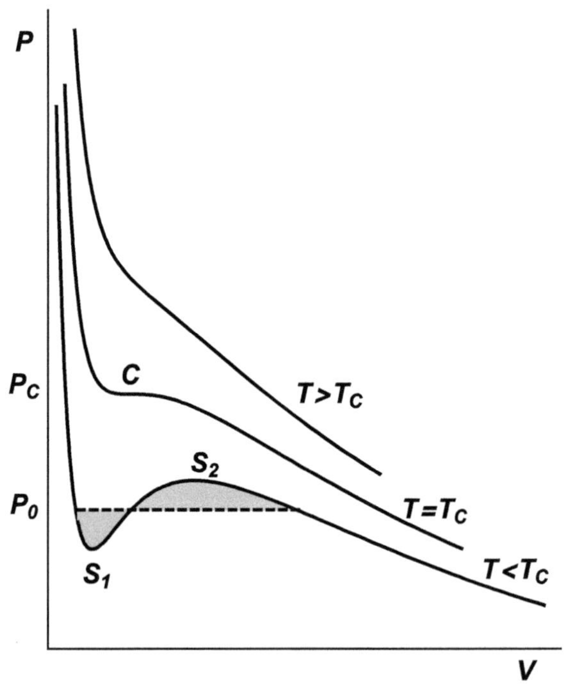

Fig. 4.2 The behavior of isotherms at various temperatures. Here P_0 is the phase transition pressure and C is the critical point. The segment S_1S_2 is absolutely unstable. The shaded regions are of equal area

ical $T < T_c$ one, each isotherm demonstrates the points of minimum and maximum. Between these points, we have a region of thermodynamic instability for the homogeneous state of substance since inequality $(\partial P/\partial V)_T < 0$ is violated. The points of minimum and maximum at the isotherm line represent the points of the spinodal.

The instability of the homogeneous state results in the appearing of the heterogeneous two-phase state of the system. We characterize such a two-phase state by the phase transition pressure $P = P_0(T)$. At the fixed temperature in Fig. 4.2, this corresponds to the horizontal line segment whose position should satisfy the equality of chemical potentials. The region of the isotherm between the transition pressure (binodal) and the spinodal points is the region of the metastable state. The phase diagram of thermodynamically stable, metastable, and absolutely unstable states is schematically shown in Fig. 4.3.

Let us consider the integral within the limits between the intersection points of the horizontal segment and the isotherm

$$\int_{V_1}^{V_2} P(T,V)\,dV = -\int_{V_1}^{V_2} \frac{\partial F(T,V)}{\partial V} dV = -(F_2 - F_1)$$

where F_1 and F_2 are the free energy of phases. Involving the relation $F_{1,2} = \Phi_{1,2} - P_0(T)V_{1,2}$ between free energies and thermodynamic potentials and using the equality of thermodynamic potentials $\Phi_1 = \Phi_2$ at the phase equilibrium line, we arrive at the following equation:

$$\int_{V_1}^{V_2} P(T,V)\,dV = P_0(T)(V_2 - V_1).$$

The geometrical interpretation of this relation is called the *Maxwell equal area rule* and means the equal areas of two shaded regions limited with the parts of the

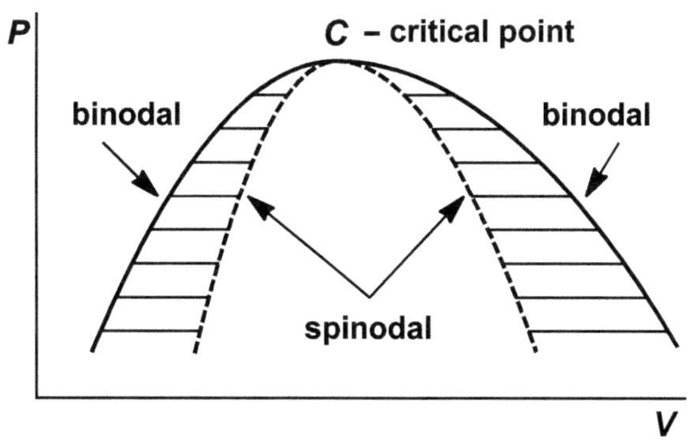

Fig. 4.3 The diagram of phase states. The dashed region is the region of metastable states for which the phenomena like overheating or overcooling are possible. The region under the spinodal line is absolutely unstable

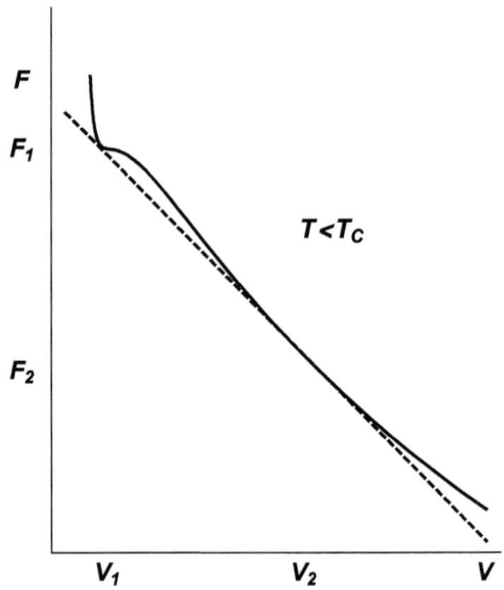

Fig. 4.4 The free energy F as a function of volume V. The slope of common tangent at points V_1 and V_2 yields the phase transition pressure P_0

isotherms lying on the different sides from the horizontal line of phase equilibrium. These shaded regions are shown in Fig. 4.2.

There exists another graphical determination of phase transition point, which is known as *Maxwell common tangent construction*. Let temperature T be lower than the critical one T_c. Let us consider the behavior of free energy $F(T, V)$ as a function of volume V, which is given in Fig. 4.4. Then we will show that the common double tangent to the free energy line determines the phase volumes V_1 and V_2 corresponding to the phase transition pressure $P_0(T)$. The derivative

$$\left(\frac{\partial F}{\partial V}\right)_T = -P$$

and the phase equilibrium condition $P(T, V_1) = P(T, V_2) = P_0(T)$ imply that the slope of line $F(T, V)$ at volume $V = V_1$ is the same as for $V = V_2$. The common tangent at points V_1 and V_2 leads to the following relation:

$$\frac{F(T, V_1) - F(T, V_2)}{V_1 - V_2} = -P_0(T),$$

entailing the necessary phase equilibrium condition as an equality of thermodynamic potentials

$$F(T, V_1) + P_0 V_1 = F(T, V_2) + P_0 V_2\,.$$

The inflection points at line $F(T, V)$, where $\left(\partial^2 F/\partial V^2\right)_T = 0$, separate the metastable states from absolutely unstable ones.

Problems

1. The van der Waals gas in the thermal equilibrium and critical conditions is in the homogeneous field of gravity. Find the distribution of gas density ρ over the altitude.

Solution. Let g be acceleration of force of gravity and z be vertical coordinate. The equation of mechanical equilibrium $dP/dz = -\rho g$ for the gas at constant temperature, $P = P(\rho)$ being the gas pressure, can readily be integrated as

$$\int\limits_{\rho_c}^{\rho} \frac{1}{\rho}\frac{dP}{d\rho}d\rho = -\int\limits_{0}^{z} g\,dz = -gz.$$

Here $\rho_c = m/3b$ is the gas density at the critical point. The substitution of the van der Waals equation of state

$$P = \frac{T}{v-b} - \frac{a}{v^2},$$

where $v = m/\rho$ is the specific volume and m is the particle mass, will give the equation for determining $v(z)$ as

$$\frac{bT}{v-b} - \frac{bT}{v_c-b} - \frac{2a}{v} + \frac{2a}{v_c} - T\ln\frac{v-b}{v_c-b} = -mgz.$$

Putting the temperature T equal to the critical one $T_c = 8a/27b$ and decomposing the left-hand side of equation at small variations of specific volume in the vicinity of critical one $v_c = 3b$, we obtain

$$-\frac{9}{16}T_c\left(\frac{v-v_c}{v_c}\right)^3 + \cdots = -mgz.$$

This results in the specific barometric behavior of gas density under critical conditions

$$\rho(z) - \rho_c = -\rho_c\left(\frac{16}{9}\frac{mgz}{T_c}\right)^{1/3} \quad \text{for} \quad z < \frac{T_c}{mg}$$

instead of usual linear behavior $\Delta\rho \sim -\rho(mgz/T_c)$. For the region of large $z \gg T_c/mg$ altitudes, the effect of closeness to the critical point reduces, and the decrease of gas density approaches the usual behavior described with the exponential barometric formula.

2. Find the latent heat $L(T)$ of gas–liquid transition for the van der Waals gas near the critical point.

Solution. Let us go over to the reduced variables as temperature $t = T/T_c$, pressure $p = P/P_c$, volume $v = V/V_c$, and chemical potential $\tilde{\mu} = \mu/T_c$. The pressure of transition $p = p(t)$, volumes of gas v_g, and liquid v_l are determined with two equations

$$p(t, v_g) = p(t, v_l) \quad \text{and} \quad \tilde{\mu}(t, v_g) = \tilde{\mu}(t, v_l).$$

To find the phase transition line, it is convenient to introduce the following parametrization:

$$s(t, v_g) - s(t, v_l) = \ln\frac{3v_g - 1}{3v_l - 1} = 2x \quad \text{or} \quad 3v_g - 1 = \frac{e^x}{f(x)} \quad \text{and} \quad 3v_l - 1 = \frac{e^{-x}}{f(x)}$$

where $s(t, v) = S/N$ is the magnitude of entropy per one particle. Unknown function $f(x)$ should be specified from two equations determining the phase transition

$$\frac{8t}{3v_g - 1} - \frac{3}{v_g^2} = \frac{8t}{3v_l - 1} - \frac{3}{v_l^2},$$

$$-t\ln(3v_g - 1) + \frac{t}{3v_g - 1} - \frac{9}{4v_g} = -t\ln(3v_l - 1) + \frac{t}{3v_l - 1} - \frac{9}{4v_l}.$$

We find from these two equations

$$t(x) = \frac{27}{4}\frac{[f(x) + \cosh x]f(x)}{[f^2(x) + 2f(x)\cosh x + 1]^2} \quad \text{and} \quad f(x) = 2\frac{x\cosh x - \sinh x}{\sinh 2x - 2x}$$

which determine implicitly the following characteristics of phase transition with the aid of function $t = t(x)$:

$$p(t) = \frac{27[1 - f^2(x)]f^2(x)}{[f^2(x) + 2f(x)\cosh x + 1]^2}, \quad \frac{dp}{dt} = \frac{16x(x\coth x - 1)}{\sinh 2x - 2x}, \quad v_g(t) - v_l(t) = \frac{2\sinh x}{3f(x)}.$$

At the critical point $t = 1$, $v_g = v_l = 1$, and near, it we can employ the decomposition of the above expressions for small $x \ll 1$

$$f(x) \approx \frac{1}{2}\left(1 - \frac{x^2}{10}\right), \quad t(x) \approx 1 - \frac{x^2}{9}, \quad v_g - v_l \approx \frac{4x}{3} \quad \text{and} \quad \frac{dp}{dt} \approx 4.$$

Next, we find according to relation $x(t) \approx 3(1 - t)^{1/2}$ that

$$v_g \approx 1 + 2(1 - t)^{1/2}, \quad v_l \approx 1 - 2(1 - t)^{1/2}, \quad s_g - s_l \approx 6(1 - t)^{1/2},$$
$$l(t) = t(v_g - v_l)\frac{dp}{dt} \approx 16(1 - t)^{1/2}.$$

Returning to the dimensional quantities according to $L(T) = P_cV_cl(t)$ or $L(T) = NT_ct(s_g - s_l)$, we get the following behavior for the latent heat of phase transition near critical temperature T_c:

$$L(T) = 6NT_c\left(1 - \frac{T}{T_c}\right)^{1/2},$$

N being the particle number in the system. The latent heat of phase transition vanishes at the critical point.

3. Find the isothermal fluctuations of the critical volume in the van der Waals gas.

Solution. At first, we calculate the following derivatives of pressure:

$$P = \frac{NT}{V - Nb} - \frac{N^2a}{V^2}$$

with respect to the gas volume V at its critical magnitude $V_c = 3Nb$:

$$\frac{\partial P}{\partial V} = \frac{2N^2a}{V^3} - \frac{NT}{(V - Nb)^2}\bigg|_{V=V_c} = \frac{6P_c}{V_c}\frac{T_c - T}{T_c},$$
$$\frac{\partial^2 P}{\partial V^2} = \frac{2NT}{(V - Nb)^3} - \frac{6N^2a}{V^4}\bigg|_{V=V_c} = \frac{18P_c}{V_c^2}\frac{T - T_c}{T_c},$$
$$\frac{\partial^3 P}{\partial V^3} = \frac{24N^2a}{V^5} - \frac{6NT}{(V - Nb)^4}\bigg|_{V=V_c} = -\frac{9P_c}{V_c^3}\frac{9T - 8T_c}{T_c}.$$

Here T_c is the critical temperature. These derivatives allow us to calculate the following fluctuations $\Delta V = V - V_c$ of critical volume:

$$\frac{\langle(\Delta V)^2\rangle}{V_c^2} = \frac{1}{6}\frac{T_c}{P_cV_c}\frac{T}{T - T_c} = \frac{4}{9N}\frac{T}{T - T_c},$$
$$\frac{\langle(\Delta V)^3\rangle}{V_c^3} = \frac{1}{12}\frac{T_c^2}{P_c^2V_c^2}\left(\frac{T}{T - T_c}\right)^2 = \frac{16}{27N^2}\left(\frac{T}{T - T_c}\right)^2,$$

$$\frac{\langle(\Delta V)^4\rangle}{V_c^4} = \frac{1}{144}\frac{T_c^3}{P_c^3 V_c^3}\frac{T^3(9T-10T_c)}{(T-T_c)^4} + \frac{1}{12}\frac{T_c^2}{P_c^2 V_c^2}\frac{T^2}{(T-T_c)^2} =$$
$$= \frac{32}{243N^3}\frac{T^3(9T-10T_c)}{(T-T_c)^4} + \frac{16}{27N^2}\frac{T^2}{(T-T_c)^2}.$$

As the temperature approaches its critical magnitude, all the volume fluctuations enhance unlimitedly. Note that fluctuation $\langle(\Delta V)^4\rangle$ becomes negative in the immediate closeness $(T-T_c)/T_c \lesssim (3\sqrt{N})^{-1}$ to the critical point. This indicates the limited application of the van der Waals equation for describing the critical properties of a gas near the critical point.

4.6 Thermodynamics of Solutions

In physics, we keep in mind that a solution is a spatially homogeneous mixture composed of two or more substances called the components. The homogeneity of the solution means that its components constitute a single-phase system and such physical characteristics as temperature, pressure, density, and concentration do not vary over the volume of the solution. In addition, diffusion and other possible energy dissipation processes are completely absent. In other words, here we will study the solutions in the complete thermodynamic equilibrium. A *solute* is a substance dissolved in another substance of larger amount, the latter being usually known as a *solvent*. There exist various aggregate states of solutions, e.g. gaseous, liquid, or solid mixtures.

The important parameter of solution is its *concentration* which is a measure of the amount of solute in a given amount of solvent. Commonly, the concentration is defined as a ratio of particle number of solute to the total number of particles in the solution. The mass concentration implies a ratio of the solute mass to the total mass of solution. The physical properties of a solution, e.g. points of melting, boiling, solidification, or phase separation, become dependent on the concentration.

Let us consider two-component or binary solution with pressure P, temperature T, and particle numbers N_1 and N_2 for each of components. We can write the differential of thermodynamic Gibbs potential $\Phi = \Phi(P, T, N_1, N_2)$ as

$$d\Phi = -S\,dT + V\,dP + \mu_1\,dN_1 + \mu_2\,dN_2.$$

The chemical potentials, introduced as

$$\mu_{1,2} = \left(\frac{\partial\Phi}{\partial N_{1,2}}\right)_{P,T},$$

depend on the pressure, temperature, and concentration. From an additivity of entropy, i.e. from equality $S(kE, kV, kN_1, kN_2) = kS(E, V, N_1, N_2)$, it follows the relation between the thermodynamic Gibbs potential and the chemical potentials of the components in the solution:

$$\Phi = \mu_1 N_1 + \mu_2 N_2\,.$$

In fact, differentiating the entropy with respect to k and putting $k = 1$, we have

$$S = \frac{\partial S}{\partial E}E + \frac{\partial S}{\partial V}V + \frac{\partial S}{\partial N_1}N_1 + \frac{\partial S_2}{\partial N_2}N_2 = \frac{E}{T} + \frac{P}{T}V + \frac{\mu_1}{T}N_1 + \frac{\mu_2}{T}N_2 \, .$$

Hence, $E - TS + PV = \Phi = \mu_1 N_1 + \mu_2 N_2$. Comparing two differentials

$$\begin{aligned} -S\,dT + V dP + \mu_1 dN_1 + \mu_2 dN_2 = \\ = d(\mu_1 N_1 + \mu_2 N_2) = \mu_1 dN_1 + \mu_2 dN_2 + N_1 d\mu_1 + N_2 d\mu_2 \end{aligned}$$

gives the *Gibbs–Duhem equation* for a solution

$$N_1\, d\mu_1 + N_2\, d\mu_2 = -S\,dT + V dP.$$

Let us call a ratio of the particle number N_1 of dissolved substance to the total number $N = N_1 + N_2$ of the particles in the solution as the *molar concentration* or simply the concentration, i.e.

$$x = \frac{N_1}{N_1 + N_2}.$$

While studying the thermodynamics of solution, we should take into account that the differentials determining different thermodynamic potentials, e.g. energy or entropy, must involve an additional variable as concentration x. For convenience, we introduce new thermodynamic potentials μ and Z. These potentials completely characterize the thermodynamic properties of solution. So, we define

$$\mu = x\mu_1 + (1 - x)\mu_2 \quad \text{and} \quad Z = \mu_1 - \mu_2.$$

The Gibbs–Duhem equation for solution, written in thermodynamic potentials μ and Z, takes the simple and clear form

$$d\mu = -sdT + vdP + Zdx.$$

Here $s = S/N$ is the entropy per one particle of solution and the specific volume $v = V/N$ is the volume per one particle of solution. This equation is a differential of chemical potential $\mu = \mu(P, T, x)$. We should define potential $Z = Z(P, T, x)$ with the aid of derivative

$$Z = \left(\frac{\partial \mu}{\partial x}\right)_{P,T}.$$

It is readily to check that the conjugate thermodynamic potential

$$\phi = \mu - xZ = x\mu_1 + (1 - x)\mu_2 - x\big(\mu_1 - \mu_2\big) = \mu_2$$

represents the chemical potential μ_2 of solvent. The corresponding differential equals

$$d\phi = d\mu - xdZ - Zdx = -sdT + vdP - xdZ$$

and, therefore, $\phi = \phi(P, T, Z)$. Because of the type of differentials, one can conclude that a pair of thermodynamic quantities x and $-Z$ is the *conjugate* thermodynamic variables and the pairs (x, T) and (x, P) are independent. In other words, the fluctuations of concentration and temperature as well as the fluctuations of concentration and pressure are statistically independent. The fluctuations of concentration x and potential Z prove to be statistically dependent. Correspondingly,[7]

$$\langle \Delta T \Delta x \rangle = 0, \quad \langle \Delta P \Delta x \rangle = 0 \quad \text{and} \quad \langle \Delta Z \Delta x \rangle = T.$$

For the variance of concentration fluctuations, we obtain that

$$\langle (\Delta x)^2 \rangle = \frac{T}{\partial Z/\partial x}.$$

The positivity of inequality $\partial Z/\partial x > 0$ is necessary for the thermodynamic stability of solution with respect to the separation into two phases and determines the line of absolute instability or spinodal in the space of parameters P, T, and x.

Let us calculate the differential of thermodynamic potential $Z = Z(P, T, x)$. We have

$$dZ = \left(\frac{\partial Z}{\partial P}\right)_{x,T} dP + \left(\frac{\partial Z}{\partial T}\right)_{x,P} dT + \left(\frac{\partial Z}{\partial x}\right)_{P,T} dx.$$

Using the thermodynamic identities for the derivatives

$$\left(\frac{\partial Z}{\partial P}\right)_{x,T} = \frac{\partial^2 \mu}{\partial P \partial x} = \left(\frac{\partial v}{\partial x}\right)_{P,T} \quad \text{and} \quad \left(\frac{\partial Z}{\partial T}\right)_{x,P} = \frac{\partial^2 \mu}{\partial T \partial x} = -\left(\frac{\partial s}{\partial x}\right)_{P,T},$$

we find straightforwardly that

$$dZ = \left(\frac{\partial v}{\partial x}\right)_{P,T} dP - \left(\frac{\partial s}{\partial x}\right)_{P,T} dT + \left(\frac{\partial Z}{\partial x}\right)_{P,T} dx.$$

On varying the concentration, temperature, and pressure, the one-phase state of solution may prove to be unstable and the solution will separate into two phases with various concentrations $x(P, T)$ and $x'(P, T)$, in general, pressure- and temperature-dependent. The conditions for the thermodynamic equilibrium between various phases of solution are the equality of pressures, temperatures, and chemical potentials

[7] Let us remind the correlations of conjugate variables $\langle \Delta v \Delta P \rangle = -T$ and $\langle \Delta s \Delta T \rangle = T$.

$$\phi'(P', T', Z') = \phi(P, T, Z), \qquad P' = P,$$
$$Z'(P', T', x') = Z(P, T, x), \qquad T' = T.$$

These equalities determine the phase equilibrium surface at the phase diagram in the following variables: concentration, temperature, and pressure. We have on the phase equilibrium surface

$$(v' - v)dP - (s' - s)dT - (x' - x)dZ = 0 \quad \text{or}$$
$$\left[v' - v - (x' - x)\frac{\partial v}{\partial x}\right]dP - \left[s' - s - (x' - x)\frac{\partial s}{\partial x}\right]dT - (x' - x)\frac{\partial Z}{\partial x}dx = 0$$

where v and v' are the specific volumes for the various phases of solution. From the last equation, we obtain the equations for the phase equilibrium lines $P = P(x, T)$ and $T = T(x, P)$

$$\left(v' - v - (x' - x)\frac{\partial v}{\partial x}\right)\left(\frac{\partial P}{\partial x}\right)_T = (x' - x)\frac{\partial Z}{\partial x},$$
$$\left(v' - v - (x' - x)\frac{\partial v}{\partial x}\right)\left(\frac{\partial P}{\partial T}\right)_x = s' - s - (x' - x)\frac{\partial s}{\partial x}.$$

In general, due to presence of interaction between the solute and solvent particles, it is a difficult problem to derive the analytical expressions for the chemical potentials of solution. *Classical ideal solution*, using the expression for the chemical potential of ideal gas

$$\mu = x\big[\mu_{10}(P, T) + T\ln x\big] + (1 - x)\big[\mu_{20}(P, T) + T\ln(1 - x)\big],$$
$$Z = \big[\mu_{10}(P, T) + T\ln x\big] - \big[\mu_{20}(P, T) + T\ln(1 - x)\big].$$

We have for the density of entropy in such a solution

$$s = -\frac{\partial \mu}{\partial T} = xs_{10} + (1 - x)s_{20} - x\ln x - (1 - x)\ln(1 - x).$$

The last two terms in the formula are called the *entropy of mixing*.

In order to demonstrate a few effects of interaction between various particles in the solution, we employ the symmetrical model of *regular solution*[8] with the following chemical potential:

$$\mu(P, T, x) = x\big[\mu_0(P, T) + T\ln x\big] +$$
$$+ (1 - x)\big[\mu_0(P, T) + T\ln(1 - x)\big] + \Theta x(1 - x).$$

[8] For example, this model is applied for solid ^{3}He-^{4}He mixtures. The volume of solid mixture is calculated as $V_{34} = xV_3 + (1 - x)V_4 - V_\Theta x(1 - x)$ where V_3 and V_4 are the volumes of solid ^{3}He and ^{4}He, volume V_Θ being the magnitude of disturbance for additivity of volumes.

The last term with $\Theta = \Theta(P) > 0$ describes the effects of interaction between the particles and represents the excess energy of solution.

We start from constructing the phase diagram for the phase separation of the solution in the temperature-concentration variables. Its construction can be performed similarly to the Maxwell double tangent construction for the gas–liquid transition, using the analogies of potential $\mu(T, x)$ with free energy $F(T, V)$, potential Z with the negative-sign pressure, and concentration x with volume V.

Let us analyze behavior $\mu(T, x)$ as a function of concentration. The general double tangent to line $\mu(T, x)$ will determine two concentrations x and x' corresponding to potential $Z_0(T)$ at which the phase transition can occur. The derivative $\left(\partial\mu/\partial x\right)_T = Z$ and phase equilibrium condition $Z(T, x) = Z(T, x') = Z_0(T)$ are independent of concentrations x and x' and, accordingly, imply that the slope of line $\mu(T, x)$ at x is the same as for x'. The general tangent at points x and x' leads to the equation

$$\frac{\mu(T, x) - \mu(T, x')}{x - x'} = Z_0(T),$$

resulting in the necessary equilibrium condition

$$\mu(T, x) - xZ_0(T) = \mu(T, x') - x'Z_0(T).$$

The derivative $\partial\mu/\partial x$ as well as the potential Z are straightforwardly calculated as

$$\partial\mu/\partial x = Z(P, T, x) = \Theta\big(1 - 2x\big) - T\ln\big[(1 - x)/x\big].$$

Due to symmetry of potential with respect to replacement $x \to (1 - x)$, the general tangent is possible only provided that $\partial\mu/\partial x = 0$, resulting in $Z_0(T) = 0$. The last condition will give the temperature

$$T_c(x) = \Theta\frac{1 - 2x}{\ln(x^{-1} - 1)}$$

of phase transition as separating the solution into two phases with different concentrations $x = x(T)$ and $x' = x'(T)$. The concentrations should be found from solving the equation $T = T_c(x)$ (Fig. 4.5). The maximum temperature of phase separation equals $\Theta/2$. Above this temperature, the solution will no longer separate at any concentration.

The inflection point at line $\mu(T, x)$, where $\partial^2\mu/\partial x^2 = \partial Z/\partial x = 0$, separates the metastable states of solution from the absolutely unstable ones and determines the spinodal line

$$T_s(x) = 2\Theta x(1 - x).$$

The maximum of temperature T_s equals $\Theta/2$ as well. The point $(T = \Theta/2,\ x = 1/2)$ in the phase diagram is the critical one (Fig. 4.5). At $T = 0$ the minimum of

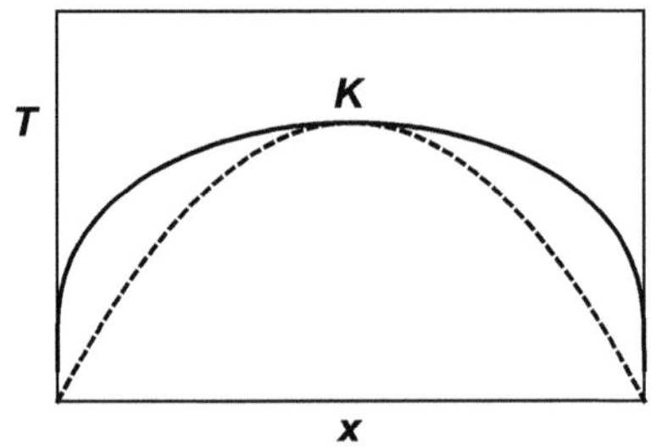

Fig. 4.5 The phase diagram for the phase separation of regular solution. The solid line, *binodal*, at which the phase separation takes place into two phases with the different concentrations $x(T)$ and $x'(T)$. The dashed line is the *spinodal*. The states between the solid and dashed lines are metastable. The region below the dashed line is absolutely unstable against the phase separation into two phases. The point K is the critical one

interaction energy realizes at concentration $x = 0$ or $x' = 1$. For $T \gg \Theta$, the energy of interaction between the components of solution becomes insignificant.

Osmosis is one of the physical phenomena that can be used to determine the chemical potential of a solvent as a function of the concentration of a solute. Let us assume that there is a semipermeable membrane separating the solution and the pure solvent into two parts since the solvent particles alone can pass through the membrane. The pressure on both sides of the membrane proves to be different. The pressure difference balancing the chemical potentials of the solution and pure solvent is called the *osmotic pressure*.

The condition for equilibrium against an exchange of solvent particles across the membrane will be an equality for the chemical potentials of solvent on both sides of membrane. Denoting the pressure in the solution as P and the pressure in the pure solvent as $P - \Pi$ where $\Pi = \Pi(P, T, x)$ is the osmotic pressure, we can write the equilibrium condition

$$\mu_2(P, T, x) = \mu_{20}(P - \Pi(P, T, x), T)$$

or

$$\phi_{20}(P - \Pi(P, T, x), T) = \phi_2(P, T, x).$$

This equation allows us to determine the chemical potential of solvent as a function of the solution concentration if we know the chemical potential of pure solvent and measure the osmotic pressure $\Pi = \Pi(P, T, x)$ as a function of P, T, and x.

Differentiating the last equation, respectively, over x, T, and P, one can obtain the following three equalities. The first reads

$$\left(\frac{\partial \Pi}{\partial x}\right)_{P,T} = \frac{x}{v_{20}}\left(\frac{\partial Z}{\partial x}\right)_{P,T},$$

$v_{20} = v_{20}(P - \Pi, T)$ being the specific volume of pure solvent at pressure $P - \Pi$. This equation is useful because it allows us to determine the derivative $\partial Z/\partial x$. Its value is essential for determining the condition for the phase separation of solution. Vanishing of the derivative specifies the spinodal, i.e. line of absolute instability.

The second equality is given by

$$\left(\frac{\partial \Pi}{\partial T}\right)_{P,x} = \frac{\bar{s} - s_{20}}{v_{20}}.$$

Here $s_{20} = s_{20}(P - \Pi, T)$ is the entropy of pure solvent per particle and $\bar{s} = s - x\partial s/\partial x$ is an "adjusted" entropy per particle. This equality allows us to estimate the temperature behavior of osmotic pressure.

The third equality reduces to

$$\left(\frac{\partial \Pi}{\partial P}\right)_{T,x} = 1 - (1 - \beta)\frac{v}{v_{20}} \quad \text{where} \quad \beta = \frac{x}{v}\frac{\partial v}{\partial x} = \frac{\partial \ln v}{\partial \ln x}.$$

As usual, derivative $\partial \Pi/\partial P$ is small at low concentrations. Provided that $v_{20}(P - \Pi) \approx v_{20}(P)$ and molar volume of solution $v(P, T, x) = v_{20}(P, T)\big(1 + x\alpha(P, T)\big)$, the derivative $\partial \Pi/\partial P$ equals zero identically.

Problems

1. Estimate the osmotic pressure for an ideal mixture of two classical gases, using the expression for the entropy of mixing.

Solution. Let us write the entropy density of ideal classical mixture with first-component concentration equal to x as

$$s = xs_{10} + (1 - x)s_{20} - x\ln x - (1 - x)\ln(1 - x).$$

Then we use that the temperature derivative of osmotic pressure equals

$$\partial \Pi/\partial T = n_{20}\big[(s - x\partial s/\partial x) - s_{20}\big] = -n_{20}\ln(1 - x).$$

Assuming the density n_{20} for the main component of mixture to be constant, we find that

$$\Pi(T) = -n_{20}T\ln(1 - x) \quad \text{and for } x \ll 1 \quad \Pi(T) \approx xn_{20}T = n_1 T,$$

n_1 being the density of first component. Here we have taken into account that the pressure of ideal gases vanishes at $T = 0$. The expression $\Pi = n_1 T$ derived at $x \ll 1$ is called the *van 't Hoff formula* and has nominally the same form as the formula for the pressure of classical ideal gas.

2. The small amount of ^{3}He atom impurities in superfluid ^{4}He can be described as an ideal Fermi gas of impuritons with dispersion $\epsilon_{\boldsymbol{p}} = -\Delta + \boldsymbol{p}^2/2M$. The dependence of the dissolution energy Δ and effective mass M on the ^{4}He density can be neglected in first approximation.

The capillary (so-called *superleak*) plays a role of membrane which passes the superfluid ^{4}He fraction of liquid mixtute though but prevents the ^{3}He impurities (nornal fluid fraction) from the penetration across the membrane. Determine the magnitude of osmotic pressure between pure superfluid ^{4}He and dilute liquid ^{3}He–^{4}He mixture at zero temperature.

Solution. Let us write chemical potential of ^{3}He atoms with density n_3

$$\mu_3 = -\Delta + p_F^2/2M, \quad p_F = \hbar(3\pi^2 n_3)^{1/3}.$$

Here p_F is the Fermi momentum of ^{3}He with density n_3. For small concentration $x \ll 1$ of ^{3}He atoms, their density equals $n_3 = xn_{40}$ where n_{40} is the density of pure ^{4}He. As it concerns the potential $Z = \mu_3 - \mu_4$, one can neglect a possible dependence μ_4 upon concentration in the $x \ll 1$ limit and put approximately $\mu_4(x) \approx \mu_4(0) = \mu_{40}$. Then,

$$\frac{\partial \Pi}{\partial x} = \frac{x}{v_{40}} \frac{\partial Z}{\partial x} = n_{40} x \frac{\partial \mu_3(x)}{\partial x} = \frac{2}{3} n_{40} \frac{p_F^2(x)}{2M} \sim x^{2/3}$$

where specific volume v_{40} and superfluid ^{4}He density n_{40} are related as $n_{40} = 1/v_{40}$. Finally, we find the osmotic pressure for $x \ll 1$

$$\Pi(x) = n_{40} \int_0^x x \frac{\partial \mu_3(x)}{\partial x} dx = \frac{2}{5} \frac{p_F^2}{2M} x n_{40} = \frac{2}{5} \frac{p_F^2}{2M} n_3 = (3\pi^2)^{2/3} \frac{\hbar^2 n_{40}^{5/3}}{5M} x^{5/3}.$$

This magnitude is approximately equal to the pressure of ideal Fermi gas with density $n_3 = xn_{40}$.

3. Find the low temperature behavior of osmotic pressure under conditions of the previous problem.

Solution. Let us employ the following relation:

$$\frac{\partial \Pi}{\partial T} = n_{40} \left(s - x \frac{\partial s}{\partial x} - s_{40} \right)$$

where s is the entropy density of liquid mixture and s_{40} is the entropy density of superfluid ^{4}He. In the low $x \ll 1$ concentration limit, we estimate the entropy of liquid mixture as a sum of pure superfluid ^{4}He and contribution from the ideal gas of *impuritons*, i.e. $s \approx s_{40} + s_3(x)$. Next, from

$$\partial \Pi / \partial T \approx (2/3) n_{40} s_3(x) \sim T x^{1/3}, \quad T \ll T_F(x),$$

we have involved that $x\, \partial s_3 / \partial x = s_3/3$ and $T_F(x)$ is the degenerate temperature for the Fermi gas with density $n_3 = xn_{40}$. Finally,

$$\Pi(T) = \Pi_0(x) + \Pi_1 T^2 \quad \text{and} \quad \Pi_1(x) \sim \Pi_0(x)/T_F^2(x).$$

4.7 Second-Order Phase Transitions

The second-order phase transitions are currently treated from the viewpoint of a spontaneous symmetry breaking. An *order parameter* should be introduced for the quantitative description of the degree of symmetry breaking. The order parameter normally vanishes in one phase (usually above the phase transition point) and increases smoothly in the other phase by getting away from the phase transition point (usually to lower temperatures). The physical properties and mathematical interpretation of order parameter φ are determined by the specific physics of second-order phase transition. From the mathematical point of view, the order parameter can be a scalar, vector, tensor, real, or complex quantity.

Here we can mention the following examples of second-order phase transitions as nonmagnetic-to-ferromagnetic transition at the *Curie temperature*, magnetization (magnetic moment per unit volume) being the order parameter, i.e. three-component vector; antiferromagnetic transition at the *Néel temperature* below which the order parameter (antiferromagnetic vector) is a difference between the magnetizations of two magnetic sublattices; difference in the densities of liquid and gas phases near the critical point. In ferroelectrics the order parameter represents the vector of spontaneous polarization arising from the mutual displacement of sublattices.

For the phase transition to superfluid or superconducting state, the order parameter is a *complex variable* associated with the wave function of condensate or electron pairs and, correspondingly, has a phase in addition to the modulus of order parameter. In superfluid ^{3}He neutral Fermi liquid with p-pairing, the order parameter becomes more complicated and is described with the complex 3×3 matrix. In the liquid–crystal phase transition, the order parameter represents an irreducible symmetrical tensor with five independent components.

For the mathematical description of second-order phase transition, the notion for the local magnitude of order parameter $\varphi(\boldsymbol{r})$ at point $\boldsymbol{r}$ should be introduced. The local order parameter is determined as a mean value taken over physically small volume $\Delta V \sim a^3$. The length a is a typical distance between particles and much smaller as compared with length ξ at which the order parameter varies noticeably. In this case one says about $\varphi(\boldsymbol{r})$ as being the field of order parameter and the field $\varphi(\boldsymbol{r})$ is a macroscopic or classical variable. The typical spatial scale at which the order parameter changes significantly must exceed the distance between the particles or the lattice constant. The order parameter $\varphi(\boldsymbol{r})$ can change or fluctuate from one point to another.

One of main characteristics for the order parameter, along with its average magnitude $\langle\varphi(\boldsymbol{r})\rangle$, is the correlation function or correlator

$$\mathcal{K}(\boldsymbol{r},\boldsymbol{r}') = \langle\varphi(\boldsymbol{r})\varphi(\boldsymbol{r}')\rangle.$$

The symbol $\langle\Lambda\rangle$ implies the thermodynamic average for variable Λ. Usually, at the temperatures above the phase transition one T_c in the disordered phase the average magnitude of order parameter vanishes $\langle\varphi\rangle = 0$. The correlator $K(\boldsymbol{r},\boldsymbol{r}')$ decays exponentially at large distances for $|\boldsymbol{r}-\boldsymbol{r}'| \to \infty$:

$$\mathcal{K}(\boldsymbol{r},\boldsymbol{r}') \sim \exp\bigl(-|\boldsymbol{r}-\boldsymbol{r}'|/\xi\bigr).$$

The correlation length $\xi = \xi(T)$ is temperature-dependent and increases unlimitedly on approaching the phase transition temperature T_c. In the ordered phase below the phase transition temperature, the correlator no longer decreases at large distances and tends to the square of average magnitude of order parameter $\langle\varphi\rangle^2$.

If the magnitude of order parameter $\varphi(\boldsymbol{r})$ is set at each point $\boldsymbol{r}$, we can represent the Helmholtz free energy F (at fixed volume) or thermodynamic Gibbs potential Φ (at fixed pressure) as a functional of variable $\varphi(\boldsymbol{r})$. As a rule, in the general theory of critical phenomena the type of external conditions or thermodynamic potentials is not specified, and the same generalized notion as an *effective Hamiltonian* $\mathcal{H}_{\text{eff}}$ is employed for calculating various thermodynamic potentials and variables.

The probability to find the system in the state with the given order parameter $\varphi(\boldsymbol{r})$ is governed with the Gibbs distribution

$$W[\varphi(\boldsymbol{r})] = Z^{-1}\exp\bigl(-E[\varphi(\boldsymbol{r})]/T\bigr).$$

Here $E[\varphi(\boldsymbol{r})]$ is the energy of the system at the given distribution of order parameter $\varphi(\boldsymbol{r})$. The partition function Z is defined as a sum or integral over all the possible configurations of field $\varphi(\boldsymbol{r})$:

$$Z = \sum_{\varphi(\boldsymbol{r})}\exp\bigl(-E[\varphi(\boldsymbol{r})]/T\bigr) = \int D\varphi(\boldsymbol{r})e^{-E[\varphi(\boldsymbol{r})]/T} = e^{-(\mathcal{H}_{\text{eff}}/T)}.$$

In essence, this is a continual integral taken over all possible field configurations $\varphi(\boldsymbol{r})$. (In what follows, we will often include factor $1/T$ to the definition of effective Hamiltonian).

From the physical point of view, the effective Hamiltonian $\mathcal{H}_{\text{eff}} = \mathcal{H}_{\text{eff}}[\varphi(\boldsymbol{r})]$ becomes temperature-dependent and represents the free energy[9] related to some distribution of order parameter in the system. From the mathematical point of view, this is a functional determined on the total set of the values of order parameter $\varphi(\boldsymbol{r})$. The effective Hamiltonian, as a functional of order parameter $\varphi(\boldsymbol{r})$, can be expanded in an infinite integral–differential series in the powers of order parameter $\varphi(\boldsymbol{r})$ and its derivatives $\nabla\varphi(\boldsymbol{r})$. The involvement of lowest powers for $\varphi(\boldsymbol{r})$ and $\nabla\varphi(\boldsymbol{r})$ represents the *Landau expansion* and, as a rule, results in the effective Landau Hamiltonian

$$\mathcal{H}_L[\varphi(\boldsymbol{r})] = \int\left[\frac{c}{2}(\nabla\varphi)^2 + \frac{a}{2}\varphi^2 + \frac{b}{4}\varphi^4 - h(\boldsymbol{r})\varphi\right]d\boldsymbol{r}.$$

In other words, we approximate the energy for the concrete configuration of order parameter with the Landau expansion, i.e. we put $E[\varphi(\boldsymbol{r})] \approx \mathcal{H}_L[\varphi(\boldsymbol{r})]$. It is natural that this may only be acceptable for the slow variations and small magnitudes of order parameter.

The type and structure of such an expansion[10] should involve the fact of phase transition existence and correspond to a number of symmetrical properties of order parameter. In general, the expansion coefficients can depend on temperature and other parameters, e.g. pressure. In order to have the ordered phase of order parameter $\langle\varphi\rangle \neq 0$ below phase transition temperature T_c and the disordered $\langle\varphi\rangle = 0$ phase above T_c, it is sufficient to assume that coefficient $a = a(T)$ is essentially temperature-dependent and changes its sign vanishing at the transition point

---

[9] More exactly, thermodynamic potential divided with temperature.

[10] For simplicity, we put the order parameter to be a scalar quantity with the number of components equal to $n = 1$. If the component number of order parameter is $n = 2, 3, \ldots$, notations φ^2, φ^4, and $(\nabla\varphi)^2$ imply the sums $\varphi_1^2 + \ldots + \varphi_n^2$, $(\varphi_1^2 + \ldots + \varphi_n^2)^2$ and $(\nabla\varphi_1)^2 + \ldots + (\nabla\varphi_n)^2$, respectively.

$$a(T) = \alpha\tau, \quad \tau = \frac{T - T_c}{T_c}, \quad \alpha > 0.$$

We call the variable τ *closeness* to the transition point or *dimensionless temperature*. Coefficients c and b are the *stiffness* and *coupling constant*, respectively. The temperature dependence of these coefficients is not so essential and can be neglected in first approximation, treating them as constant ones. The term with φ^4 is referred to as the coupling of order parameter fluctuations. The parameter $h(\boldsymbol{r})$ can be interpreted as the generalized *external field*.

Depending on the type of phase transition and symmetry of order parameter, it may be possible to include the cubic term φ^3 in the expansion of effective Hamiltonian. Such a situation arises in the critical phenomena near the gas–liquid critical point. The cubic term in the effective Hamiltonian leads to the first-order phase transition. At the first-order phase transition point the continuity breaks down, entailing a jump for the order parameter in the course of transition from one phase to another.

In some physical systems, for example, to describe the phase transitions in ferroelectrics, the expansion to sixth-order terms has often to be used, resulting from the smallness and the possible simultaneous vanishing of coefficients b and a at the tricritical point. The description of phase transitions under violating the translational symmetry and appearing of the long-range periodic structures, e.g. charge density waves in ferroelectrics, spin density waves, helicoidal structures in antiferromagnets, or spatially inhomogeneous phases in superconductors, may require to augment the expansion of effective Hamiltonian with the spatial derivatives of higher order than second, e.g. $\sim (\nabla^2\varphi)^2$. Such a situation occurs at the *Lifshitz point* where the stiffness coefficient c vanishes.

4.8 Self-Consistent Field Approximation

Let us turn now to considering the properties of phase transition in the framework of the Landau approximation, assuming the expansion of the effective Hamiltonian in a series over the powers of order parameter and slowness of its spatial variation. We write the probability $W[\varphi(\boldsymbol{r})]$ for the given field configuration of order parameter as

$$\begin{aligned} W[\varphi(\boldsymbol{r})] &= Z^{-1}\exp\bigl(-\mathcal{H}_L[\varphi(\boldsymbol{r})]/T\bigr), \\ Z &= \int D\varphi(\boldsymbol{r})\exp\bigl(-\mathcal{H}_L[\varphi(\boldsymbol{r})]/T\bigr). \end{aligned}$$

The calculation of partition function represents an evaluation of functional (continual) integral meaning a complicated mathematical problem due to presence of non-quadratic $\varphi(\boldsymbol{r})$ expansion terms in the exponent. The self-consistent or mean-field approximation reduces to a choice of such a field configuration which provides us with the maximum probability of its realization. The maximum probability has the configuration for which the magnitude $\mathcal{H}_L[\varphi(\boldsymbol{r})]$ is minimum. The necessary condition is the following:

$$\frac{\delta\mathcal{H}_L[\varphi(\boldsymbol{r})]}{\delta\varphi(\boldsymbol{r})} = 0.$$

In the self-consistent or mean-field approximation, the fluctuations of order parameter φ are completely neglected, and the equilibrium of the system is determined with the condition of minimum for $\mathcal{H}_L[\varphi(\boldsymbol{r})]$.

Let us analyze the properties of second-order phase transition in the self-consistent field approximation for the effective Landau Hamiltonian

$$\mathcal{H}_L = \int\left[\frac{c}{2}(\nabla\varphi)^2 + \frac{a}{2}\varphi^2 + \frac{b}{4}\varphi^4 - h(\boldsymbol{r})\varphi\right]d\boldsymbol{r}, \quad a(\tau) = \alpha\tau \quad (\alpha > 0).$$

We start from the case of zero external field $h(\boldsymbol{r}) = 0$. In this case the condition of the minimum Hamiltonian reaches the homogeneous distribution of order parameter $\langle\varphi(\boldsymbol{r})\rangle = \varphi$, satisfying the equation

$$(a + b\varphi^2)\varphi = 0.$$

Here, two states are possible such as a disordered phase $\varphi = 0$ available at all temperatures and an ordered phase with $\varphi^2 = -a/b$ available only if $T < T_c$. For the disordered phase, we have $\mathcal{H}_L = 0$ and for the ordered one: $\mathcal{H}_L = -a^2/4b < 0$. Thus, at low $T < T_c$ temperature there will exist an ordered state energetically more favorable as compared with the disordered one. As a result, we have for the equilibrium magnitude of order parameter $\varphi = \varphi_0$

$$\varphi_0(\tau) = \begin{cases} 0 & \text{if } \tau > 0, \\ \pm(\alpha|\tau|/b)^{1/2} \sim |\tau|^{1/2} & \text{if } \tau < 0. \end{cases}$$

Below T_c, as the temperature lowers, the order parameter magnitude increases and reaches the saturation away from the phase transition point in the low temperature limit.

For the specific heat $C = -T\partial^2\mathcal{H}_L[\phi_0]/\partial T^2$, we find

$$C(\tau) = \begin{cases} 0 & \text{if } \tau > 0, \\ \frac{T}{T_c}\frac{\alpha^2}{2bT_c} & \text{if } \tau < 0. \end{cases}$$

The specific heat increases in the course of the disordered–ordered phase transition and experiences a finite-magnitude jump directly at the phase transition point.

Let us consider the effect of external homogeneous field $h(\boldsymbol{r}) = h$ on the properties of phase transition. Most important characteristic for the behavior of the system in external field is the generalized *susceptibility* defined as $\chi = \partial\langle\varphi\rangle/\partial h$. The condition for the minimum of effective Hamiltonian yields the equation for determining φ

$$a\varphi + b\varphi^3 = h.$$

For $h \neq 0$, this equation has no solution with $\varphi = 0$ at any temperature and, strictly speaking, the phase transition is absent. The solution with the maximum modulus of order parameter corresponds to the energy minimum.

Differentiating over the field results in the equation for the susceptibility

$$\chi = \frac{1}{a + 3b\varphi^2(h)} .$$

We analyze these two equations in the limits of weak $h \ll h_\tau$ and high $h \gg h_\tau$ external field. The critical field h_τ can be estimated from the condition $3b\varphi^2(h_\tau) = -a$ and equation for order parameter. This gives approximately

$$h_\tau \sim \frac{2\alpha^{3/2}}{3\sqrt{3b}}|\tau|^{3/2}.$$

So, at the phase transition point the effect of external field is always strong, regardless of its magnitude.

Let us start from the low field limit. In this case we can put $\varphi \approx \varphi_0(\tau)$ and find the temperature behavior divergent at the phase transition point and described by the *Curie law*

$$\chi(\tau) = \begin{cases} \frac{1}{\alpha\tau} & \text{if } \tau > 0, \\ & \qquad (h \ll h_\tau) \\ \frac{1}{2\alpha|\tau|} & \text{if } \tau < 0. \end{cases}$$

In the high $h \gg h_\tau$ external field, the nonlinear term in φ becomes predominant and the closeness to the phase transition has no essential effect on the behavior of order parameter as a function of external field

$$\varphi = \left(\frac{h}{b}\right)^{1/3} \quad \text{and} \quad \chi = \frac{1}{3b^{1/3}h^{2/3}} \quad \text{for} \quad h \gg h_\tau .$$

Problem

1. In the ordered ($\tau < 0$) phase, an emergence of the domain walls or plane defects becomes possible as a result of degenerating the states with $+\varphi_0$ and $-\varphi_0$ in energy. Let $\varphi(z)$ depend on coordinate z normal to the domain wall plane and $\varphi(\pm\infty) = \pm\varphi_0$ where φ_0 is the equilibrium magnitude of order parameter.

Find the behavior $\varphi(z)$ and the domain wall energy E.

Solution. Minimizing the effective Hamiltonian, according to $\delta\mathcal{H}_L/\delta\varphi = 0$, leads to the equation

$$-c\varphi''(z) + a\varphi(z) + b\varphi(z) = 0.$$

This equation should be solved for the following boundary conditions: $\varphi(\pm\infty) = \pm\varphi_0$ and $\varphi'(\pm\infty) = 0$. Multiplying the equation by $\varphi'(z)$, we arrive at first integral

$$-\frac{c}{2}\varphi'^2(z) + \frac{a}{2}\varphi^2(z) + \frac{b}{4}\varphi^4(z) = \text{const} = -\frac{b}{4}\varphi_0^4$$

and then at the equation

$$\xi^2\varphi'^2(z) = \frac{(\varphi_0^2 - \varphi^2)^2}{2\varphi_0^2} \quad \text{where} \quad \xi = \left(\frac{c}{|a|}\right)^{1/2}.$$

The solution of equation can be represented as

$$\varphi(z) = \varphi_0 \tanh \frac{z - z_0}{\sqrt{2}\,\xi}$$

where z_0 is an arbitrary constant determining the position of domain wall.

The energy of domain wall per unit area equals the integral

$$E = \int_{-\infty}^{\infty} dz\left(\frac{c}{2}\varphi'^2(z) + \frac{a}{2}\varphi^2(z) + \frac{b}{4}\varphi^4(z) - \frac{b}{4}\varphi_0^4\right) = \int_{-\infty}^{\infty} c\varphi'^2(z)dz = \frac{8\sqrt{2}}{3}\xi\frac{b\varphi_0^4}{4}.$$

4.9 Critical Exponents

The behavior of physical quantities near continuous phase transitions is usually extrapolated with the power laws. These power-like properties are well supported by experimental data. The *critical exponent* is defined as an exponent in the limiting behavior at $T \to T_c$ and, if the physical quantity behaves as $f(\tau) \sim \tau^\lambda$, its exponent reads

$$\lambda = \lim_{\tau\to 0} \frac{\ln|f(\tau)|}{\ln|\tau|}.$$

The case $\lambda = 0$ means either finite jump of quantity $f(\tau)$ or logarithmic divergence at $\tau = 0$. The critical exponent shows only the limiting behavior at the phase transition point. In general, there are less divergent corrections, e.g.

$$f(\tau) = A|\tau|^\lambda(1 + B|\tau|^\epsilon + \ldots) \quad \text{where} \quad \epsilon > 0.$$

The proportionality coefficients A and B do not necessarily coincide on both sides of the phase transition. Though the correction term vanishes in limit $\tau \to 0$, it may be essential however and noticeable for small but finite values τ, imitating another critical exponent at $|\tau| > |B|^{-1/\epsilon}$.

The following conventional symbols are accepted for the critical exponents. For specific heat, this is exponent α in accordance with the relation

$$C(\tau) \sim |\tau|^{-\alpha}.$$

The critical exponent β is introduced for describing the critical behavior of order parameter in the ordered phase

$$\varphi_0(\tau) \sim (-\tau)^\beta \quad (\tau \leqslant 0).$$

The critical exponents γ and δ are related with the behavior of order parameter as a function of external field h. Exponent γ determines the temperature behavior of susceptibility in the low field

$$\chi \sim |\tau|^{-\gamma}$$

and the exponent δ does the behavior of susceptibility in the high external field limit or just at the phase transition point $T = T_c$

$$\varphi(h) \sim h^{\frac{1}{\delta}} \quad \text{and} \quad \chi(h) \sim h^{\frac{1}{\delta}-1}.$$

In the mean-field approximation we have $\alpha = 0$, $\beta = 1/2$, $\gamma = 1$, and $\delta = 3$. Other critical indices will be introduced below.

4.10 Fluctuations of Order Parameter

The physical quantities in the thermodynamic systems are subjected to thermal fluctuations due to openness of the system. This is fully applicable to the order parameter fluctuations which we have neglected so far, treating the order parameter unchangeable from one point to another in the space.

Let us start to analyze the scalar order parameter fluctuations from the high temperature region, treating the magnitude of order parameter φ to be sufficiently small and neglecting the coupling term φ^4. Thus we study the following quadratic effective Hamiltonian:

$$\mathcal{H}_0 = \int \left[\frac{c}{2}(\nabla\varphi)^2 + \frac{a}{2}\varphi^2\right] d\boldsymbol{r}.$$

This approximation is called the *free-field model* and plays a key role in the theory of critical phenomena as an exactly solvable model. The free-field model is a specific type of effective Landau Hamiltonian at $b = 0$ and, in particular, originates from calculating the fluctuating corrections in the Landau expansion. Here we can treat $\varphi(\boldsymbol{r})$ as a deviation of order parameter from its mean value $\varphi_0 = \langle\varphi(\boldsymbol{r})\rangle$.

Below we consider the spatial correlation of scalar order parameter φ, studying the behavior of *irreducible correlator* $K(\boldsymbol{r}, \boldsymbol{r}')$

$$K(\boldsymbol{r}, \boldsymbol{r}') = \langle\Delta\varphi(\boldsymbol{r})\Delta\varphi(\boldsymbol{r}')\rangle = \langle\varphi(\boldsymbol{r}) - \langle\varphi(\boldsymbol{r})\rangle\,\rangle\langle\varphi(\boldsymbol{r}') - \langle\varphi(\boldsymbol{r}')\rangle\,\rangle.$$

The angular brackets mean the thermodynamic average. In what follows, for definiteness, we imply the temperature above the critical one $T \geqslant T_c$ and, correspondingly, we put $\langle\varphi(\boldsymbol{r})\rangle = 0$. According to the definition of correlation function, we have

$$K(\boldsymbol{r},\boldsymbol{r}') = \langle \varphi(\boldsymbol{r})\varphi(\boldsymbol{r}')\rangle = \frac{\sum\limits_{\varphi(\boldsymbol{r})} \varphi(\boldsymbol{r})\varphi(\boldsymbol{r}')e^{-\mathcal{H}_0/T}}{\sum\limits_{\varphi(\boldsymbol{r})} e^{-\mathcal{H}_0/T}} = $$

$$= \frac{\int D\varphi(\boldsymbol{r})\,\varphi(\boldsymbol{r})\varphi(\boldsymbol{r}')e^{-\mathcal{H}_0/T}}{\int D\varphi(\boldsymbol{r})\,e^{-\mathcal{H}_0/T}}.$$

Here the sums (integrals) must be calculated over all the field configurations of order parameter $\varphi(\boldsymbol{r})$. For the calculation in the framework of the free-field model, we reduce Hamiltonian $\mathcal{H}_0$ to the simple diagonal form

$$\mathcal{H}_0 = \frac{1}{2}\sum_{\boldsymbol{k}}\left[(ck^2+a)|\varphi_{\boldsymbol{k}}|^2\right]$$

with the aid of transformation to the Fourier representation

$$\varphi(\boldsymbol{r}) = \frac{1}{\sqrt{V}}\sum_{\boldsymbol{k}}\varphi_{\boldsymbol{k}}e^{i\boldsymbol{k}\boldsymbol{r}}, \quad \varphi^*(\boldsymbol{r}) = \frac{1}{\sqrt{V}}\sum_{\boldsymbol{k}}\varphi^*_{\boldsymbol{k}}e^{-i\boldsymbol{k}\boldsymbol{r}} \quad \text{and} \quad \varphi_{-\boldsymbol{k}} = \varphi^*_{\boldsymbol{k}}.$$

The expression for the correlation function $K(\boldsymbol{r},\boldsymbol{r}')$ reads

$$K(\boldsymbol{r},\boldsymbol{r}') = \langle \varphi(\boldsymbol{r})\varphi(\boldsymbol{r}')\rangle = \sum_{\boldsymbol{k},\boldsymbol{k}'}\langle \varphi_{\boldsymbol{k}}\varphi_{\boldsymbol{k}'}\rangle e^{i\boldsymbol{k}\boldsymbol{r}+i\boldsymbol{k}'\boldsymbol{r}'} = \sum_{\boldsymbol{k}}\langle|\varphi_{\boldsymbol{k}}|^2\rangle e^{i\boldsymbol{k}(\boldsymbol{r}-\boldsymbol{r}')}$$

where we have taken into account that the harmonics with different wave vectors $\boldsymbol{k}$ fluctuate independently from each other, i.e. $\langle \varphi_{\boldsymbol{k}}\varphi_{\boldsymbol{k}'}\rangle = \langle \varphi_{\boldsymbol{k}}\varphi^*_{\boldsymbol{k}}\rangle\delta_{\boldsymbol{k},-\boldsymbol{k}'}$.

To find the average, we employ the method of the generating functional $\mathcal{Z}$

$$\mathcal{Z}[h_{\boldsymbol{k}}] = \frac{\int D\varphi_{\boldsymbol{k}}\,e^{-\sum\limits_{\boldsymbol{k}}\left(\frac{1}{2}\varphi_{\boldsymbol{k}}K_{\boldsymbol{k}}^{-1}\varphi_{-\boldsymbol{k}}-\varphi_{\boldsymbol{k}}h_{-\boldsymbol{k}}\right)}}{\int D\varphi_{\boldsymbol{k}}\,e^{-\frac{1}{2}\sum\limits_{\boldsymbol{k}}\varphi_{\boldsymbol{k}}K_{\boldsymbol{k}}^{-1}\varphi_{-\boldsymbol{k}}}}, \qquad K_{\boldsymbol{k}} = \frac{T}{ck^2+a}.$$

To calculate the generating functional, we perform the transformation of homogeneous shift for the field $\varphi_{\boldsymbol{k}}$ representing the variable for integrating

$$\varphi_{\boldsymbol{k}} \to \varphi_{\boldsymbol{k}} + \xi_{\boldsymbol{k}} \quad \text{and} \quad \varphi_{-\boldsymbol{k}} \to \varphi_{-\boldsymbol{k}} + \xi_{-\boldsymbol{k}}\,.$$

Under such a replacement of variables, the value of integrals included in definition $\mathcal{Z}$ remains unchanged since the integration over $\varphi_{\boldsymbol{k}}$ is performed in the infinite limits and $D(\varphi_{\boldsymbol{k}}+\xi_{\boldsymbol{k}}) = D\varphi_{\boldsymbol{k}}$. Next,

$$\int D(\varphi_{\boldsymbol{k}}+\xi_{\boldsymbol{k}})\,e^{-\sum\limits_{\boldsymbol{k}}[\frac{1}{2}(\varphi_{-\boldsymbol{k}}+\xi_{-\boldsymbol{k}})K_{\boldsymbol{k}}^{-1}(\varphi_{\boldsymbol{k}}+\xi_{\boldsymbol{k}})-h_{-\boldsymbol{k}}(\varphi_{\boldsymbol{k}}+\xi_{\boldsymbol{k}})-(\varphi_{-\boldsymbol{k}}+\xi_{-\boldsymbol{k}})h_{\boldsymbol{k}}]} =$$

$$= e^{\frac{1}{2}\sum\limits_{\boldsymbol{k}}h_{-\boldsymbol{k}}K_{\boldsymbol{k}}h_{\boldsymbol{k}}}\int D\varphi_{\boldsymbol{k}}\,e^{-\frac{1}{2}\sum\limits_{\boldsymbol{k}}\varphi_{-\boldsymbol{k}}K_{\boldsymbol{k}}^{-1}\varphi_{\boldsymbol{k}}}$$

where we have chosen the magnitude of shift as $\xi_k = K_k h_k$ and $\xi_{-k} = K_k h_{-k}$ in order to get rid of linear terms over φ_k. We have also involved that $\sum_k h_{-k}\varphi_k = \sum_k h_k\varphi_{-k}$. Hence we obtain the following representation for the generating functional $\mathcal{Z}[h_k]$:

$$\mathcal{Z}[h_k] = \frac{\int D\varphi_k \, e^{-\sum\limits_k \left(\frac{1}{2}\varphi_k K_k^{-1}\varphi_{-k} - \varphi_k h_{-k}\right)}}{\int D\varphi_k \, e^{-\frac{1}{2}\sum\limits_k \varphi_k K_k^{-1}\varphi_{-k}}} = e^{\frac{1}{2}\sum\limits_k h_k K_k h_{-k}}.$$

Decomposing the left-hand and right-hand sides of equality to second order in h_k, we have

$$1 + \frac{1}{2}\sum_k h_k K_k h_{-k} = 1 + \sum_k \langle\varphi_k\rangle h_{-k} + \sum_{k,k'} \frac{1}{2}\langle\varphi_k\varphi_{k'}\rangle h_{-k}h_{-k'}.$$

Since h_k is arbitrary, the comparison of the expansion terms with the same powers results in the obvious equality $\langle\varphi_k\rangle = 0$ and the equality which we have sought for

$$\langle\varphi_k\varphi_{k'}\rangle = K_k\delta_{k',-k} \quad \text{and} \quad \langle\varphi_{-k}\varphi_k\rangle = \langle|\varphi_k|^2\rangle = \frac{T}{ck^2 + a}.$$

Eventually, the correlation function is expressed with the following integral generalized to an arbitrary spatial dimensionality d:

$$K(\boldsymbol{r}, \boldsymbol{r}') = \int \frac{d^dk}{(2\pi)^d} \frac{T}{ck^2 + a} e^{ik(r-r')} = \\ = \frac{T}{2\pi c}\left(2\pi|\boldsymbol{r} - \boldsymbol{r}'|\xi\right)^{1-d/2} K_{\frac{d-2}{2}}\left(\frac{|\boldsymbol{r} - \boldsymbol{r}'|}{\xi}\right).$$

Here $K_\nu(x)$ is the modified Bessel function of the second kind[11] with index ν. The correlation length $\xi = (c/a)^{1/2} \sim |\tau|^{-1/2}$ is a typical size of region where a new phase develops. In fact, for $|\boldsymbol{r} - \boldsymbol{r}'| \gg \xi$, the correlation function decays exponentially

$$K(\boldsymbol{r} - \boldsymbol{r}') \sim \exp(-|\boldsymbol{r} - \boldsymbol{r}'|/\xi).$$

In the phase transition point $\tau = 0$, the correlation length diverges and a new phase starts to occupy the whole volume of the system. For usual dimensionality $d = 3$, the correlation function equals

$$K(\boldsymbol{r}, \boldsymbol{r}') = \frac{T}{4\pi c|\boldsymbol{r} - \boldsymbol{r}'|} e^{-|\boldsymbol{r}-\boldsymbol{r}'|/\xi}.$$

[11] Rarely: Macdonald function.

At the very phase transition point $T = T_c$, the typical and natural length scale is absent since $\xi = \infty$ and correlation function $K(\boldsymbol{r}, \boldsymbol{r}')$ behaves in the power-like manner

$$K(\boldsymbol{r}, \boldsymbol{r}') \sim |\boldsymbol{r} - \boldsymbol{r}'|^{2-d}.$$

In the two-dimensional case at $|\boldsymbol{r} - \boldsymbol{r}'| \ll \xi$ we have a logarithmic behavior $K(\boldsymbol{r}, \boldsymbol{r}') \approx (T/2\pi c)\ln(\xi/|\boldsymbol{r} - \boldsymbol{r}'|)$.

In order to describe the anomalous behavior of correlation function in the immediate vicinity of phase transition point, two critical exponents are necessary to introduce more. The exponent ν determines the degree of divergence for the correlation length within the vicinity of phase transition point

$$\xi(\tau) \sim |\tau|^{-\nu}.$$

The *critical Fisher exponent* η characterizes the behavior of correlation function $K(\boldsymbol{r}, \boldsymbol{r}')$ at the very transition point according to

$$K(\boldsymbol{r}, \boldsymbol{r}') \sim \frac{1}{|\boldsymbol{r} - \boldsymbol{r}'|^{d-2+\eta}}.$$

In the free-field model these critical exponents are as follows: $\nu = 1/2$ and $\eta = 0$.

Let us analyze the external field effect on the properties of order parameter $\varphi(\boldsymbol{r})$ within the free-field model. In other words, we study the effective Hamiltonian

$$\mathcal{H}_0 = \int \left[\frac{c}{2}(\nabla\varphi)^2 + \frac{a}{2}\varphi^2 - h(\boldsymbol{r})\varphi(\boldsymbol{r}) \right] d\boldsymbol{r}.$$

The equilibrium magnitude of order parameter is determined by the minimum of effective Hamiltonian. Varying $\mathcal{H}_0$ yields

$$\frac{\delta \mathcal{H}_0}{\delta \varphi(\boldsymbol{r})} = -c\nabla^2\varphi(\boldsymbol{r}) + a\varphi(\boldsymbol{r}) - h(\boldsymbol{r}) = 0.$$

The solution of this equation can conveniently be written with the aid of the *Green function* or *response function* as

$$\varphi(\boldsymbol{r}) = \int G_0(\boldsymbol{r} - \boldsymbol{r}')h(\boldsymbol{r}')d\boldsymbol{r}',$$

the Green function being the solution of equation

$$\left(-c\frac{\partial^2}{\partial \boldsymbol{r}^2} + a\right) G_0(\boldsymbol{r}, \boldsymbol{r}') = \delta(\boldsymbol{r} - \boldsymbol{r}').$$

Since $G_0(\boldsymbol{r}, \boldsymbol{r}') = G_0(\boldsymbol{r} - \boldsymbol{r}')$, the solution can be found using the transformation to the Fourier representation for the Green function $G_0(\boldsymbol{k})$. So, we have

$$(ck^2 + a)G_0(\mathbf{k}) = 1 \quad \text{and} \quad G_0^{-1}(\mathbf{k}) = ck^2 + a.$$

Hence we disclose the direct relation between the correlation and Green functions: $K(\mathbf{r}) = TG_0(\mathbf{r})$ and

$$G_0(\mathbf{r}) = \frac{(2\pi r\xi)^{1-d/2}}{2\pi c} K_{\frac{d-2}{2}}(r/\xi).$$

Thus, the Green function differs from the correlation function by a multiplier alone.

Problem

Find the relation between the generalized susceptibility χ and irreducible correlator $K(\mathbf{r})$ for the fluctuations of order parameter field $\varphi(\mathbf{r})$.

Solution. Let us introduce external field $h(\mathbf{r})$ by augmenting the effective Hamiltonian $H_0[\varphi]$ with the following term:

$$\delta H[\varphi] = -\int d\mathbf{r}\, h(\mathbf{r})\varphi(\mathbf{r}) \quad \text{and} \quad H = H_0[\varphi] + \delta H[\varphi].$$

Let external field be uniform: $h(\mathbf{r}) = \text{const}$. Then we write down the general definition of partition function Z and the relations for the averages:

$$Z = \sum_{\varphi} e^{-H/T}, \quad \langle\varphi(\mathbf{r})\rangle = \frac{1}{Z}\sum_{\varphi}\varphi(\mathbf{r})e^{-H/T},$$

$$\int d\mathbf{r}\,\langle\varphi(\mathbf{r})\rangle = T\frac{1}{Z}\frac{\partial Z}{\partial h}, \quad \iint d\mathbf{r}\,d\mathbf{r}'\,\langle\varphi(\mathbf{r})\varphi(\mathbf{r}')\rangle = T^2\frac{1}{Z}\frac{\partial^2 Z}{\partial h^2}.$$

Due to constancy of external field and homogeneity of system, the mean value of order parameter $\langle\varphi(\mathbf{r})\rangle$ is independent of coordinate $\mathbf{r}$ and the mean value $\langle\varphi(\mathbf{r})\varphi(\mathbf{r}')\rangle$ depends on the coordinate difference alone. Therefore, one of the integrations will give us the volume V of the system and, in particular,

$$\langle\varphi\rangle = \frac{T}{V}\frac{\partial \ln Z}{\partial h}.$$

Then we obtain from the definition of susceptibility

$$\chi = \frac{\partial\langle\varphi\rangle}{\partial h} = \frac{T}{V}\frac{\partial^2 \ln Z}{\partial h^2}.$$

The double differentiation of logarithm $\ln Z$ results in

$$\frac{1}{T^2}\frac{1}{Z}\frac{\partial^2 Z}{\partial h^2} = \iint d\mathbf{r}\,d\mathbf{r}'\,\langle\Delta\varphi(\mathbf{r})\Delta\varphi(\mathbf{r}')\rangle = \iint d\mathbf{r}\,d\mathbf{r}'\,K(\mathbf{r}, \mathbf{r}').$$

Then, involving the homogeneity of the system $K(\mathbf{r}, \mathbf{r}') = K(\mathbf{r} - \mathbf{r}')$, we find the relation desired

$$\chi = \frac{1}{T}\int K(\mathbf{r})d\mathbf{r} = \frac{1}{T}\int \langle\Delta\varphi(\mathbf{r})\Delta\varphi(0)\rangle\, d\mathbf{r}.$$

This formula represents the general relation between the irreducible correlator for the fluctuations of order parameter φ and its response χ to the conjugate external field h.

4.11 The Ginzburg–Levanyuk Criterion

So far we have neglected the effect of possible spatial order parameter fluctuations associated directly with the presence of gradient term in the effective free-field Hamiltonian which the Fourier transform reads as

$$\mathcal{H}_0 = \frac{1}{2}\sum_{\boldsymbol{k}}\Big[(ck^2 + a)|\varphi_{\boldsymbol{k}}|^2 - h^*_{\boldsymbol{k}}\varphi_{\boldsymbol{k}} - \varphi^*_{\boldsymbol{k}}h_{\boldsymbol{k}}\Big].$$

Let us start the calculation of partition function in the lack of external field, i.e. at $h = 0$. The thermodynamic potential is $F = -T \ln Z$ where partition function Z represents a sum over the infinite set of all possible spatial realizations for the order parameter field $\varphi(\boldsymbol{r})$. Running over all the possible Fourier transforms of field $\varphi_{\boldsymbol{k}}$ is completely equivalent to that over all the possible realizations of field $\varphi(\boldsymbol{r})$:

$$Z_0 = \sum_{\mathrm{Re}\,\varphi_{\boldsymbol{k}},\,\mathrm{Im}\,\varphi_{\boldsymbol{k}}} e^{-\sum_{\boldsymbol{k}}\frac{ck^2+a}{2T}|\varphi_{\boldsymbol{k}}|^2} = \prod_{\boldsymbol{k}} \sum_{\mathrm{Re}\,\varphi_{\boldsymbol{k}},\,\mathrm{Im}\,\varphi_{\boldsymbol{k}}} e^{-\frac{ck^2+a}{2T}|\varphi_{\boldsymbol{k}}|^2}.$$

Then, in the infinite product we calculate the value for the sum with the fixed wave vector $\boldsymbol{k}$ as

$$\sum_{\mathrm{Re}\,\varphi_{\boldsymbol{k}},\,\mathrm{Im}\,\varphi_{\boldsymbol{k}}} e^{-\frac{ck^2+a}{2T}|\varphi_{\boldsymbol{k}}|^2} = \frac{1}{2}\int_{-\infty}^{\infty} d(\mathrm{Re}\,\varphi_{\boldsymbol{k}}) \int_{-\infty}^{\infty} d(\mathrm{Im}\,\varphi_{\boldsymbol{k}}) \times$$

$$\times \exp\left(-\frac{ck^2+a}{2T}\left[(\mathrm{Re}\,\varphi_{\boldsymbol{k}})^2 + (\mathrm{Im}\,\varphi_{\boldsymbol{k}})^2\right]\right) = \frac{\pi T}{ck^2+a}.$$

Here we have introduced factor 1/2. Otherwise, the states with wave vectors $\pm\boldsymbol{k}$ would be counted twice for the scalar real order parameter field due to relation $\varphi^*_{\boldsymbol{k}} = \varphi_{-\boldsymbol{k}}$. Finally, we arrive at the following expression[12] for the partition function at $h = 0$:

$$Z_0 = \prod_{\boldsymbol{k}} \frac{\pi T}{ck^2+a}.$$

Then the thermodynamic potential equals

$$F = -T\sum_{\boldsymbol{k}} \ln\frac{\pi T}{ck^2+a} \approx -T_c\sum_{\boldsymbol{k}} \ln\frac{\pi T_c}{ck^2+a}.$$

---

[12] For the real order parameter, its Fourier transforms are not independent and the integration should be limited to the half-space of the wave vector $\boldsymbol{k}$. As for the complex order parameter, the components $\mathrm{Re}\,\varphi_{\boldsymbol{k}}$ and $\mathrm{Im}\,\varphi_{\boldsymbol{k}}$ are completely independent and factor 1/2 is superfluous. The numerator in Z_0 will be $2\pi T$ instead of πT.

Since, as we will see, the singular behavior of thermodynamic variables is related with the region of small wave vectors $\boldsymbol{k}$ and smallness $a = \alpha\tau$, it is possible to neglect the difference between T and T_c in the regular places in the expression for thermodynamic potential F. Then, the singular correction to specific heat $C = -T\partial^2 F/\partial T^2$ equals

$$\delta C_{\text{sing}}(\tau) = -T_c^{-1}\frac{\partial^2 F}{\partial \tau^2} = \sum_{\boldsymbol{k}} \frac{\alpha^2}{(\alpha\tau + ck^2)^2} = \int \frac{V_d\, d^d k}{(2\pi)^d} \frac{(\alpha/c)^2}{(k^2 + \xi^{-2})^2},$$

resulting from the spatial order parameter fluctuations. Here $V_d = L^d$ is the volume of the system in the space of dimensionality d and $\xi = (c/\alpha\tau)^{1/2}$ is the correlation length. Calculating the integral gives the following answer:

$$\delta C_{\text{sing}}(\tau) = \frac{V_d S_d}{(2\pi)^d} \frac{\pi(2-d)}{4\sin(\pi d/2)} \left(\frac{\alpha}{c}\right)^2 \xi^{4-d} \sim \frac{1}{\tau^{(4-d)/2}}$$

where S_d is the surface area of unit d-dimensional sphere. The divergence originating from $\xi \to \infty$ is associated with the singular contribution to the integral from the region of small wave vectors or from the long-wave fluctuations of order parameter. The less the spatial dimensionality d, the stronger the order parameter fluctuations will manifest by approaching the phase transition point.

In the usual dimensionality $d = 3$ one has

$$\delta C_{\text{sing}}(\tau) = \frac{V}{8\pi}\left(\frac{\alpha}{c}\right)^2 \xi \sim \frac{1}{\tau^{1/2}}.$$

For $d = 4$, the singularity of correction to the specific heat is logarithmic $\delta C_{\text{sing}} \sim \ln(1/\tau)$. No specific singular contribution to the specific heat behavior appears for dimensions $d > 4$. The critical exponents, predicted by the self-consistent Landau theory, remain unchanged. In this sense the space dimensionality $d = 4$ proves to be boundary and is called the *upper critical dimensionality* in this case. In the dimensions above the upper critical one, the critical exponents coincide with those predicted by the mean-field theory.

If we compare the fluctuation correction δC_{sing} with the specific heat jump ΔC found in the Landau theory approximation, we arrive at the *Ginzburg–Levanyuk criterion* of applicability for the Landau theory in which the spatial fluctuations of order parameter are completely neglected. In the $d = 3$ case the fulfillment of the following strong inequality is required:

$$\frac{\alpha^2}{2bT_c} \gg \frac{1}{8\pi}\left(\frac{\alpha}{c}\right)^{3/2} \frac{1}{|\tau|^{1/2}} \sim \frac{1}{|\tau|^{1/2}}.$$

Hence we find[13] the criterion of applicability for the Landau theory (self-consistent field theory):

$$1 \gg |\tau| \gg \mathrm{Gi}, \quad \mathrm{Gi} = \frac{b^2 T_c^2}{\alpha c^3}.$$

The left-hand side of inequality expresses simply the closeness to the phase transition point, which is necessary for the Landau expansion. The right-hand side of inequality ensures the smallness of order parameter fluctuations. The immediate vicinity of phase transition point $|\tau| \lesssim \mathrm{Gi}$, where the strong spatial fluctuations of order parameter develop, is called the *fluctuation* or *critical region* of phase transition. For $\mathrm{Gi} \gtrsim 1$, no applicability region exists for the theory of self-consistent field (Landau expansion).

We can also obtain the Ginzburg–Levanyuk criterion from the following speculations. The mean-field theory is justified provided that the order parameter fluctuations in volume $V_\xi \sim \xi^d$ with linear size $\sim \xi$ are small as compared with the typical magnitude of order parameter $\varphi_0 \sim (\alpha|\tau|/b)^{1/2}$ in volume V_ξ. In the opposite case, it is necessary to take the order parameter fluctuations into account. For the conventional spatial dimension $d = 3$ and from the condition

$$\langle(\Delta\varphi)^2\rangle = \frac{T\chi}{V_\xi} \sim \frac{T_c\chi}{V_\xi} \sim \frac{T_c}{\xi^3\alpha|\tau|} \ll \varphi_0^2 \sim \frac{\alpha|\tau|}{b}$$

where $\chi \sim 1/(\alpha|\tau|)$ is the generalized susceptibility, we arrive at the same Ginzburg–Levanyuk criterion.

The dimensionless parameter Gi, temperature-independent and characterizing the specific condensed matter, is referred to as the *Ginzburg–Levanyuk number*. To clarify the physical meaning of this parameter, let us rewrite it as follows:

$$\mathrm{Gi} = \frac{b^2 T_c^2}{\alpha c^3} \sim \left(\frac{T_c}{\epsilon_c a^3}\right)^2 \left(\frac{a}{\xi_0}\right)^6.$$

Here length $\xi_0 = (c/\alpha)^{1/2}$ represents approximately the magnitude of correlation length (coherence length) far from the transition point, roughly speaking, at zero temperature when $|\tau| \sim 1$. Parameter $\epsilon_c = \alpha^2/b$ by the order of magnitude can be identified with the condensation energy of the ordered phase far from the phase transition temperature as well. Since parameter a means the typical spacing between the particles of medium, the quantity $\epsilon_c a^3$ determines the typical scale of condensation energy per particle or energy gain as compared with the disordered phase as a result of emerging the low-temperature ordered phase. The main condition for the smallness of the Ginzburg–Levanyuk number is the large correlation length as compared with the interatomic spacing, i.e. $\xi \gg a$.

[13] We omit the small numerical factor $1/(4\pi)^2$.

The condensed media with small numbers Gi $\ll 1$, first of all, include superconductors[14] and superfluid helium ^{3}He for which Gi $\sim (a/\xi_0)^4$ or $\sim (T_c/\varepsilon_F)^4$. The substances with relatively small numbers Gi ~ 0.1 can be presented by ferromagnets and ferroelectrics with the Curie temperature about several kilokelvins. The normal liquid–superfluid transition in ^{4}He serves as an example of phase transition which has no temperature region for applicability of Landau expansion since Gi ~ 1. The coherence length ξ_0 here has really one or two interatomic distances, i.e. $\xi_0 \sim a$. The condensation energy per particle is of the same order of the magnitude as the transition temperature $T_c \sim 2.17$ K into the superfluid state.

Problems

1. Find the critical exponent μ and correlation radius ξ for the fluctuations of order parameter φ in the external field h at the phase transition point $\tau = 0$.

Solution. We expand the effective Landau Hamiltonian

$$H_L[\varphi] = \frac{c}{2}(\nabla\varphi)^2 + \frac{b\varphi^4}{4} - h\varphi$$

at phase transition point $\tau = 0$ over the deviations of order parameter from its equilibrium value $\varphi_0(h) = (h/b)^{1/3}$ to quadratic terms

$$H_L[\varphi] = H_L[\varphi_0] + \frac{c}{2}(\nabla\varphi)^2 + \frac{3b\varphi_0^2}{2}(\varphi - \varphi_0)^2 + \dots$$

This Hamiltonian is analogous to that of free field treated above. Thus we write straightforwardly the answer

$$\xi = \sqrt{\frac{c}{3b^{1/3}}\,\frac{1}{h^{1/3}}}, \quad \text{i.e.}\quad \xi(h) \sim h^{-\mu}$$

and critical exponent μ equals 1/3.

2. Determine the singular correction to specific heat $\delta C_{\text{sing}} \sim h^{-\epsilon}$ in the external field h at the phase transition point $\tau = 0$.

Solution. Let us expand the effective Landau Hamiltonian $H_L[\varphi]$ over the deviations of order parameter φ from its equilibrium value φ_0, satisfying the equation

$$a(\tau)\varphi_0 + b\varphi_0^3 = h, \quad a(\tau) = \alpha\tau.$$

We restrict ourselves with the second-order terms in deviations. We have

$$H_L[\varphi] = H_L[\varphi_0] + \frac{c}{2}(\nabla\varphi)^2 + \frac{a + 3b\varphi_0^2}{2}(\varphi - \varphi_0)^2 + \dots$$

The thermodynamic potential determining the singular contribution due to order parameter fluctuations will be equal to

$$F = -T_c \sum_{\boldsymbol{k}} \ln \frac{\pi T_c}{ck^2 + a + 3b\varphi_0^2}\,.$$

Next,

$$\delta C_{\text{sing}} = -T_c^{-1}\frac{\partial^2 F}{\partial\tau^2} = \sum_{\boldsymbol{k}} \frac{\left[\partial(a + 3b\varphi_0^2)/\partial\tau\right]^2}{(ck^2 + a + 3b\varphi_0^2)^2} - \sum_{\boldsymbol{k}} \frac{\partial^2(a + 3b\varphi_0^2)/\partial\tau^2}{ck^2 + a + 3b\varphi_0^2}.$$

[14] The condensation energy per particle is $\epsilon_c a^3 \sim T_c^2/\varepsilon_F$ and coherence length is $\xi_0 \sim a\varepsilon_F/T_c$.

Since the main contribution to the singular behavior of thermodynamic variables is connected with the long-wave fluctuations of order parameter or small wave vectors $\boldsymbol{k}$, we analyze the first term alone as most divergent at $h \to 0$. So,

$$\delta C_{\text{sing}} = \sum_{\boldsymbol{k}} \frac{\left[\partial(a+3b\varphi_0^2)/\partial\tau\right]^2}{(ck^2+a+3b\varphi_0^2)^2} = \int \frac{V d^3k}{(2\pi)^3} \frac{\left[\alpha+6b\varphi_0\partial\varphi_0/\partial\tau\right]^2}{(ck^2+a+3b\varphi_0^2)^2}.$$

To calculate the numerator of a fraction at $\tau = 0$, we differentiate the equation determining the order parameter. We obtain

$$\alpha\varphi_0 + \left(\alpha\tau + 3b\varphi_0^2\right)\frac{\partial\varphi_0}{\partial\tau} = 0.$$

Hence for $\tau = 0$ we have $\partial\varphi_0/\partial\tau = -\alpha/(3b\varphi_0)$ where $\varphi_0 = (h/b)^{1/3}$. The result at $\tau = 0$ reads

$$\delta C_{\text{sing}}(h) = \int \frac{V d^3k}{(2\pi)^3} \frac{\alpha^2}{(ck^2+3b\varphi_0^2)^2} = \int \frac{V d^3k}{(2\pi)^3} \frac{(\alpha/c)^2}{(k^2+\xi^{-2})^2} \quad \text{and} \quad \xi = \sqrt{\frac{c}{3b^{1/3}}}\,\frac{1}{h^{1/3}}.$$

The analysis of integral leads to the following singular behavior of specific heat at phase transition point $\tau = 0$ as $h \to 0$:

$$\delta C_{\text{sing}}(h) = \frac{V}{8\pi}\left(\frac{\alpha}{c}\right)^2 \xi = \frac{V}{8\pi}\frac{\alpha^2}{\sqrt{3c^3 b^{1/3}}}\frac{1}{h^{1/3}}, \quad \text{i.e.} \quad \delta C_{\text{sing}}(h) \sim h^{-\epsilon},$$

the critical exponent being $\epsilon = 1/3$.

4.12 Critical Point

A series of thermodynamic systems with the phase transitions can be described by the effective Hamiltonian having the cubic term in the expansion over order parameter φ

$$\mathcal{H}(\varphi) = \frac{\tau}{2}\varphi^2 + \frac{a}{3}\varphi^3 + \frac{b}{4}\varphi^4 - h\varphi \quad (b > 0).$$

The parameter h represents an external field. As an example of such a system, we can indicate the vapor–liquid phase transition near the critical point. Here the deviation of density from its value at the critical point plays a role of order parameter. Another example is the phase transition between the isotropic and anisotropic nematic phases in the liquid crystal nematic. Below for the definiteness, we put $a \geqslant 0$. (The case $a \leqslant 0$ reduces to replacing $\varphi \to -\varphi$).

Let external field at first be absent $h = 0$. The selected point $\tau = 0$ and $a = 0$ represents the *critical point*. The magnitude $\mathcal{H} = 0$ of effective Hamiltonian corresponds to the disordered phase $\varphi = 0$. Accordingly, the condition $\mathcal{H} \leqslant 0$ meets the energy gain of the ordered phase. The phase transition line or binodal τ_0 and the value of order parameter φ_0 at the binodal

$$\tau_0 = \frac{2}{9}\frac{a^2}{b} \quad \text{and} \quad \varphi_0 = -\frac{2}{3}\frac{a}{b}$$

are found from the conditions

$$\mathcal{H} = \frac{\tau}{2}\varphi_0^2 + \frac{a}{3}\varphi_0^3 + \frac{b}{4}\varphi_0^4 = 0 \quad \text{and} \quad \frac{\partial \mathcal{H}}{\partial \varphi} = \tau\varphi_0 + a\varphi_0^2 + b\varphi_0^3 = 0.$$

The finite magnitude of the order parameter discontinuity at the transition point for $a \neq 0$ means the first-order phase transition. For $\tau < \tau_0$, the order parameter is determined by the minimum of effective Hamiltonian

$$\partial \mathcal{H}/\partial \varphi = (\tau + a\varphi + b\varphi^2)\varphi = 0.$$

The absolute minimum of effective Hamiltonian corresponds to the root with the maximum value of order parameter φ. As a result, we have

$$\varphi(\tau) = \begin{cases} 0 & \text{for } \tau > \tau_0, \\ -\frac{\left(a+\sqrt{a^2-4b\tau}\right)}{2b} & \text{for } \tau < \tau_0. \end{cases}$$

Since the phase transition is the first-order one for $a \neq 0$, the latent heat of phase transition acquires the finite magnitude proportional to the jump of first derivative of effective Hamiltonian with respect to temperature

$$\Delta\big(\partial \mathcal{H}/\partial \tau\big) = -\varphi_0^2/2 = -2a^2/9b^2.$$

The spinodal line or the line of absolute phase instability is found from breaking down the minimum of effective Hamiltonian, i.e. when the second derivative changes its sign from positive to negative.

$$\left.\frac{\partial^2 \mathcal{H}}{\partial \varphi^2}\right|_{\varphi=\varphi_0} = 0.$$

Hence we determine the spinodal line (maximum overcooling line) for the disordered phase $\tau_{s_-} = 0$ and the spinodal line (maximum overheating line) for the ordered phase $\tau_{s_+} = a^2/4b$. It is obvious that $\tau_{s_-} < \tau_0 < \tau_{s_+}$ and the temperature regions $\tau_{s_-} < \tau < \tau_0$ and $\tau_0 < \tau < \tau_{s_+}$ correspond to the metastable regions of phases (Fig. 4.6).

Let us turn to analyzing the behavior of susceptibility χ in the low external field. First, we differentiate the equation

$$\tau\varphi + a\varphi^2 + b\varphi^3 = h,$$

determining the order parameter φ, with respect to external field h, and substitute the value $\varphi \approx \varphi(0)$ in the case of low external field. Then we find the inverse susceptibility

$$\chi^{-1} = \tau + 2a\varphi(0) + 3b\varphi^2(0) = \begin{cases} \tau, & \tau > \tau_0\,, \\ -2\tau + a\frac{a+\sqrt{a^2-4b\tau}}{2b}, & \tau < \tau_0\,. \end{cases}$$

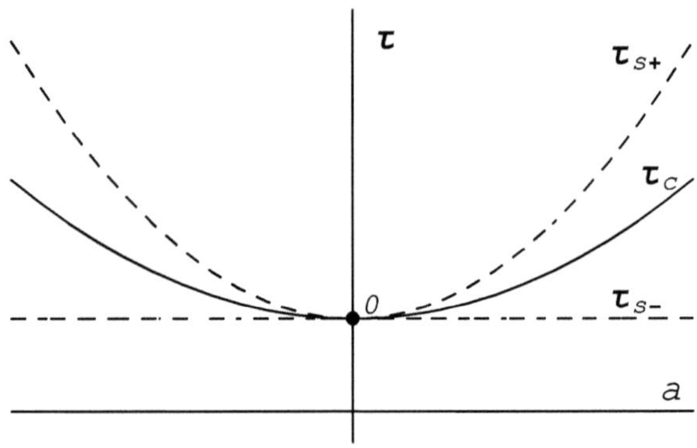

Fig. 4.6 The diagram of phase states. The solid line is the binodal. The dashed lines are the spinodals

The susceptibility exhibits the cusp-like maximum $\chi(\tau_0) = 9b/2a^2$ at the first-order phase transition line $\tau = \tau_0$. Its derivative $\partial\chi/\partial\tau$ has a discontinuity. At both spinodals $\tau_{s_-} = 0$ and $\tau_{s_+} = a^2/4b$, the magnitudes of susceptibility would be infinite.

The value of coefficient $a = 0$ is specific. At $a = 0$ the phase transition to the disordered phase occurs at temperature $\tau = 0$ with $\varphi_0 = 0$ and becomes the second-order phase transition. Thus, point $a = 0$ and $\tau = 0$ is that where the phase transition line terminates and thus *critical*. The binodal and the spinodal merge at the critical point since the regions of metastable phases are absent for the second-order phase transition.

4.13 Multicritical Point

In the thermodynamic system having the second-order phase transitions, the region of their existence may depend on a number of physical parameters such as temperature, pressure, volume, concentration, and electric or magnetic field. The *multicritical point* is usually understood as a specific point in the region of physical parameters at which the phase transition takes place. As a rule, at the multicritical point we have a coexistence of several ordered phases by intersecting, merging or branching the phase transition lines under possible change in the type of phase transition. The classification of multicritical point depends on the number of thermodynamic parameters necessary for describing the state of the thermodynamic system.

For emerging the multicritical point,[15] it is necessary to have, at least, one line of second-order phase transition depending on two physical parameters at any rate, e.g. temperature and pressure. The presence of multicritical point leads to appearing of a number of *crossover phenomena* in the critical behavior of thermodynamic variables

[15] The critical point of gas–liquid transition is not usually classified as a multicritical point since the gas–liquid transition is a first-order one and the critical behavior takes place at the single point alone.

in the vicinity of this point, such as changing the critical exponents and the upper critical dimension which determines the applicability of mean-field theory.

Let us consider the conditions for emerging the multicritical point within the framework of the mean-field approximation in which the effective Hamiltonian (thermodynamic potential) is formulated as an expansion into a series over the powers of order parameter

$$H(\varphi) = H_0 + \frac{1}{2}A\varphi^2 + \frac{1}{3}B\varphi^3 + \frac{1}{4}C\varphi^4 + \frac{1}{5}D\varphi^5 + \frac{1}{6}E\varphi^6 + \dots$$

In the general case the coefficients of expansion depend on the temperature and other physical parameters, e.g. pressure, magnetic or electric field.

Let us start our consideration from the symmetrical case when the terms with the odd powers are absent due to symmetry reasons resulting from invariance $H(-\varphi) = H(\varphi)$. Below we study the following effective Hamiltonian:

$$H(\varphi) = \frac{c}{2}(\nabla\varphi)^2 + \frac{1}{2}\tau\varphi^2 + \frac{1}{4}b\varphi^4 + \frac{1}{6}d\varphi^6 - h\varphi \quad (d > 0).$$

The expansion coefficients depend on temperature T and in addition, say, on pressure P. Let coefficient $\tau(T, P)$ vanish at some line[16] $\tau(T, P) = 0$ of physical parameters. Provided that coefficient $b(T, P) > 0$, the second-order phase transition occurs at that line $\tau(T, P) = 0$. If, otherwise, coefficient $b(T, P)$ proves to be negative, the phase transition becomes the first-order one. In general, in such a situation there may exist a point where both coefficients τ and b vanish simultaneously at the phase transition line $\tau(T, P) = 0$

$$\tau(T_t, P_t) = 0 \quad \text{and} \quad b(T_t, P_t) = 0, \quad d(P_t, T_t) > 0.$$

The point (T_t, P_t) in the plane of parameters T and P is referred to as *tricritical*. If the system along with two parameters is more characterized by additional thermodynamic variables, the set of tricritical points may compose the line of triple points.

If the coefficients in the expansion of effective Hamiltonian depend on the three parameters, the point where all the three coefficients τ, b, and d vanish at the same time will represent the *tetracritical point*. In this case it is necessary to involve the next term φ^8 of expansion into consideration. Additional singularities will appear in the behavior of thermodynamic variables near such points.

So, first of all, the transition to the ordered phase $\varphi \neq 0$ at $\tau = 0$ is the second-order phase transition for $b > 0$ and $h = 0$. The role of term φ^6 is inessential since the type of phase transition and qualitative behavior of the thermodynamic system remain unchanged within the immediate vicinity of second-order phase transition

[16] As above, coefficient $\tau(T, P)$ means the relative closeness to the transition $\tau = 1 - T/T_c(P)$.

line $\tau_c = 0$. In fact, while $|\tau| \ll b^2/d$ the term φ^6 can simply be neglected due to inequality $d\varphi_0^6 \ll b\varphi_0^4$ since $\varphi_0^2 \sim |\tau|/b$.

In the region of values $b < 0$, the behavior of thermodynamic variables changes significantly and the first-order phase transition occurs at the line determined with two equations

$$H(\varphi) = \left(\frac{\tau}{2} + \frac{b\varphi^2}{4} + \frac{d\varphi^4}{6}\right)\varphi^2 = 0 \quad \text{and} \quad \frac{\partial H}{\partial \varphi} = (\tau + b\varphi^2 + d\varphi^4)\varphi = 0.$$

Hence we find the line of phase transition which proves to be the first-order one

$$\tau_c = \frac{3}{16}\frac{b^2}{d}$$

and determines the finite magnitude of order parameter in the ordered phase, appearing with the jump immediately below the phase transition line

$$\varphi_c^2 = -\frac{3}{4}\frac{b}{d}.$$

As a result, we obtain the full line of phase transition from the disordered phase $\varphi = 0$ to the ordered one with nonzero order parameter $\varphi \neq 0$ as a function of coefficient b

$$\tau_c(b) = \begin{cases} 0 & \text{if } b \geqslant 0, \\ \frac{3b^2}{16d} & \text{if } b \leqslant 0. \end{cases}$$

Thus, there is a change of the type for phase transition at the tricritical point determined with condition $\tau = b = 0$ (Fig. 4.7). The lines of first- and second-order phase

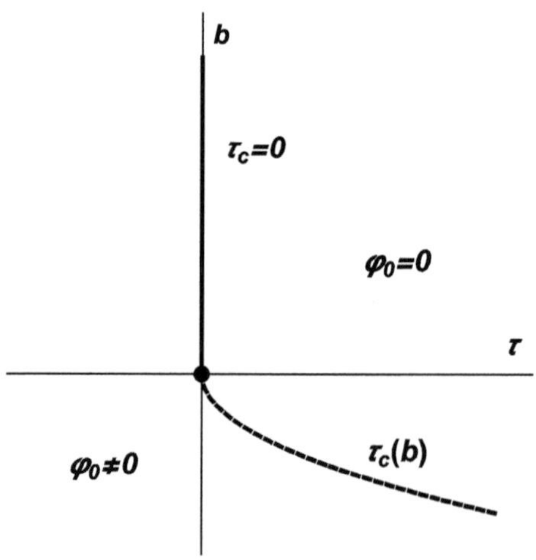

Fig. 4.7 The diagram for the phase states. The solid line is the second-order phase transition. The dashed line is the first-order phase transition

transitions will have the common tangent in the assumption of linear behavior for coefficient $b(T, P)$ as a function of deviations T and P from values T_t and P_t.

Within the whole region where the ordered phase exists, the order parameter is determined by the relation

$$\varphi_0^2(\tau) = \frac{-b + \sqrt{b^2 - 4d\tau}}{2d}.$$

(The negative sign in front of square root corresponds either to the lack of real solutions for the order parameter or to the larger magnitude of effective Hamiltonian). The value of effective Hamiltonian $H(\varphi_0)$ in the disordered phase at $h = 0$ is given by the formula

$$H(\varphi_0) = \frac{b^3 - 6bd\tau - (b^2 - 4d\tau)^{3/2}}{24d^2}.$$

We will mean the specific heat as $C(\tau) = -\partial^2 H(\varphi_0)/\partial\tau^2$, neglecting the factor T_t^{-1} insignificant for our purpose. The specific heat vanishes in the disordered phase and equals

$$C(\tau) = \frac{1}{2\sqrt{b^2 - 4d\tau}}$$

in the ordered phase. The specific heat is characterized by various behaviors in two regions. One is close to the tricritical point when $b^2 \lesssim |\tau|d$ and the other, if $b^2 \gtrsim |\tau|d$, is distant. For the phase transition far from the tricritical point, the specific heat experiences a finite jump of magnitude $\Delta C = 1/(2|b|)$ at the second-order transition line and magnitude of $\Delta C = 1/|b|$ at the first-order transition line. In the region $b^2 \lesssim |\tau|d$ close to the tricritical region, we discover an enhancement of singularity in the specific heat like $\Delta C \sim |\tau|^{-1/2}$ by approaching the tricritical point where the specific heat tends simply to infinity. Thus, on crossing the *crossover line* $|\tau| \sim b^2/d$, we come across the transition from one critical behavior to another.

In the finite external field $h \neq 0$ the picture of possible phase states becomes more complicated. For $b < 0$, the second-order phase transition at the line determined with the simultaneous fulfillments of the following conditions:

$$\frac{\partial H}{\partial\varphi} = \tau\varphi + b\varphi^3 + d\varphi^5 - h = 0,$$

$$\frac{\partial^2 H}{\partial\varphi^2} = \tau + 3b\varphi^2 + 5d\varphi^4 = 0 \quad \text{and} \quad \frac{\partial^3 H}{\partial\varphi^3} = 6b\varphi + 20d\varphi^3 = 0.$$

The simple solution yields the following result determining the phase transition line as a function of coefficient $b < 0$:

$$\tau_{c2} = \frac{9}{20}\frac{b^2}{d}, \quad \varphi_h^2 = \frac{3}{10}\frac{|b|}{d}, \quad h = \frac{6}{25}\frac{b^2}{d}\varphi_h = \frac{3\sqrt{6}}{25\sqrt{5}}\frac{|b|^{5/2}}{d^{3/2}}.$$

In the temperature region $\tau_c(b) < \tau < \tau_{c2}(b)$ $(b < 0)$, there remains a possibility for the first-order phase transition as a function of external field.

Let us consider the behavior of susceptibility $\chi = \partial\varphi/\partial h$ satisfying the equation

$$\chi = \left(\tau + 3b\varphi^2 + 5d\varphi^4\right)^{-1}$$

where $\varphi = \varphi(h)$. In the low field limit $h \to 0$, we can put the order parameter equal to its value φ_0 in the lack of external field. As a result, we have

$$\chi^{-1}(\tau, b) = \begin{cases} \tau, & \text{the disordered phase } (\tau > 0), \\ \frac{-b+\sqrt{b^2-4d\tau}}{d}\sqrt{b^2 - 4d\tau}, & \text{the ordered phase.} \end{cases}$$

Similar to the behavior of specific heat, we can distinguish two vicinities separated by the crossover line $\tau \sim b^{\varphi}/d$ demonstrating the *crossover exponent* $\varphi = 2$ in the behavior of susceptibility. The crossover line specifies the transient region from one critical behavior to another. It is seen that for $|\tau| \gtrsim b^2/d$, we disclose the typical behavior corresponding to the Curie law $\chi \sim |\tau|^{-1}$. In the vicinity $|\tau| \lesssim b^2/d$, there appears a stronger singular behavior $\chi \sim d/b^2$ in the ordered phase as $|b| \to 0$.

It is interesting to compare the magnitudes of susceptibility at the first-order phase transition line $\tau_c = 3b^2/(16d)$ from the sides of disordered and ordered phases. So, we have

$$\chi(\tau_c) = \begin{cases} \frac{16}{3}\frac{d}{b^2} \text{ (the disordered phase)}, \\ \frac{4}{3}\frac{d}{b^2} \text{ (the ordered phase)}. \end{cases}$$

We see that there is a finite jump-like discontinuity of susceptibility at the first-order phase transition line. The jump magnitude equals $\Delta\chi_c = 4d/b^2$ and increases unlimitedly on approaching the tricritical point $b = 0$ and $\tau = 0$. As for the line of second-order phase transition, the susceptibility turns always to be infinite regardless of the external field magnitude.

In conclusion, we will discuss the limits for applicability of the mean-field approximation in order to describe the critical behavior at the tricritical point. First of all, we are interested in the upper critical dimension of the tricritical point where coefficient $b = 0$. Let us employ the Ginzburg–Levanyuk criterion requiring the smallness of order parameter fluctuations in volume $V_\xi \sim \xi^D$ with the linear size of about the correlation length ξ. Accordingly,

$$\langle(\Delta\varphi)^2\rangle = \frac{T_t\chi}{V_\xi} \sim \frac{T_t\chi}{\xi^D} \ll \varphi_0^2$$

where $\chi \sim |\tau|^{-1}$ is the susceptibility and D is the spatial dimensionality.

For estimating the correlation length, it is necessary to involve the inhomogeneity energy equal to $c(\nabla\varphi)^2/2$ and compare it with energy $|\tau|\varphi^2/2$. Then this gives the correlation length $\xi \sim (c/|\tau|)^{1/2}$. At the tricritical point $b = 0$, as we have seen above, the equilibrium value of order parameter corresponds approximately to relation $|\tau|\varphi_0^2 \sim d\varphi_0^6$. Hence we have $\varphi_0^2 \sim (|\tau/d)^{1/2}$ and we arrive at the inequality

$$T_t \frac{1}{|\tau|}\left(\frac{|\tau|}{c}\right)^{D/2} \ll \left(\frac{|\tau|}{d}\right)^{1/2} \quad \text{or} \quad \frac{1}{|\tau|^{D-3}} \gg \frac{dT_t^2}{c^D} = (\mathrm{Gi})^{3-D}.$$

In the limit $\tau \to 0$, the last inequality is always satisfied for the space dimensionality $D > 3$ and breaks down for $D < 3$. Thus, for the tricritical point $b = 0$ and interaction energy $\sim \varphi^6$, the upper critical dimension equals three. The predictions of mean-field theory remain valid for the critical exponents at $D > 3$. For $D = 3$, the consideration of the order parameter fluctuations can only result in the appearing of the logarithmic corrections to the results of mean-field theory. Two crossover lines $\tau \sim \pm b^{\varphi}/d$ with the exponent $\varphi = 2$ separate the regions of different critical behavior.

Problem

Estimate the fluctuation correction to the specific heat δC_{sing} at the tricritical point $b = 0$ on the side of disordered phase $\tau < 0$.

Solution. Let us decompose the effective Hamiltonian to the quadratic deviations $\Delta\varphi = \varphi - \varphi_0$ near the equilibrium value of order parameter $\varphi_0 = (|\tau|/d)^{1/2}$:

$$H(\varphi) = H(\varphi_0) + \frac{c}{2}(\nabla\varphi)^2 + |\tau|(\Delta\varphi)^2 + \dots$$

Since this expression for the effective Hamiltonian is analogous to the ones considered in the previous sections, we can readily write the answer

$$\delta C_{\text{sing}} = \frac{V}{8\pi}\left(\frac{2}{c}\right)^2 \xi = \frac{V}{8\pi}\left(\frac{2}{c}\right)^2 \sqrt{\frac{c}{2|\tau|}} \sim \frac{1}{c^{3/2}|\tau|^{1/2}}.$$

This result should be compared with that $C = V(4T_t\sqrt{d|\tau|})^{-1}$ of mean-field approximation. The equally divergent temperature dependences $|\tau|^{-1/2}$ of both contributions are evidence for a smaller role of fluctuations of the order parameter field near the multicritical point.

4.14 Phases with the Incommensurate Periodicity: The Lifshitz Point

There exists a vast number of condensed media in which the formation of spatially inhomogeneous structures occurs as a result of the phase transition. From the general point of view, such structures can be characterized by the spatial period which may be commensurate or incommensurate with the original spatial structure. Accordingly, such phases are referred to as *commensurate* or *incommensurate*. Below we analyze the transitions by violating the translational symmetry and emerging the long-period structures. The examples of such transitions are the formation of charge density waves in ferroelectrics, spin density waves or helicoidal structures in antiferromagnets, and spatially inhomogeneous Fulde–Ferrell–Larkin–Ovchinnikov (FFLO) phase in superconductors.

In what follows, we study an existence of long-period modulated spatial structures within the framework of the mean-field (self-consistent) model. We imply that the wave vectors associated with the spatial structure period are much smaller as compared with the inverse interatomic distances. On the whole, the thermodynamics of phase transitions with the formation of the modulated structures has a number of specific features. One of most striking and instructive examples are the transitions associated with an existence of the *Lifshitz points*.

The spatially modulated phases, as a solution of the problem on the minimum of some effective Hamiltonian (thermodynamic potential), may appear when additional terms with the spatial derivatives of order parameter augment the effective Hamiltonian. Next, we consider the simplest example of such an effective Hamiltonians:

$$H[\varphi(\boldsymbol{r})] = \int d\boldsymbol{r} \left(\frac{d}{4}(\nabla^2\varphi)^2 + \frac{c}{2}(\nabla\varphi)^2 + \frac{\tau}{2}\varphi^2 + \frac{b}{4}\varphi^4 \right), \quad (b,\, d > 0).$$

We naturally assume the coefficients b and d for higher powers of order parameter to be positive. The sign of coefficient c can be arbitrary and change from positive to negative under variation of thermodynamic variable additional to the temperature, e.g. pressure or magnetic field. Coefficient τ is the relative closeness to the phase transition temperature.

It is clear that for coefficient $c > 0$, the emergence of spatially inhomogeneous state is energetically unfavorable. As for $c < 0$, the minimum of effective Hamiltonian can correspond to the state with the spatially varying order parameter $\varphi(\boldsymbol{r})$. In this case the point where coefficients c and τ vanish together, i.e. $\tau = 0$ and $c = 0$, is referred to as[17] the *Lifshitz point*.

For analyzing the phase states of the given effective Hamiltonian, it is convenient to represent the order parameter as an expansion into a Fourier series

$$\varphi(\boldsymbol{r}) = \sum_{\boldsymbol{k}} \varphi_{\boldsymbol{k}} e^{i\boldsymbol{k}\boldsymbol{r}}.$$

Then the effective Hamiltonian reads

$$H[\varphi] = \sum_{\boldsymbol{k}} \frac{\tau(k)}{2} |\varphi_{\boldsymbol{k}}|^2 + \frac{b}{4} \int d\boldsymbol{r}\, \varphi^4(\boldsymbol{r})$$

where coefficient $\tau(k)$ is determined with the simple relation

$$\tau(k) = \tau + ck^2 + \frac{d}{2}k^4.$$

[17] To specify the type singular point, one may indicate the tricritical (in our case) or tetracritical Lifshitz point, etc.

The instability of disordered phase with order parameter $\varphi(\boldsymbol{r})$ will be due to appearance of negative value $\tau(k)$ at some k. For $c > 0$, the minimum $\tau(k)$, depending on wave vector k, realizes at $k = 0$ and, for the first time, coefficient $\tau(k)$ changes the sign at $\tau = 0$. As a consequence, in the region $c > 0$ at $\tau = 0$ there occurs a usual second-order phase transition to the homogeneous ordered phase with zero wave vector $k = 0$ and order parameter $\varphi(\boldsymbol{r}) = \varphi_0 = (|\tau|/b)^{1/2}$ constant in space. For this homogeneous ordered phase, the value of effective Hamiltonian will equal

$$H_0 = H(\varphi_0) = -\frac{\tau^2}{4b}\,.$$

In the region $c < 0$ the minimum $\tau(k)$ is achieved at the finite value of wave vector

$$k_0 = \left(|c|/d\right)^{1/2}$$

and the sign $\tau(k)$ changes from positive to negative at the transition temperature

$$\tau_{m_+} = \frac{c^2}{2d}\,.$$

For $\tau < \tau_{m_+}$, the spatially modulated and ordered phase becomes energetically more favorable. The order parameter varies in the wave-like manner in space according to equation

$$\varphi(\boldsymbol{r}) = \varphi_m \cos \boldsymbol{kr}$$

with some amplitude φ_m and period L of spatial structure

$$L = \frac{2\pi}{k_0} = 2\pi\left(\frac{d}{|c|}\right)^{1/2}.$$

We find the value of effective Hamiltonian for the modulated ordered phase from the relation

$$H_m = H(\varphi_m) = \frac{\tau}{2}\langle\varphi^2(\boldsymbol{r})\rangle + \frac{ck^2}{2}\langle\varphi^2(\boldsymbol{r})\rangle + \frac{dk^4}{4}\langle\varphi^2(\boldsymbol{r})\rangle + \frac{b}{4}\langle\varphi^4(\boldsymbol{r})\rangle.$$

Here notation $\langle\cdots\rangle$ implies averaging over the period of spatial structure. Taking $\langle\cos^2 \boldsymbol{kr}\rangle = 1/2$ and $\langle\cos^4 \boldsymbol{kr}\rangle = 3/8$ into account, we have

$$H_m = \frac{\tau(k)}{4}\varphi_m^2 + \frac{3}{32}b\varphi_m^4$$

where φ_m is the amplitude of order parameter modulations. The condition of minimum H_m leads us to the amplitude of modulations and the value of effective Hamiltonian

$$\varphi_m^2 = -\frac{4}{3}\frac{\tau - \tau_{m_+}}{b} \quad \text{and} \quad H_m = -\frac{(\tau - \tau_{m_+})^2}{6b} \quad \text{if} \quad \tau \leqslant \tau_{m_+}.$$

The transition to the modulated ordered phase from homogeneous disordered phase represents the second-order phase transition since the order parameter φ_m or modulation amplitude changes continuously for such a transition.

Comparing the values H_m and H_0, we see that value H_0 becomes smaller as compared with value H_m for sufficiently large negative values τ. The transition from the modulated ordered phase to the homogeneous ordered one will take place at equality $H_m(\tau_{m_-}) = H_0(\tau_{m_-})$. Hence we arrive readily at the transition temperature τ_{m_-} equal to

$$\tau_{m_-} = -\frac{\sqrt{2}}{\sqrt{3} - \sqrt{2}}\tau_{m_+}.$$

The phase transition at temperature τ_{m_-} proves to be first-order one. The magnitude of the order parameter modulus and spatial structure period change in the jump-like manner. The mean-field theory gives the jumps of finite magnitude for the specific heat at all phase transition lines.

The specific point, where parameters are $\tau = 0$ and $c = 0$ at the same time, represents the tricritical Lifshitz point and, in addition, is the point of phase states where all the phase transition lines merge, two lines being second-order and one line being first-order. For $c < 0$, the modulated ordered phase exists in the temperature region $\tau_{m_-} < \tau < \tau_{m_+}$. The region for an existence of homogeneous ordered phase is $\tau < \tau_{m_-}$ at $c < 0$ and $\tau < 0$ at $c > 0$. The homogeneous disordered phase remains in the region $\tau > \tau_{m_+}$ at $c < 0$ and $\tau > 0$ at $c > 0$. All three phases coexist at the Lifshitz point (Fig. 4.8).

The tendency to the formation of spatially modulated structure can be seen by studying the dispersion of susceptibility or linear response $\chi(\boldsymbol{r})$ to the inhomogeneous external field $h(\boldsymbol{r})$. Let us write the definition of linear response

$$\varphi(\boldsymbol{r}) = \int \chi(\boldsymbol{r} - \boldsymbol{r}')h(\boldsymbol{r}')d\boldsymbol{r}'$$

which reduces to the simple algebraic relation in the Fourier transform

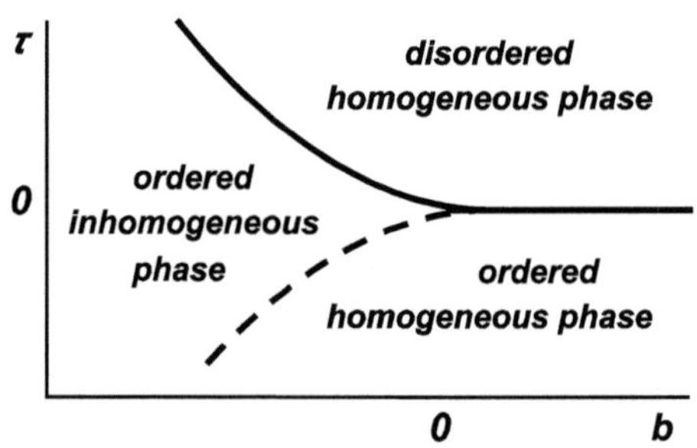

Fig. 4.8 The diagram of phase states. The solid line is second-order phase transition. The dashed line is first-order phase transition

$$\varphi_k = \chi(k)h_k\,.$$

In what follows, we consider the sufficiently high temperatures when order parameter φ is sufficiently small and we can neglect the fourth-order term $b\varphi^4/4$ as compared with quadratic ones. Varying the effective Hamiltonian

$$H[\varphi(\boldsymbol{r})] = \int d\boldsymbol{r}\left(\frac{d}{4}(\nabla^2\varphi)^2 + \frac{c}{2}(\nabla\varphi)^2 + \frac{\tau}{2}\varphi^2 - h(\boldsymbol{r})\varphi\right)$$

results in equation determining the order parameter field

$$\frac{\delta H}{\delta\varphi(\boldsymbol{r})} = \frac{d}{2}\nabla^2\left(\nabla^2\varphi\right) - c\nabla^2\varphi + \tau\varphi - h(\boldsymbol{r}) = 0.$$

Due to linearity of this equation, we readily get susceptibility $\chi(k)$

$$\left(\frac{d}{2}k^4 + ck^2 + \tau\right)\varphi_k = h_k \quad \text{or} \quad \chi^{-1}(k) = \tau - \tau_{m_+} + \frac{d}{2}\left(k^2 + \frac{c}{d}\right)^2.$$

Evidently, the most interesting situation arises for $c < 0$. In this case the linear response as a function of wave vector k has the maximum at $k = k_0 = (|c|/d)^{1/2}$, which enhances on approaching the transition point $\tau = \tau_{m_+}$. This indicates the instability of homogeneous state for the order parameter field.

The correlation length ξ at $c = 0$ near transition can be estimated by comparing the energy $\tau\varphi^2/2$ and energy of inhomogeneity $d(\nabla^2\varphi)^2/4$ on the neglect of term $b\varphi^4$ due to its smallness. Putting $\tau\varphi^2/2 \sim d(\nabla^2\varphi)^2/4$ and estimating $d(\nabla^2\varphi)^2 \sim d\varphi^2/\xi^2$, we have $\xi \sim (d/\tau)^{1/4}$. This gives the critical exponent $\nu = 1/4$ instead of 1/2 far away from the Lifshitz point.

The specific feature of Lifshitz point $c = 0$ is a trend to the infinite period $L \sim \tau^{-1/4}$ for the spatial structure modulation of order parameter field. On the whole, this promotes for increasing the order parameter fluctuations and enlarging the region of critical fluctuations near the Lifshitz point at $c = 0$. The upper critical dimensionality becomes equal to $d = 8$.

Problems

1. Find the fluctuation correction δC_{sing} to the specific heat in the approximation of effective free-field Hamiltonian with the Lifshitz point

$$\mathcal{H}_{Lif}[\varphi(\boldsymbol{r})] = \int d\boldsymbol{r}\left(\frac{d}{4}(\nabla^2\varphi)^2 + \frac{c}{2}(\nabla\varphi)^2 + \frac{\tau}{2}\varphi^2\right), \quad (c < 0,\ d > 0)$$

in the vicinity of the second-order phase transition line $\tau = \tau_{m_+} = c^2/2d$.

Solution. Let us write the effective Hamiltonian in the Fourier representation

$$\mathcal{H}_{Lif} = \frac{1}{2}\sum_{\boldsymbol{k}}(\tau + ck^2 + dk^4/2)|\varphi_{\boldsymbol{k}}|^2 = \frac{1}{2}\sum_{\boldsymbol{k}}(\tau - \tau_{m+} + d(k^2 - k_0)^2/2)|\varphi_{\boldsymbol{k}}|^2$$

where $k_0 = (|c|/d)^{1/2}$ is the wave vector of the modulated ordered structure. On the analogy with the previous considerations, we write the thermodynamic potential

$$F = -T\sum_k \ln\frac{2\pi T}{\tau - \tau_{m_+} + d(k^2 - k_0^2)^2/2} \approx -T_{m+}\sum_k \ln\frac{2\pi T_{m+}}{\tau - \tau_{m_+} + d(k^2 - k_0^2)^2/2}.$$

Accordingly, resulting from the order parameter fluctuations, the singular correction to the specific heat will be equal to

$$\delta C_{\text{sing}}(\tau) = -T_{m+}^{-1}\frac{\partial^2 F}{\partial \tau^2} = \sum_k \frac{1}{(\tau - \tau_{m_+} + d(k^2 - k_0^2)^2/2)^2} =$$

$$= \int \frac{V d^3k}{(2\pi)^3}\frac{1}{(\tau - \tau_{m_+} + d(k^2 - k_0^2)^2/2)^2} = \frac{2V}{\pi^2 d^2 k_0^5}\int_0^\infty \frac{x^2\,dx}{\left(\frac{2(\tau-\tau_{m+})}{dk_0^4} + (x^2 - 1)^2\right)^2}.$$

After integration we obtain the correction

$$\delta C_{\text{sing}}(\tau) = \frac{V}{4\pi\sqrt{2}}\frac{1}{\left(c^2 + 2d(\tau - \tau_{m_+})\right)^{1/2}}\frac{d^{1/2}}{\left(\left[c^2 + 2d(\tau - \tau_{m_+})\right]^{1/2} - |c|\right)^{3/2}}$$

and its limiting behavior

$$\delta C_{\text{sing}}(\tau) = \frac{V}{4\pi\sqrt{2}}\begin{cases} \frac{|c|^{1/2}}{d^{1/2}}\frac{1}{(\tau-\tau_{m_+})^{3/2}}, & (\tau - \tau_{m_+}) \ll c^2/2d, \\ \frac{1}{d^{1/4}}\frac{1}{\left(2(\tau-\tau_{m_+})\right)^{5/4}}, & (\tau - \tau_{m_+}) \gg c^2/2d. \end{cases}$$

The presence of Lifshitz point $c = 0$ within its vicinity results in the appearing of the crossover phenomena separated by the crossover lines $\tau - \tau_{m+} \sim c^2/2d$ with the crossover exponent $\varphi = 2$.

2. Find the correlation function $K(x)$ for the fluctuations of order parameter field $\varphi(x)$ described with the effective Hamiltonian of the previous problem in the case of one-dimensional space $d = 1$.

Solution. The correlation function is determined with the relation

$$K(x) = \sum_k \langle|\varphi_k|^2\rangle e^{ikx} = \int_{-\infty}^{\infty}\frac{dk}{2\pi}\frac{2T\,e^{ikx}}{\tau + ck^2 + dk^4/2} = \frac{T}{\pi}\int_{-\infty}^{\infty}\frac{\cos(kx)\,dk}{\tau - \tau_{m+} + d(k^2 - k_0^2)^2/2}.$$

As a above, we imply here $c < 0$, $d > 0$, $k_0^2 = |c|/d$, and $\tau > \tau_{m_+}$. Calculating the integral, e.g. with the aid of residue theorem, we obtain

$$K(x) = \frac{T}{\sqrt{c^2 + 2d(\tau - \tau_{m_+})}}\left(\xi\cos\frac{|x|}{\eta} + \eta\sin\frac{|x|}{\eta}\right)e^{-|x|/\xi}$$

where

$$\xi = \left(\frac{2d}{\sqrt{c^2 + 2d(\tau - \tau_{m+})} - |c|}\right)^{1/2} \quad\text{and}\quad \eta = \left(\frac{2d}{\sqrt{c^2 + 2d(\tau - \tau_{m+})} + |c|}\right)^{1/2}.$$

First, according to the expressions obtained, we see that the correlation function decays exponentially at the distances exceeding the correlation length ξ which becomes infinite at the phase transition point $\tau = \tau_{m_+}$. Second, the existence of the Lifshitz point, inducing the transition to the modulated phase, results in the oscillations of correlation function with the period equal to $2\pi\eta$. The correlation function also reflects the crossover phenomena in the region of line $\tau - \tau_{m_+} \sim c^2/2d$. In fact,

$$\xi \sim \left(\frac{|c|}{\tau - \tau_{m_+}}\right)^{1/2} \quad \text{and} \quad \eta \sim \left(\frac{d}{|c|}\right)^{1/2} \quad \text{at} \quad \tau - \tau_{m_+} \ll c^2/2d,$$
$$\xi \sim \left(\frac{2d}{\tau - \tau_{m_+}}\right)^{1/4} \quad \text{and} \quad \eta \sim \left(\frac{2d}{\tau - \tau_{m_+}}\right)^{1/4} \quad \text{at} \quad \tau - \tau_{m_+} \gg c^2/2d.$$

4.15 Fundamentals of Critical Phenomena

When approaching the point of second-order phase transition, the correlation radius ξ for the fluctuations of order parameter field $\varphi(\boldsymbol{r})$ increases unlimitedly and becomes the single largest parameter of length in the system, exceeding by many times all the other possible lengths such as interparticle distances (lattice constants) or radii of interparticle interactions. In each region or cell with size ξ and volume $V_\xi \sim \xi^d$, the value of order parameter remains approximately the same and the total order parameter

$$\Phi_\xi = \sum_{\boldsymbol{r}\in V_\xi} \varphi(\boldsymbol{r})$$

grows directly proportional to V_ξ as correlation radius ξ increases. Near the phase transition, we can imagine the thermodynamic system as a set of macroscopic regions or cells, the order parameter being ordered singly and independently in each of the cells. However, as a whole, the system does not look as an ordered one since these cells are disordered with respect to each other.

If all the linear sizes are measured in units of correlation radius ξ and the total order parameter is done in units of Φ_ξ, the change of closeness to transition τ will not result in varying the thermodynamic quantities and functions. Within the framework of such a *scaling hypothesis* or *scaling invariance*, the behavior of all thermodynamic variables is expressed in terms of power laws as a function of τ as well as the behavior of correlators as a function of distances.

As is noted above, in the theory of critical phenomena the singular behavior of physical quantities in the immediate vicinity of phase transition is characterized with the *critical exponents*. The latter ones describe the power-like behavior of physical quantities as a function of their closeness τ to the transition point. The main critical exponents α, β, γ, and δ are introduced for specific heat C, order parameter φ at $\tau \leqslant 0$, and susceptibility χ in the low and high external field h:

$$C \sim |\tau|^{-\alpha}, \quad \varphi \sim (-\tau)^{\beta} \quad (\tau \leqslant 0), \quad \chi \sim |\tau|^{-\gamma}, \quad \chi \sim h^{1/\delta - 1}.$$

To specify the correlations of order parameter field φ in the disordered phase for $\tau > 0$ and at the very point of phase transition, two more critical exponents are introduced such as exponent of correlation length ν according to

$$\langle\varphi(\boldsymbol{r})\varphi(\boldsymbol{r}')\rangle_{|\boldsymbol{r}-\boldsymbol{r}'|\to\infty} \sim \exp\big(-|\boldsymbol{r}-\boldsymbol{r}'|/\xi\big), \quad \xi\sim\tau^{-\nu} \quad (\tau>0)$$

and the exponent of anomalous dimensionality or Fisher exponent η according to

$$\langle\varphi(\boldsymbol{r})\varphi(\boldsymbol{r}')\rangle_{|\boldsymbol{r}-\boldsymbol{r}'|\to\infty} \sim |\boldsymbol{r}-\boldsymbol{r}'|^{-(d-2+\eta)} \quad (\tau=0)$$

where d is the dimensionality of the space.

We first accept the fluctuation field of order parameter near the phase transition point to be described with the single macroscopic length scale referred to as the correlation length ξ. Since the correlation length becomes infinite at the very phase transition point and thus the natural length unit or length scale vanishes, we may assume an existence of *scale invariance* or *scaling*. A similar change of distances, i.e. scaling, in the lack of typical size or scale

$$\boldsymbol{r}\to\boldsymbol{r}'=\lambda\boldsymbol{r}$$

cannot change the state of thermodynamic system since this transformation reduces to varying the length. For such a scaling transformation, the physical quantity $A(\boldsymbol{r})$ of thermodynamic system, e.g. order parameter φ, energy density ε, temperature τ, and external field h, should be multiplied by some power λ according to the law

$$A(\boldsymbol{r})\to A'(\boldsymbol{r}')=\lambda^{-\Delta_A}A(\boldsymbol{r}).$$

The exponent Δ_A is called the *scaling dimensionality*.[18] The hypothesis of scale invariance or scaling hypothesis incorporates the statement that the scale change of quantities A_1, A_2, ... cannot affect the relations in the theory. In other words, the factors λ should vanish from all relations after the scaling transform.

The scaling dimensionality of coordinate, say x, is obviously equal to $\Delta_x=-1$. Accordingly, the scaling dimensionality for volume element $dV=dx_1dx_2\ldots dx_d$ will equal $\Delta_V=-d$, d being the number of spatial dimensions for the thermodynamic system. Accordingly, the dimensionality for effective Hamiltonian density is $\Delta_{\mathcal{H}}=d$. The dimensionality of the thermodynamically conjugated quantities equals d in a sum. For example, requiring the invariance for contribution $\varphi h\,dV$ of external field h to the effective Hamiltonian, we readily arrive at equality $\Delta_\varphi+\Delta_h=d$. The important role is assigned to the quantity $s(\boldsymbol{r})$ thermodynamically conjugate to temperature τ. This quantity can be called the entropy density. Following the general rule for the conjugate quantities, we have $\Delta_s+\Delta_\tau=d$.

Assuming the validity of the scaling hypothesis, we can relate the scaling dimensions with the critical exponents and derive a number of scaling relations between the critical exponents. In fact, since $\Delta_\xi=-1$ for correlation length and $\xi(\tau)\sim\tau^{-\nu}$, we find straightforwardly that $\Delta_\tau=1/\nu$. Then, because of $\varphi(\tau)\sim|\tau|^\beta$, we have $\Delta_\varphi=\beta\Delta_\tau=\beta/\nu$.

[18] In general, the scaling dimensionality does not coincide with the usual dimensionality of physical quantity.

Let us determine the scaling relations for the critical exponents characterizing the singular behavior of thermodynamic system near the critical point in the low external field. According to the definition, the singular behavior of specific heat $C(\tau)$ is given by the second derivative of effective Hamiltonian density with respect to temperature

$$C(\tau) = -\frac{\partial^2 \mathcal{H}}{\partial \tau^2} \quad \text{and} \quad C(\tau) \sim \tau^{-\alpha}.$$

Hence, employing the relations for dimensions

$$\Delta_C = \Delta_{\mathcal{H}} - 2\Delta_\tau = d - 2\Delta_\tau \quad \text{and} \quad \Delta_C = -\alpha\Delta_\tau\,,$$

and $\Delta_\tau = 1/\nu$, we obtain the first equality

$$2 - \alpha = d\nu.$$

To determine the second relation, we write the following expressions for susceptibility:

$$\chi(\tau) = \frac{\partial \varphi}{\partial h} \quad \text{and} \quad \chi(\tau) \sim \tau^{-\gamma}.$$

Next, taking $\Delta_\varphi + \Delta_h = d$ and $\Delta_\varphi = \beta\Delta_\tau$, we find

$$\Delta_\chi = \Delta_\varphi - \Delta_h = 2\Delta_\varphi - d = 2\beta\Delta_\tau - d \quad \text{and} \quad \Delta_\chi = -\gamma\Delta_\tau\,.$$

As a result, we have $2\beta + \gamma = d/\Delta_\tau = d\nu$. Recalling the relation $d\nu = 2 - \alpha$ found before, we arrive at the following important equality relating three critical exponents:

$$\alpha + 2\beta + \gamma = 2.$$

In the spatially isotropic phases, the fluctuations of thermodynamic quantities should also be isotropic. This means that the correlators of scalar quantities must depend only on the mutual distance $|\boldsymbol{r}_1 - \boldsymbol{r}_2|$ between the points. In the important case of the pair correlator

$$K_{AB}(\boldsymbol{r}_1, \boldsymbol{r}_2) = \langle A(\boldsymbol{r}_1)B(\boldsymbol{r}_2)\rangle$$

for two scalar quantities A and B, the symmetrical properties of isotropy, homogeneity, and scaling invariance yield the power law at the phase transition point $\tau = 0$

$$K_{AB}(\boldsymbol{r}_1, \boldsymbol{r}_2) = \frac{k_{AB}}{|\boldsymbol{r}_1 - \boldsymbol{r}_2|^{\Delta_A + \Delta_B}}.$$

The constant k_{AB} depends on the properties of quantities A, B, and microscopic characteristics of the phases. In the free-field model, the pair correlator $\langle\varphi(\boldsymbol{r}_1)\varphi(\boldsymbol{r}_2)\rangle$

for the order parameter fluctuations has the same power form with $2\Delta_\varphi = d - 2$. In the general case, the scaling dimension of order parameter field $\varphi(\boldsymbol{r})$ is associated with the anomalous dimension exponent or Fisher exponent η as

$$\Delta_\varphi = \big(d - 2 + \eta\big)/2 \quad \text{or} \quad \beta = \nu\big(d - 2 + \eta\big)/2.$$

Thus, the scaling hypothesis allows us to determine the structure of pair correlator within the accuracy to numerical factor. Comparing the last expression and $2\beta + \gamma = d\nu$ results in the equality

$$\gamma = \nu(2 - \eta).$$

For the many-particle correlator, the scaling hypothesis gives the condition of homogeneity alone

$$\langle A_1(\lambda \boldsymbol{r}_1) \dots A_n(\lambda \boldsymbol{r}_n)\rangle = \lambda^{-\Delta_{A_1} - \dots - \Delta_{A_n}} \langle A_1(\boldsymbol{r}_1) \dots A_n(\boldsymbol{r}_n)\rangle.$$

Hence, in particular, $\Delta_{\varphi^n} = n\Delta_\varphi$, n being the positive integer.

Similar scaling relations can also be established for the critical exponents characterizing the singular behavior at the critical point in the high external field h. If the behavior of order parameter as a function of external field is described by the power law $\varphi(h) \sim h^{1/\delta}$ with exponent δ, we have $\delta = \Delta_h/\Delta_\varphi = (d - \Delta_\varphi)/\Delta_\varphi$. Then, taking $\Delta_\varphi = (d - 2 + \eta)/2$ into account, we arrive at the following scaling relation:

$$\delta = \frac{d + 2 - \eta}{d - 2 + \eta}\,.$$

Using this relation, we can represent relation $\beta + \gamma = d\nu - \beta$ with the aid of exponent δ as

$$\beta + \gamma = \beta\delta.$$

On the whole, the scaling hypothesis or *Widom scaling* assumes that the density of effective Hamiltonian $\mathcal{H}(\tau, h)$ or singular part of free energy can be written as the following homogeneous function:

$$\mathcal{H}\big(\lambda^{\Delta_\tau}\tau,\, \lambda^{\Delta_h} h\big) = \lambda^d \mathcal{H}\big(\tau,\, h\big) \quad \text{where} \quad d = \Delta_\tau + \Delta_h\,.$$

By means of two exponents, e.g. ν and η, the Widom scaling allows us to express all the other critical exponents.

Similar scaling relations can be established for the critical exponents of parameter order φ, specific heat C, and correlation length ξ in the high magnetic field

$$\varphi(h) \sim h^{1/\delta}, \quad C(h) \sim h^{-\epsilon} \quad \text{and} \quad \xi(h) \sim h^{-\mu}.$$

Correspondingly, we have $\Delta_\varphi = \Delta_h/\delta$, $\Delta_h = 1/\mu$, and $\Delta_C = -\epsilon\Delta_h$. Recalling that $\Delta_\varphi = \beta/\nu$ and $\Delta_C = -\alpha/\nu$ as well, we arrive at the following equalities: $\beta\delta\mu = \nu$

and $\alpha\mu = \epsilon\nu$. Using that $\beta\delta = \beta + \gamma$, we rewrite them in the equivalent form

$$(\beta + \gamma)\mu = \nu \quad \text{and} \quad \epsilon(\beta + \gamma) = \alpha.$$

In the three-dimensional case the experimental data and numerical simulations show that, as a rule, the Fisher exponent and critical exponent for specific heat have relatively small magnitudes: $\alpha \lesssim 0.1$ and $\eta \lesssim 0.05$. If we put approximately $\alpha = 0$ and $\eta = 0$, the estimate for the other exponents gives $\beta = 1/3 \sim 0.3, \gamma = 4/3 \sim 1.3$, $\nu = 2/3 \sim 0.7$, and $\delta = 5$ instead of those in the Landau theory of self-consistent field as $\beta = 1/2$, $\gamma = 1$, $\nu = 1/2$, and $\delta = 3$. It can be readily seen that the Landau theory of self-consistent field does not satisfy all the relations of scaling.

The magnitudes of critical exponents depend on the number of spatial variables as well as on the symmetry and number of components for the order parameter field. However, in spite of large variety of thermodynamic systems the critical phenomena have the *property of universality*, meaning that the limiting and singular properties for a wide class of the thermodynamic systems are independent of microscopic details in these systems. Accordingly, the thermodynamic systems, demonstrating the same critical behavior, are unified in the same *universality class*.

4.16 Approximate Calculation of Critical Exponents: The Renormalization-Group Method

The most important problem in the theory of critical phenomena is to calculate the critical exponents and determine the equation of state. The critical exponents depend on the number of spatial variables d. In the space $d \geqslant 4$ the critical exponents in the self-consistent field approximation satisfy the scaling relations but in the dimensions $d < 4$ this is invalid. Therefore, the spatial dimension $d = 4$ is the upper critical dimension and can be taken as initial zero approximation for constructing the perturbation theory or *ϵ-expansion* with the continuous mathematical transition to the space of smaller dimension $d = 4 - \epsilon$. Naturally, the physical sense in such expansions has the integer values $\epsilon = 1, 2, 3$ alone.

The typical scale for the fluctuations of order parameter $\varphi(\boldsymbol{r})$ grows unlimitedly by approaching the phase transition point. The long-wave fluctuations of order parameter can be described with the aid of effective Landau Hamiltonian[19]

$$H = H[\varphi_{\boldsymbol{q}};\, q_0] = \int d\boldsymbol{r} \left[\frac{(\nabla\varphi)^2}{2} + \frac{\tau_0\varphi^2}{2} + g_0\varphi^4 \right]$$

in which the field of order parameter $\varphi(\boldsymbol{r})$ is smoothed. The *smoothness* means that, in the Fourier expansion of order parameter field

---

[19] Here we select the normalization of the order parameter with the stiffness coefficient equal to unity.

$$\varphi(\boldsymbol{r}) = \sum_{q<q_0} \varphi_{\boldsymbol{q}} e^{i\boldsymbol{q}\boldsymbol{r}},$$

the harmonics are absent for the large wave vectors $q \geqslant q_0$. The choice for the magnitude q_0 is conditional to some extent. However, the linear scale $1/q_0$ is implied to be macroscopic.

The solution for the problem on continuous phase transition reduces in essence to analyzing the partition function

$$Z = \operatorname{tr} e^{-H} = \int e^{-H[\varphi_{\boldsymbol{q}};\, q_0]} \prod_{q<q_0} d\varphi_{\boldsymbol{q}}$$

for the order parameter field φ described by the effective Landau Hamiltonian. The partition function is the functional (infinite-dimensional) integral over all possible configurations of order parameter field $\varphi_{\boldsymbol{q}}$. The direct calculation of such a functional integral is a very complicated problem.

The effective Hamiltonian $H[\varphi;\, q_0]$ depends on the cutting parameter q_0. In order to comprehend this behavior, we integrate $\exp(-H)$ over the wave vectors within interval $\lambda q_0 < q < q_0$ $(\lambda < 1)$ and determine a new effective Hamiltonian $H_\lambda = S_\lambda H$. The result of such an action can be represented as an integral

$$e^{-S_\lambda H} = \int e^{-H[\varphi_{\boldsymbol{q}};\, q_0]} \prod_{\lambda q_0<q<q_0} d\varphi_{\boldsymbol{q}}\,.$$

The transform from Hamiltonian $H[\varphi_{\boldsymbol{q}};\, q_0]$ to a new one $S_\lambda H = H[\varphi_{\boldsymbol{q}};\, \lambda q_0]$ is referred to as the *smoothing operation*. The smoothing operation separates the field $\varphi(\boldsymbol{r})$ into the slow $\varphi_0(\boldsymbol{r})$ and rapid $\varphi_1(\boldsymbol{r})$ parts

$$\varphi(\boldsymbol{r}) = \varphi_0(\boldsymbol{r}) + \varphi_1(\boldsymbol{r}),$$
$$\varphi_0(\boldsymbol{r}) = \sum_{q<\lambda q_0} \varphi_{\boldsymbol{q}} e^{i\boldsymbol{q}\boldsymbol{r}} \quad \text{and} \quad \varphi_1(\boldsymbol{r}) = \sum_{\lambda q_0<q<q_0} \varphi_{\boldsymbol{q}} e^{i\boldsymbol{q}\boldsymbol{r}}.$$

The smoothing operation is nonlinear, and its successive application reads $S_{\lambda_1} S_{\lambda_2} = S_{\lambda_1\lambda_2}$.

According to the scaling hypothesis, the smoothed Hamiltonian $S_\lambda H$ should have the same structure as the initial one H and remain unvaried at the very phase transition point. To compare the smoothed Hamiltonian $S_\lambda H$ with the initial one H, it is necessary to perform the inverse transformation of spatial scale to the previous one

$$q \to q' = \lambda^{-1} q$$

and the simultaneous scale transformation or *dilation* D_λ of order parameter field $\varphi(\boldsymbol{r})$:

$$\varphi_{\boldsymbol{q}} \to \varphi'_{\boldsymbol{q}'} = Z_\lambda^{-1} \varphi'_{\lambda \boldsymbol{q}'} = Z_\lambda^{-1} \varphi_{\boldsymbol{q}}\,.$$

The successive application of the smoothing S_λ and dilation D_λ is called the *renormalization* of Hamiltonian and $R_\lambda = D_\lambda S_\lambda$. The renormalized Hamiltonian

$$R_\lambda H[\varphi_{\boldsymbol{q}};\, q_0] = H[Z_\lambda \varphi_{\lambda^{-1}\boldsymbol{q}};\, \lambda^{-1}q_0] = H'[\varphi'_{\boldsymbol{q}};\, \lambda^{-1}q_0]$$

describes the fluctuations of a new field in the same region of wave vectors $q < q_0$. Provided that the thermodynamic system is at the critical point, the renormalized Hamiltonian should remain unchanged and cease to depend on the choice of wave vector. Otherwise, if the thermodynamic system does not reside at the critical point, we come to sufficiently large scales after the multiple usage of renormalization. The large scales mean the small λq_0 and thus exceed the correlation length, corresponding to the uncorrelated Gaussian fluctuations of order parameter field.

Let us start from renormalizing the effective Hamiltonian of free field H_0 in the space of dimension d

$$H_0[\varphi_{\boldsymbol{q}};\, q_0] = \frac{1}{2}\int\limits_{q<q_0} (q^2+\tau_0)|\varphi_{\boldsymbol{q}}|^2 \frac{d^dq}{(2\pi)^d}\,.$$

We turn first to the smoothing operation S_λ. Since the Fourier transforms with various wave vectors $\boldsymbol{q}$ do not interact with each other, the smoothing of Hamiltonian leads to the appearing of the constant term as a result of integration over the Fourier transform $\varphi_{\boldsymbol{q}}$ for the wave vectors between λq_0 and q_0. This term has a sense of the simple shift of energy reference point

$$S_\lambda H_0 = \frac{1}{2}\int\limits_{q<\lambda q_0} (q^2+\tau_0)|\varphi_{\boldsymbol{q}}|^2 \frac{d^dq}{(2\pi)^d} + \frac{1}{2}\sum_{\lambda q_0<q<q_0} \ln\frac{q^2+\tau_0}{\pi}\,.$$

We omit this constant term as insignificant and independent of order parameter field $\varphi_{\boldsymbol{q}}$. Performing the operation dilation D_λ, we arrive at the renormalized Hamiltonian

$$R_\lambda H_0[\varphi_{\boldsymbol{q}};\, q_0] = \frac{\lambda^d Z_\lambda^2}{2}\int\limits_{q<q_0} (\lambda^2 q^2+\tau_0)|\varphi_{\boldsymbol{q}}|^2 \frac{d^dq}{(2\pi)^d}\,.$$

For the free-field Hamiltonian, the temperature point $\tau_0 = 0$ is the critical point of phase transition. The renormalized Hamiltonian $R_\lambda H_0$ does not change if we take the scale

$$Z_\lambda = \lambda^{-(d+2)/2}.$$

In this case the temperature τ_0 will be transformed according to the law

$$\tau_0 \to \tau_0' = \lambda^{-2}\tau_0$$

corresponding to the scaling dimension $\Delta_\tau^0 = 2$.

Let us turn to the analysis of effective Landau Hamiltonian $H = H_0 + H_i$ where H_i is the interacting Hamiltonian

$$H_i = g_0 \int \varphi^4(\boldsymbol{r})\, d\boldsymbol{r} = g_0 \underset{q_1, q_2, q_3}{\iiint} \varphi_{q_1} \varphi_{q_2} \varphi_{q_3} \varphi_{-(q_1+q_2+q_3)} \frac{d^d q_1}{(2\pi)^d} \frac{d^d q_2}{(2\pi)^d} \frac{d^d q_3}{(2\pi)^d} .$$

The direct integration of the Landau Hamiltonian is a mathematically difficult problem due to presence of the coupling term with $\varphi^4(\boldsymbol{r})$. Thus, we perform the calculation of functional integral using the free-field Hamiltonian H_0 as zero approximation. Let us employ the equality

$$e^{-S_\lambda H} = \frac{\int e^{-H_i} e^{-H_0} \prod\limits_{\lambda q_0 < q < q_0} d\varphi_{\boldsymbol{q}}}{\int e^{-H_0} \prod\limits_{\lambda q_0 < q < q_0} d\varphi_{\boldsymbol{q}}} \int e^{-H_0} \prod_{\lambda q_0 < q < q_0} d\varphi_{\boldsymbol{q}} = \langle e^{-H_i} \rangle_0 \, e^{-S_\lambda H_0}$$

where the brackets mean the thermodynamic average with respect to the free-field Hamiltonian. Then we represent the smoothed Hamiltonian as

$$S_\lambda H = S_\lambda H_0 - \ln \langle e^{-H_i} \rangle_0 .$$

The calculation of the last term will be performed as an expansion into a series over the coupling constant g_0, assuming it sufficiently small. So, we have

$$\langle e^{-H_i} \rangle_0 = \sum_0^\infty (-1)^n \frac{g_0^n}{n!} \left\langle \left[\int \left(\varphi_0(\boldsymbol{r}) + \varphi_1(\boldsymbol{r})\right)^4 d\boldsymbol{r} \right]^n \right\rangle_0 .$$

The smoothing operation S_λ does not disturb the slow component $\varphi_0(\boldsymbol{r})$ of order parameter field having the Fourier transforms with wave vectors $q < \lambda q_0$ alone. The component $\varphi_0(\boldsymbol{r})$ will play a role of external parameter in calculations. As it concerns the rapidly changing component $\varphi_1(\boldsymbol{r})$ with harmonics $\lambda q_0 < q < q_0$, the average $\langle \cdots \rangle$ means the thermodynamic one with the free-field Hamiltonian H_0.

Let us restrict ourselves only with first and second order in the coupling constant g_0. Then, we have

$$\begin{aligned} \langle e^{-H_i} \rangle_0 =& 1 - g_0 \langle \int d\boldsymbol{r} \left[\varphi_0(\boldsymbol{r}) + \varphi_1(\boldsymbol{r})\right]^4 \rangle_0 + \\ &+ \frac{g_0^2}{2} \langle \iint d\boldsymbol{r}\, d\boldsymbol{r}' \left[\varphi_0(\boldsymbol{r}) + \varphi_1(\boldsymbol{r})\right]^4 \left[\varphi_0(\boldsymbol{r}') + \varphi_1(\boldsymbol{r}')\right]^4 \rangle_0 + \ldots \end{aligned}$$

It is more convenient to perform the calculation of averages in the Fourier representation, e.g. for the real field of the order parameter $\varphi(\boldsymbol{r})$

$$\langle\int d\boldsymbol{r}\varphi_1^2(\boldsymbol{r})\rangle_0 = \langle\varphi_1^2\rangle_0 = \int\limits_{\lambda q_0}^{q_0}\frac{d^dq}{(2\pi)^d}\langle|\varphi_1(\boldsymbol{q})|^2\rangle \text{ where } \langle|\varphi_1(\boldsymbol{q})|^2\rangle = \frac{1}{q^2+\tau_0},$$

and we take into account that $\varphi_1(-\boldsymbol{q}) = \varphi_1^*(\boldsymbol{q})$ and the average is $\langle\varphi_1(\boldsymbol{q})\varphi_1(\boldsymbol{q}')\rangle_0 = \langle|\varphi_1(\boldsymbol{q})|^2\rangle_0\delta_{\boldsymbol{q},-\boldsymbol{q}'}$. In what follows, we omit the integral signs for brevity.

Let us start from consideration of first-order corrections

$$\begin{aligned}\langle[\varphi_0+\varphi_1]^4\rangle_0 &= \varphi_0^4 + 4\varphi_0^3\langle\varphi_1\rangle_0 + 6\varphi_0^2\langle\varphi_1^2\rangle_0 + 4\varphi_0\langle\varphi_1^3\rangle_0 + \langle\varphi_1^4\rangle_0 = \\ &= \varphi_0^4 + 6\varphi_0^2G_\lambda + G_\lambda^{(2)}, \quad \text{where} \quad G_\lambda = \langle\varphi_1^2\rangle_0 \quad \text{and} \quad G_\lambda^{(2)} = \langle\varphi_1^4\rangle_0.\end{aligned}$$

Obviously, the terms with the odd number of φ_1 vanish and we find for first-order correction in the coupling constant

$$S_\lambda H = S_\lambda H_0 - \ln\langle e^{-H_i}\rangle_0 \approx H_0 + g_0\int\left[\varphi_0^4(\boldsymbol{r}) + 6\varphi_0^2(\boldsymbol{r})G_\lambda + G_\lambda^{(2)}\right]d\boldsymbol{r}.$$

We have used here that $\ln(1-x)\approx -x$ at $|x|\ll 1$. The last term independent of slow component $\varphi_0(\boldsymbol{r})$, i.e. constant, represents the insignificant shift of energy level reference and can be dismissed. The physical sense for the term with $\varphi_0^2(\boldsymbol{r})$ is a shift of the critical temperature of phase transition

$$\tau_0 \to S_\lambda\tau_0 = \tau_0 + \delta\tau_0 = \tau_0 + 12g_0G_\lambda + O(g_0^2).$$

It is natural that relation $\tau^* = \tau_0 + \delta\tau_0 = 0$ must be satisfied at the critical point of phase transition.

Let us now pay attention to the coupling constant. It can be readily seen that nothing takes place with the coupling constant g_0 in first order. The coupling constant remains the same

$$g_0 \to S_\lambda g_0 = g_0 + O(g_0^2).$$

Therefore, for determining the shift of the coupling constant, it is necessary to use the second order of the perturbation theory and take all the terms proportional to $g_0^2\varphi_0^4$ in $S_\lambda H$. Below we need the following approximation in expanding the logarithm $\ln\langle e^{-H_0}\rangle_0$

$$\ln\left(1 - g_0x + \frac{g_0^2}{2}y^2\right) \approx -g_0x + \frac{g_0^2}{2}\left(y^2 - x^2\right)$$

in limit $g_0 \ll 1$. Then, selecting the terms proportional to $g_0^2\varphi_0^4$ in the decomposition of logarithm $\ln\langle e^{-H_0}\rangle_0$, we find how the coupling constant transforms

$$g_0 \to S_\lambda g_0 = g_0 - 36g_0^2K_\lambda\,.$$

The irreducible correlator K_λ is given by the relation

$$K_\lambda = \frac{1}{2}\big[\langle\varphi_1^2(\boldsymbol{r})\varphi_1^2(\boldsymbol{r}')\rangle_0 - \langle\varphi_1^2(\boldsymbol{r})\rangle_0\langle\varphi_1^2(\boldsymbol{r}')\rangle_0\big].$$

Employing the operation of dilation D_λ yields the following transformations for temperature τ_0:

$$D_\lambda\tau_0 = Z_\lambda^2\lambda^d = \lambda^{-2}\tau_0$$

and for the coupling constant g_0:

$$D_\lambda g_0 = Z_\lambda^4\lambda^{3d} = \lambda^{d-4}g_0 = \lambda^{-\epsilon}g_0 \quad (d = 4 - \epsilon).$$

The appearance of factor λ^{3d} can be seen by representing the coupling term as a triple integral over wave vectors. Here it is also clear that the space dimension $d = 4$ is outlined in the sense $D_\lambda g_0 = g_0$. So, we accept always $\epsilon \ll 1$.

In what follows, we are interested in the wave vectors of the slow field components φ_0 much smaller as compared with the rapid components ($\lambda \ll 1$). For simplification, we can treat the wave vectors of slow field as zero ones, i.e. consider component $\varphi_0(\boldsymbol{r})$ as a constant quantity. The following relation for the Gaussian integrals is used to calculate correlator K_λ:

$$\int d\boldsymbol{r}\, d\boldsymbol{r}'\, \langle\varphi_1^2(\boldsymbol{r})\varphi_1^2(\boldsymbol{r}')\rangle_0 = \int d\boldsymbol{r}\, d\boldsymbol{r}'\, \langle\varphi_1^2(\boldsymbol{r})\rangle_0\langle\varphi_1^2(\boldsymbol{r}')\rangle_0 + \\ +2\int d\boldsymbol{r}\, d\boldsymbol{r}'\, \langle\varphi_1(\boldsymbol{r})\varphi_1(\boldsymbol{r}')\rangle_0\langle\varphi_1(\boldsymbol{r}')\varphi_1(\boldsymbol{r})\rangle_0.$$

This relation can straightforwardly be comprehended from the identities for the mean values of the Fourier harmonics

$$\langle\varphi_{\boldsymbol{q}}\varphi_{-\boldsymbol{q}}\varphi_{\boldsymbol{q}'}\varphi_{-\boldsymbol{q}'}\rangle_0 = \\ = \langle\varphi_{\boldsymbol{q}}\varphi_{-\boldsymbol{q}}\rangle_0\langle\varphi_{\boldsymbol{q}'}\varphi_{-\boldsymbol{q}'}\rangle_0 + \langle\varphi_{\boldsymbol{q}}\varphi_{-\boldsymbol{q}}\rangle_0^2\,\delta_{\boldsymbol{q}',-\boldsymbol{q}} + \langle\varphi_{\boldsymbol{q}}\varphi_{-\boldsymbol{q}}\rangle_0^2\,\delta_{\boldsymbol{q}',\boldsymbol{q}}$$

if we use an independence of Fourier harmonics

$$\langle\varphi_{\boldsymbol{q}}\varphi_{\boldsymbol{q}'}\rangle_0 = \langle|\varphi_{\boldsymbol{q}}|^2\rangle_0\,\delta_{\boldsymbol{q}',-\boldsymbol{q}} \quad (\varphi_{-\boldsymbol{q}} = \varphi_{\boldsymbol{q}}^*)$$

and equality $\langle|\varphi_{\boldsymbol{q}}|^4\rangle_0 = 3\langle|\varphi_{\boldsymbol{q}}|^2\rangle_0^2$. Finally, we have for the irreducible correlator K_λ

$$K_\lambda = \int d\boldsymbol{r}\, d\boldsymbol{r}'\langle\varphi_1(\boldsymbol{r})\varphi_1(\boldsymbol{r}')\rangle_0\langle\varphi_1(\boldsymbol{r}')\varphi_1(\boldsymbol{r})\rangle_0 = \\ = \int_{\lambda q_0}^{q_0} \frac{d^d q}{(2\pi)^d}\langle|\varphi_{1,\boldsymbol{q}}|^2\rangle_0\langle|\varphi_{1,-\boldsymbol{q}}|^2\rangle_0.$$

In order to return the thermodynamic system to the phase transition point, we must choose τ_0 so that the renormalized value τ_0 vanishes. Accordingly, we have $\langle|\varphi_1(\boldsymbol{q})|^2\rangle_0 = 1/q^2$ and the correlator K_λ reads

$$K_\lambda = \int\limits_{\lambda q_0}^{q_0} \frac{d^d q}{(2\pi)^d} \langle|\varphi_{1,\boldsymbol{q}}|^2\rangle_0^2 = \int\limits_{\lambda q_0}^{q_0} \frac{1}{q^4}\frac{d^d q}{(2\pi)^d} = \frac{S_d}{(2\pi)^d}\int\limits_{\lambda q_0}^{q_0} \frac{1}{q^{1+\epsilon}} =$$
$$= \frac{S_d}{(2\pi)^d} q_0^{-\epsilon}\frac{\lambda^{-\epsilon}-1}{\epsilon} \quad (\epsilon = 4-d).$$

Here S_d is the surface area of unit radius sphere in the d-dimensional space

$$S_d = 2\pi^{d/2}/\Gamma(d/2).$$

In first lowest approximation it is sufficient to put $\epsilon = 0$. Taking into account that $S_4 = 2\pi^4$, we obtain in this limit

$$K_\lambda = -\frac{1}{8\pi^2}\ln\lambda \quad \text{and} \quad S_\lambda g_0 = g_0 + \frac{9}{2\pi^2} g_0^2 \ln\lambda + O(g_0^3).$$

The small parameter that justifies our expansion equals $g_0 \ln\lambda \ll 1$.

After dilation D_λ, the renormalized value for the coupling constant $g(\lambda) = R_\lambda g_0$ will equal

$$g(\lambda) \approx \lambda^{-\epsilon}\left(g_0 + \frac{9g_0^2}{2\pi^2}\ln\lambda\right) \quad \text{or} \quad \ln g(\lambda) \approx \ln g_0 - \epsilon\ln\lambda + \frac{9g_0}{2\pi^2}\ln\lambda.$$

Differentiating with respect to $\ln\lambda$ gives us the following:

$$\frac{d\,\ln g}{d\,\ln\lambda} = -\epsilon + \frac{9}{2\pi^2}g_0 + O(g_0^2).$$

Staying within our initial approximation of small coupling constant $g_0 \ll 1$, we can replace the "bare" constant g_0 with its renormalized value g and derive the *renormalization-group equation*

$$\frac{d\,\ln g}{d\,\ln\lambda} = f(g) \quad \text{and} \quad f(g) = -\epsilon + \frac{9}{2\pi^2}g + \ldots$$

The function $f(g)$ is called the Gell-Mann–Low function. In the quantum field theory, instead of function $f(g)$, it is usually adopted to define the *beta function* $\beta(g) = gf(g)$ in accordance with equation

$$\frac{dg}{d\,\ln\lambda} = \beta(g).$$

The function $f(g)$ has been determined in lowest order in expanding into a series over the coupling constant g and, in general, it should be subjected to the further calculation and determination by analyzing the next expansion terms in the renormalized Hamiltonian. The expansion in the powers of coupling constant g corresponds to that in the ϵ-powers.

The *fixed point* $Rg^* = g^*$ of our renormalization, i.e. the coupling constant remains unchangeable, is found from the condition $f(g^*) = 0$

$$g^* = \frac{2\pi^2}{9}\epsilon \ll 1.$$

The smallness $\epsilon = 4 - d \ll 1$ agrees with the approximation of small magnitude for the coupling constant.

The scaling dimension Δ_τ of reduced temperature τ means that the quantity τ with renormalization R_λ is governed by the scaling law

$$\tau \to \tau(\lambda) = R_\lambda \tau = \lambda^{-\Delta_\tau}\tau \quad \text{or} \quad \frac{d\,\ln\tau(\lambda)}{d\,\ln\lambda} = -\Delta_\tau\,.$$

Let us determine the scaling dimension of reduced temperature τ at the fixed point g^*. For this purpose, it is necessary to calculate the renormalized quantity τ within the accuracy of first order in ϵ. We start from the mean values G_λ

$$G_\lambda = \langle \int d\boldsymbol{r}\,\varphi_1^2\boldsymbol{r}\rangle_0 = \int\limits_{\lambda q_0}^{q_0} \frac{d^d q}{(2\pi)^d}\langle|\varphi_1(\boldsymbol{q})|^2\rangle_0 = \frac{S_d}{(2\pi)^d}\int\limits_{\lambda q_0}^{q_0}\frac{q^{d-1}dq}{q^2+\tau_0}\,.$$

For the calculations within first-order accuracy in ϵ, it is sufficient to put $\epsilon = 0$ or $d = 4$. Then, new temperature of phase transition or critical point τ^* will be determined from equation $\tau^* = 0$

$$0 = \tau^* + \frac{3}{2\pi^2}g_0\int\limits_{\lambda q_0}^{q_0}\frac{q^3dq}{q^2+\tau^*} = \tau^* + \frac{3}{2\pi^2}g_0\int\limits_{\lambda q_0}^{q_0} q\left(1 - \frac{\tau^*}{q^2+\tau^*}\right)dq.$$

Renormalizing the quantity τ_0 yields the following equation:

$$R_\lambda\tau_0 = \lambda^{-2}\left[\tau^* + \frac{3}{2\pi^2}g_0\int\limits_{\lambda q_0}^{q_0} q\left(1 - \frac{\tau^*}{q^2+\tau^*}\right)dq\right].$$

Analyzing this equation, we should take into account that there appears a shift of critical point or phase transition point from $\tau_0 = 0$ to new point $\tau_0 = \tau^*$ as a result of interacting with the fluctuations of order parameter field. Therefore, we must determine the temperature from new phase transition point τ^*. In other words, it is necessary to analyze how quantity $\tau' = \tau_0 - \tau^*$ renormalizes. As a result, we obtain the following relations, using first-order approximation in the coupling constant g_0

$$\tau(\lambda) = R_\lambda \tau_0' = \lambda^{-2}\left(\tau_0' - \frac{3g_0}{2\pi^2}\tau_0' \int\limits_{\lambda q_0}^{q_0} \frac{q\,dq}{q^2}\right) = \lambda^{-2}\left(1 + \frac{3g_0}{2\pi^2}\ln\lambda\right)\tau_0',$$

$$\ln\tau = \ln\tau_0 - 2\ln\lambda + \ln\left(1 + \frac{3g_0}{2\pi^2}\ln\lambda\right) \approx \ln\tau_0 - \left(2 - \frac{3g_0}{2\pi^2}\right)\ln\lambda.$$

Replacing the constant g_0 with g, we arrive at the renormalization-group equation for the variable τ

$$\frac{d\,\ln\tau}{d\,\ln\lambda} = -\Delta_\tau(g) \quad \text{where} \quad \Delta_\tau = 2 - \frac{3}{2\pi^2}g + \ldots$$

And finally, substituting $g = g^* = 2\pi^2\epsilon/9$, we find the magnitude of scaling dimension for temperature τ at the fixed point, i.e. at the phase transition point

$$\Delta_\tau = 2 - \frac{\epsilon}{3} + O(\epsilon^2).$$

Correspondingly, critical exponent ν for the correlation length $\xi \sim |\tau|^{-\nu}$ equals

$$\nu = \frac{1}{\Delta_\tau} = \frac{1}{2} + \frac{\epsilon}{12} + O(\epsilon^2).$$

For the physically interesting case $\epsilon = 1$, this delivers the following approximate value: $\nu = 7/12 \approx 0.6$.

To determine the exponent α in the behavior of specific heat $C \sim |\tau|^{-\alpha}$, one can apply the scaling relation $d\nu = 2 - \alpha$ for the dimension $d = 4 - \epsilon$. This entails the following correction for α in the linear approximation in ϵ:

$$\alpha = \frac{\epsilon}{6}.$$

The critical exponent β, describing the behavior of order parameter $\varphi(\tau < 0) \sim (-\tau)^\beta$, can be determined from relation $\beta = \nu\Delta_\varphi$. As we have seen above, the Fourier transform $\varphi_{\boldsymbol{q}}$ of order parameter should be multiplied by factor $Z_\lambda^{-1} = \lambda^{(d+2)/2}$ by performing the scale transformation. Accordingly, the order parameter $\varphi(\boldsymbol{r})$ crosses over to $Z_\lambda^{-1}\lambda^{-d}\varphi(\boldsymbol{r}) = \lambda^{-(d-2)/2}\varphi(\boldsymbol{r})$ since obviously $\varphi(\boldsymbol{r}) = \int \varphi_{\boldsymbol{q}} d^d q/(2\pi)^d$ and $q \to \lambda^{-1}q$. Thus, in the linear ϵ-approximation the scaling dimension of order parameter φ equals

$$\Delta_\varphi = \frac{d-2}{2} = 1 - \frac{\epsilon}{2}.$$

Hence the critical exponent β is given by

$$\beta = \Delta_\varphi \nu = \left(1 - \frac{\epsilon}{2}\right)\left(\frac{1}{2} + \frac{\epsilon}{12}\right) = \frac{1}{2} - \frac{\epsilon}{6} + O(\epsilon^2)$$

and, for $\epsilon = 1$, the approximate value is $\beta = 1/3$.

The critical exponent γ in the behavior of susceptibility $\chi \sim |\tau|^{-\gamma}$ can be found by using the scaling relation $\alpha + 2\beta + \gamma = 2$. Hence the critical exponent γ equals approximately

$$\gamma = 1 + \frac{\epsilon}{6} + O(\epsilon^2).$$

For $\epsilon = 1$, this leads to value $\gamma = 7/6$.

The critical exponent δ, specifying the order parameter behavior in the high external field $\varphi(h) \sim h^{1/\delta}$, can be found from the scaling relation $\beta\delta = \beta + \gamma$. Appropriately, critical exponent δ equals approximately

$$\delta = 3 + \epsilon + O(\epsilon^2).$$

For $\epsilon = 1$, this entails the value $\delta = 4$.

The critical Fisher exponent or anomalous dimension η in the linear ϵ-pproximation proves to be zero, i.e. $\eta = O(\epsilon^2)$. The point here is the following. As we have seen above, the coefficient before $q^2|\varphi_q|^2$ after the smoothing operation S_λ does not change within the linear approximation in the coupling constant g proportional to ϵ. Along with that the scaling coefficient Z_λ will be the same, i.e. Z_λ. As a result, the spatial behavior for correlator $K(\boldsymbol{r}, \boldsymbol{r}') = \langle \varphi(\boldsymbol{r})\varphi(\boldsymbol{r}' \rangle$ at the phase transition point remains unvaried, i.e. $K(\boldsymbol{r}, \boldsymbol{r}') \sim |\boldsymbol{r} - \boldsymbol{r}'|^{-(d-2)}$ as for the case of free-field Hamiltonian. Omitting the narrative of more complicated calculations in the second-order ϵ-expansion, we will cite the result for the critical Fisher exponent

$$\eta = \frac{\epsilon^2}{54} + \ldots$$

The critical exponents calculated with the aid of ϵ-expansion agree well with experiment and numerical simulation data.

Problem

1. Show that the following relation: $\langle|\varphi_q|^4\rangle_0 = 3\langle|\varphi_q|^2\rangle_0^2$ is valid for the mean values of order parameter in the free-field model.

Solution. Let us employ the formula derived above for the generating functional of the free-field Hamiltonian

$$e^{\frac{1}{2}\sum\limits_q h_q K_q h_{-q}} = \frac{\int D\varphi_q e^{-\frac{1}{2}\sum\limits_q \varphi_q K_q^{-1}\varphi_{-q}} e^{\sum\limits_q \varphi_q h_{-q}}}{\int D\varphi_q e^{-\frac{1}{2}\sum\limits_q \varphi_q K_q^{-1}\varphi_{-q}}}.$$

Then we decompose the left-hand and right-hand sides of equality to fourth-order terms in h_q

$$1+\frac{1}{2}\sum_q h_q K_q h_{-q}+\frac{1}{2!}\frac{1}{2^2}\sum_{q,q'} h_q h_{q'} K_q K_{q'} h_{-q} h_{-q'}+\ldots=1+\sum_q \langle\varphi_q\rangle_0 h_{-q}+$$
$$+\frac{1}{2!}\sum_{q,q'}\langle\varphi_q\varphi_{q'}\rangle_0 h_{-q}h_{-q'}+\frac{1}{3!}\sum_{q,q',q''}\langle\varphi_q\varphi_{q'}\varphi_{q''}\rangle_0 h_{-q}h_{-q'}h_{-q''}+$$
$$+\frac{1}{4!}\sum_{q,q',q'',q'''}\langle\varphi_q\varphi_{q'}\varphi_{q''}\varphi_{q'''}\rangle_0 h_{-q}h_{-q'}h_{-q''}h_{-q'''}+\ldots$$

It is clear that the average $\langle\varphi_q\varphi_{q'}\cdots\varphi_{q''}\rangle_0=0$ vanishes if the number of elements φ_q is odd. The comparison of second-order terms in h entails $\langle\varphi_q\varphi_{q'}\rangle_0=K_q\delta_{q',-q}$.

The analogous comparison of fourth-order terms in h leads to the following answer:

$$\langle\varphi_q\varphi_{q'}\varphi_{q''}\varphi_{q'''}\rangle_0=K_qK_{q''}\delta_{q,-q'}\delta_{q'',-q'''}+$$
$$+K_qK_{q'}\delta_{q,-q''}\delta_{q',-q'''}+K_qK_{q'}\delta_{q,-q'''}\delta_{q',-q''}\,.$$

In fact, the direct substitution into the last sum above yields

$$\sum_{q,q',q'',q'''}\langle\varphi_q\varphi_{q'}\varphi_{q''}\varphi_{q'''}\rangle_0=3\sum_{q,q'}h_qh_{q'}K_qK_{q'}h_{-q}h_{-q'}$$

by identifying the terms on the left-hand and right-hand sides of expansion at arbitrary h_q. Next, we find that

$$\langle\varphi_q\varphi_{-q}\varphi_{q'}\varphi_{-q'}\rangle_0=K_qK_{q'}+K_qK_{-q}\left(\delta_{q,q'}+\delta_{q,-q'}\right),$$

and finally

$$\langle\varphi_q\varphi_{-q}\varphi_q\varphi_{-q}\rangle_0=\langle|\varphi_q|^4\rangle_0=3K_q^2=3\langle|\varphi_q|^2\rangle_0^2$$

by taking $\varphi_{-q}=\varphi_q^*$ and $K_{-q}=K_q$ into account.

In the general case the average $\langle\varphi_q\varphi_{q'}\cdots\varphi_{q''}\rangle_0$ at the even number of elements φ_q can be represented as a sum of possible pair averages $\langle\varphi_q\varphi_{-q}\rangle_0$, and all the possible pairings of elements φ_q should be summed. Each pairing gives the multiplier $K_q=\langle\varphi_q\varphi_{-q}\rangle_0$. It can be shown that

$$\langle\varphi_q^n\varphi_{-q}^n\rangle_0=\langle|\varphi_q|^{2n}\rangle_0=(2n-1)!!\langle|\varphi_q|^2\rangle_0^n\,.$$

The multiplier $(2n-1)!!=(2n)!/(2^n)$ is the total number of partitions of $2n$-element set into $2n$ pairs.

4.17 Two-Dimensional Degenerate System

The two-dimensional plane degenerate systems have a particular interest in the theory of phase transitions. Unlike the three-dimensional systems, the mean value of order parameter is zero in the degenerate two-dimensional system. The most interesting case here is when the number of order parameter components equals two. We consider below the classical XY-model as an example.

The *classical XY-model* represents the two-dimensional square lattice whose sites are occupied with the two-component $\boldsymbol{S}_r=(S_r^x,\,S_r^y)$ spins[20] of unit magni-

[20] The XY-model is a special case of n-vector model or $O(n)$-model. For the case of one-component $S\equiv S^z=\pm1$ spins, this is the *Ising model* (n=1). The case of three-component (n=3) spins $\boldsymbol{S}=(S^x,\,S^y,\,S^z)$ is called the *Heisenberg model*.

tude $|\boldsymbol{S}_{\boldsymbol{r}}| = 1$. The square lattice vector period equals $\boldsymbol{a}$. Let each spin at the lattice site interact with the spins at the closest neighbor sites alone. The Hamiltonian of such a system can be written as follows:

$$\mathcal{H} = -J \sum_{\langle \boldsymbol{r}, \boldsymbol{a} \rangle} \left(S_{\boldsymbol{r}}^x S_{\boldsymbol{r}+\boldsymbol{a}}^x + S_{\boldsymbol{r}}^y S_{\boldsymbol{r}+\boldsymbol{a}}^y \right).$$

The sign of exchange constant $J > 0$ corresponds to the ferromagnetic parallel ordering of two neighboring spins.

It is obvious that the interaction between the spins can completely be neglected in the limit of very high $T \gg J$ temperatures. In this limit the spin system, in essence, will be an ideal paramagnet with the lack of any correlation in the mutual orientation of spins. To comprehend what happens for the spin orientation by lowering the temperature, we analyze the temperature behavior of correlation function $K_{\boldsymbol{r},\boldsymbol{r}'} = \langle \boldsymbol{S}_{\boldsymbol{r}} \boldsymbol{S}_{\boldsymbol{r}'} \rangle$ between the spins at the lattice sites $\boldsymbol{r}$ and $\boldsymbol{r}'$.

It is convenient to treat the problem in the angular variables introduced according to

$$S_{\boldsymbol{r}}^x = \cos \varphi_{\boldsymbol{r}} \quad \text{and} \quad S_{\boldsymbol{r}}^y = \sin \varphi_{\boldsymbol{r}} \, .$$

The Hamiltonian reads in these variables

$$\mathcal{H} = -J \sum_{\langle \boldsymbol{r}, \boldsymbol{a} \rangle} \cos(\varphi_{\boldsymbol{r}} - \varphi_{\boldsymbol{r}+\boldsymbol{a}}).$$

The spin correlator will be represented in terms of the functional integral which should be taken over the whole region of the possible variations of angle $\varphi_{\boldsymbol{r}}$ for each spin in the lattice

$$\langle \boldsymbol{S}_{\boldsymbol{r}} \boldsymbol{S}_{\boldsymbol{r}'} \rangle = \langle e^{i(\varphi_{\boldsymbol{r}} - \varphi_{\boldsymbol{r}'})} \rangle = \frac{1}{Z} \int_0^{2\pi} \cdots \int_0^{2\pi} \Big(\prod_{\boldsymbol{r}} d\varphi_{\boldsymbol{r}}\Big) e^{i(\varphi_{\boldsymbol{r}} - \varphi_{\boldsymbol{r}'})} e^{\frac{J}{T} \sum_{\boldsymbol{r},\boldsymbol{a}} \cos(\varphi_{\boldsymbol{r}} - \varphi_{\boldsymbol{r}+\boldsymbol{a}})} .$$

Here Z is the partition function equal to the integral

$$Z = \int_0^{2\pi} \cdots \int_0^{2\pi} \Big(\prod_{\boldsymbol{r}} d\varphi_{\boldsymbol{r}}\Big) e^{\frac{J}{T} \sum_{\boldsymbol{r},\boldsymbol{a}} \cos(\varphi_{\boldsymbol{r}} - \varphi_{\boldsymbol{r}+\boldsymbol{a}})} .$$

Let us start our consideration from the high $T \gg J$ temperature region and determine the magnitude of correlator at the distance equal to the spacing between the nearest neighbor spins at sites $\boldsymbol{r}$ and $\boldsymbol{r} + \boldsymbol{a}$. For $T \gg J$, in the above formulas we expand the exponential function into a series over the powers of ratio $J/T \ll 1$ and restrict ourselves with terms not higher than linear ones. As a result of identities

$$\int_0^{2\pi} d\varphi_\rho \sin\varphi_\rho = 0 \quad \text{and} \quad \int_0^{2\pi} d\varphi_\rho \cos\varphi_\rho = 0,$$

from the total sum over the lattice sites the only term with $\boldsymbol{\rho} = \boldsymbol{r}$ delivers nonzero contribution to the correlator between the nearest neighbor spins at sites $\boldsymbol{r}$ and $\boldsymbol{r} + \boldsymbol{a}$. Then we have

$$\langle e^{i(\varphi_r - \varphi_{r+a})} \rangle \approx \frac{J}{T} \int_0^{2\pi} d\varphi_r e^{i(\varphi_r - \varphi_{r+a})} \cos(\varphi_r - \varphi_{r+a}) = \frac{J}{2T} \,.$$

For the high temperatures, it is natural to suppose a drastic decay of correlation in the mutual orientation of spins as the spacing between the spins increases. In this case in first approximation, we can neglect any correlation in the directions for the initial spin and the spin following the nearest neighboring spin. Accordingly, we can approximately write

$$\langle e^{i(\varphi_r - \varphi_{r+na})} \rangle \approx \langle e^{i(\varphi_r - \varphi_{r+a})} \rangle \langle e^{i(\varphi_r - \varphi_{r+2a})} \rangle \dots \langle e^{i(\varphi_r - \varphi_{r+na})} \rangle.$$

Here we should take the shortest path from site $\boldsymbol{r}$ to site $\boldsymbol{r} + n\boldsymbol{a}$, providing us the maximum result since each transition from one site to the neighboring one gives an additional factor $J/2T \ll 1$. We have finally

$$\langle \boldsymbol{S}_r \boldsymbol{S}_{r+na} \rangle \approx (J/2T)^n = e^{\;n \ln(2T/J)} \quad (T \gg J).$$

Hence we see that the spin orientation correlator decays exponentially with the spacing between the spins. The correlation radius, expressed in the lattice constant a, equals approximately $\xi/a \sim 1/\ln(2T/J) \ll 1$.

For the low temperature $T \ll J$ region, it is energetically favorable to have the maximum value $\cos(\varphi_r - \varphi_{r+a})$ or, in other words, angle φ_r would have a negligibly small variation at the spacing between two neighboring sites. This means that the correlation radius should significantly exceed the lattice constant a. In such a situation one can employ the following estimate:

$$\cos(\varphi_r - \varphi_{r+a}) \approx 1 - \frac{(\varphi_r - \varphi_{r+a})^2}{2} \,.$$

We can omit the unit from our consideration since it entails the shift of energy level reference. Next, the following approximation is used

$$\sum_{\boldsymbol{a}} (\varphi_r - \varphi_{r+a})^2 \approx a^2 \left(\frac{\partial \varphi}{\partial x} \right)^2 + a^2 \left(\frac{\partial \varphi}{\partial y} \right)^2$$

where a is the lattice constant. The rule of crossover from the lattice site sum to the integral over area

$$a^2\sum_{\boldsymbol{r}}\longrightarrow\int dx\,dy$$

completes our transition to the continual approximation with the equivalent effective Hamiltonian governed with the continuous field $\varphi(x, y)$

$$H=\frac{J}{2}\int\left[\left(\frac{\partial\varphi}{\partial x}\right)^2+\left(\frac{\partial\varphi}{\partial y}\right)^2\right]dx\,dy=\int d^2\boldsymbol{r}\,\frac{J(\nabla\varphi)^2}{2}.$$

For the parametrization $n_x=\cos\varphi$ and $n_y=\sin\varphi$, we have $\boldsymbol{n}^2=1$ and the following representation:

$$H=\frac{J}{2}\int(\nabla\boldsymbol{n})^2dx\,dy.$$

Here $\boldsymbol{n}(x, y)$ is the two-dimensional unit vector determined with the single angle $\varphi(x, y)$ and pointing the terminal position of vector at the unit circumference.

The Gibbs distribution, proportional to $\exp(-H/T)$ with effective Hamiltonian H, reproduces the Gaussian one. So, we can use the familiar relation for the mean values under the Gaussian distribution of probabilities

$$\langle e^{i[\varphi(\boldsymbol{r})-\varphi(\boldsymbol{r}')]}\rangle=e^{-\langle[\varphi(\boldsymbol{r})-\varphi(\boldsymbol{r}')]^2\rangle/2}.$$

The mean value of the exponent

$$\langle\left[\varphi(\boldsymbol{r})-\varphi(\boldsymbol{r}')\right]^2\rangle/2=\langle\varphi^2(\boldsymbol{r})\rangle-\langle\varphi(\boldsymbol{r})\varphi(\boldsymbol{r}')\rangle$$

can be calculated with the aid of Fourier transformation

$$\varphi(\boldsymbol{r})=\sum_{\boldsymbol{q}}e^{i\boldsymbol{q}\boldsymbol{r}}\varphi_{\boldsymbol{q}}\quad(\varphi_{\boldsymbol{q}}^*=\varphi_{-\boldsymbol{q}})$$

in which the mean value required reads

$$\langle\varphi(\boldsymbol{r})\varphi(\boldsymbol{r}')\rangle=\sum_{\boldsymbol{q},\boldsymbol{q}'}\langle\varphi_{\boldsymbol{q}}\varphi_{\boldsymbol{q}'}\rangle e^{i\boldsymbol{q}\boldsymbol{r}+i\boldsymbol{q}'\boldsymbol{r}'}.$$

The effective Hamiltonian H in the Fourier transforms is already diagonal in $\varphi_{\boldsymbol{q}}$

$$H=\frac{1}{2}\sum_{\boldsymbol{q}}Jq^2|\varphi_{\boldsymbol{q}}|^2.$$

The mean values, similar to $\langle\varphi_{\boldsymbol{q}}\varphi_{\boldsymbol{q}'}\rangle$, have already been calculated for such fluctuating quantities in the previous sections. So, using the above-derived expressions

$$\langle\varphi_{\boldsymbol{q}}\varphi_{\boldsymbol{q}'}\rangle = \langle|\varphi_{\boldsymbol{q}}|^2\rangle\delta_{-\boldsymbol{q},\boldsymbol{q}'} \quad \text{where} \quad \langle|\varphi_{\boldsymbol{q}}|^2\rangle = \frac{T}{Jq^2},$$

we obtained the final answer

$$\frac{1}{2}\langle\left[\varphi(\boldsymbol{r}) - \varphi(\boldsymbol{r}')\right]^2\rangle = \sum_{\boldsymbol{q}} \frac{T}{Jq^2}\left(1 - e^{i\boldsymbol{q}(\boldsymbol{r}-\boldsymbol{r}')}\right) =$$

$$= \frac{T}{J}\int_0^\infty \frac{q\,dq}{2\pi q^2}\int_0^{2\pi}\frac{d\vartheta}{2\pi}\left(1 - e^{iq|\boldsymbol{r}-\boldsymbol{r}'|\cos\vartheta}\right) = \frac{T}{2\pi J}\int_0^\infty dq\,\frac{1 - J_0(q|\boldsymbol{r}-\boldsymbol{r}'|)}{q}.$$

Integrating over angle ϑ leads us to the Bessel function of the first kind $J_0(x)$. Analyzing the last integrand, we see that the integral over q diverges logarithmically at the large values of wave vector. The divergency results from inapplicability of the continual approximation at the small distances of the order of lattice constant a or at large wave vectors $q \sim 1/a$. The latter vector magnitudes are associated with the rapid variations of field φ at the scales about spin lattice constant a. For this reason, in order to eliminate non-physical divergency at wave vectors $q \gtrsim 1/a$, we cut the upper limit of integration at wave vector $q \sim 1/a$. Then, we get the following estimate of the mean value within the logarithmic accuracy

$$\frac{1}{2}\langle\left[\varphi(\boldsymbol{r}) - \varphi(\boldsymbol{r}')\right]^2\rangle \approx \frac{T}{2\pi J}\ln\frac{|\boldsymbol{r}-\boldsymbol{r}'|}{a}.$$

As a result, in the low $T \ll J$ temperature region we disclose the power-like behavior for the spin correlation function

$$\langle\boldsymbol{S}_{\boldsymbol{r}}\boldsymbol{S}_{\boldsymbol{r}'}\rangle = \langle e^{i(\varphi_{\boldsymbol{r}}-\varphi_{\boldsymbol{r}'})}\rangle = \left(\frac{a}{|\boldsymbol{r}-\boldsymbol{r}'|}\right)^{T/2\pi J}$$

with the infinite correlation radius.

In accordance with such a behavior of spin correlator we can conclude that, strictly speaking, the spontaneous far-distant spin ordering is absent in the XY-model. In fact, since the spin correlator decouples $\langle\boldsymbol{S}_{\boldsymbol{r}}\boldsymbol{S}_{\boldsymbol{r}'}\rangle \to \langle\boldsymbol{S}_{\boldsymbol{r}}\rangle\langle\boldsymbol{S}_{\boldsymbol{r}'}\rangle$ and vanishes at $|\boldsymbol{r}-\boldsymbol{r}'| \to \infty$, we have zero mean value $\langle\boldsymbol{S}_{\boldsymbol{r}}\rangle = 0$. On the other hand, we see the exponential decay of spin correlator at high $T \gg J$ temperatures but for the low $T \ll J$ temperatures we find another power-like behavior different in kind. So, we can expect some kind of phase transition in the region $T \sim J$ as the temperature lowers.

According to the scaling hypothesis, the correlation function at the phase transition point should have a power-like behavior as a function of distance. For spatial dimension $d = 2$, we expect

$$\langle \boldsymbol{S_r S_{r'}} \rangle \sim |\boldsymbol{r} - \boldsymbol{r}'|^{-\eta}.$$

This corresponds to the critical Fischer exponent $\eta = T/2\pi J$, however, temperature-dependent. The highest temperature $T_{\text{BKT}} \sim J$, when the exponential decay of the correlator crosses over to the power-like behavior, proposes the specific phase transition referred to as the *Berezinskii–Kosterlitz–Thouless transition*.

Problems

1. The electrically neutral two-dimensional system of positive and neutral point charges occupies the plane surface of area A. According to the Poisson equation in the two-dimensional space, two point charges of magnitude e_i and e_j at the points $\boldsymbol{r}_i$ and $\boldsymbol{r}_j$ interact logarithmically as a function of distance $|\boldsymbol{r}_i - \boldsymbol{r}_j|$ between them

$$U_{ij} = -q_i q_j \ln |\boldsymbol{r}_i - \boldsymbol{r}_j|, \quad q_{i,j} = \sqrt{2}\, e_{i,j}\,.$$

In other words, the point charges in the two-dimensional space interact like the charged and infinitely long parallel filaments. (An arbitrary constant is chosen in the interaction potential so that $U_{ij} = 0$ at $|\boldsymbol{r}_i - \boldsymbol{r}_j| = 1$). The total energy for the system of charges is given by the following sum:

$$U = -\frac{1}{2}\sum_{i \neq j} q_i q_j \ln |\boldsymbol{r}_i - \boldsymbol{r}_j| = -\sum_{i<j} q_i q_j \ln |\boldsymbol{r}_i - \boldsymbol{r}_j|.$$

Find the equation of state for such a two-dimensional system and analyze its thermodynamic stability in the low-temperature limit.

Solution. Let us write down the configurational part of the partition function for the system of N charges occupying the area A

$$\begin{aligned} Z = \frac{1}{N_+!\,N_-!}\int d^2r_1 d^2r_2 \ldots d^2r_N\, e^{-U/T} = \\ = \frac{1}{N_+!\,N_-!}\int d^2r_1 d^2r_2 \ldots d^2r_N\, \exp\Bigl(\sum_{i<j}\frac{q_i q_j}{T}\ln |\boldsymbol{r}_i - \boldsymbol{r}_j|\Bigr) = \\ = \frac{1}{N_+!\,N_-!}\int d^2r_1 d^2r_2 \ldots d^2r_N \prod_{i<j} |\boldsymbol{r}_i - \boldsymbol{r}_j|^{\frac{q_i q_j}{T}} \end{aligned}$$

where N_+ and N_- are the number of positive and negative charges, respectively. Here we mean the integration performed over the whole area A for each charge.

The pressure P is given by differentiating the free energy F with respect to volume, i.e. to area A occupied with the charges

$$P = -\frac{\partial F}{\partial A} = T\frac{\partial \ln Z}{\partial A}\,.$$

Hence we see that the dependence Z versus A alone is significant for determining the pressure. For this purpose, in the expression for Z we change the scale of length unit, introducing new variables $\boldsymbol{\xi}_i$ according to $\boldsymbol{r}_i = \boldsymbol{\xi}_i A^{1/2}$. Then we obtain

$$Z(A, T, N_+, N_-) = A^{(N + \sum_{i<j}\frac{q_i q_j}{2T})} z(T, N_+, N_-).$$

Here $N = N_+ + N_-$ is the total number of charges and the z factor is A-independent. The pressure is readily found and equal to

$$P = \frac{T}{A}\left(N + \frac{1}{2T}\sum_{i<j} q_i q_j\right) = \frac{T}{A}\left[N + \frac{1}{4T}\left(\left(\sum_i q_i\right)^2 - \sum_i q_i^2\right)\right].$$

Taking electroneutrality of system $\sum_i q_i = 0$ into account, we arrive at the following equation of state:

$$PA = NT\left[1 - \frac{1}{4T}\left(\frac{N_+}{N}q_+^2 + \frac{N_-}{N}q_-^2\right)\right] = NT\left(1 - \frac{q^2}{4T}\right).$$

The last equality is written for the symmetrical case when $q_+ = q$ and $q_- = -q$.

The condition of thermodynamic stability requires that an expansion of volume or area in our case should be accompanied by decreasing the pressure under constant temperature. So,

$$\left(\frac{\partial P}{\partial A}\right)_T < 0.$$

This condition breaks down at the temperatures below critical temperature $T_c = q^2/4$ and, therefore, two-dimensional neutral system of charges cannot exist in the homogeneous state at $T < T_c$. Below the critical temperature T_c the bound pairs (molecules) appear and consist of positive and negative charges.

2. Estimate the energy for a single vortex excitation of circulation $\varkappa$ in the classical plane XY-model.

Solution. The vortex excitation is governed with the following gradient of continuous field $\varphi(x, y)$:

$$\nabla\varphi = \frac{\varkappa}{2\pi r}\boldsymbol{e}_\theta .$$

Here $\boldsymbol{e}_\theta$ is unit vector in the direction of rotation angle φ and radius vector is $r = \sqrt{x^2 + y^2}$. The corresponding circulation of field $\varphi(x, y)$ equals

$$\oint_C d\boldsymbol{l} \cdot \nabla\varphi = \varkappa$$

where C is an arbitrary closed contour encircling the vortex center $\boldsymbol{r} = 0$. The energy of vortex excitation is given by the logarithmic integral

$$E = \int d^2r \frac{J(\nabla\varphi)^2}{2} = \frac{J}{2}\int_a^L \frac{\varkappa^2}{4\pi^2 r^2} 2\pi r\, dr = \frac{\varkappa^2}{2\pi}\ln\frac{L}{a} .$$

Here, we have put the lattice constant a as a lower limit since the continual approximation breaks down on this scale. As it concerns the upper limit, we take the size of the plane $L \gg a$, being the macroscopic size.

3. In the thin (quasi-two-dimensional) superfluid film, the interaction energy $U(r)$ of vortex–antivortex pair, i.e. two parallel vortex filaments with the opposite directed circulations of the same magnitude $\varkappa$, depends logarithmically on the distance r between the filaments

$$U(r) = U_0 \ln\frac{r}{a}, \quad U_0 = \rho d\frac{\varkappa^2}{2\pi} .$$

Here a is the minimally possible distance between the vortex filaments, ρ is the liquid density, and d is the film thickness or filament length.

Find the mean squared distance $\langle r^2 \rangle$ between the filaments as a function of temperature. Determine the *Berezinskii–Kosterlitz–Thouless temperature* T_{BKT} above which the dissociation of vortex–antivortex pair should occur.

Solution: Since the probability for fluctuation is proportional to $\exp[-U(r)/T]$, the mean square of distance between the vortex filaments can be calculated using the formula

$$\langle r^2\rangle = \frac{\int_a^\infty 2\pi r dr\, r^2 e^{-U(r)/T}}{\int_a^\infty 2\pi r dr\, e^{-U(r)/T}} = \frac{\int_a^\infty dr\, r^3 \exp(-\frac{U_0}{T}\ln\frac{r}{a})}{\int_a^\infty dr\, r \exp(-\frac{U_0}{T}\ln\frac{r}{a})} = a^2\frac{U_0-2T}{U_0-4T}.$$

For $T = 0$, the mean square of distance between the filaments is minimum $\langle r^2\rangle = a^2$. As the temperature increases, the mean square $\langle r^2\rangle$ grows and at the temperature equal to

$$T_{\mathrm{BKT}} = \frac{U_0}{4} = \frac{\rho d\varkappa^2}{8\pi}$$

becomes infinite, entailing the destruction of bound state or dissociation of vortex–antivortex pair.

4. Find the free energy $F(T)$ of classical XY-model for the one-dimensional chain with the free ends, N two-component spins of unit length being located at the chain sites. The spins interact with the nearest neighbors alone. The exchange coupling constant $J > 0$ has the ferromagnetic sign.

Solution. The Hamiltonian of spin chain reads

$$H = -J\sum_{k=1}^{N-1}\cos(\varphi_{k+1}-\varphi_k) = -J\big[\cos(\varphi_1-\varphi_2)+\dots+\cos(\varphi_{N-1}-\varphi_N)\big]$$

where φ_k is the rotation angle of kth spin. The corresponding partition function will equal the N-dimensional integral

$$Z = \int_0^{2\pi} d\varphi_1 \int_0^{2\pi} d\varphi_2 \cdots \int_0^{2\pi} d\varphi_N \exp\Big(-\beta J\sum_{k=1}^{N-1}\cos(\varphi_{k+1}-\varphi_k)\Big), \quad \beta = T^{-1}.$$

Performing the following replacement of variables $\varphi_k = \phi_k - \varphi_{k-1}$ and involving that the integration should be fulfilled over the whole period, we obtain

$$Z = \int_0^{2\pi} d\varphi_1 \int_0^{2\pi} d\phi_2 \cdots \int_0^{2\pi} d\phi_N \exp\Big(-\beta J\sum_{k=1}^{N-1}\cos\phi_{k+1}\Big) = 2\pi\Big[\int_0^{2\pi} d\phi\, e^{-\beta J\cos\phi}\Big]^{N-1}.$$

The last integral can be expressed in terms of modified Bessel function of the first kind $I_0(x)$. Thus we have for the partition function Z and free energy F

$$Z = 2\pi\big[2\pi I_0(\beta J)\big]^{N-1} \quad\text{and}\quad F = -NT\ln(2\pi) - (N-1)T\ln I_0(J/T).$$

The specific heat, calculated according to $C(T) = -T\partial^2F/\partial^2T$, is given by the formula

$$C(T) = (N-1)x^2\Big[\frac{I_0(x)+I_2(x)}{2I_0(x)} - \Big(\frac{I_1(x)}{I_0(x)}\Big)^2\Big], \quad x = \frac{J}{T}$$

where I_0, I_1, and I_2 are the modified Bessel functions of the first kind. At $T = 0$ the specific heat is finite and equals $(N-1)/2$. As the temperature grows, the specific heat increases linearly as $\Delta C(T) \sim T/4J$, runs across the maximum at $T \sim 0.4J$, and then decreases as $(N-1)J^2/2T^2$ in the region $T \gg J$.

Chapter 5
Normal Fermi Liquid

Unlike ideal gases where the energy of a gas is a simple sum of the energies of individual particles, in the system of interacting particles the determination of energy levels is very difficult and often impossible task. As a consequence, though the basic ideas and methods for describing the condensed matter can be the same, nevertheless the concrete implementation of these ideas and methods may differ in various cases and types of condensed matter.

5.1 Creation and Annihilation Operators

The occupation number representation (or second quantization) is a convenient formalism for analyzing and describing the quantum many-particle systems with interaction. The description employs the *basis of single-particle states* as an occupation of each single-particle state with certain number of identical particles. The formalism is wholly based on the *principle of indistinguishability* of identical particles in quantum mechanics.

The occupation number representation means the crossover from describing the system of N particles with the aid of the orthogonal and complete set of wave functions $\varphi_1(\xi)$, $\varphi_2(\xi)$, ... of a particle, where ξ is a set of variables (coordinate, momentum, spin) identifying the particle state, to the occupation numbers when the number of particles N_i is given for each of possible states with wave function $\varphi_i(\xi)$.

Let the first particle occupy the state with wave function φ_1, second particle occupy the state with wave function φ_2, and so on. Then the total wave function for the whole system of N particles can be written as a product of wave functions

$$\Phi(\xi_1, \xi_2, \dots, \xi_N) = \varphi_1(\xi_1)\varphi_2(\xi_2)\dots\varphi_N(\xi_N).$$

S. N. Burmistrov, *Statistical and Condensed Matter Physics*,
https://doi.org/10.1007/978-981-97-9072-2_5

Such wave function, however, does not satisfy the symmetry principle with respect to permutation for any pair of particles. In other words, if we want to describe the interacting particle system in the terms of total wave function, the latter must satisfy some definite symmetry conditions. For the case of the Fermi–Dirac statistics, the total wave function Φ_F should be antisymmetric with respect to permutation of any two variables. It is usually convenient to represent this as the *Slater determinant*

$$\Phi_F(\xi_1, \xi_2, \ldots, \xi_N) = \mathcal{N}\det \begin{Vmatrix} \varphi_1(\xi_1) & \varphi_1(\xi_1) & \ldots & \varphi_1(\xi_N) \\ \vdots & \vdots & \ddots & \vdots \\ \varphi_N(\xi_1) & \varphi_N(\xi_2) & \ldots & \varphi_N(\xi_N) \end{Vmatrix}$$

where $\mathcal{N}$ is the normalization factor.

In the case of the Bose–Einstein statistics, the total wave function Φ_B should be symmetrical with respect to the permutations of all variables. This can briefly be written with the aid of *permanent*[1] square $N \times N$ matrix

$$\Phi_B = \Phi(\xi_1, \xi_2, \ldots, \xi_N) = \mathcal{N}\mathrm{perm} \begin{Vmatrix} \varphi_1(\xi_1) & \varphi_1(\xi_2) & \ldots & \varphi_1(\xi_N) \\ \vdots & \vdots & \ddots & \vdots \\ \varphi_N(\xi_1) & \varphi_N(\xi_2) & \ldots & \varphi_N(\xi_N) \end{Vmatrix}$$

and $\mathcal{N}$ is the normalization factor.

As a rule, usage of such symmetrical (antisymmetrical) wave function is associated with a number of computational disadvantages. To characterize the state of many-particle system, instead the total wave function and due to principle of indistinguishability between particles, we can simply indicate how many particles N_i are in the state with wave function φ_i, i.e. specify the occupation of this state. The wave function $|N_1, N_2, \ldots\rangle$ denotes that there are N_1 particles in the state with wave function φ_1 and so on. The states in which the number of particles is completely determined are called the *Fock states*. The state $|0, 0 \ldots\rangle$ which has no particles is referred to as *vacuum*.

Since in the physical processes the number of particles in one state or another can vary, we need to introduce the operators changing the particle number in the given state, i.e. transforming the occupation numbers according to $N \leftrightarrows N \pm 1$. An operator a^+, increasing the number of particles in the given state with wave function φ by one, is called the *creation operator*. An operator a, decreasing the number of particles in the given state by one, is called the *annihilation operator*. The successive application of several creation or annihilation operators allows us to change the occupation numbers by several units.

[1] Unlike the determinant, the definition for the permanent of square matrix implies that any permutation of matrix elements does not change its sign. For example, $\mathrm{perm}\begin{Vmatrix} a_1 & b_1 \\ a_2 & b_2 \end{Vmatrix} = a_1b_2 + a_2b_1$ and $\mathrm{perm}\begin{Vmatrix} a_1 & b_1 & c_1 \\ a_2 & b_2 & c_2 \\ a_3 & b_3 & c_3 \end{Vmatrix} = a_1b_2c_3 + a_1c_2b_3 + b_1c_2a_3 + b_1a_2c_3 + c_1a_2b_3 + c_1b_2a_3$. Whereas, $\det\begin{Vmatrix} a_1 & b_1 \\ a_2 & b_2 \end{Vmatrix} = a_1b_2 - a_2b_1$ and $\det\begin{Vmatrix} a_1 & b_1 & c_1 \\ a_2 & b_2 & c_2 \\ a_3 & b_3 & c_3 \end{Vmatrix} = a_1b_2c_3 - a_1c_2b_3 + b_1c_2a_3 - b_1a_2c_3 + c_1a_2b_3 - c_1b_2a_3$.

Let us start from determining the creation and annihilation operators for the Bose–Einstein statistics, assuming a possibility for an arbitrary occupation of the same state with bosons. So, we put

$$a|N\rangle = \alpha_N|N-1\rangle \quad \text{and} \quad a^+|N\rangle = \beta_N|N+1\rangle.$$

It is convenient to determine as $a^+a|N\rangle = N|N\rangle$, i.e. particle number operator $\hat{N} = a^+a$ and its action on the given state shows the number of particles in the given state. Then, using the Hermitian property of the following matrix elements: $\langle N|a^+|N-1\rangle = \langle N-1|a|N\rangle^*$, we have $\beta_{N-1} = \alpha_N$. Next from

$$\langle N|a^+a|N-1\rangle = \beta_{N-1}\alpha_N = |\alpha_N|^2 = N$$

we find[2] $\alpha_N = \sqrt{N}$ and $\beta_N = \sqrt{1+N}$. Finally, we arrive at the following relations for the bosonic operators of creating and annihilating the particles

$$a^+|N\rangle = \sqrt{1+N}|N+1\rangle \quad \text{and} \quad a|N\rangle = \sqrt{N}|N-1\rangle,$$
$$a^+a|N\rangle = N|N\rangle \quad \text{and} \quad aa^+|N\rangle = (1+N)|N\rangle.$$

The following commutation relations for the bosonic creation and annihilation operators can readily be obtained from the above equations:

$$[a^+,\, a^+] = 0, \quad [a,\, a] = 0, \quad [a,\, a^+] = 1.$$

These commutation relations can be used as the initial ones to determine the bosonic operators of creating and annihilating the particles. The state having N particles can be written by the N-time application of creation operator a^+ to the vacuum state

$$|N\rangle = \frac{1}{\sqrt{N!}}\left(a^+\right)^N|0\rangle.$$

All the formulas above are generalized to an arbitrary number of the Fock states

$$a_k^+|\ldots N_i, N_k, N_l\ldots\rangle = \sqrt{1+N_k}|\ldots N_i, N_k+1, N_l\ldots\rangle,$$
$$a_k|\ldots N_i, N_k, N_l\ldots\rangle = \sqrt{N_k}|\ldots N_i, N_k-1, N_l\ldots\rangle.$$

The operators related to various states commute

$$[a_i^+,\, a_k^+] = 0, \quad [a_i,\, a_k] = 0, \quad [a_i,\, a_k^+] = \delta_{ik}\,.$$

In contrast to the Bose–Einstein statistics, the Fermi–Dirac statistics does not allow more than one particle in the same state. The occupation numbers N take the

[2] For the reasons of convenience, we choose α_N and β_N to be real without complex phase multipliers.

values 0 or 1 alone. If state $|1\rangle$ is occupied, the further action of creation operator c^+ must destroy this state or transfer it to the unoccupied state $|0\rangle$ in the full accordance with the Pauli principle forbidding two identical fermions to be in the same state. A fermion from the occupied state $|1\rangle$ is removed with the annihilation operator c. In fact, it is necessary to determine only two relations for the creation and annihilation operators

$$c|1\rangle = \alpha|0\rangle \quad \text{and} \quad c^+|1\rangle = \beta|1\rangle$$

and we require the fulfillment of relation $c^+c|N\rangle = N|N\rangle$ which sets the particle number operator $\hat{N} = c^+c$. Hence, it follows that $\alpha\beta = N = 0$ or 1. Using that $\langle 1|c^+|0\rangle = \langle |c|\rangle^*$, we have $\beta = \alpha^*$ and, correspondingly, $|\alpha|^2 = N$. As a result, coefficients α and β take values[3] either 0 or 1. We give below all possible cases for the action of operators

$$c^+|0\rangle = 1|1\rangle, \quad c|0\rangle = 0|1\rangle,$$
$$c^+|1\rangle = 0|0\rangle, \quad c|1\rangle = 1|0\rangle.$$

These relations can be written as follows:

$$c|N\rangle = \sqrt{N}|1-N\rangle \quad \text{and} \quad c^+|N\rangle = \sqrt{1-N}|1-N\rangle.$$

Hence, one sees that $c^+c^+|N\rangle = 0$ and $cc|N\rangle = 0$. Since $cc^+|N\rangle = (1-N)|N\rangle$, the anticommutator $\{c,\, c^+\} = cc^+ + c^+c$ satisfies the equality

$$\{c,\, c^+\} = 1.$$

In some applications it is convenient to introduce the so-called the *Majorana fermion operators*:

$$\chi = \frac{c + c^+}{\sqrt{2}} \quad \text{and} \quad \bar{\chi} = \frac{c - c^+}{i\sqrt{2}}$$

with the anticommutative relations

$$\{\chi, \chi\} = 1, \quad \{\bar{\chi}, \bar{\chi}\} = 1 \quad \text{and} \quad \{\chi, \bar{\chi}\} = 0.$$

The results obtained above can be generalized to an arbitrary number of Fock states, as follows:

$$c_k|\ldots N_i, N_k, N_l \ldots\rangle = (-1)^L\sqrt{1-N_k}|\ldots N_i, 1-N_k, N_l \ldots\rangle,$$
$$c_k^+|\ldots N_i, N_k, N_l \ldots\rangle = (-1)^L\sqrt{N_k}|\ldots N_i, 1-N_k, N_l \ldots\rangle.$$

[3] For the reasons of convenience, we put α and β to be real without any complex phase multipliers.

Here factor $(-1)^L$ keeps in view the number of fermionic states preceding the kth state. This reflects the fact that the fermionic many-particle wave function changes its sign with permutating two adjacent states, i.e. $|N_i, N_k\rangle = -|N_k, N_i\rangle$, and displays in the anticommutative properties of fermionic operators

$$\{c_i^+, c_k^+\} = 0, \quad \{c_i, c_k\} = 0, \quad \{c_i, c_k^+\} = \delta_{ik}.$$

Let operators a_i^+ and a_i be creation and annihilation ones for the single-particle states with wave function $\varphi_i(\boldsymbol{r})$. Then the corresponding operators of creation $\psi^+(\boldsymbol{r})$ and annihilation $\psi(\boldsymbol{r})$ in the coordinate space are given by the formulas

$$\psi(\boldsymbol{r}) = \sum_i \varphi_i(\boldsymbol{r})a_i \quad \text{and} \quad \psi^+(\boldsymbol{r}) = \sum_i \varphi_i^*(\boldsymbol{r})a_i^+$$

with the summing all the possible states of the system. Since coefficients $\varphi_i(\boldsymbol{r})$ and $\varphi_i^*(\boldsymbol{r})$ in the sum are the usual wave functions, the matrix operators for these ψ-operators will coincide with the same matrix elements for the conventional wave functions. The commutative and anticommutative rules conserve

$$\left[\psi(\boldsymbol{r}), \psi^+(\boldsymbol{r}')\right] = \delta(\boldsymbol{r} - \boldsymbol{r}') \quad \text{for bosons,}$$
$$\left\{\psi(\boldsymbol{r}), \psi^+(\boldsymbol{r}')\right\} = \delta(\boldsymbol{r} - \boldsymbol{r}') \quad \text{for fermions.}$$

If operators a_p^+ and a_p are the creation and annihilation of a particle with momentum $\boldsymbol{p}$, the operators $\psi^+(\boldsymbol{r})$ and $\psi(\boldsymbol{r})$ have the sense of creating and annihilating the particle at the coordinate point $\boldsymbol{r}$. The relation between these operators is achieved with the aid of the conventional Fourier transformation

$$\psi(\boldsymbol{r}) = \frac{1}{V^{3/2}}\int \frac{Vd^3p}{(2\pi\hbar)^3} a_p e^{i\boldsymbol{pr}/\hbar} \quad \text{and} \quad \psi^+(\boldsymbol{r}) = \frac{1}{V^{3/2}}\int \frac{Vd^3p}{(2\pi\hbar)^3} a_p^+ e^{-i\boldsymbol{pr}/\hbar}$$

where V is the volume[4] of the system. Then the particle number operator reads in the coordinate representation

$$\hat{N} = \int \psi^+(\boldsymbol{r})\psi(\boldsymbol{r})d^3r.$$

Let total energy of the system be $E = \sum_i \varepsilon_i N_i$ where ε_i is the energy of single-particle with wave function φ_i. Since the particle number operator in the ith state is $\hat{N}_i = a_i^+ a_i$, the corresponding Hamiltonian operator equals $\hat{H} = \sum_i \varepsilon_i a_i^+ a_i$. If the states are classified by the momentum of a particle as, for example, in an ideal gas $\varepsilon(\boldsymbol{p}) = \boldsymbol{p}^2/2m$, the Hamiltonian is given by

---

[4] To write the formulas more concise and less cumbersome, the volume $V = 1$ and Planck constant $\hbar = 1$ are usually supposed.

$$\hat{H} = \sum_{p} \frac{\boldsymbol{p}^2}{2m} a_{\boldsymbol{p}}^+ a_{\boldsymbol{p}}.$$

In the coordinate representation, this expression corresponds to the integral determining the kinetic energy of the particles

$$\hat{H} = \int \frac{\hbar^2 \nabla\psi^+(\boldsymbol{r}) \cdot \nabla\psi(\boldsymbol{r})}{2m} d^3r = \int \frac{\hbar^2 |\nabla\psi(\boldsymbol{r})|^2}{2m} d^3r.$$

For the classical expressions of energy E_{ext} of the system in the external potential field $V(\boldsymbol{r})$

$$E_{\text{ext}} = \int n(\boldsymbol{r}) V(\boldsymbol{r}) d^3r$$

and for the energy[5] of two-particle interaction E_{int} with the coupling potential $U(\boldsymbol{r} - \boldsymbol{r}')$

$$E_{\text{int}} = \frac{1}{2} \iint n(\boldsymbol{r}) U(\boldsymbol{r} - \boldsymbol{r}') n(\boldsymbol{r}') d^3r\, d^3r',$$

the corresponding terms in the Hamiltonian via ψ-operators[6] are composed as follows:

$$\hat{H}_{\text{ext}} = \int \psi^+(\boldsymbol{r}) V(\boldsymbol{r}) \psi(\boldsymbol{r}) d^3r,$$

$$\hat{H}_{\text{int}} = \frac{1}{2} \iint \psi^+(\boldsymbol{r}) \psi^+(\boldsymbol{r}') U(\boldsymbol{r} - \boldsymbol{r}') \psi(\boldsymbol{r}') \psi(\boldsymbol{r}) d^3r\, d^3r'.$$

Note here that operator $\hat{n}(\boldsymbol{r}) = \psi^+(\boldsymbol{r})\psi(\boldsymbol{r})$ plays a role of particle density operator at point $\boldsymbol{r}$.

The current density operator $\hat{\boldsymbol{j}}(\boldsymbol{r})$, expressed via creation $\psi^+(\boldsymbol{r})$ and annihilation $\psi(\boldsymbol{r})$ operators for the spinless particles

$$\hat{\boldsymbol{j}}(\boldsymbol{r}) = \frac{ie\hbar}{2m} \Big(\nabla\psi^+(\boldsymbol{r})\psi(\boldsymbol{r}) - \psi^+(\boldsymbol{r})\nabla\psi(\boldsymbol{r}) \Big) - \frac{e^2}{mc} \boldsymbol{A}(\boldsymbol{r}) \psi^+(\boldsymbol{r}) \psi(\boldsymbol{r}),$$

corresponds to the usual quantum mechanical expression.

---

[5] The coefficient with the interaction term should be written as $1/2V$ but, for simplicity of the formulas, volume $V = 1$ is usually put.

[6] It is customary to place the creation operators on the left-hand side from the annihilation operators.

5.2 Normal (Nonsuperfluid) Fermi-Liquid

The condensed system of interacting particles obeying the Fermi–Dirac statistics is commonly referred to as a *Fermi liquid.* As most important examples, we note electrons in most metals, liquid ^{3}He being the helium isotope with two protons, one neutron and two electrons in an atom, and nucleons (protons and neutrons) composing the nuclear matter. So, below we always imply the Fermi particle spin equal to 1/2. The temperatures that are of greatest interest here are sufficiently low as compared with the typical degeneracy temperature of the Fermi system or when the de Broglie wavelength corresponding to the thermal motion of particles becomes larger than the interparticle distance. For such low temperatures, the quantum effects begin essentially to affect the macroscopic properties of the Fermi liquid.

Since the fermions obey the Pauli principle which forbids two fermions to occupy the same state, the concept of the *Fermi surface* can be introduced in the momentum space, separating the occupied fermionic states from the unoccupied ones in the ground state of Fermi liquid at zero temperature. The Fermi surface is a sphere in isotropic ^{3}He Fermi liquid.

In metals, as an example of anisotropic Fermi liquids, the shape of Fermi surface is anisotropic and corresponds completely to the symmetry of crystal lattice and its periodicity. The Fermi surface in the momentum space is governed by certain equation $E(\boldsymbol{p}) = \mu$ where μ is the chemical potential at zero temperature. The Fermi surfaces in metals may have very complicated geometrical shape and usually they are divided into the *closed* and *open* Fermi surfaces. The Fermi surface is referred to as the *closed* one if it nowhere reaches the boundary of the first Brillouin zone. In other words, the magnitude of any vector lying inside the closed Fermi surface is always less than the reciprocal lattice period.

In the case of *open* surface the Fermi one reaches the boundary of the first Brillouin zone. The topology of the Fermi surface to great extent determines the dynamics of electrons in electric and magnetic fields and, therefore, influences the galvanomagnetic properties of metals.

In what follows, we consider thoroughly the normal liquid ^{3}He representing neutral isotropic Fermi liquid. The background for the phenomenological description of normal (non-superconducting) Fermi liquid incorporates the concepts of the spectrum of elementary excitations or quasiparticles with momentum $\boldsymbol{p}$ and energy $\varepsilon(\boldsymbol{p})$. The structure of low energy excitations in Fermi liquid is completely similar to the energy spectrum of ideal Fermi gas. At zero temperature corresponding to the ground state of Fermi liquid, all quasiparticles in the momentum space fill entirely the region of momenta $|\boldsymbol{p}| < p_F$ or *Fermi sphere* limited with the Fermi momentum p_F. The magnitude of the Fermi momentum is independent of interaction between particles.[7] The volume of momentum space limited with the Fermi surface is proportional to the Fermi particle density and determined with the particle density n of Fermi liquid. The magnitude of the Fermi momentum coincides with that for the ideal Fermi gas of the same density n.

---

[7] This statement is the essence of general Luttinger's theorem.

The number of quasiparticles equals that of particles of Fermi liquid. It is necessary to distinguish the momentum distribution $N(\boldsymbol{p})$ of genuine particles in the liquid from the similar distribution like the Fermi step $n(\boldsymbol{p}) = \vartheta(p_F - p)$. The magnitude of distribution $N(\boldsymbol{p})$ at zero momentum $p = 0$ is somewhat smaller than unity and decreases with increasing the momentum p from zero value to the Fermi momentum p_F. For the boundary value $p = p_F$, the distribution $N(\boldsymbol{p})$ has a jump $Z = N(p_F - 0) - N(p_F + 0)$ of finite magnitude $(0 < Z < 1)$ dependent on the interaction between particles of the liquid. In the region $p > p_F$ the distribution $N(\boldsymbol{p})$ continues to decrease down to zero as the momentum magnitude p increases.

The ground state energy of the Fermi liquid at zero temperature does not reduce to a simple sum of energies $\varepsilon(\boldsymbol{p})$ for all quasiparticles. The general statement can only be formulated, meaning the fact that the ground state energy $\mathcal{E}$ for the Fermi liquid at zero temperature is some functional depending on the distribution function $n(\boldsymbol{r}, \boldsymbol{p})$ alone[8]

$$\mathcal{E} = \mathcal{E}[n(\boldsymbol{r}, \boldsymbol{p}] = \int E[n(\boldsymbol{r}, \boldsymbol{p})](2s+1)\frac{d^3p}{(2\pi\hbar)^3}d^3r.$$

The ground state energy, i.e. energy minimum, realizes at the genuine distribution function[9] $n(\boldsymbol{r}, \boldsymbol{p})$. For spin $s = 1/2$, the number of spin states is $2s + 1 = 2$. When it is necessary to take the spin and magnetic field effects into account, distribution function $n_\sigma(\boldsymbol{r}, \boldsymbol{p}) = n_{\alpha\beta}(\boldsymbol{r}, \boldsymbol{p})$ will be spin-dependent as well and represent the 2×2 matrix. In addition to integration, taking the matrix trace is also implied.

The equilibrium state at nonzero temperature and, in general, any excited state in the Fermi liquid can be obtained by the successive transfer of one or several quasiparticles from the space region inside the Fermi sphere to outside it. Every such transfer is treated as a simultaneous creation of a quasiparticle outside the Fermi sphere and a hole inside it. Thus, a hole means directly a non-occupied state inside the Fermi sphere. The energy of single quasiparticle $\varepsilon[n(\boldsymbol{r}, \boldsymbol{p})]$ is determined as a variational derivative of the total energy with respect to the distribution function according to

$$\delta\mathcal{E} = \mathcal{E}[n + \delta n] - \mathcal{E}[n] = \int \delta E[n(\boldsymbol{r}, \boldsymbol{p})]\frac{2d^3p}{(2\pi\hbar)^3}d^3r =$$
$$= \int \varepsilon[n(\boldsymbol{r}, \boldsymbol{p})]\delta n(\boldsymbol{r}, \boldsymbol{p})]\frac{2d^3p}{(2\pi\hbar)^3}d^3r, \quad \text{i.e.} \quad \varepsilon[n(\boldsymbol{r}, \boldsymbol{p})] = \frac{\delta\mathcal{E}[n(\boldsymbol{r}, \boldsymbol{p})]}{\delta n(\boldsymbol{r}, \boldsymbol{p})}.$$

---

[8] Determining a specific type of the functional is a subject of microscopic Fermi-liquid theory. This problem will be discussed on the example of a dipole Fermi gas.

[9] The similar statement, representing the variational principle for the interacting electron system in the external electric field, is called the *Hohenberg–Kohn theorem*. The latter provides us one of arguments for applying the density functional theory as a method of computing the electronic structure of molecules and metals.

The sense of this definition is transparent. In fact, if a quasiparticle is created with momentum $\boldsymbol{p} = \boldsymbol{Q}$ at point $\boldsymbol{r} = \boldsymbol{R}$ and the distribution function varies as $\delta n(\boldsymbol{r}, \boldsymbol{p}) = (1/2)(2\pi\hbar)^3\delta(\boldsymbol{p} - \boldsymbol{Q})\delta(\boldsymbol{r} - \boldsymbol{R})$, the energy of Fermi liquid increases by the magnitude equal to $\varepsilon[n(\boldsymbol{R}, \boldsymbol{Q})]$. In the spatially homogeneous isotropic Fermi liquid the quasiparticle distribution function at zero temperature equals the step-like function $n_0(\boldsymbol{p}) = \vartheta(p_F - p)$, the quasiparticle excitation energy depending on the momentum magnitude alone, i.e. $\varepsilon\big(n_0(\boldsymbol{p})\big) = \varepsilon_0(p)$. In the close vicinity of the Fermi sphere, the quasiparticle energy can be expanded in the powers of difference $p - p_F$

$$\xi(p) = \varepsilon_0(p) - \varepsilon_0(p_F) \approx v_F(p - p_F), \quad |p - p_F| \ll p_F\,.$$

Coefficient v_F is called the *Fermi velocity*. It can also be expressed in the terms of *effective quasiparticle mass* m^* according to definition

$$v_F = \frac{p_F}{m^*}\,.$$

The magnitude of effective mass[10] depends on the interaction between the particles in the Fermi liquid and differs from the mass of the genuine particles composing the Fermi liquid.

The dependence of quasiparticle energy on the distribution function is an important and specific feature of the phenomenological approach for describing the Fermi liquid with the aid of the functional for the distribution function. In fact, let us expand the functional for the total energy of a homogeneous Fermi liquid into the small deviations of distribution function $\delta n = n - n_0$ from the equilibrium one n_0

$$\mathcal{E} = \mathcal{E}[n_0(\boldsymbol{p})] + \int \varepsilon_0(\boldsymbol{p})\delta n(\boldsymbol{p})V d\tau_p + \\ + \frac{1}{2}\int f(\boldsymbol{p}, \boldsymbol{p}')\delta n(\boldsymbol{p})\,\delta n(\boldsymbol{p}')V d\tau_p\, d\tau_{p'} + \ldots$$

Here $V = \int d^3r$ is the volume of Fermi liquid and $d\tau_p = 2d^3p/(2\pi\hbar)^3$ is the volume element of momentum space. Hence we see that the quasiparticle excitation energy depends on the distribution function and equals

$$\varepsilon(\boldsymbol{p}) = \varepsilon_0(\boldsymbol{p}) + \int f(\boldsymbol{p}, \boldsymbol{p}')\delta n(\boldsymbol{p}')d\tau_{p'} + \ldots = \varepsilon_0(\boldsymbol{p}) + \delta\varepsilon[n(\boldsymbol{p})].$$

The second variational derivative of energy functional with respect to the distribution function, which appears here,

$$f(\boldsymbol{p}, \boldsymbol{p}') = \left.\frac{\delta^2\mathcal{E}}{\delta n(\boldsymbol{p})\,\delta n(\boldsymbol{p}')}\right|_{n=n'=n_0(\boldsymbol{p})}$$

---

[10] The calculation of the effective mass can be performed within the framework of the microscopic theory of Fermi liquid.

is symmetrical due to its definition, i.e. $f(\boldsymbol{p}, \boldsymbol{p}') = f(\boldsymbol{p}', \boldsymbol{p})$, and referred to as the *Landau function*. The Landau function $f(\boldsymbol{p}, \boldsymbol{p}')$ governs the coupling of quasiparticles and determines the most important low temperature properties of Fermi liquid such as effective mass magnitude, specific heat, and sound velocity.

There is an integral relation connecting the effective mass m^*, parameters of the Landau function f, and genuine mass m of the Fermi particles in a liquid. In what follows, we use that the momentum per unit volume $\mathcal{P}$ equals the mass flow density $\boldsymbol{j}$ and that the number of quasiparticles equals the number of genuine Fermi-particles of mass m. Since the velocity of a quasiparticle $\boldsymbol{v}$ is $\partial\varepsilon/\partial\boldsymbol{p}$, the density of quasiparticle flow is given by

$$\boldsymbol{i} = \int \frac{\partial\varepsilon}{\partial\boldsymbol{p}} n(\boldsymbol{p}) d\tau_p \quad \text{where} \quad d\tau_p = 2\frac{d^3 p}{(2\pi\hbar)^3}\,.$$

Accordingly, due to equality between the numbers of quasiparticles and genuine particles, the quasiparticle number flow $\boldsymbol{i}$ is identical to the genuine particle flow equal to the mass flow density $\boldsymbol{j}$ divided by the Fermi particle mass m, i.e. $\boldsymbol{i} = \boldsymbol{j}/m$. Since the momentum per unit volume of liquid equals the integral

$$\mathcal{P} = \int \boldsymbol{p} n\, d\tau_p\,,$$

we arrive at the following equality:

$$\int \boldsymbol{p} n\, d\tau_p = m \int \frac{\partial\varepsilon}{\partial\boldsymbol{p}} n(\boldsymbol{p})\, d\tau_p\,.$$

Varying this equality with respect to the distribution function gives

$$\begin{aligned}
\int \boldsymbol{p}\, \delta n\, d\tau_p &= m \int \frac{\partial\varepsilon}{\partial\boldsymbol{p}}\, \delta n\, d\tau_p + m \int n(\boldsymbol{p}) \frac{\partial(\delta\varepsilon)}{\partial\boldsymbol{p}} d\tau_p = \\
&= m \int \frac{\partial\varepsilon}{\partial\boldsymbol{p}} \delta n\, d\tau_p + m \int d\tau_p\, n(\boldsymbol{p}) \frac{\partial}{\partial\boldsymbol{p}} \left(\int f(\boldsymbol{p}, \boldsymbol{p}')\, \delta n(\boldsymbol{p}')\, d\tau_{p'} \right) = \\
&= m \int \frac{\partial\varepsilon}{\partial\boldsymbol{p}} \delta n\, d\tau_p + m \int d\tau_p\, \delta n(\boldsymbol{p}) \int d\tau_{p'}\, n(\boldsymbol{p}') \frac{\partial}{\partial\boldsymbol{p}'} f(\boldsymbol{p}', \boldsymbol{p}).
\end{aligned}$$

Due to arbitrariness of the variation δn and due to symmetry of the Landau function $f(\boldsymbol{p}, \boldsymbol{p}') = f(\boldsymbol{p}', \boldsymbol{p})$, we obtain the equality

$$\frac{\boldsymbol{p}}{m} = \frac{\partial\varepsilon}{\partial\boldsymbol{p}} + \int \frac{\partial f(\boldsymbol{p}, \boldsymbol{p}')}{\partial\boldsymbol{p}'} n(\boldsymbol{p}')\, d\tau_{p'}\,.$$

The integration by parts and the vanishing of the integrand at the infinitely far surface $p' = \infty$ lead us to the integral relation desired

$$\frac{\boldsymbol{p}}{m} = \frac{\partial \varepsilon}{\partial \boldsymbol{p}} - \int f(\boldsymbol{p}, \boldsymbol{p}') \frac{\partial n(\boldsymbol{p}')}{\partial \boldsymbol{p}'} \, d\tau_{p'} \, .$$

For the equilibrium distribution function at zero temperature, the derivative $\partial n/\partial \boldsymbol{p}$ reads

$$\frac{\partial n}{\partial \boldsymbol{p}} = \frac{\partial n}{\partial \varepsilon}\frac{\partial \varepsilon}{\partial \boldsymbol{p}} = -v_F \frac{\boldsymbol{p}}{p_F}\delta(\varepsilon - \varepsilon_F) = -\frac{\boldsymbol{p}}{p_F}\delta(p - p_F).$$

When finding the effective mass m^* due to the last equality, we are interested in the values of the momentum at the Fermi-sphere surface $|\boldsymbol{p}| = p_F$. So, let us choose direction of vector $\boldsymbol{p}$ as a polar axis and introduce angle χ equal to that between vectors $\boldsymbol{p} = p_F\boldsymbol{n}$ and $\boldsymbol{p}' = p_F\boldsymbol{n}'$

$$\frac{\boldsymbol{p}}{m} = \left.\frac{\partial \varepsilon}{\partial \boldsymbol{p}}\right|_{p=p_F} + \int f(p_F\boldsymbol{n}, p_F\boldsymbol{n}')\delta(p' - p_F)\frac{\boldsymbol{p}'}{p_F}\frac{2p'^2 dp'\, d\Omega_\chi}{(2\pi\hbar)^3} =$$

$$= v_F\frac{\boldsymbol{p}}{p_F} + \int f(\chi)\boldsymbol{n}' \frac{p_F^2}{\pi^2\hbar^3}\frac{d\Omega_\chi}{4\pi} = v_F\frac{\boldsymbol{p}}{p_F} + \frac{p_F^2}{\pi^2\hbar^3}\boldsymbol{n}\int f(\chi)\cos\chi\frac{d\Omega_\chi}{4\pi}\, .$$

Here $d\Omega_\chi = 2\pi \sin\chi\, d\chi$ is the solid angle element. Then we use relation $v_F = p_F/m$ and define the dimensionless *Landau parameter* F_1 according to

$$F_1 = 3\frac{p_F^2}{\pi^2\hbar^3 v_F}\int f(\chi)\cos\chi\frac{d\Omega_\chi}{4\pi} = 3\frac{m^* p_F}{\pi^2\hbar^3}\int f(\chi)\cos\chi\frac{d\Omega_\chi}{4\pi}\, .$$

The Landau parameter F_1 characterizes the quasiparticle-quasiparticle interaction averaged over the Fermi surface with the cosine of angle.[11] Finally, we express the effective mass via the Landau parameter F_1

$$\frac{1}{m} = \frac{1}{m^*}\left(1 + \frac{F_1}{3}\right) \quad \text{or} \quad m^* = m\left(1 + \frac{F_1}{3}\right).$$

In liquid ^{3}He the Landau parameter is $F_1 = 6.25$ and, correspondingly, $m^* = 3.08m$, m being the ^{3}He atom mass. In metals or semiconductors, the effective mass of electron excitations can be both larger or smaller than the electron one. In the so-called heavy fermion metallic alloys based on cerium or uranium the effective mass m^* can exceed the electron mass by a factor of 100 or more.

The effective mass of quasiparticles determines entropy S and specific heat of a Fermi liquid in the low temperature limit $T \ll \varepsilon_F$. The specific heat will be found by differentiating the energy of a Fermi liquid with respect to the temperature at the fixed particle number N. The variation of equilibrium distribution function $\delta n = n(T) - n_0$ with temperature changes the energy of a liquid per unit volume according

[11] From the mathematical point of view the parameter F_1 is a coefficient in the expansion of function $F(\chi) = \sum_{l=0}^{\infty}(2l+1)F_l P_l(\cos\chi)$ over the Legendre polynomials $P_l(\cos\chi)$.

to

$$\delta\mathcal{E} = \int \varepsilon\, \delta n\, d\tau_p = \int \varepsilon_0\, \delta n\, d\tau_p + \int \delta\varepsilon\, \delta n\, d\tau_p \approx \int \varepsilon_0\, \delta n\, d\tau_p\,.$$

Note here the following important aspect. We can neglect the second term provided that two conditions are satisfied. First, the contribution of the term retained is linear in temperature T in the low temperature limit. Second, the contribution from the discarded term, resulting from the interaction of quasiparticles, is of lower in magnitude than the first power of temperature. So, we write the derivative

$$\frac{\partial(\delta\varepsilon)}{\partial T} = \int f \frac{\partial(\delta n')}{\partial T} d\tau_{p'}$$

and employ the following expansion valid as $T \to 0$:

$$\frac{\partial n}{\partial T} = \frac{\partial \mu}{\partial T}\delta(\varepsilon - \mu) - \frac{\pi^2}{3} T \frac{\partial}{\partial \varepsilon}\delta(\varepsilon - \mu) + \dots$$

The derivative $\partial\mu/\partial T$ equals the entropy which, like the specific heat, is linear in temperature. Therefore, quantity $\delta\varepsilon$ will be of second order in temperature and the approximation[12] above is entirely correct. As a result, for the fixed particle number $N = \text{const}$, we have

$$C(T) \approx \int \varepsilon_0(\boldsymbol{p}) \frac{\partial n}{\partial T} d\tau_p \quad \text{and} \quad \int \frac{\partial n}{\partial T} d\tau_p = 0.$$

In the lowest order in temperature, we can replace the exact values of energy ε with the approximate ones ε_0 in the derivative $\partial n/\partial T$ above. After these steps, the further calculation becomes identical to that in the case of an ideal Fermi gas if we replace the particle mass m with the effective one m^* in the density of states. Thus, we readily obtain the following answer for the low temperature behavior of specific heat in a normal Fermi liquid:

$$C(T) = \gamma T, \quad \gamma = \frac{\pi^2}{3} g(\varepsilon_F), \quad g(\varepsilon_F) = \frac{m^* p_F}{\pi^2 \hbar^3}\,.$$

Here γ is the *Sommerfeld constant* and $g(\varepsilon_F)$ is the density of quasiparticle states at the Fermi surface. Because of the linear temperature behavior the specific heat and the entropy equal each other. The requirement of positiveness for the specific heat entails one of the necessary conditions for the thermodynamic stability of a Fermi liquid as an inequality $F_1 > -3$ for the Landau parameter. Condition $F_1 = -3$ and in the general[13] case $F_l = -(2l + 1)$ are often referred to as the *Pomeranchuk instability*

[12] The calculation of the next expansion term, proportional to $T^3 \ln(\varepsilon_F/T)$, may be an inappropriate problem within the framework of the energy functional depending on the distribution function alone.

[13] Let us recall the expansion $F(\chi) = \sum_{l=0}^{\infty}(2l + 1) F_l P_l(\cos\chi)$ over the Legendre polynomials.

criterion. The point at which $F_l = -(2l+1)$ is of interest as it indicates a quantum phase transition from a Fermi liquid to a different state of matter.

5.3 Paramagnetic Susceptibility of Normal Fermi Liquid

Liquid ^{3}He displays magnetic properties because the ^{3}He nucleus has spin 1/2 and, therefore, magnetic moment. The magnitude of magnetic moment is determined by the product of gyromagnetic ratio and magnetic nuclear Bohr magneton. An existence of magnetic moment in the particles results in paramagnetism of a Fermi liquid.

Since the Fermi particles have spin, the Landau function $f_{\sigma,\sigma'}(\boldsymbol{p}, \boldsymbol{p}')$, in general, should include the term corresponding to the exchange interaction between quasiparticles

$$f_{\sigma,\sigma'}(\boldsymbol{p}, \boldsymbol{p}') = f(\boldsymbol{p}, \boldsymbol{p}') + \zeta(\boldsymbol{p}, \boldsymbol{p}')(\boldsymbol{\sigma}\boldsymbol{\sigma}').$$

In the external magnetic field H both the energy of quasiparticles in a Fermi liquid and the distribution function $\delta n_\sigma = n_\sigma - n_0(\boldsymbol{p})$ vary simultaneously

$$\delta\varepsilon_\sigma(\boldsymbol{p}) = -\beta\boldsymbol{\sigma}\boldsymbol{H} + \mathrm{tr}_{\sigma'}\int f_{\sigma,\sigma'}(\boldsymbol{p}, \boldsymbol{p}')\delta n_{\sigma'}(\boldsymbol{p}')\, d\tau_{p'}\,, \quad d\tau_p = \frac{d^3p}{(2\pi\hbar)^3}\,.$$

Here β is the effective Bohr magneton and matrices $\boldsymbol{\sigma}$ are the Pauli ones. The paramagnetic susceptibility will be calculated from the relation

$$\chi = \frac{\partial \boldsymbol{M}}{\partial \boldsymbol{H}} = \frac{\partial}{\partial \boldsymbol{H}}\mathrm{tr}_\sigma \int \beta\boldsymbol{\sigma} n_\sigma(\boldsymbol{p})\, d\tau_p = \beta\frac{\partial}{\partial \boldsymbol{H}}\mathrm{tr}_\sigma \int \boldsymbol{\sigma}\delta n_\sigma(\boldsymbol{p})\, d\tau_p$$

where $\boldsymbol{M}$ is the magnetic moment per unit volume. Here we take into account that the magnetization of a liquid vanishes in the absence of magnetic field. Formally, this follows from identity $\mathrm{tr}\,\boldsymbol{\sigma} = 0$.

We can neglect the chemical potential variation in the linear approximation in the magnetic field. The chemical potential, as a scalar quantity, changes only in second order with respect to magnetic field being a vector quantity. Thus we write approximately

$$\chi \approx \beta\frac{\partial}{\partial H}\mathrm{tr}_\sigma \int \boldsymbol{\sigma}\frac{\partial n_0(\varepsilon)}{\partial \varepsilon}\delta\varepsilon_\sigma(\boldsymbol{p})\, d\tau_p\,,$$

$$\delta\varepsilon_\sigma(\boldsymbol{p}) \approx -\beta\boldsymbol{\sigma}\boldsymbol{H} + \mathrm{tr}_{\sigma'}\int f_{\sigma,\sigma'}(\boldsymbol{p}, \boldsymbol{p}')\frac{\partial n_0(\varepsilon_0(\varepsilon'))}{\partial \varepsilon'}\delta\varepsilon_{\sigma'}(\boldsymbol{p}')\, d\tau_{p'}\,.$$

Since derivative $n_0'(\varepsilon) = -\delta(\varepsilon - \varepsilon_F) = -\delta(p - p_F)/v_F$, integration in the both integrals is performed over the Fermi surface and the essential momenta are only those which magnitudes equal to the Fermi momentum, i.e. $|\boldsymbol{p}| = |\boldsymbol{p}'| = p_F$. Next, assuming the structure of solution as $\delta\varepsilon_\sigma(\boldsymbol{p}) = -\xi(\boldsymbol{p})\beta\boldsymbol{\sigma}\boldsymbol{H}$, we obtain two equations for determining the unknown ξ and χ

$$\xi(\boldsymbol{p}) = 1 + 2\int \zeta(\boldsymbol{p}, \boldsymbol{p}')\frac{\partial n_0(\varepsilon')}{\partial \varepsilon'}\xi(\boldsymbol{p}')\,d\tau_{p'}\,,$$

$$\chi = -2\beta^2\int \xi(\boldsymbol{p})\frac{\partial n_0(\varepsilon)}{\partial \varepsilon}\,d\tau_p = \beta^2\frac{m^* p_F}{\pi^2\hbar^3}\xi(p_F) = \beta^2 g(\varepsilon_F)\xi(p_F),$$

where $g(\varepsilon_F)$ is the density of quasiparticle states at the Fermi surface. We have used here the following identities: $\mathrm{tr}_{\sigma'}(\boldsymbol{\sigma\sigma}')\boldsymbol{\sigma}' = 2\boldsymbol{\sigma}$ and $\mathrm{tr}_{\sigma}(\boldsymbol{\sigma H})\boldsymbol{\sigma} = 2\boldsymbol{H}$. The spin-independent part of the Landau function, as expected, cancels from the final result.

Let us choose direction of vector $\boldsymbol{p}$ as a polar axis z and introduce angle ϑ equal to that between vectors $\boldsymbol{p} = p_F\boldsymbol{n}$ and $\boldsymbol{p}' = p_F\boldsymbol{n}'$. Then, we get

$$\xi(p_F) = 1 - \int \zeta(\vartheta)\frac{m^* p_F}{\pi^2\hbar^3}\xi(p_F)\frac{d\Omega_\vartheta}{4\pi}\,, \quad \zeta(\vartheta) = \zeta(p_F\boldsymbol{n}, p_F\boldsymbol{n}')$$

where the solid angle element equals $d\Omega_\vartheta = 2\pi\sin\vartheta\,d\vartheta$. We find after calculation

$$\xi(p_F) = \frac{1}{1+Z_0} \quad \text{and} \quad Z_0 = g(\varepsilon_F)\int \zeta(\vartheta)\frac{d\Omega_\vartheta}{4\pi}\,.$$

Here we have introduced the dimensionless *exchange Landau parameter* Z_0 equal to the magnitude of the exchange interaction averaged over the whole the Fermi surface and multiplied by the density of states. Finally, the paramagnetic (spin) susceptibility of a Fermi liquid is given by the formula

$$\chi = \frac{\beta^2 g(\varepsilon_F)}{1+Z_0}\,.$$

The stability condition for the paramagnetic state of a Fermi liquid is governed by the relation $Z_0 > -1$. If $Z_0 = -1$, the paramagnetic susceptibility diverges. The latter means the appearance of a ferromagnetic instability[14] and inapplicability of the Landau theory to a Fermi liquid. In liquid ^{3}He the Landau parameter is $Z_0 \approx -0.67$. The palladium metal has $Z_0 \approx -0.9$ and, thus, is close to the ferromagnetic instability.

5.4 Dilute Fermi Gas with the Dipolar Interaction

The experimental realization of laser traps has allowed us to study the physical properties of rarified atom gases in the degenerate state at ultralow temperatures. As

[14] The similar condition for originating the ferromagnetic ordering in the system of collective electrons or in metals is called the *Stoner criterion of instability*.

one of the interesting cases, we consider below a dilute (weakly non-ideal) Fermi gas with the dipole interaction between the particles as an example of an anisotropic Fermi system.

Let *dipole Fermi gas* be a gas of fermions with mass m, in which the fermion spins are completely polarized, e.g. with magnetic field, in the z-axis direction. The interaction $U(\boldsymbol{r})$ between two particles is anisotropic and has the dipole-dipole type

$$U(\boldsymbol{r}) = \frac{d^2}{r^3}\left(1 - 3\frac{z^2}{r^2}\right)$$

where d is the dipole electric or magnetic moment of a fermion.

First of all, we start from determining the Fermi surface shape and use the Hartree–Fock approximation, assuming the dipole-dipole interaction to be small, i.e. $nd^2/\varepsilon_F \ll 1$ where n is the particle number and ε_F is the Fermi energy. Let us express the energy of a gas per unit volume in terms of the Fermi–Dirac distribution function $n_{\boldsymbol{p}}$ and Fourier transform $U(\boldsymbol{p})$ of interparticle interaction, employing the Hartree–Fock approximation

$$\begin{aligned} E &= \sum_{\boldsymbol{p}} \frac{\boldsymbol{p}^2}{2m} n_{\boldsymbol{p}} + \frac{1}{2}\sum_{\boldsymbol{p}} n_{\boldsymbol{p}} \int U(\boldsymbol{r})\, d^3r \sum_{\boldsymbol{p}'} n_{\boldsymbol{p}'} - \frac{1}{2}\sum_{\boldsymbol{p},\boldsymbol{p}'} n_{\boldsymbol{p}} U(\boldsymbol{p}-\boldsymbol{p}') n_{\boldsymbol{p}'} = \\ &= \sum_{\boldsymbol{p}} \varepsilon_0(\boldsymbol{p}) n_{\boldsymbol{p}} + \frac{1}{2} n^2 U(0) - \frac{1}{2}\sum_{\boldsymbol{p},\boldsymbol{p}'} n_{\boldsymbol{p}} U(\boldsymbol{p}-\boldsymbol{p}') n_{\boldsymbol{p}'}. \end{aligned}$$

Here the first term of a sum is the kinetic energy, the second is the energy of direct dipole-dipole interaction of particles distributed homogeneously in the space, and the third is the exchange interaction of fermions. In these expressions, there is no summing over the fermion spin since we consider the direction of fermion spins to be fixed. The gas density n is thus given by a sum over momentum alone

$$n = \sum_{\boldsymbol{p}} n_{\boldsymbol{p}}.$$

The Fourier transform $U(\boldsymbol{r})$ can be found with the aid of the equalities for the Coulomb potential derivatives

$$-\frac{\partial^2}{\partial x_\alpha \partial x_\beta}\left(\frac{1}{r}\right) = \frac{r^2\delta_{\alpha\beta} - 3x_\alpha x_\beta}{r^5} + \frac{4\pi}{3}\delta_{\alpha\beta}\delta(\boldsymbol{r}).$$

Hence it is readily to see for the Fourier transform at $\alpha = \beta = z$ that

$$-(-ip_z)^2\frac{4\pi}{p^2} = \big(\frac{r^2-3z^2}{r^5}\big)_p + \frac{4\pi}{3} \quad \text{and} \quad \big(\frac{r^2-3z^2}{r^5}\big)_p = -\frac{4\pi}{3}\big(1-3\frac{p_z^2}{p^2}\big).$$

In conclusion, the Fourier transform of dipole-dipole interaction $U(\boldsymbol{r})$ equals

$$U(\boldsymbol{p}) = -\frac{4\pi d^2}{3}\left(1-3\frac{p_z^2}{p^2}\right).$$

The Landau function $f(\boldsymbol{p}, \boldsymbol{p}')$, defined as second variational derivative of energy with respect to the distribution function, equals

$$f(\boldsymbol{p}, \boldsymbol{p}') = \delta^2 E/\delta n_{\boldsymbol{p}}\delta n_{\boldsymbol{p}'} = U(0) - U(\boldsymbol{p}-\boldsymbol{p}').$$

The quasiparticle energy $\varepsilon(\boldsymbol{p})$ is given by the first variational derivative $\delta E/\delta n_{\boldsymbol{p}}$ and equal to

$$\varepsilon(\boldsymbol{p}) = \varepsilon_0(\boldsymbol{p}) + \sum_{\boldsymbol{p}'} f_{\boldsymbol{p}\boldsymbol{p}'}n_{\boldsymbol{p}'} = \varepsilon_0(\boldsymbol{p}) - \sum_{\boldsymbol{p}'} U(\boldsymbol{p}-\boldsymbol{p}')n_{\boldsymbol{p}'} = \varepsilon_0(\boldsymbol{p}) + \delta\varepsilon(\boldsymbol{p}).$$

Here we have taken into account that the Hartree contribution from the direct interaction $U(0)$ vanishes as a result of integrating over the angles, i.e. $U(0) = 0$.

For calculating $\delta\epsilon(\boldsymbol{p})$, we put $\hbar = 1$ for a moment and use the following technique. Since $\delta\epsilon(\boldsymbol{p})$ is a convolution, the quasiparticle energy in the coordinate representation can straightforwardly be obtained

$$\epsilon(\boldsymbol{r}) = \epsilon_0(\boldsymbol{r}) + \delta\epsilon(\boldsymbol{r}) = -\frac{\nabla^2}{2m} - U(\boldsymbol{r})n(\boldsymbol{r}),$$

$n(\boldsymbol{r})$ being the Fourier transform of distribution function

$$n(\boldsymbol{r}) = \sum_{\boldsymbol{p}} n(\boldsymbol{p})e^{i\boldsymbol{p}\boldsymbol{r}}.$$

We put $n(\boldsymbol{p}) \approx n_0(\boldsymbol{p}) = \vartheta(p_F - p)$ as a first approximation and then find

$$n(\boldsymbol{r}) \approx n_0(\boldsymbol{r}) = \sum_{|\boldsymbol{p}|\leqslant p_F} e^{i\boldsymbol{p}\boldsymbol{r}} = 3n\frac{\sin(p_F r) - p_F r\cos(p_F r)}{(p_F r)^3}, \quad n = \frac{p_F^3}{6\pi^2}$$

where p_F is the Fermi momentum in the gas without dipole interaction and n is the gas density. Then, determining the variation $\delta\epsilon(\boldsymbol{p})$ reduces to the integral

$$\delta\epsilon(\boldsymbol{p}) = \int d^3r U(\boldsymbol{r}) n_0(r) e^{-i\boldsymbol{pr}} = d^2 \int d^3r \big(1 - \frac{3z^2}{r^2}\big) \frac{n_0(r)}{r^3} e^{-i\boldsymbol{pr}}.$$

Let us analyze the following integral:

$$J_{ik} = d^2 \int d^3r \left(\delta_{ik} - \frac{3x_i x_k}{r^2} \right) \frac{n_0(r)}{r^3} e^{-i\boldsymbol{pr}} = A(p)\delta_{ik} + B(p) e_i e_k, \quad e_i = \frac{p_i}{p}.$$

The answer can be expressed via two possible tensors of second-rank. The first is unit one δ_{ik}. The second $e_i e_k$ is composed of the projections of unit vector $\boldsymbol{e}$, directed along the momentum. It is obvious that convolution is $J_{ii} = 3A + B = 0$ and, correspondingly,

$$J_{ik} = A(p) \left(\delta_{ik} - 3\frac{p_i p_k}{p^2} \right).$$

We are interested in the components $i = k = z$ or $J_{zz} = A(p)(1 - 3p_z^2/p^2)$. Let us direct vector $\boldsymbol{p}$ along axis z, i.e. $\boldsymbol{p} = (0,\ 0,\ p)$. Then,

$$A(p) = -\frac{1}{2} J_{zz} = -\frac{d^2}{2} \int d^3r \left(1 - \frac{3z^2}{r^2} \right) \frac{n_0(r)}{r^3} e^{-i\boldsymbol{pr}} =$$

$$= -\frac{3nd^2}{2} \int_0^\infty \frac{2\pi r^2\, dr}{r^3} \frac{\sin(p_F r) - p_F r \cos(p_F r)}{(p_F r)^3} \int_{-1}^{1} dx (1 - 3x^2) e^{iprx} =$$

$$= 12\pi n d^2 \int_0^\infty \frac{dr}{r} \frac{\sin(p_F r) - p_F r \cos(p_F r)}{(pr)^3} \frac{(p^2 r^2 - 3)\sin(pr) + 3pr\cos(pr)}{(pr)^3}.$$

Using that

$$F(x) = \int \frac{dx}{x} \frac{\sin(bx) - bx\cos(bx)}{b^3 x^3} \frac{(3 - x^2)\sin x - 3x\cos x}{x^3} =$$

$$= \frac{1}{96 b^3 x^6} \Big[(b^2 - 1)^3 x^6 \big[\mathrm{ci}\,\big(x(b-1)\big) - \mathrm{ci}\,\big(x(b+1)\big)\big] +$$

$$+ 4bx^2\big(-12 + (b^2 - 1)x^2\big) \sin bx \cos x +$$

$$+ bx\big(24 - 2(3 + b^2)x^2 + (b^2 - 1)x^4\big) \sin x \cos bx -$$

$$- \big(24 + 6(b^2 - 1)x^2 + (b^4 - 1)x^4\big) \sin bx \Big]$$

where $\mathrm{ci}\,(x) = -\int_x^\infty \cos t\, dt/t$ is the cosine integral and that

$$F(0) = -\frac{1}{48} \left(\frac{1}{b^2} + \frac{8}{3} - b^2 + \frac{(b^2 - 1)^3}{2b^3} \ln\Big|\frac{b+1}{b-1}\Big| \right), \quad F(\infty) = 0,$$

we arrive at the following answer:

$$A(p) = nd^2 I\left(\frac{p}{p_F}\right),$$

$$I(x) = \frac{\pi}{4}\left(x^2 + \frac{8}{3} - \frac{1}{x^2} + \frac{(1-x^2)^3}{2x^3}\ln\left|\frac{1+x}{1-x}\right|\right) \approx \begin{cases} 4\pi x^2/5, x \ll 1 \\ 4\pi/3, \quad x \gg 1 \end{cases}.$$

Note that function $I'''(x)$ has a logarithmic singularity at point $x = 1$. Finally, we obtain the following quasiparticle spectrum in first approximation in smallness of dipole-dipole interaction:

$$\varepsilon(\boldsymbol{p}) = \varepsilon_0(p) + nd^2 I\left(\frac{p}{p_F}\right)\left(1 - 3\frac{p_z^2}{p^2}\right).$$

The spectrum has an anisotropic character in the momentum as a result of spatial anisotropy for the interaction between particles. The behavior of function $I(x)$ is shown in Fig. 5.1.

The energy spectrum anisotropy leads to the anisotropic shape of the Fermi surface governed by equation $\varepsilon(\boldsymbol{p}) = \mu$. Involving that the distortion of the Fermi sphere is small and the momenta $|\boldsymbol{p}| \sim p_F$ close to the Fermi momentum are only essential, we can approximately write the following:

$$\frac{p_x^2 + p_y^2 + p_z^2}{2m} + nd^2 I(1)\left(1 - \frac{3p_z^2}{p_F^2}\right) = \mu,$$

$$p_x^2 + p_y^2 + p_z^2(1 - 2\pi nd^2/\varepsilon_F) = 2m(\mu - 2\pi nd^2/3)$$

where $\varepsilon_F = p_F^2/2m$ is the Fermi energy in the non-interacting gas. The Fermi surface acquires the shape of spheroid elongated in the direction of the dipole polarization axis z with a ratio of axes as

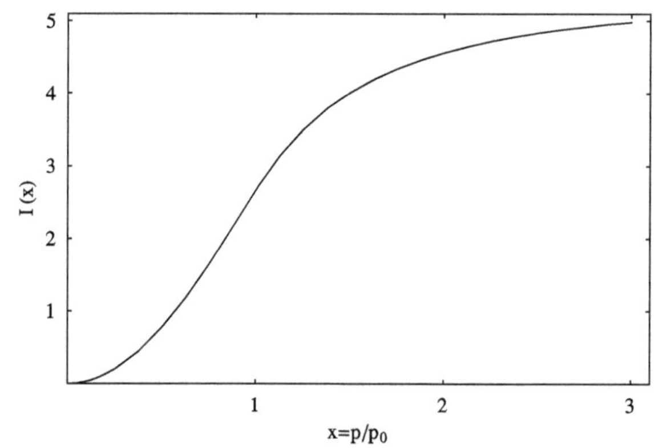

Fig. 5.1 The behavior of function $I(x)$

$$\frac{(p_z)_{\max}}{(p_x)_{\max}} = \frac{(p_z)_{\max}}{(p_y)_{\max}} \approx 1 + \frac{\pi n d^2}{\varepsilon_F}.$$

The chemical potential $\mu \approx p_F^2/2m$ remains unchanged in first order in interaction. As the magnitude of the dipole-dipole interaction increases and approaches about $nd^2 \gtrsim \varepsilon_F$, the homogeneous gas state becomes unstable and the spatially inhomogeneous structures appear.

Using the formula obtained above, we calculate the renormalization of chemical potential at zero temperature in second order in the smallness of dipole-dipole interaction $nd^2/\varepsilon_F \ll 1$. In the thermal equilibrium the quasiparticle distribution function $n_{\rm p}$ looks like the Fermi–Dirac distribution and is a step-like function $n_p = n_0(\varepsilon_p) = \vartheta(\mu - \varepsilon_p)$ at zero temperature $T = 0$. The constancy of particle number N means that the variation δn_p of distribution function satisfies the following condition:

$$\delta N = \sum_p \delta n_p = 0.$$

Expanding $\delta n_p = n_0(\varepsilon_0 + \delta\varepsilon_p - \varepsilon_F - \delta\mu) - n_0(\varepsilon_0 - \varepsilon_F)$ to second order in $(\delta\varepsilon_p - \delta\mu)$

$$\delta n_p = n_0'(\varepsilon_0 - \varepsilon_F)(\delta\varepsilon_p - \delta\mu) + \frac{1}{2} n_0''(\varepsilon_0 - \varepsilon_F)(\delta\varepsilon_p - \delta\mu)^2 + \ldots,$$

we arrive at the following relation:

$$\delta\mu \sum_p n_0' = \sum_p n_0' \delta\varepsilon_p + \frac{1}{2} \sum_p n_0''(\delta\varepsilon_p - \delta\mu)^2.$$

Next, we represent the quasiparticle energy variation $\delta\varepsilon_p$ as a sum of corrections of first $\delta\varepsilon_p^{(1)} \sim d^2$ and second $\delta\varepsilon_p^{(2)} \sim d^4$ orders in interaction

$$\delta\varepsilon_p = \delta\varepsilon_p^{(1)} + \delta\varepsilon_p^{(2)}.$$

The sum $\sum n_0' \delta\varepsilon_p^{(1)}$ gives zero contribution due to integration over the angle sector $\delta\varepsilon_p^{(1)}$. Thus, changing the chemical potential occurs only in second order in interaction, i.e. $\delta\mu \sim d^4$. As a result, we have in second order in interaction

$$\delta\mu = \frac{\sum_p n_0' \delta\varepsilon_p^{(2)} + \frac{1}{2} \sum_p n_0'' \left(\delta\varepsilon_p^{(1)}\right)^2}{\sum_p n_0'}.$$

The density of states $g(\varepsilon_F)$ of an ideal gas equals

$$g(\varepsilon_F) = -\sum_p n_0' = \int \frac{d^3 p}{(2\pi)^3} \delta\big(\varepsilon_0(\boldsymbol{p}) - \varepsilon_F\big) = \frac{m p_F}{2\pi^2} = \frac{3}{2} \frac{n}{\varepsilon_F}.$$

To calculate the second term in the numerator

$$\frac{1}{2}\sum_{p} n_0''\big(\delta\varepsilon_{p}^{(1)}\big)^2 = -\frac{1}{2}\int \frac{d^3p}{(2\pi)^3}\,\delta'\big(\varepsilon_0(p)-\varepsilon_F\big)\big(\delta\varepsilon_{p}^{(1)}\big)^2 =$$
$$= -\frac{(nd^2)^2}{4\pi^2}\int_0^\infty dp\, p^2 I^2(p/p_F)\delta'(\varepsilon_0(\boldsymbol{p})-\varepsilon_F))\int \frac{d\Omega}{4\pi}(1-3\cos^2\theta)^2 =$$
$$= \frac{(nd^2)^2}{4\pi^2}\,\frac{4}{5}\,m\sqrt{2m}\,\frac{\partial}{\partial\varepsilon}\big[\sqrt{\varepsilon}\, I^2(\sqrt{\varepsilon/\varepsilon_F}\,)\big]\Big|_{\varepsilon=\varepsilon_F} = \frac{16\pi^2}{45}\,\frac{(nd^2)^2}{\varepsilon_F}\,g(\varepsilon_F),$$

we have used the previous results and took here into account that $I(1) = 2\pi/3$ and $I'(1) = \pi$.

It is more complicated to calculate the first term in the nominator and one requires to determine $\delta\varepsilon_{p}^{(2)}$ first of all. For this purpose, in order to solve the equation

$$\varepsilon_{\boldsymbol{p}} = \varepsilon_0(\boldsymbol{p}) - \sum_{p'} U_{\boldsymbol{p}-\boldsymbol{p}'}\, n_0(\varepsilon_{\boldsymbol{p}'}-\mu),$$

we will apply the method of successive approximations, taking into account that the first correction $\delta\mu_1$ for the chemical potential vanishes. The energy of second approximation in interaction U equals

$$\varepsilon_{p}^{(2)} = \varepsilon_0 - \sum_{p'} U_{\boldsymbol{p}-\boldsymbol{p}'}\, n_0\big(\varepsilon_0(\boldsymbol{p}') + \delta\varepsilon_{p'}^{(1)} - \varepsilon_F - \delta\mu_1\big) \approx$$
$$\approx \varepsilon_{p}^{(1)} - \sum_{p'} U_{\boldsymbol{p}-\boldsymbol{p}'}\, n_0'\big(\varepsilon_0(\boldsymbol{p}')-\varepsilon_F\big)\delta\varepsilon_{p'}^{(1)} = \varepsilon_{p}^{(1)} + \delta\varepsilon_{p}^{(2)}, \quad \delta\mu_1 = 0.$$

The next calculation $\delta\varepsilon_{p}^{(2)}$ is performed with two steps. This is analogous to calculating $\delta\varepsilon_{p}^{(1)}$ above. First, we introduce $\delta n_2(\boldsymbol{r})$ according to

$$\delta n_2(\boldsymbol{r}) = \sum_{p} n_0'\big(\varepsilon_0(\boldsymbol{p})-\varepsilon_F\big)\delta\varepsilon_{p}^{(1)} e^{i\boldsymbol{pr}} =$$
$$= -nd^2 I(1)\frac{m}{p_F}\int \frac{d^3p}{(2\pi)^3}\,\delta(p-p_F)\big(1-3\frac{p_z^2}{p^2}\big)e^{i\boldsymbol{pr}} =$$
$$= -nd^2 I(1)\frac{mp_F}{2\pi^2}\,\frac{(p_F^2r^2-3)\sin p_Fr + 3p_Fr\cos p_Fr}{(p_Fr)^3}\left(1-3\frac{z^2}{r^2}\right).$$

Then, using the properties of convolution in the coordinate space, we find

$$\delta\varepsilon^{(2)}(\boldsymbol{r}) = -U(\boldsymbol{r})\delta n_2(\boldsymbol{r}) =$$
$$= \frac{2\pi nd^2}{3}\,g(\varepsilon_F)\,\frac{(p_F^2r^2-3)\sin p_Fr + 3p_Fr\cos p_Fr}{(p_Fr)^3}\,\frac{d^2}{r^3}\left(1-3\frac{z^2}{r^2}\right)^2.$$

The sum $\sum_{p} n_0' \delta\varepsilon_p^{(2)}$ is calculated with the aid of transformation to the coordinate space

$$\sum_{p} n_0' \delta\varepsilon_p^{(2)} = \int d^3r\, n_0'(-\boldsymbol{r})\delta\varepsilon(\boldsymbol{r})^{(2)}$$

where $n_0'(\boldsymbol{r})$ is the Fourier transform for the derivative of distribution function

$$n_0'(\boldsymbol{r}) = -\int \frac{d^3p}{(2\pi)^3}\delta\big(\varepsilon_0(\boldsymbol{p}) - \varepsilon_F\big)e^{i\boldsymbol{pr}} = -\frac{mp_F}{2\pi^2}\frac{\sin p_F r}{p_F r} = -\frac{3}{2}\frac{n}{\varepsilon_F}\frac{\sin p_F r}{p_F r}.$$

So, we obtain the following result:

$$\begin{aligned}\sum_{p} n_0' \delta\varepsilon_p^{(2)} &= -\frac{\pi(nd^2)^2}{\varepsilon_F}g(\varepsilon_F)\times \\ &\times \int d^3r \frac{\sin p_F r}{p_F r}\frac{(p_F^2 r^2 - 3)\sin p_F r + 3p_F r\cos p_F r}{(p_F r)^3}\frac{1}{r^3}\left(1 - 3\frac{z^2}{r^2}\right)^2 = \\ &= -\frac{\pi(nd^2)^2}{\varepsilon_F}g(\varepsilon_F)\frac{16\pi}{5}\int_0^\infty dx \frac{\sin x}{x^2}\frac{(x^2-3)\sin x + 3x\cos x}{x^3} = \\ &= \frac{4\pi^2}{15}\frac{(nd^2)^2}{\varepsilon_F}g(\varepsilon_F).\end{aligned}$$

Summing up two fractions 16/45 and 4/15 and then dividing the sum of two contributions in the nominator by $-g(\varepsilon_F)$, we come to the final answer for the change of chemical potential

$$\mu = \varepsilon_F - \frac{28\pi^2}{45}\frac{(nd^2)^2}{\varepsilon_F}.$$

At temperature $T = 0$ the chemical potential depends on density $n = N/V$ alone. The ground state energy E_0 of a gas can be determined using the relations $\mu = \partial E_0/\partial N$ and $\varepsilon_F \sim N^{2/3}$:

$$E_0 = \frac{3}{5}N\varepsilon_F\left(1 - \frac{4\pi^2}{9}\frac{(nd^2)^2}{\varepsilon_F^2}\right) = V\frac{3}{5}n\varepsilon_F\left(1 - \frac{4\pi^2}{9}\frac{(nd^2)^2}{\varepsilon_F^2}\right).$$

Taking the previous results and that $\varepsilon_F \sim V^{-2/3}$, we can determine the pressure

$$P = -\frac{\partial E_0}{\partial V} = \frac{2}{5}n\varepsilon_F - \frac{16\pi^2}{45}n\frac{(nd^2)^2}{\varepsilon_F}.$$

To find the density of states at the Fermi surface $g(\mu)$, we use that derivative $\partial E/\partial\mu$ at $T = 0$ equals

$$\frac{\partial E}{\partial \mu} = \sum_{p} \varepsilon_p \, \delta(\mu - \varepsilon_p) = \mu \sum_{p} \delta(\mu - \varepsilon_p) = \mu g(\mu),$$

and that

$$\frac{\partial E}{\partial \mu} = \frac{\partial E/\partial n}{\partial \mu/\partial n} \approx \frac{3}{2} n \left[1 - \frac{28\pi^2}{45} \left(\frac{nd^2}{\varepsilon_F} \right)^2 \right].$$

Hence it follows that the density of states at the Fermi surface does not change within the accuracy of approximations made

$$g(\mu) = \frac{1}{\mu} \frac{\partial E}{\partial \mu} \approx \frac{3}{2} \frac{n}{\varepsilon_F} = g(\varepsilon_F).$$

To conclude this section, we find the variation of quasiparticle excitation spectrum within the linear approximation in interaction and gas flow velocity $\boldsymbol{u}$. In addition, we show that the momentum $\boldsymbol{\mathcal{P}}$ per gas volume unit coincides with the gas flow density $\boldsymbol{j} = \rho \boldsymbol{u}$ where ρ is the mass density of a gas.

So, let dipole Fermi gas flow as a whole at small constant velocity $\boldsymbol{u}$. Correspondingly, the equilibrium Fermi–Dirac distribution function changes due to the Doppler shift as $n_0(\varepsilon_p) \to n_0(\varepsilon_p - \boldsymbol{pu})$.

This result can also be derived from the condition of entropy maximum S under constancy of energy $E = \sum_p \varepsilon_p n_p$, particle number $N = \sum_p n_p$, and gas momentum $\boldsymbol{\mathcal{P}} = \sum_p \boldsymbol{p} n_p$. In fact, let us first introduce the following Lagrange multipliers: $1/T$, μ/T, and $\boldsymbol{u}/T$. The requirement for the extremum of function

$$\widetilde{S} = -\sum_{p} [n_p \ln n_p + (1 - n_p) \ln(1 - n_p)] - \frac{1}{T} \sum_{p} [\varepsilon_p - \mu - \boldsymbol{up}] n_p$$

with respect to n_p leads us to the quasiparticle distribution desired

$$n_0 = [e^{(\varepsilon_p - \mu - \boldsymbol{pu})/T} + 1]^{-1}.$$

Let us denote energy of quasiparticles in the flowing Fermi gas as H_p. Then, energy H_p satisfies the equation

$$H_p = \varepsilon_0(\boldsymbol{p}) - \boldsymbol{pu} + \sum_{p'} f(\boldsymbol{p}, \boldsymbol{p}') n_0(H_p) = \varepsilon_0(\boldsymbol{p}) - \boldsymbol{pu} - \sum_{p'} U_{p-p'} n_0(H_p).$$

Here $\varepsilon_0(\boldsymbol{p}) = \boldsymbol{p}^2/2m$, $f(\boldsymbol{p}, \boldsymbol{p}')$ is the Landau function equal to $U_0 - U_{p-p'}$ in the Hartree–Fock approximation. Besides, we have already taken $U_0 = 0$ into account. Solving the equation in the linear approximation in interaction U and velocity $\boldsymbol{u}$, we find

$$H_{\boldsymbol{p}} \approx \varepsilon_0(\boldsymbol{p}) - \boldsymbol{pu} - \sum_{\boldsymbol{p}'} U_{\boldsymbol{p}-\boldsymbol{p}'} n_0\big(\varepsilon_0(\boldsymbol{p}')\big) + \sum_{\boldsymbol{p}'} U_{\boldsymbol{p}-\boldsymbol{p}'} (\boldsymbol{p}'\boldsymbol{u}) n_0'\big(\varepsilon_0(\boldsymbol{p}')\big) =$$

$$= \varepsilon_{\boldsymbol{p}} - \boldsymbol{pu} + m\boldsymbol{u} \cdot \int \frac{d^3p'}{(2\pi\hbar)^3} U_{\boldsymbol{p}-\boldsymbol{p}'} \frac{\partial}{\partial \boldsymbol{p}'} n_0\big(\varepsilon_0(\boldsymbol{p}')\big)$$

where $\varepsilon_{\boldsymbol{p}}$ is the quasiparticle excitation spectrum in the gas at rest. We use below the relation

$$U_{\boldsymbol{p}-\boldsymbol{p}'} \nabla_{\boldsymbol{p}'} n_{\boldsymbol{p}'} = \nabla_{\boldsymbol{p}'} [U_{\boldsymbol{p}-\boldsymbol{p}'} n_{\boldsymbol{p}'}] + n_{\boldsymbol{p}'} \nabla_{\boldsymbol{p}} U_{\boldsymbol{p}-\boldsymbol{p}'}$$

for the transformation of the integrand function. Then, involving that the first term with the complete derivative gives zero contribution after integration over the infinitely distant surface, we arrive at the following relations:

$$H_{\boldsymbol{p}} = \varepsilon_{\boldsymbol{p}} - \boldsymbol{pu} + m\boldsymbol{u} \frac{\partial}{\partial \boldsymbol{p}} \sum_{\boldsymbol{p}'} U_{\boldsymbol{p}-\boldsymbol{p}'} n_0\big(\varepsilon_0(\boldsymbol{p}')\big) =$$

$$= \varepsilon_{\boldsymbol{p}} - \boldsymbol{pu} + m\boldsymbol{u} \frac{\partial}{\partial \boldsymbol{p}} \big(\varepsilon_0(\boldsymbol{p}) - \varepsilon_{\boldsymbol{p}}\big).$$

Finally, the approximate formula is obtained

$$H_{\boldsymbol{p}} = \varepsilon_{\boldsymbol{p}} - m\boldsymbol{u} \frac{\partial \varepsilon_{\boldsymbol{p}}}{\partial \boldsymbol{p}} = \varepsilon_{\boldsymbol{p}} - m(\boldsymbol{u}\boldsymbol{v}_{\boldsymbol{p}}).$$

The momentum per volume unit $\boldsymbol{\mathcal{P}}$ is determined with the expression

$$\boldsymbol{\mathcal{P}} = \int \boldsymbol{p} n_0(H_{\boldsymbol{p}}) \frac{d^3p}{(2\pi\hbar)^3} = \int \boldsymbol{p} n_0(\varepsilon_{\boldsymbol{p}} - m\boldsymbol{u}\boldsymbol{v}_{\boldsymbol{p}}) \frac{d^3p}{(2\pi\hbar)^3}.$$

For zero flow velocity $\boldsymbol{u} = 0$ due to parity $\varepsilon_{\boldsymbol{p}} = \varepsilon_{-\boldsymbol{p}}$, it is obvious that $\boldsymbol{\mathcal{P}} = 0$. Expanding n_0 in $\boldsymbol{u}$ entails nonzero contribution

$$\boldsymbol{\mathcal{P}} = -\int \frac{d^3p}{(2\pi\hbar)^3} \boldsymbol{p} \left(m\boldsymbol{u} \frac{\partial \varepsilon_{\boldsymbol{p}}}{\partial \boldsymbol{p}} \right) \frac{\partial n_0}{\partial \varepsilon_{\boldsymbol{p}}} = -m \int \frac{d^3p}{(2\pi\hbar)^3} \boldsymbol{p} (\boldsymbol{u}\nabla_{\boldsymbol{p}}) n_0(\boldsymbol{p}) =$$

$$= m \int \frac{d^3p}{(2\pi\hbar)^3} \big[\boldsymbol{u} n_0 - (\boldsymbol{u}\nabla_{\boldsymbol{p}}) \boldsymbol{p} n_0\big] = m\boldsymbol{u} \int \frac{d^3p}{(2\pi\hbar)^3} n_0(\boldsymbol{p}) = mn\boldsymbol{u}.$$

The second term in the integrand results in nonzero contribution since it transforms into integration over the infinitely distant surface. The integral from the first integrand term coincides with the particle density n. Thus, we are convinced, at least, in first order in interaction that the momentum per unit volume $\boldsymbol{\mathcal{P}}$ equals the gas flow density $\boldsymbol{j} = \rho \boldsymbol{u}$ in spite of the Fermi surface anisotropy originating from the anisotropic character of interaction between the Fermi gas particles.

Chapter 6
Phenomenon of Superconductivity. The BCS Theory

The phenomenon when the electrical resistance of a conductor vanishes completely with the simultaneous expulsion of an external magnetic field from the conductor bulk has been called the *superconductivity phenomenon*. The conductor having such properties is called a *superconductor*. The expulsion of a magnetic field from the superconductor bulk is referred to as the *Meissner–Ochsenfeld effect*. The superconducting properties of a conductor appear when its temperature becomes below the *critical temperature* at which the phase transition occurs from the normal state to the superconducting one. The transition to the superconducting state is accompanied with changing the thermodynamic properties of a conductor. The phenomenon of superconductivity should be attributed to the *macroscopic quantum phenomena*.

The critical temperature of superconducting transition for pure chemical elements changes within a wide range. The highest critical temperature is recorded for niobium $T_c = 9.25$ K and, for example, tungsten has a rather low temperature as $T_c = 15.4$ mK. The elements that do not exhibit the superconducting properties should primarily include the transition (Cr, Fe, Co, Ni) and rare earth (Gd, Dy, Ho, Er, etc.) metals with the unfilled d and f-shells, demonstrating spontaneous magnetic ordering.

The synthesis of various alloys and chemical compounds makes it possible to increase the upper limit of the known critical temperatures. The critical temperature in such materials can significantly exceed that of their constituent components. Superconductivity with the critical temperature $T_c \gtrsim 30$ K is usually referred to as *high-temperature superconductivity*. Examples of high-temperature superconductors include the family of iron oxypnictides with the transition temperatures within the 30–55 K range and the family of cuprates with temperatures in the region 90 - 130 K. The external pressure is one more parameter which can change the superconducting transition temperature. A striking example is hydrogen sulfide transiting into the metal state at the pressures of about 150 GPa with the critical temperature $T_c \sim 200$ K.

S. N. Burmistrov, *Statistical and Condensed Matter Physics*,
https://doi.org/10.1007/978-981-97-9072-2_6

Concerning the behavior in the external magnetic field, the superconductors can be divided into two main types, namely type-I and type-II. Type-I superconductors have one *thermodynamic critical field* H_c above which the superconducting state breaks completely down and below which the magnetic field is completely expelled from the superconductor bulk, demonstrating the Meissner effect.

Type-II superconductors are characterized by two critical fields called the *lower critical field* H_{c1} and *upper critical field* H_{c2}. In the magnetic $H < H_{c1}$ fields the type-II superconductors demonstrate the Meissner effect similar to that in the type-I superconductors. In the intermediate $H_{c1} < H < H_{c2}$ magnetic fields the type-II superconductor is in the *mixed state*[1] with the partial and gradual penetration of magnetic field into the superconductor bulk as the magnetic field increases. When achieving the upper critical field H_{c2}, there occurs a complete destruction of the superconducting state with the full penetration of magnetic field into the superconductor bulk.

From the viewpoint of theoretical aspects, we may discern conventional and unconventional superconductors. We imply that the properties of conventional superconductors can be described within the framework of the traditional *BCS (Bardeen–Cooper–Schrieffer) model* or *electron-phonon Eliashberg model*. The unconventional superconductors are those which properties have no explanation within these two traditional models. As an example of superconductors with the unconventional properties, we can mention the so-called heavy-fermion metals, e.g. $CeCu_2Si_2$, UPt_3, UBe_{13}, UCoGe. The latter compound is also interesting because it demonstrates a coexistence of superconductivity and ferromagnetism in the temperature region below 0.6 K.

The emergence of a superconducting state is associated with the presence of some effective attraction between the electrons which energies lie within the sufficiently close vicinity of the Fermi surface. In the traditional models such effective attraction appears as a result of the *electron–phonon interaction*. For appearing the attraction between the electrons which should repel each other due to the Coulomb interaction, the electrons must interact with another system. In a metal the positive ion lattice is such system. An electron in the lattice tends to attract the nearby positive ions to itself and thus the density of positive charges becomes slightly higher than on average. A small excess of positive charge, in turn, enables additionally to attract another electron, thereby creating an effective attraction between the electrons. The description of such electron–electron interaction looks like an exchange of the same phonon between two electrons.

Let us consider the following two events shown schematically in Fig. 6.1. An electron with momentum $\boldsymbol{p}$ and energy $\varepsilon_{\boldsymbol{p}}$ emits a phonon with momentum $\boldsymbol{k}$ and frequency $\omega_{\boldsymbol{k}}$ and transits to the state with momentum $\boldsymbol{p} - \boldsymbol{k}$ and energy $\varepsilon_{\boldsymbol{p}-\boldsymbol{k}}$. The second electron with momentum $\boldsymbol{p}'$ and energy $\varepsilon_{\boldsymbol{p}'}$ absorbs the phonon emitted and crosses over to the state with momentum $\boldsymbol{p}' + \boldsymbol{k}$ and energy $\varepsilon_{\boldsymbol{p}'+\boldsymbol{k}}$.

---

[1] The mixed state is often said as a vortex state. The magnetic field flux passes through the superconductor by means of quantized vortex lines composing the Abrikosov or flux line lattice.

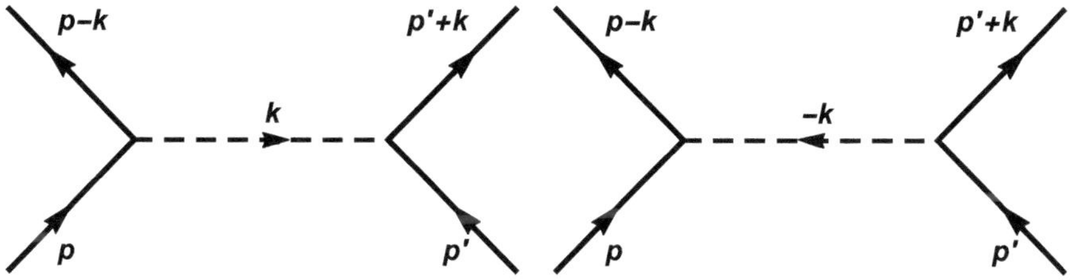

Fig. 6.1 The interaction between the electron with momentum $\boldsymbol{p}$ and the electron with momentum $\boldsymbol{p}'$ via emitting (left-hand figure) and absorbing a phonon (right-hand figure)

The second event is described as follows. The electron with momentum $\boldsymbol{p}$ and energy ε_p absorbs a phonon with momentum $-\boldsymbol{k}$ emitted with another electron with momentum $\boldsymbol{p}'$ and energy $\varepsilon_{p'}$. Absorbing the phonon, the electron acquires the momentum $\boldsymbol{p}-\boldsymbol{k}$ and energy ε_{p-k}. Accordingly, the electron emitting a phonon will transit to the state with momentum $\boldsymbol{p}'+\boldsymbol{k}$ and energy $\varepsilon_{p'+k}$. In second order of the perturbation theory, these two transitions are characterized with the following amplitudes:

$$\frac{|V_{p-k,p}|^2}{\varepsilon_p-\varepsilon_{p-k}-\hbar\omega_k} \quad \text{and} \quad \frac{|V_{p'+k,p'}|^2}{\varepsilon_{p'}-\varepsilon_{p'+k}-\hbar\omega_{-k}}\,.$$

Unifying these two processes and denoting $V_{p-k,p'}=V_k$ and $V_{p'+k,p'}=V_{-k}$ for brevity, we obtain in the total

$$\frac{|V_k|^2}{\varepsilon_p-\varepsilon_{p-k}-\hbar\omega_k}+\frac{|V_{-k}|^2}{\varepsilon_{p'}-\varepsilon_{p'+k}-\hbar\omega_{-k}}=\frac{2\hbar\omega_k}{(\varepsilon_p-\varepsilon_{p-k})^2-(\hbar\omega_k)^2}|V_k|^2\,.$$

Here we have involved the conservation law for the total energy of electrons

$$\varepsilon_p+\varepsilon_{p'}=\varepsilon_{p-k}+\varepsilon_{p'+k}$$

and invariance V_k and ω_k against replacing $\boldsymbol{k}$ with $-\boldsymbol{k}$.

Let us discuss this formula representing the *Fröhlich interaction*. For the small difference $|\varepsilon_p-\varepsilon_{p-k}|<\hbar\omega_k$, we have the negative magnitude which sign corresponds to an effective attraction between electrons with the close energies. The maximum typical frequencies of phonon spectrum are of the order of the Debye frequency ω_D corresponding to the maximum typical momentum $\hbar k_D\sim\pi\hbar/a$. For $|\varepsilon_p-\varepsilon_{p-k}|\gtrsim\hbar\omega_D$, the electron–phonon interaction becomes only repulsive and decreases rapidly as the difference in the electron energies grows. Since at low temperatures the main contribution to the thermodynamic quantities is provided with the electrons from the close vicinity of the Fermi surface, we can roughly say that the effective electron-electron attraction realizes within the immediate vicinity of the Fermi surface, the width being about $\hbar\omega_D$. In other words, while the energies of electrons are within the region $|\varepsilon_p-\varepsilon_F|\leqslant\hbar\omega_D\ll\varepsilon_F$.

6.1 The BCS Model. The Thermodynamic Properties of Superconductors

Based on the arguments above, we can use the simplest approximation, namely *BCS model*. In this model, it is assumed that the interaction potential $V_{\boldsymbol{p},\boldsymbol{p}'}$ between the electrons with momenta $\boldsymbol{p}$ and $\boldsymbol{p}'$ is isotropic and constant within the narrow energy layer beside the Fermi surface $\hbar\omega_D \ll \epsilon_F$

$$V_{\boldsymbol{p},\boldsymbol{p}'} = \begin{cases} g, \ |\xi_{\boldsymbol{p}}| \text{ and } |\xi_{\boldsymbol{p}'}| \leqslant \hbar\omega_D \ (g < 0) \\ 0, \ |\xi_{\boldsymbol{p}}| \text{ or } \ \ |\xi_{\boldsymbol{p}'}| > \hbar\omega_D \end{cases}$$

where $\xi_{\boldsymbol{p}} = \varepsilon_{\boldsymbol{p}} - \varepsilon_F$ and $\xi_{\boldsymbol{p}'} = \varepsilon_{\boldsymbol{p}'} - \varepsilon_F$ are the energy of electrons with momenta $\boldsymbol{p}$ and $\boldsymbol{p}'$, taken from the Fermi energy. We emphasize in advance that the adequate description of the superconducting transition within the framework of the BCS model assumes implicitly the magnitude of the electron–electron interaction to be sufficiently small (*weak coupling approximation*). This means that the superconducting transition temperature T_c determined in the model would be much less than the Debye temperature, i.e. $T_c \ll \hbar\omega_D$.

Let us write Hamiltonian of the electron system[2] with the two-particle interaction $U(\boldsymbol{r} - \boldsymbol{r}')$ independent of particle spins

$$\hat{H} = \int d^3r \psi_\sigma^+(\boldsymbol{r})\left(\frac{-\hbar^2\nabla^2}{2m}\right)\psi_\sigma(\boldsymbol{r}) + \\ + \frac{1}{2}\int d^3r\, d^3r' \psi_\sigma^+(\boldsymbol{r})\psi_{\sigma'}^+(\boldsymbol{r}')U(\boldsymbol{r} - \boldsymbol{r}')\psi_{\sigma'}(\boldsymbol{r}')\psi_\sigma(\boldsymbol{r}).$$

Indices σ and σ' mean the spin projections and we imply summing over the pairs of identical spins. Next, in accordance with the model the electron–electron interaction has a point-like character, i.e. $U(\boldsymbol{r} - \boldsymbol{r}') = g\delta(\boldsymbol{r} - \boldsymbol{r}')$. Therefore, in this case the electrons of the same spin projections do not interact. In other words, the Pauli principle forbids two identical Fermi particles at the same state to be at one point. Thus, the electrons should necessarily have the spins directed in the opposite directions.

As is mentioned above, it is more reasonable to calculate the thermodynamic quantities in the interacting systems without fixing the particle number N but by studying the dependence of thermodynamic quantities as a function of chemical potential μ. Therefore, we will calculate the thermodynamic averages for the following Hamiltonian:

[2] The coefficient in the interaction term should be written as $1/2V$. To simplify the view of formulas, we put volume $V = 1$.

$$\hat{\mathcal{H}} = \hat{H} - \mu\hat{N} = \int d^3r \sum_{\sigma} \psi_{\sigma}^{+}(\boldsymbol{r})\left(-\frac{\hbar^2\nabla^2}{2m} - \mu\right)\psi_{\sigma}(\boldsymbol{r}) +$$
$$+ g\int d^3r\, d^3r' \psi_{\uparrow}^{+}(\boldsymbol{r})\psi_{\downarrow}^{+}(\boldsymbol{r}')\psi_{\downarrow}(\boldsymbol{r}')\psi_{\uparrow}(\boldsymbol{r}).$$

Let us turn from the coordinate representation to the momentum one. Then the Hamiltonian reads

$$\hat{\mathcal{H}} = \hat{H} - \mu\hat{N} = \sum_{p,\sigma} \xi_p a_{p,\sigma}^{+} a_{p,\sigma} + g \sum_{p'+q'=p+q} a_{p'\uparrow}^{+} a_{q'\downarrow}^{+} a_{q\downarrow} a_{p\uparrow}$$

where $\xi_p = \varepsilon_p - \mu = \boldsymbol{p}^2/2m - \mu$ is the energy taken from the chemical potential level. We imply a sum over the momenta within the narrow intervals $|\xi_p|$, $|\xi_q|$, $|\xi_{p'}|$ and $|\xi_q'| \leqslant \hbar\omega_D$ near the Fermi surface. The further action should be the calculation of thermodynamic averages, i.e. determination of grand potential $\Omega(\mu) = \langle\hat{\mathcal{H}}\rangle = \operatorname{tr}(\hat{\rho}\hat{\mathcal{H}})$ where $\hat{\rho}$ is the equilibrium density matrix corresponding to the Gibbs distribution.

For the negative electron-phonon coupling constant $g < 0$, the ground state of normal Fermi liquid becomes unstable and the qualitative change of ground state structure takes place. In a superconductor, the excited state proves to be separated with an energy gap from the ground state. The calculation of thermodynamic quantities is most simple if the Hamiltonian has a diagonal form such as $\sum \varepsilon_p a_{p,\sigma}^{+} a_{p,\sigma}$ and, for example, thermodynamic average $\langle a_{p,\sigma}^{+} a_{p,\sigma}\rangle$ is a usual Fermi distribution $\left(e^{\varepsilon_{p,\sigma}/T} + 1\right)^{-1}$.

The *variational metod* is employed to find the energy of a superconductor and its elementary excitation spectrum. In this method we introduce the auxiliary Fermi operators $\alpha_{p\sigma}^{+}$ for creation and $\alpha_{p\sigma}$ for annihilation of quasiparticles playing a role of elementary excitations in the superconducting state. For this purpose, we apply the formulas of the *Bogoliubov transformation*

$$\alpha_{p\uparrow} = u_p a_{p\uparrow} - v_p a_{-p\downarrow}^{+},\ \alpha_{p\uparrow}^{+} = u_p a_{p\uparrow}^{+} - v_p a_{-p\downarrow};$$
$$\alpha_{p\downarrow} = u_p a_{p\downarrow} + v_p a_{-p\uparrow}^{+},\ \alpha_{p\downarrow}^{+} = u_p a_{p\downarrow}^{+} + v_p \alpha_{-p\uparrow}.$$

Here u_p and v_p are the amplitudes[3] of the *uv*-transform, which determine the fractions of mixing the creation and annihilation operators of particles. In this case, one says that two electrons with the opposite momenta and spins constitute the bound state or *Cooper pair.*

The operators $a_{p,\sigma}^{+}$ and $a_{p,\sigma}$, having a sense of particle creation and annihilation, satisfy the anticommutative Fermi relations

$$\{a_{p,\sigma},\, a_{p',\sigma'}^{+}\} = \delta_{p,p'}\delta_{\sigma,\sigma'} \quad \text{and} \quad \{a_{p,\sigma}^{+},\, a_{p',\sigma'}^{+}\} = \{a_{p,\sigma},\, a_{p',\sigma'}\} = 0.$$

[3] In the general case the amplitudes are the complex quantities satisfying the relations $u_{-p} = u_p^*$ and $v_{-p} = v_p^*$.

In order the quasiparticle operators $\alpha^+_{p\sigma}$ and $\alpha_{p\sigma}$ would have the same meaning as usual creation and annihilation particle operators, it is necessary to require the fulfillment of the same conditions

$$\{\alpha_{p,\sigma},\, \alpha^+_{p',\sigma'}\} = \delta_{p,p'}\delta_{\sigma,\sigma'} \quad \text{and} \quad \{\alpha^+_{p,\sigma},\, \alpha^+_{p',\sigma'}\} = \{\alpha_{p,\sigma},\, \alpha_{p',\sigma'}\} = 0.$$

So, we write the following anticommutator for quasiparticles:

$$\begin{aligned}\{\alpha_{p\uparrow}\,, \alpha^+_{p\uparrow}\} &= u_p^2\{a_{p\uparrow}\,, a^+_{p\uparrow}\} - u_p v_p\{a^+_{-p\downarrow}\,, a^+_{p\uparrow}\} + \\ &+ v_p^2\{a^+_{-p\downarrow}\,, a_{-p\downarrow}\} - u_p v_p\{a_{p\uparrow}\,, a^+_{-p\downarrow}\} = u_p^2 + v_p^2.\end{aligned}$$

Thus we should require the following condition for the amplitude of uv-transform:

$$u_p^2 + v_p^2 = 1$$

and, in fact, we have one free parameter v_p or u_p which should be chosen from the condition of thermodynamic grand potential Ω minimum.

We turn to calculating the thermodynamic average $\langle\hat{\mathcal{H}}\rangle$ for Hamiltonian $\hat{\mathcal{H}} = \hat{H} - \mu\hat{N}$, substituting the following inverted formulas into Hamiltonian:

$$\begin{aligned} a_{p\uparrow} &= u_p\alpha_{p\uparrow} + v_p\alpha^+_{-p\downarrow},\ a^+_{p\uparrow} = u_p\alpha^+_{p\uparrow} + v_p\alpha_{-p\downarrow};\\ a_{p\downarrow} &= u_p\alpha_{p\downarrow} - v_p\alpha^+_{-p\uparrow},\ a^+_{p\downarrow} = u_p\alpha^+_{p\downarrow} - v_p\alpha_{-p\uparrow}.\end{aligned}$$

Calculating the averages, we keep only nonzero terms $\langle\alpha^+_{p,\sigma}\alpha_{p,\sigma}\rangle = n_{p,\sigma}$ and $\langle\alpha_{p,\sigma}\alpha^+_{p,\sigma}\rangle = 1 - n_{p,\sigma}$ where $n_{p,\sigma}$ is the mean occupation number of quasiparticle states or Fermi distribution for quasiparticles with energy $\varepsilon_{p,\sigma}$ which should be determined from the condition of the minimum $\langle\hat{\mathcal{H}}\rangle$.

Most complicated thing is to average the term with interaction. For the system of non-interacting electrons, the average for the product of four Fermi operators would decouple into a sum of the averages for all possible products of the operator pairs. In the BCS model implying the weak electron–electron interaction, we can use the approximation of free electrons or that of self-consistent field.[4] So we take an advantage of the following approximation:

$$\begin{aligned}\langle a^+_{p'\uparrow}a^+_{q'\downarrow}a_{q\downarrow}a_{p\uparrow}\rangle =& \langle a^+_{p'\uparrow}a^+_{q'\downarrow}\rangle\langle a_{q\downarrow}a_{p\uparrow}\rangle - \\ &- \langle a^+_{p'\uparrow}a_{q\downarrow}\rangle\langle a^+_{q'\downarrow}a_{p\uparrow}\rangle + \langle a^+_{p'\uparrow}a_{p\uparrow}\rangle\langle a^+_{q'\downarrow}a_{q\downarrow}\rangle,\end{aligned}$$

in which the *anomalous averages* $\langle a^+_{p\uparrow}a^+_{p\downarrow}\rangle$ and $\langle a_{p\downarrow}a_{p\uparrow}\rangle$ appear in addition to the *normal average* $\langle a^+_{p\sigma}a_{p\sigma}\rangle$.

---

[4] The latter approximation can be justified with the large correlation length ξ_0 in superconductor, exceeding significantly the interatomic distance $a \sim \hbar/p_F$ or, correspondingly, with the very small Ginzburg–Levanyuk number $\mathrm{Gi} \sim (a/\xi_0)^4 \ll 1$.

Before continuing the calculation further, we will make the following remark. The second term vanishes, and the last term's contribution to the thermodynamic potential

$$g\sum_p \langle a^+_{p\uparrow}a_{p\uparrow}\rangle \sum_q \langle a^+_{q\downarrow}a_{q\downarrow}\rangle$$

is of no interest since it has the free term structure and results in a small renormalization of chemical potential to the extent of small electron-electron coupling constant g.

Performing the calculation for the contributions from the free and interaction terms to the thermodynamic grand potential $\Omega(\mu)$, we find finally

$$\begin{aligned}\Omega(\mu) =& \langle \hat{H} - \mu\hat{N}\rangle = \sum_p \xi_p\left[u_p^2(n_{p\uparrow}+n_{p\downarrow}) + v_p^2(2-n_{-p\uparrow}-n_{-p\downarrow}\right]+ \\ &+ g\sum_p u_p v_p(1-n_{p\uparrow}-n_{-p\downarrow})\sum_q u_q v_q(1-n_{-q\uparrow}-n_{q\downarrow}).\end{aligned}$$

Here $n_{p,\sigma}$ is the occupation number of quasiparticle excitations and $n_{p\sigma}=n_{-p\sigma}$ due to even symmetry. For varying the potential $\Omega(\mu)$ in order to find its minimum, it is convenient to parametrize the amplitudes u_p and v_p as

$$u_p^2 = \frac{1+z_p}{2}, \quad v_p^2 = \frac{1-z_p}{2}$$

and introduce new self-consistent parameter

$$\Delta = -g\sum_q u_q v_q(1-n_{q\uparrow}-n_{q\downarrow}) = -\frac{g}{2}\sum_q \sqrt{1-z_q^2}(1-n_{q\uparrow}-n_{q\downarrow}).$$

This parameter Δ will be called the *superconducting gap*. So, we derive the equation resulting from condition $\delta\Omega/\delta z_p = 0$:

$$-\xi_p(1-n_{p\uparrow}-n_{p\downarrow}) + \frac{z_p}{\sqrt{1-z_p^2}}(1-n_{p\uparrow}-n_{p\downarrow})\Delta = 0 \quad \text{or} \quad \frac{z_p}{\sqrt{1-z_p^2}} = \frac{\xi_p}{\Delta}.$$

Hence we determine readily that

$$z_p = \xi_p/\varepsilon_p \quad \text{where} \quad \varepsilon_p = \sqrt{\xi_p^2+\Delta^2}.$$

Accordingly, the *self-consistent equation* for determining the superconducting gap takes the form

$$\Delta = -\frac{g}{2}\sum_p \frac{1-n_{p\uparrow}-n_{p\downarrow}}{\sqrt{\xi_p^2+\Delta^2}}\Delta.$$

It is obvious that there is always a trivial solution $\Delta = 0$ meaning the normal state. Since $n_{p\uparrow} + n_{p\downarrow} \leqslant 1$, nontrivial solution $\Delta \neq 0$ is only possible for the attraction interaction $g < 0$ and can be realized for sufficiently low temperatures as we will see further.

The energy of elementary excitations or quasiparticles can be determined as a variational derivative of thermodynamic potential $\Omega(\mu)$ with respect to the distribution function

$$\delta\Omega(\mu) = \sum_{p} \varepsilon_{p\uparrow}\delta n_{p\uparrow} + \sum_{p} \varepsilon_{p\downarrow}\delta n_{p\downarrow} .$$

In this case the equilibrium distribution function $n_{p,\sigma}$ for elementary excitations proves to be the ordinary Fermi–Dirac distribution with $\varepsilon_{p,\sigma}$ as a energy of elementary excitation

$$n_{p,\sigma} = \left[\exp(\varepsilon_{p,\sigma}/T) + 1\right]^{-1}.$$

Varying the potential $\Omega(\mu)$, e.g. with respect to $n_{p\uparrow}$, yields

$$\varepsilon_{p,\sigma} = \delta\Omega(\mu)/\delta n_{p\uparrow} = \xi_p(u_p^2 - v_p^2) + 2u_p v_p \Delta = \sqrt{\xi_p^2 + \Delta^2} = \varepsilon_p .$$

Thus, we can ascribe energy $\varepsilon_p = \pm\sqrt{\xi_p^2 + \Delta^2}$, where $\xi_p = (\boldsymbol{p}^2/2m) - \mu$, to the quasiparticle excitations in the superconducting state. The excitations related to the minus sign are customary called the holes and those related to the positive sign are referred to as quasiparticles or simply the particles. As we see in Fig. 6.2, there appears an energy gap of magnitude 2Δ in the elementary excitation spectrum.[5]

6.2 Temperature Behavior of the Superconducting Gap

We start our consideration of the thermodynamic properties in the superconducting state of a metal from studying the temperature behavior of superconducting gap $\Delta(T)$. If there is no magnetic field or spin–orbit interaction, the distribution functions coincide for various spin directions, i.e. $n_{p\uparrow} = n_{p\downarrow} = n(\varepsilon_p)$. Let us write the self-consistent equation which implicitly determines the temperature behavior of superconducting gap $\Delta(T)$

$$1 = \frac{|g|}{2} \int\limits_{|\xi_p| \leqslant \hbar\omega_D} \frac{d^3p}{(2\pi\hbar)^3} \frac{1 - 2n(\varepsilon_p)}{\varepsilon_p}$$

[5] The spectrum found is the simplest expression in order to describe an emergence of energy gap at the Fermi surface.

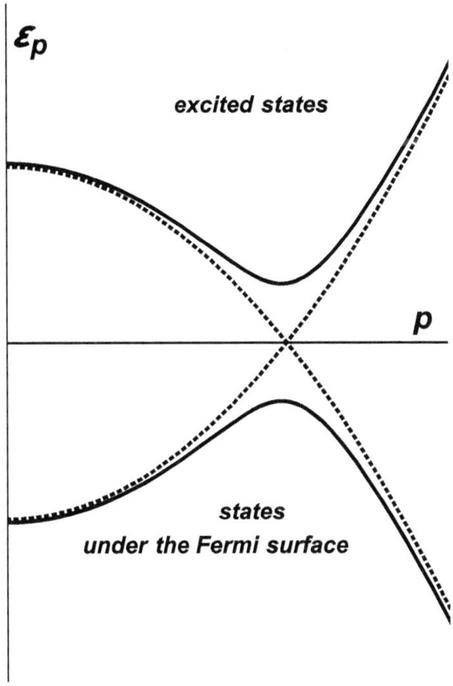

Fig. 6.2 The elementary excitation spectrum in a superconductor. The dashed lines correspond to the quasiparticle and hole excitations in a normal nonsuperconducting conductor

where $\varepsilon_p = \sqrt{\xi_p^2 + \Delta^2}$. Here we have taken explicitly into account that the interaction takes place at the energies $|\xi_p| \leqslant \hbar\omega_D$ alone. Substituting the Fermi–Dirac distribution $n(\varepsilon_p) = \left[e^{\varepsilon_p/T} + 1\right]^{-1}$, we go over to integration with respect to ξ_p

$$1 = |g| \int\limits_{|\xi_p| \leqslant \hbar\omega_D} \frac{p^2 dp}{4\pi^2\hbar^3} \frac{1 - 2n(\varepsilon)}{\varepsilon} \approx |g| \frac{p_F^2}{4\pi^2 v_F \hbar^3} \int\limits_{-\hbar\omega_D}^{\hbar\omega_D} d\xi \frac{1 - 2n(\sqrt{\xi^2 + \Delta^2})}{\sqrt{\xi^2 + \Delta^2}}.$$

Here we have performed an approximate replacement $p^2 dp$ with $(p_F^2/v_F)d\xi$. The point is that the integration region considered corresponds to the momentum region $|p - p_F| \ll p_F$ lying within the narrow band near the Fermi surface.

Let us introduce both the density of electron states $N(0)$ at the Fermi surface with the fixed spin projection and the dimensionless coupling constant λ according to

$$N(0) = \frac{p_F^2}{2\pi^2 v_F \hbar^3} = \frac{m p_F}{2\pi^2\hbar^3} \quad \text{and} \quad \lambda = |g| N(0).$$

Then we obtain the equation

$$\frac{1}{\lambda} = \int_0^{\hbar\omega_D} \frac{\tanh\left(\sqrt{\xi^2+\Delta^2}/2T\right)}{\sqrt{\xi^2+\Delta^2}} d\xi.$$

For zero temperature $T = 0$, the integral can straightforwardly be estimated[6] as $\ln\left(2\hbar\omega_D/\Delta(0)\right)$. This yields the exponential behavior as a function of coupling constant λ

$$\Delta(0) = 2\hbar\omega_D e^{-1/\lambda} \quad (\lambda \ll 1).$$

To find the critical superconducting transition T_c, it is necessary to put $\Delta(T_c) = 0$. Integrating by parts[7] results in

$$\frac{1}{\lambda} = \int_0^{\hbar\omega_D} \frac{\tanh(\xi/2T_c)}{\xi} d\xi = \ln\frac{\hbar\omega_D}{2T_c} - \int_0^{\infty} \frac{\ln x}{\cosh^2 x} dx$$
$$= \ln\left(\frac{\gamma}{\pi}\frac{2\hbar\omega_D}{T_c}\right)$$

where $\ln\gamma = C = 0.577\ldots$ is Euler's constant. Thus the superconducting transition temperature equals

$$T_c = \frac{\gamma}{\pi} 2\hbar\omega_D e^{-1/\lambda}.$$

The next ratio

$$\frac{2\Delta(0)}{T_c} = \frac{2\pi}{\gamma} \approx 3.52$$

is universal in the BCS model or in the weak coupling limit $\lambda \ll 1$. In the general case of strong ($\lambda \sim 1$) coupling this ratio depends on the coupling constant λ.

The magnitude of *isotopic effect*

$$\alpha = -\frac{d(\ln T_c)}{d(\ln M)},$$

where M is the ion mass, proves to be $\alpha = 1/2$ in the BCS model. In fact, let superconductors be composed of the isotopes with the same electronic properties but different masses M. Such superconductors have different Debye frequencies varying as $\omega_D \sim M^{-1/2}$. This entails the typical isotopic effect for the phonon mechanism of superconductivity with $T_c \sim M^{-1/2}$.

---

[6] Estimating the integrals, we keep always in mind the initial approximation of weak coupling, i.e. smallness of coupling constant g and, therefore, $\Delta \ll \hbar\omega_D$.

[7] Since the second integral is convergent, the upper limit $\hbar\omega_D/2T_c$ can be replaced with the infinity. The integral equals $-\ln(2\gamma/\pi)$, $\ln\gamma = C = 0.577\ldots$ being Euler's constant.

Let us describe the superconducting gap in the vicinity $T = 0$ and $T = T_c$. Near $T = 0$ the exponential temperature behavior is observed

$$\Delta(T) = \Delta(0) - \sqrt{2\pi T \Delta(0)}\, e^{-\Delta(0)/T} \quad (T \ll T_c).$$

The exponential behavior is a direct consequence of creating the quasiparticles (elementary excitations) with the temperature growth. The number of quasiparticles is proportional to $\exp(-\Delta(0)/T)$, resulting from the energy gap in the excitation spectrum.

If $T \to T_c$, we have $\Delta(T) \to 0$ and

$$\Delta(T) = \pi \left(\frac{8}{7\zeta(3)}\right)^{1/2} T_c \left(\frac{T_c - T}{T_c}\right)^{1/2} \approx 3.06 T_c \left(\frac{T_c - T}{T_c}\right)^{1/2}$$

where $\zeta(3) \approx 1.202$ is the Riemann zeta-function.

If the superconducting gap $\Delta(T)$ is treated as an order parameter of the normal metal-superconductor phase transition, i.e. $\Delta(T > T_c) = 0$ and $\Delta(T < T_c) \neq 0$, the critical exponent $\beta = 1/2$ of order parameter corresponds completely to the mean-field approximation for the theory of second-order phase transitions. The temperature behavior $\Delta(T)$ is shown in Fig. 6.3.

Problems

1. Estimate the behavior of the superconducting gap near zero temperature $T = 0$.
Solution. We have from the self-consistent equation for the superconducting gap

$$\ln \frac{2\hbar\omega_D}{\Delta(0)} = \frac{1}{\lambda} = \int_0^{\hbar\omega_D} \frac{d\xi}{\sqrt{\xi^2 + \Delta^2}} - 2\int_0^{\infty} \frac{d\xi}{\sqrt{\xi^2 + \Delta^2}} \frac{1}{\exp(\sqrt{\xi^2 + \Delta^2}/T) + 1}.$$

We obtain expanding in the large exponential in the second integrand and transforming to the integration over ε

$$\ln \frac{\Delta}{\Delta(0)} \approx -2\int_\Delta^\infty d\varepsilon \frac{e^{-\varepsilon/T}}{\sqrt{\varepsilon^2 - \Delta^2}} \approx -\frac{2}{\sqrt{2\Delta}} \int_\Delta^\infty d\varepsilon \frac{e^{-\varepsilon/T}}{\sqrt{\varepsilon - \Delta}} = -\sqrt{\frac{2\pi T}{\Delta}} e^{-\Delta/T}.$$

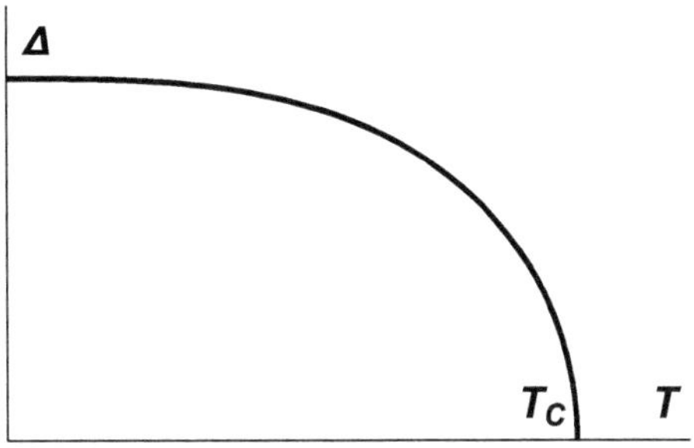

Fig. 6.3 The temperature behavior of energy gap Δ

Representing approximately the logarithm as $(\Delta - \Delta(0))/\Delta(0)$ and putting $\Delta \approx \Delta(0)$ on the right-hand side, we arrive at the answer desired

$$\Delta \approx \Delta(0) - \sqrt{2\pi T \Delta(0)} e^{-\Delta(0)/T}.$$

2. Estimate the behavior of the superconducting gap near the transition temperature $T = T_c$.
Solution. Let us use the following values of integrals:

$$\frac{1}{\lambda} = \int_0^{\hbar\omega_D} d\xi \frac{\tanh(\xi/2T_c)}{\xi} \quad \text{and} \quad \int_0^{\hbar\omega_D} d\xi \frac{\tanh(\xi/2T)}{\xi} = \ln\left(\frac{\gamma}{\pi}\frac{2\hbar\omega_D}{T}\right)$$

and transform the self-consistent equation for the superconducting gap to the form

$$\ln\frac{T_c}{T} = \int_0^{\infty} d\xi \left[\frac{\tanh(\xi/2T)}{\xi} - \frac{\tanh(\sqrt{\xi^2+\Delta^2}/2T)}{\sqrt{\xi^2+\Delta^2}}\right].$$

(Due to convergence of the integral the upper limit $\hbar\omega_D$ of the integral is replaced with the infinity). Next, we apply the formula

$$\tanh\frac{z}{2T} = 4T\sum_{n=0}^{\infty}\frac{z}{\omega_n^2 + z^2}, \quad \omega_n = \pi T(2n+1)$$

and obtain

$$\ln\frac{T_c}{T} = 4T\sum_{n=0}^{\infty}\int_0^{\infty} d\xi \left[\frac{1}{\omega_n^2+\xi^2} - \frac{1}{\omega_n^2+\xi^2+\Delta^2}\right].$$

Expanding in Δ and integrating over ξ, we find

$$\ln\frac{T_c}{T} = 4T\sum_{n=0}^{\infty}\left[\frac{\pi}{4}\frac{\Delta^2}{\omega_n^3} - \frac{3\pi}{16}\frac{\Delta^4}{\omega_n^5} + \ldots\right] = \frac{\Delta^2}{(\pi T)^2}\frac{7}{8}\zeta(3) - \frac{3}{4}\frac{\Delta^4}{(\pi T)^4}\frac{31}{32}\zeta(5) + \ldots$$

Here we have used the sum series formula

$$\sum_{n=0}^{\infty}\frac{1}{(2n+1)^z} = \left(1 - \frac{1}{2^z}\right)\left(1 + \frac{1}{2^z} + \frac{1}{3^z} + \frac{1}{4^z} + \ldots\right) = \left(1 - \frac{1}{2^z}\right)\zeta(z)$$

where $\zeta(z)$ is the Riemann zeta-function: $\zeta(3) \approx 1.202$ and $\zeta(5) \approx 1.037$. To get the answer, it is sufficient to involve the first term of expansion alone

$$\Delta(T) \approx \pi\sqrt{\frac{8}{7\zeta(3)}}\sqrt{T_c(T_c - T)}.$$

6.3 Thermodynamic Functions of a Superconductor

In the previous section we have studied the solutions of the self-consistent equation for the superconducting gap and found that, in addition to trivial solution $\Delta = 0$, there appears a nontrivial solution $\Delta \neq 0$ below some temperature. The nontrivial solution will describe new physical state of a conductor provided only that the superconducting state with $\Delta \neq 0$ is energetically more favorable than the normal state $\Delta = 0$. Below, first of all, we should compare the thermodynamic potentials in the normal and superconducting states.

The grand thermodynamic potential $\Omega(\mu)$ is expressed via grand partition function $\mathcal{Z}$ as

$$\Omega(\mu) = -T \ln \mathcal{Z} \quad \text{and} \quad \mathcal{Z} = \operatorname{tr} e^{-\hat{\mathcal{H}}/T}$$

where $\hat{\mathcal{H}} = \hat{H} - \mu \hat{N}$ is the BCS Hamiltonian. Let us differentiate potential $\Omega(\mu)$ with respect to the coupling constant g

$$\frac{\partial \Omega}{\partial g} = -\frac{T}{\mathcal{Z}}\frac{\partial \mathcal{Z}}{\partial g} = \frac{1}{g}\frac{\operatorname{tr}\left(\hat{H}_{\text{int}} e^{-\hat{\mathcal{H}}/T}\right)}{\operatorname{tr}\left(e^{-\hat{\mathcal{H}}/T}\right)} = \frac{1}{g}\langle \hat{H}_{\text{int}} \rangle$$

where $\hat{H}_{\text{int}}$ is the term responsible for the interaction of electrons. Then, in the thermodynamic average $\langle \hat{H}_{\text{int}} \rangle$ we retain the product of anomalous averages $\sum_p \langle a^+_{p\uparrow} a^+_{-p\downarrow} \rangle \sum_q \langle a_{q\downarrow} a_{-q\uparrow} \rangle$ different from zero only in the superconducting state. Accordingly, we can write the following:

$$\frac{\partial \Omega}{\partial g} = \frac{\Delta^2}{g^2}.$$

The relation between g and Δ is already known for the temperature given. Therefore, the difference between the magnitudes of potential Ω in the superconducting and normal states will be equal to

$$\Omega_s - \Omega_n = \int_0^g \frac{\Delta^2}{g^2} dg = -\int_0^\Delta \Delta^2 \frac{d}{d\Delta}\left(\frac{1}{g}\right) d\Delta$$

after transition from variable g to variable Δ.

Let us give asymptotic values of difference $\Omega_s - \Omega_n$ per unit volume at low $T \to 0$ temperatures and near the transition $T \to T_c$. We recall first the value of potential Ω in the normal state at low temperatures $\Omega_n = -\pi^2 N(0) T^2/3$. For $T = 0$, the potential is $\Omega_n = 0$ and, correspondingly, in the superconducting state the potential Ω_s equals

$$\Omega_s(T = 0) = -N(0)\frac{\Delta^2(0)}{2}.$$

The qualitative interpretation of the formula above is the following. The electrons within the energy band $\sim \Delta(0)$ near the Fermi surface, whose number is $\sim N(0)\Delta(0)$, are bound into the Cooper pairs with an energy gain about $\Delta(0)$. For the finite but low temperature, this results in the exponential behavior

$$\Omega_s(T) - \Omega_s(0) \approx N(0)\left[8\pi T^3 \Delta(0)\right]^{1/2} e^{-\Delta(0)/T}.$$

Such behavior leads to the exponential freezing of entropy and specific heat due to existence of energy gap for the elementary excitation spectrum in the superconductor

$$S_s = -\partial\Omega_s/\partial T \approx N(0)\big[8\pi\Delta^3(0)/T\big]^{1/2}e^{-\Delta(0)/T},$$
$$C_s = T\partial S_s/\partial T \approx N(0)\big[8\pi\Delta^5(0)/T^3\big]^{1/2}e^{-\Delta(0)/T}.$$

To derive the asymptotic formula near the transition temperature T_c, we start from the expansion obtained above in Problem 2:

$$\ln\frac{T_c}{T} = \frac{\Delta^2}{(\pi T)^2}\frac{7}{8}\zeta(3) + \ldots$$

Hence we find for the differential dT_c

$$\frac{dT_c}{T_c} = \frac{2\Delta}{(\pi T)^2}\frac{7}{8}\zeta(3)d\Delta + \ldots$$

On the other hand, we have from $T_c \sim \exp(-1/\lambda)$

$$\frac{dT_c}{T_c} = \frac{d\lambda}{\lambda^2} = -d\left(\frac{1}{\lambda}\right) = \frac{1}{N(0)}d\left(\frac{1}{g}\right).$$

The comparison gives

$$\frac{d}{d\Delta}\left(\frac{1}{g}\right) = \frac{7}{4}\zeta(3)N(0)\frac{\Delta}{(\pi T)^2} + \ldots$$

and we find involving the first term alone

$$\Omega_s - \Omega_n = -\int_0^{\Delta}\Delta^2 d\left(\frac{1}{g}\right)d\Delta =$$

$$= -\frac{7\zeta(3)}{4}\frac{N(0)}{(\pi T)^2}\int_0^{\Delta}\Delta^3\,d\Delta = -\frac{7\zeta(3)}{16}N(0)\frac{\Delta^4}{(\pi T)^2}.$$

Substituting the value Δ at $T \to T_c$, we have

$$\Omega_s - \Omega_n = -\frac{4\pi^2}{7\zeta(3)}N(0)(T_c - T)^2.$$

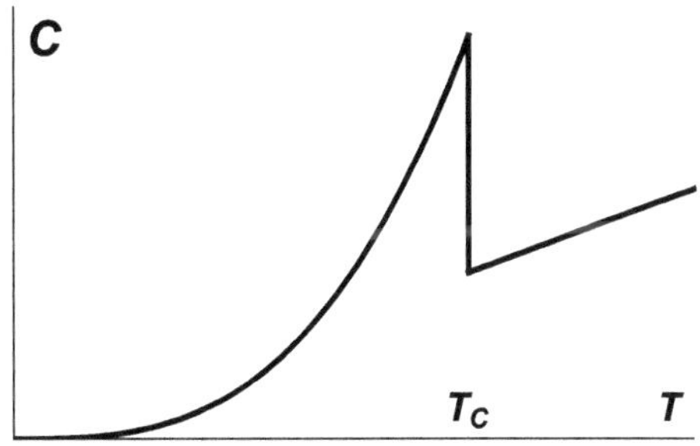

Fig. 6.4 The temperature behavior of specific heat $C(T)$

Hence we readily find the difference for the entropies in the superconducting and normal states[8]

$$S_s - S_n = -\frac{8\pi^2}{7\zeta(3)} N(0)(T_c - T).$$

After differentiating with respect to temperature once more and retaining the main terms, we arrive at the following specific heat for a superconductor at the transition point

$$C_s(T_c) = C_n(T_c) + \frac{8\pi^2}{7\zeta(3)} N(0) T_c \,.$$

Thus, the specific heat of a conductor is subjected to the finite-magnitude jump at the transition to the superconducting state. Reminding that the specific heat in the normal state equals $C_n(T) = (2\pi^2/3)N(0)T$, we obtain the following ratio of the specific heats at the transition point

$$\frac{C_s(T_c)}{C_n(T_c)} = 1 + \frac{12}{7\zeta(3)} \approx 2.43.$$

In the BCS model, this ratio is independent of the coupling constant and represents a universal quantity. The temperature behavior of specific heat is shown in Fig. 6.4.

The continuity of thermodynamic potentials Ω_n and Ω_s, as well as their first derivatives at $T = T_c$, indicate that the normal-superconducting state transition is second-order phase transition.

Problems

1. Estimate the temperature behavior of specific heat in the superconducting state near zero temperature $T = 0$.

Solution. We use the relation

$$\frac{1}{g} = -N(0)\left(\ln \frac{2\hbar\omega_D}{\Delta} - 2\int\limits_0^\infty \frac{d\xi}{\sqrt{\xi^2+\Delta^2}} \frac{1}{\exp(\sqrt{\xi^2+\Delta^2}/T)+1} \right)$$

[8] The superconducting state proves to be more ordered than the normal one if entropy is interpreted as a measure of disorder.

in which we expand the integrand in a series in the exponential and go over to integration over ε. A series of integrals can be expressed via modified Bessel function $K_0(x)$ (Macdonald function)[9]

$$\frac{1}{g} = -N(0)\left[\ln\frac{2\hbar\omega_D}{\Delta} + 2\sum_{n=1}^{\infty}(-1)^n K_0\left(\frac{n\Delta}{T}\right)\right].$$

Differentiating this formula with respect to Δ yields

$$\frac{d}{d\Delta}\left(\frac{1}{g}\right) = N(0)\left[\frac{1}{\Delta} + 2\sum_{n=1}^{\infty}(-1)^n\frac{n}{T}K_1\left(\frac{n\Delta}{T}\right)\right].$$

Then we find for the difference $\Omega_s - \Omega_n$ in the thermodynamic potentials

$$\Omega_s - \Omega_n = -N(0)\int_0^{\Delta} d\Delta\left[\Delta + 2\sum_{n=1}^{\infty}(-1)^n\frac{n\Delta^2}{T}K_1\left(\frac{n\Delta}{T}\right)\right] =$$

$$= -N(0)\left[\frac{\Delta^2}{2} + 2\sum_{n=1}^{\infty}(-1)^n\frac{T^2}{n^2}\int_0^{n\Delta/T} dx\, x^2 K_1(x)\right] =$$

$$= -N(0)\left[\frac{\Delta^2}{2} + 2\sum_{n=1}^{\infty}(-1)^n\frac{T^2}{n^2}\left[2 - \left(\frac{n\Delta}{T}\right)^2 K_2\left(\frac{n\Delta}{T}\right)\right)\right].$$

The sum

$$4T^2\sum_{n=1}^{\infty}\frac{(-1)^n}{n^2} = -\frac{\pi^2}{3}T^2$$

is analytically exact, leading to the term $(\pi^2/3)N(0)T^2$ equal to the opposite-sign potential $\Omega_n = -(\pi^2/3)N(0)T^2$ in the normal state. As a result, we arrive at the following expansion for the thermodynamic potential in the superconducting state:

$$\Omega_s = -N(0)\left[\frac{\Delta^2}{2} - 2\Delta^2\sum_{n=1}^{\infty}(-1)^n K_2\left(\frac{n\Delta}{T}\right)\right].$$

For $T \to 0$, we involve the temperature correction to $\Delta(0)$ and the term $n = 1$ in a sum. Eventually, we have

$$\Omega_s(T) - \Omega_n(0) \approx N(0)\sqrt{8\pi T^3\Delta(0)}\, e^{-\Delta(0)/T}.$$

Hence we have the entropy and specific heat as $T \to 0$

$$S_s(T) = -\partial\Omega_s/\partial T \approx N(0)\sqrt{8\pi\Delta^3(0)/T}\, e^{-\Delta(0)/T},$$

$$C_s(T) = T\partial S_s/\partial T \approx N(0)\sqrt{8\pi\Delta^5(0)/T^3}\, e^{-\Delta(0)/T}.$$

2. The Ginzburg–Landau theory interprets the normal-superconducting phase transition within the framework of self-consistent field with the superconducting gap Δ as an order parameter and

---

[9] Let us remind the definition of function $K_\nu(x)$ and a number of useful identities as $K_\nu(x) = \int_0^\infty e^{-x\cosh t}\cosh(\nu t)\,dt$, $K_\nu(x\to\infty) \approx (\pi/2x)^{1/2}e^{-x}$, $K_0'(x) = -K_1(x)$, $\left(x^2K_2(x)\right)' = -x^2K_1(x)$ and $\int_0^\infty x^2K_1(x)\,dx = 2$.

represents the thermodynamic potential difference $\Omega_s - \Omega_n$ as an expansion in the powers of gap Δ near $T = T_c$

$$\Omega_s - \Omega_n = a(T)|\Delta|^2 + \frac{b}{2}|\Delta|^4, \quad a(T) = a_0(T - T_c) \quad (a_0 > 0).$$

Find the expansion coefficients a and b, using the known behavior for $\Delta(T)$.

Solution. For $T < T_c$, we have

$$|\Delta|^2 = -\frac{a(T)}{b} \quad \text{and} \quad \Omega_s - \Omega_n = -\frac{a^2(T)}{2b}.$$

Hence we find the expansion coefficients

$$b = -2\frac{\Omega_s - \Omega_n}{|\Delta|^4} = \frac{7\zeta(3)}{8}\frac{N(0)}{(\pi T_c)^2} \quad \text{and} \quad a(T) = 2\frac{\Omega_s - \Omega_n}{|\Delta|^2} = \frac{N(0)}{T_c}(T - T_c).$$

In the Ginzburg–Landau theory, it is customary to introduce a superconducting order parameter ψ differing by the multiplier $C^{1/2}$ from the superconducting gap according to

$$|\Delta|^2 = C|\psi|^2 \quad \text{where} \quad C = \frac{6(\pi T_c)^2}{7\zeta(3)}\frac{1}{\mu N(0)} = \frac{8\pi^2 T_c^2}{7\zeta(3)n}.$$

Here n is the electron density. Accordingly, the expansion for the thermodynamic potential difference can be written as

$$\Omega_s - \Omega_n = \alpha(T)|\psi|^2 + \frac{\beta}{2}|\psi|^4,$$
$$\alpha(T) = Ca(T) = \frac{6\pi^2}{7\zeta(3)}\frac{T_c}{\mu}(T - T_c) \quad \text{and} \quad \beta = C^2 b = \frac{9\pi^2}{14\zeta(3)}\frac{T_c^2}{\mu^2 N(0)}.$$

6.4 Method of Self-Consistent Field. The Bogoliubov–de Gennes Equations

We have considered the superconducting properties of Fermi particles with attraction in the spatially homogeneous state in the lack of any external field. In the general case, the electrons in a superconductor can be in the external magnetic field $\boldsymbol{B}(\boldsymbol{r})$. Let us turn to constructing the description of superconducting state in the external magnetic field.

As we have seen above, the description of the superconducting state reduces in essence to the self-consistent decoupling of the electron–electron interaction

$$U_{\text{int}} = g\int d^3r\, \psi(\boldsymbol{r})^+_\uparrow \psi(\boldsymbol{r})^+_\downarrow \psi(\boldsymbol{r})_\downarrow \psi(\boldsymbol{r})_\uparrow$$

with the aid of introducing the *pairing potential* $\Delta(\boldsymbol{r}) = g\langle\psi(\boldsymbol{r})_\downarrow \psi(\boldsymbol{r})_\uparrow\rangle$ and $\Delta^*(\boldsymbol{r}) = g\langle\psi(\boldsymbol{r})^+_\uparrow \psi(\boldsymbol{r})^+_\downarrow\rangle$. The pairing potential is nonzero in the superconducting state and should be determined from the condition of the thermodynamic potential minimum. In the homogeneous state it is obvious that $\Delta(\boldsymbol{r}) = \Delta^*(\boldsymbol{r}) = \text{const}$.

We start from the usual Hamiltonian for an electron in the external magnetic field. The term responsible for the electron–electron interaction U_{int} will be written in accordance with the self-consistent decoupling. Then the effective Hamiltonian reads

$$\hat{\mathcal{H}}_{\text{eff}} = \hat{H} - \mu\hat{N} =$$
$$= \int d^3r \sum_{\sigma} \psi_{\sigma}^{+}(\boldsymbol{r}) \left[\frac{1}{2m} \left(-i\hbar\nabla - \frac{e}{c}\boldsymbol{A}(\boldsymbol{r}) \right)^2 - \mu \right] \psi_{\sigma}(\boldsymbol{r}) +$$
$$+ \int d^3r \left[\Delta(\boldsymbol{r})\psi_{\uparrow}^{+}(\boldsymbol{r})\psi_{\downarrow}^{+}(\boldsymbol{r}) + \Delta^*(\boldsymbol{r})\psi_{\downarrow}(\boldsymbol{r})\psi_{\uparrow}(\boldsymbol{r}) \right]$$

where $\boldsymbol{A}(\boldsymbol{r})$ is the vector potential determining the magnetic field $\boldsymbol{B} = \nabla \times \boldsymbol{A}$.

To find the eigenvalues of effective Hamiltonian $\hat{\mathcal{H}}_{\text{eff}}$ and the eigenfunctions, we diagonalize it with the aid of the Bogoliubov transformation

$$\psi_{\uparrow}(\boldsymbol{r}) = \sum_{p} \left[u_p(\boldsymbol{r})\alpha_{p\uparrow} - v_p^*(\boldsymbol{r})\alpha_{-p\downarrow}^{+} \right],$$
$$\psi_{\uparrow}^{+}(\boldsymbol{r}) = \sum_{p} \left[u_p^*(\boldsymbol{r})\alpha_{p\uparrow}^{+} - v_p(\boldsymbol{r})\alpha_{-p\downarrow} \right].$$

And the analogous transformation is given for the spin down

$$\psi_{\downarrow}(\boldsymbol{r}) = \sum_{p} \left[u_p(\boldsymbol{r})\alpha_{p\downarrow} + v_p^*(\boldsymbol{r})\alpha_{-p\uparrow}^{+} \right],$$
$$\psi_{\downarrow}^{+}(\boldsymbol{r}) = \sum_{p} \left[u_p^*(\boldsymbol{r})\alpha_{p\downarrow}^{+} + v_p(\boldsymbol{r})\alpha_{-p\uparrow} \right].$$

Here $\alpha_{p\sigma}^{+}$ and $\alpha_{p\sigma}$ are auxiliary creation and annihilation operators of elementary excitations (quasiparticles) in the state $|p\sigma\rangle$ obeying the conventional Fermi anticommutative relations

$$\alpha_{p\sigma}\alpha_{p'\sigma'} + \alpha_{p'\sigma'}\alpha_{p\sigma} = 0, \quad \alpha_{p\sigma}^{+}\alpha_{p'\sigma'}^{+} + \alpha_{p'\sigma'}^{+}\alpha_{p\sigma}^{+} = 0,$$
$$\alpha_{p\sigma}\alpha_{p'\sigma'}^{+} + \alpha_{p'\sigma'}^{+}\alpha_{p\sigma} = \delta_{pp'}\delta_{\sigma\sigma'}.$$

In addition, the amplitudes $u_p(\boldsymbol{r})$ and $v_p(\boldsymbol{r})$ should satisfy the relations of completeness and orthogonality

$$\sum_{p} u_p^*(\boldsymbol{r})u_p(\boldsymbol{r}') + v_p^*(\boldsymbol{r})v_p(\boldsymbol{r}') = \delta(\boldsymbol{r} - \boldsymbol{r}'),$$
$$\sum_{p} u_p^*(\boldsymbol{r})v_p(\boldsymbol{r}') + v_p(\boldsymbol{r})u_p^*(\boldsymbol{r}') = 0.$$

The aim of such transformation is to determine the amplitudes $u_p(\boldsymbol{r})$ and $v_p(\boldsymbol{r})$ which could transfer Hamiltonian $\hat{\mathcal{H}}_{\text{eff}}$ to the diagonal form

$$\hat{\mathcal{H}}_{\text{eff}} = \Omega_0(\mu) + \sum_{p\sigma} \varepsilon_{p\sigma} \alpha^+_{p\sigma} \alpha_{p\sigma} \,.$$

Here $\Omega_0(\mu)$ is the thermodynamic potential in the ground state and $\varepsilon_{p\sigma}$ has the sense of elementary excitation energy in the state $|p\sigma\rangle$. It is possible to check directly for the diagonal Hamiltonian that its commutators with the creation $\alpha^+_{p\sigma}$ and annihilation $\alpha_{p\sigma}$ operators will be equal to

$$\begin{aligned}
\left[\alpha_{p\sigma}, \hat{\mathcal{H}}_{\text{eff}}\right] &= \alpha_{p\sigma}\hat{\mathcal{H}}_{\text{eff}} - \hat{\mathcal{H}}_{\text{eff}}\alpha_{p\sigma} = \varepsilon_{p\sigma}\alpha_{p\sigma} \,, \\
\left[\alpha^+_{p\sigma}, \hat{\mathcal{H}}_{\text{eff}}\right] &= \alpha^+_{p\sigma}\hat{\mathcal{H}}_{\text{eff}} - \hat{\mathcal{H}}_{\text{eff}}\alpha^+_{p\sigma} = -\varepsilon_{p\sigma}\alpha^+_{p\sigma} \,.
\end{aligned}$$

It is more convenient to find the necessary conditions by choosing the amplitudes $u_p(\boldsymbol{r})$ and $v_p(\boldsymbol{r})$ via preliminary calculations[10] of commutating Hamiltonian $\hat{\mathcal{H}}_{\text{eff}}$ with $\psi_\sigma(\boldsymbol{r})$ and $\psi^+_\sigma(\boldsymbol{r})$

$$\begin{aligned}
\left[\psi_\uparrow(\boldsymbol{r}), \hat{\mathcal{H}}_{\text{eff}}\right] &= \hat{H}_e(\boldsymbol{r})\psi_\uparrow(\boldsymbol{r}) + \Delta(\boldsymbol{r})\psi^+_\downarrow(\boldsymbol{r}), \\
\left[\psi_\downarrow(\boldsymbol{r}), \hat{\mathcal{H}}_{\text{eff}}\right] &= \hat{H}_e(\boldsymbol{r})\psi_\downarrow(\boldsymbol{r}) - \Delta(\boldsymbol{r})\psi^+_\uparrow(\boldsymbol{r})
\end{aligned}$$

and

$$\begin{aligned}
\left[\psi^+_\uparrow(\boldsymbol{r}), \hat{\mathcal{H}}_{\text{eff}}\right] &= -\hat{H}_e(\boldsymbol{r})\psi^+_\uparrow(\boldsymbol{r}) - \Delta^*(\boldsymbol{r})\psi_\downarrow(\boldsymbol{r}), \\
\left[\psi^+_\downarrow(\boldsymbol{r}), \hat{\mathcal{H}}_{\text{eff}}\right] &= -\hat{H}_e(\boldsymbol{r})\psi^+_\downarrow(\boldsymbol{r}) + \Delta^*(\boldsymbol{r})\psi_\uparrow(\boldsymbol{r}).
\end{aligned}$$

For brevity, we have denoted

$$\hat{H}_e = \frac{1}{2m}\left(-i\hbar\nabla - \frac{e}{c}\boldsymbol{A}(\boldsymbol{r})\right)^2 - \mu.$$

Expressing $\psi^+_\sigma(\boldsymbol{r})$ and $\psi_\sigma(\boldsymbol{r})$ via $\alpha^+_{p\sigma}$ and $\alpha_{p\sigma}$ and using the commutators found above, we arrive at the equalities which should identically be satisfied with coinciding the coefficients before $\alpha^+_{p\sigma}$ and $\alpha_{p\sigma}$ on the left- and right-hand side of the resulting equalities. In fact, from one side

$$\begin{aligned}
&\left[\psi_\uparrow(\boldsymbol{r}), \hat{\mathcal{H}}_{\text{eff}}\right] = \hat{H}_e(\boldsymbol{r})\psi_\uparrow(\boldsymbol{r}) + \Delta(\boldsymbol{r})\psi^+_\downarrow(\boldsymbol{r}) = \\
&= \hat{H}_e(\boldsymbol{r})\sum_p\left(u_p\alpha_{p\uparrow} - v^*_p\alpha^+_{-p\downarrow}\right) + \Delta(\boldsymbol{r})\sum_p\left(u^*_p\alpha^+_{p\downarrow} + v_p\alpha_{-p\uparrow}\right),
\end{aligned}$$

[10] We use the anticommutation rules for the Fermi operators as $\psi_\sigma(\boldsymbol{r})\psi_{\sigma'}(\boldsymbol{r}') + \psi_{\sigma'}(\boldsymbol{r}')\psi_\sigma(\boldsymbol{r}) = 0$, $\psi^+_\sigma(\boldsymbol{r})\psi^+_{\sigma'}(\boldsymbol{r}') + \psi^+_{\sigma'}(\boldsymbol{r}')\psi'_\sigma(\boldsymbol{r}) = 0$, and $\psi_\sigma(\boldsymbol{r})\psi^+_{\sigma'}(\boldsymbol{r}') + \psi^+_{\sigma'}(\boldsymbol{r}')\psi_\sigma(\boldsymbol{r}) = \delta_{\sigma\sigma'}\delta(\boldsymbol{r} - \boldsymbol{r}')$.

and on the other side

$$[\psi_\uparrow(\boldsymbol{r}), \hat{\mathcal{H}}_{\text{eff}}] = \sum_p [(u_p\alpha_{p\uparrow} - v_p^*\alpha_{-p\downarrow}^+), \hat{\mathcal{H}}_{\text{eff}}] = \\ = \sum_p (u_p\varepsilon_p\alpha_{p\uparrow} + v_p^*\varepsilon_{-p}\alpha_{-p\downarrow}^+).$$

Comparing[11] the coefficients before $\alpha_{p\uparrow}$ and $\alpha_{p\downarrow}^+$, we derive the Bogoliubov-de Gennes (BdG) equations for determining the amplitudes $u_p(\boldsymbol{r})$ and $v_p(\boldsymbol{r})$

$$\varepsilon_p u_p(\boldsymbol{r}) = \hat{H}_e u_p(\boldsymbol{r}) + \Delta(\boldsymbol{r}) v_p(\boldsymbol{r}), \\ \varepsilon_p v_p(\boldsymbol{r}) = -\hat{H}_e^* v_p(\boldsymbol{r}) + \Delta^*(\boldsymbol{r}) u_p(\boldsymbol{r}).$$

The equations above are customary to write in the matrix form as well

$$\begin{pmatrix} \hat{H}_e(\boldsymbol{r}) & \Delta(\boldsymbol{r}) \\ \Delta^*(\boldsymbol{r}) & -\hat{H}_e^*(\boldsymbol{r}) \end{pmatrix} \begin{pmatrix} u_p \\ v_p \end{pmatrix} = \varepsilon_p \begin{pmatrix} u_p \\ v_p \end{pmatrix}.$$

Eventually, the amplitudes $u_p(\boldsymbol{r})$ and $v_p(\boldsymbol{r})$ satisfy a set of linear homogeneous differential equations.

For the spatially homogeneous case under zero magnetic field when $\boldsymbol{A}(\boldsymbol{r}) = 0$ and $\Delta(\boldsymbol{r}) = \text{const}$, the Bogoliubov–de Gennes equations can be solved with the aid of substitution

$$u_p(\boldsymbol{r}) = u_p e^{i\boldsymbol{pr}/\hbar} \quad \text{and} \quad v_p(\boldsymbol{r}) = v_p e^{i\boldsymbol{pr}/\hbar},$$

resulting in the following equations:

$$\xi_p u_p + \Delta v_p = \varepsilon_p u_p\,, \\ \Delta^* u_p - \xi_p v_p = \varepsilon_p v_p\,, \text{where} \quad \xi_p = \frac{\boldsymbol{p}^2}{2m} - \mu.$$

Nontrivial solution of this system takes place for the eigenvalue $\varepsilon_p = \pm\sqrt{\xi_p^2 + |\Delta|^2}$. Taking the condition for amplitudes $u_p^2 + v_p^2 = 1$ into account, we have

$$u_p^2 = \frac{1}{2}\left(1 + \frac{\xi_p}{\sqrt{\xi_p^2 + |\Delta|^2}}\right), \\ v_p^2 = \frac{1}{2}\left(1 - \frac{\xi_p}{\sqrt{\xi_p^2 + |\Delta|^2}}\right).$$

[11] For simplicity and clarity, we do not consider a possible dependence of the elementary excitation energy ε_p upon the spin projection $\sigma = \uparrow$ or $\downarrow$.

6.5 Superconducting Current

In this section, we turn to investigating the Bogoliubov–de Gennes equations when the pairing potential $\Delta(\boldsymbol{r})$ varies slowly in the space. Such situation may appear in the presence of magnetic field strength $\boldsymbol{h} = \text{curl}\,\boldsymbol{A}(\boldsymbol{r})$ where $\boldsymbol{A}$ is the vector potential in the presence of the current states. Let the pairing potentials $\Delta(\boldsymbol{r}) = \Delta\, e^{i\varphi(\boldsymbol{r})}$ and $\Delta^*(\boldsymbol{r}) = \Delta\, e^{-i\varphi(\boldsymbol{r})}$ have the phase $\varphi(\boldsymbol{r})$ varying slowly in the space. Nonzero phase of the pairing potential will result in the current state.

The solution of the Bogoluibov-de Gennes equations will be sought as

$$u(\boldsymbol{r}) = u_{\boldsymbol{p}} e^{\frac{i\boldsymbol{p}\boldsymbol{r}}{\hbar} + \frac{i\varphi(\boldsymbol{r})}{2}},$$
$$v(\boldsymbol{r}) = v_{\boldsymbol{p}} e^{\frac{i\boldsymbol{p}\boldsymbol{r}}{\hbar} - \frac{i\varphi(\boldsymbol{r})}{2}}.$$

A possible dependence of preexponential factors $u_{\boldsymbol{p}}$ and $v_{\boldsymbol{p}}$ on the coordinates will be neglected. In the first approximation, substituting the amplitudes $u(\boldsymbol{r})$ and $v(\boldsymbol{r})$ into the Bogoluibov-de Gennes equations, we neglect also the derivatives of vector potential $\boldsymbol{A}(\boldsymbol{r})$ and second derivatives of phase $\varphi(\boldsymbol{r})$ due to assumption about sufficiently slow variation of these quantities in the space. Under these assumptions, it is easy to check that

$$\hat{H}_e u(\boldsymbol{r}) = \left[\frac{\left(-i\hbar\nabla - \frac{e\boldsymbol{A}(\boldsymbol{r})}{c}\right)^2}{2m} - \mu\right]u(\boldsymbol{r}) = \left[\frac{\left(\boldsymbol{p} + \frac{\hbar\nabla\varphi}{2} - \frac{e\boldsymbol{A}(\boldsymbol{r})}{c}\right)^2}{2m} - \mu\right]u(\boldsymbol{r}),$$
$$\hat{H}_e^* v(\boldsymbol{r}) = \left[\frac{\left(i\hbar\nabla - \frac{e}{c}\boldsymbol{A}(\boldsymbol{r})\right)^2}{2m} - \mu\right]v(\boldsymbol{r}) = \left[\frac{\left(\boldsymbol{p} - \frac{\hbar\nabla\varphi}{2} + \frac{e\boldsymbol{A}(\boldsymbol{r})}{c}\right)^2}{2m} - \mu\right]v(\boldsymbol{r}).$$

As a result, we arrive at the following system of equations:

$$\left(\varepsilon - \xi_{\boldsymbol{P}_+}\right)u(\boldsymbol{r}) - \Delta(\boldsymbol{r})v(\boldsymbol{r}) = 0, \quad \xi_{\boldsymbol{P}_\pm} = \frac{\boldsymbol{P}_\pm^2}{2m} - \mu,$$
$$-\Delta^*(\boldsymbol{r})u(\boldsymbol{r}) + \left(\varepsilon + \xi_{\boldsymbol{P}_-}\right)u(\boldsymbol{r}) = 0, \quad \boldsymbol{P}_\pm = \boldsymbol{p} \pm \frac{\hbar\nabla\varphi}{2} \mp \frac{e}{c}\boldsymbol{A}.$$

The compatibility condition for this system of equations, i.e. zero determinant,

$$\left(\varepsilon_{\boldsymbol{p}} - \xi_{\boldsymbol{P}_+}\right)\left(\varepsilon_{\boldsymbol{p}} + \xi_{\boldsymbol{P}_-}\right) - \Delta^2 = 0$$

gives us the energy spectrum of elementary excitations in the superconducting state

$$\varepsilon_{\boldsymbol{p}} = \frac{1}{2}\left[\left(\xi_{\boldsymbol{P}_+} - \xi_{\boldsymbol{P}_-}\right) \pm \sqrt{\left(\xi_{\boldsymbol{P}_+} + \xi_{\boldsymbol{P}_-}\right)^2 + 4\Delta^2}\right].$$

In what follows, we consider the phase gradient $\nabla\varphi$ and magnitude of vector potential $\boldsymbol{A}$ to be small as compared with the typical electron momentum of about

Fermi one. Next, we can write the following expansions restricted with the lowest terms alone:

$$\xi_{P_+} - \xi_{P_-} \approx \frac{p}{m}\left(\hbar\nabla\varphi - \frac{2e}{c}A\right) \quad \text{and} \quad \xi_{P_+} + \xi_{P_-} \approx \frac{p^2}{m} - 2\mu = 2\xi_p\,.$$

Then we come to the following elementary excitation spectrum in the presence of small gradient phase and magnetic field:

$$\varepsilon_p \approx \pm\sqrt{\xi_p^2 + \Delta^2} + \frac{1}{m}p\cdot\left(\frac{\hbar\nabla\varphi}{2} - \frac{e}{c}A(r)\right).$$

Thus, as we see in the presence of magnetic field and phase gradient, the elementary excitation energy is shifted by the quantity pv_s where

$$v_s = \frac{\hbar}{2m}\left(\nabla\varphi - \frac{2e}{\hbar c}A(r)\right) = \frac{\hbar}{2m}\nabla\Phi$$

is called the *superfluid velocity* and the phase Φ is referred to as the *gauge phase*. The magnitude $v_{cr} = \Delta/p_F$, where $p_F = (2m\mu)^{1/2}$ is the Fermi momentum, serves as a *critical magnitude* of superfluid velocity. For the magnitudes of velocity v_s larger than the critical one $v_s > v_{cr}$, the energy gap in the elementary excitation spectrum will vanish.

This criterion allows us to estimate what spatial variations of gauge phase Φ can be considered as sufficiently slow. Let phase Φ vary at the typical distance δ so that $\nabla\Phi \sim 1/\delta$. Requiring small variation for the energy spectrum of excitations, i.e. $v_s \ll v_{cr}$, we have

$$\frac{\hbar}{m}\nabla\Phi \sim \frac{\hbar}{m\delta} \ll v_{cr} \sim \frac{\Delta}{p_F} = \frac{\Delta}{mv_F} \quad \text{or} \quad \delta \gg \frac{\hbar v_F}{\Delta} = \xi$$

where v_F is the Fermi velocity. The length $\xi(T) = \hbar v_F/\Delta(T)$ will be called the *correlation length* of a superconductor, which[12] is temperature-dependent together with the energy gap $\Delta = \Delta(T)$.

Returning to the Bogoluibov-de Gennes equations, we can see that the amplitudes u_p and v_p in our approximation depend on $\nabla\varphi$ and A by means of a sum $\xi_{P_+} - \xi_{P_-} = 2\xi_p$. In fact, it is readily to see that $\varepsilon_p - \xi_{P_\pm} \approx \varepsilon_p \mp \xi_p$. Thus, in our initial approximation the assumption about coordinate independence of amplitudes u_p and v_p is completely justified. The amplitudes remain the same as in the case of the spatially independent potential of pairing $\Delta(r)$.

Now let us turn to calculating the current in a superconductor. The magnitude of current density $j(r)$ at the given point r is the thermodynamic average of the familiar quantum mechanical expression for the current density operator $\hat{j}(r)$

[12] The notation ξ is common. It should not be confused with the energy of electrons $\xi_p = p^2/2m - \mu$ taken from chemical potential μ.

$$\hat{j}(\boldsymbol{r}) = \frac{ie\hbar}{2m}\sum_{\sigma}\Big(\nabla\psi_{\sigma}^{+}(\boldsymbol{r})\psi_{\sigma}(\boldsymbol{r}) - \psi_{\sigma}^{+}(\boldsymbol{r})\nabla\psi_{\sigma}(\boldsymbol{r})\Big)$$
$$- \frac{e^2}{mc}\sum_{\sigma}\boldsymbol{A}(\boldsymbol{r})\psi_{\sigma}^{+}(\boldsymbol{r})\psi_{\sigma}(\boldsymbol{r})$$

stated in terms of creation $\psi^{+}(\boldsymbol{r})$ and annihilation $\psi(\boldsymbol{r})$ operators. Thus, we should be able to calculate the following thermodynamic averages: $\langle\nabla\psi^{+}\psi\rangle$, $\langle\psi^{+}\nabla\psi\rangle$, and $\langle\psi^{+}\psi\rangle$.

To find the averages, we employ the Bogoluibov transformation formulas expressing the operators ψ^{+} and ψ via elementary excitation operators α^{+} and α. For calculating the derivatives $\nabla\psi^{+}$ and $\nabla\psi$, it is sufficient only to differentiate the exponential factors $e^{i\boldsymbol{pr}/\hbar \pm i\varphi(\boldsymbol{r})/2}$ in the amplitudes $u_{\boldsymbol{p}}(\boldsymbol{r})$ and $v_{\boldsymbol{p}}(\boldsymbol{r})$, i.e.

$$\nabla u_{\boldsymbol{p}} \approx i(\boldsymbol{p}/\hbar + \nabla\varphi/2)u_{\boldsymbol{p}}\,, \qquad \nabla u_{\boldsymbol{p}}^{*} \approx -i(\boldsymbol{p}/\hbar + \nabla\varphi/2)u_{\boldsymbol{p}}^{*}\,,$$
$$\nabla v_{\boldsymbol{p}} \approx i(\boldsymbol{p}/\hbar - \nabla\varphi/2)v_{\boldsymbol{p}}\,, \qquad \nabla v_{\boldsymbol{p}}^{*} \approx -i(\boldsymbol{p}/\hbar - \nabla\varphi/2)v_{\boldsymbol{p}}^{*}\,.$$

We calculate the current $\boldsymbol{j}(\boldsymbol{r}) = \langle\hat{\boldsymbol{j}}(\boldsymbol{r})\rangle$ as a thermodynamic average for the current density operator $\hat{\boldsymbol{j}}$

$$\boldsymbol{j}(\boldsymbol{r}) = \frac{e}{m}\sum_{\boldsymbol{p}\sigma}\left[\frac{i\hbar}{2}\big(u_{\boldsymbol{p}}\nabla u_{\boldsymbol{p}}^{*} - u_{\boldsymbol{p}}^{*}\nabla u_{\boldsymbol{p}}\big) - \frac{e}{c}\boldsymbol{A}u_{\boldsymbol{p}}u_{\boldsymbol{p}}^{*}\right]\langle\alpha_{\boldsymbol{p}\sigma}^{+}\alpha_{\boldsymbol{p}\sigma}\rangle +$$
$$+ \frac{e}{m}\sum_{\boldsymbol{p}\sigma}\left[\frac{i\hbar}{2}\big(v_{\boldsymbol{p}}^{*}\nabla v_{\boldsymbol{p}} - v_{\boldsymbol{p}}\nabla v_{\boldsymbol{p}}^{*}\big) - \frac{e}{c}\boldsymbol{A}v_{\boldsymbol{p}}v_{\boldsymbol{p}}^{*}\right]\langle\alpha_{\ \boldsymbol{p}\sigma}\alpha_{-\boldsymbol{p}\sigma}^{+}\rangle =$$
$$= \frac{e}{m}\sum_{\boldsymbol{p}\sigma}(\boldsymbol{p} + \hbar\nabla\Phi/2)\Big[|u_{\boldsymbol{p}}|^2\langle\alpha_{\boldsymbol{p}\sigma}^{+}\alpha_{\boldsymbol{p}\sigma}\rangle + |v_{\boldsymbol{p}}|^2\langle\alpha_{-\boldsymbol{p}\sigma}\alpha_{-\boldsymbol{p}\sigma}^{+}\rangle\Big].$$

Here the thermodynamic average $\langle\alpha_{\boldsymbol{p}\sigma}^{+}\alpha_{\boldsymbol{p}\sigma}\rangle = n_{\boldsymbol{p}\sigma}$ represents the occupation numbers of elementary quasiparticle excitations with energy $\varepsilon_{\boldsymbol{p}}$, i.e. Fermi distribution

$$n_{\boldsymbol{p}\sigma} = \big[\exp(\varepsilon_{\boldsymbol{p}\sigma}/T) + 1\big]^{-1}.$$

Accordingly,[13] we have $\langle\alpha_{\boldsymbol{p}\sigma}\alpha_{\boldsymbol{p}\sigma}^{+}\rangle = 1 - n_{\boldsymbol{p}\sigma}$ resulting from the anticommutation relations for the Fermi operators. In our case we will not take a possible dependence

[13] If we perform the Bogoliubov transformation for the electron density operator

$$\hat{n}_e(\boldsymbol{r}) = \sum_{\boldsymbol{p}\sigma}\psi_{\sigma}^{+}(\boldsymbol{r})\psi_{\sigma}(\boldsymbol{r})$$

and then accomplish the thermodynamic averaging for the operators of quasiparticle excitations $\alpha_{\boldsymbol{p}\sigma}^{+}$ and $\alpha_{\boldsymbol{p}\sigma}$, we find the following relations for the electron density:

$$n_e(\boldsymbol{r}) = \sum_{\boldsymbol{p}\sigma}\Big[|u_{\boldsymbol{p}}(\boldsymbol{r})|^2\langle\alpha_{\boldsymbol{p}\sigma}^{+}\alpha_{\boldsymbol{p}\sigma}\rangle + |v_{\boldsymbol{p}}(\boldsymbol{r})|^2\langle\alpha_{\boldsymbol{p}\sigma}\alpha_{\boldsymbol{p}\sigma}^{+}\rangle\Big].$$

of excitation energy on the spin into account, i.e.

$$\varepsilon_{p\sigma} = \varepsilon_p = \pm\sqrt{\xi_p^2 + |\Delta|^2} + \hbar \boldsymbol{p}\nabla\Phi/2m = \varepsilon_{p0} + \boldsymbol{p}\boldsymbol{v}_s.$$

Thus, the current density reads

$$\boldsymbol{j}(\boldsymbol{r}) = \frac{2e}{m}\sum_p (\boldsymbol{p} + \hbar\nabla\Phi/2)\Big[|u_p|^2 n(\varepsilon_p) + |v_p|^2\big(1 - n(\varepsilon_{-p})\big)\Big].$$

Factor 2 in front of the sum appears after summing over the spin projections.

We obtain in the linear approximation in smallness $\nabla\Phi$ or $\boldsymbol{v}_s$ with involving the dependence ε_p upon $\nabla\Phi$ and the parity $\varepsilon_{-p0} = \varepsilon_{p0}$

$$\begin{aligned}\boldsymbol{j}(\boldsymbol{r}) = &\frac{e\hbar}{m}\sum_p \nabla\Phi\Big[|u_p|^2 n(\varepsilon_{p0}) + |v_p|^2\big(1 - n(\varepsilon_{p0})\big)\Big] + \\ &+\frac{e\hbar}{m}\sum_p \frac{1}{m}\boldsymbol{p}(\boldsymbol{p}\nabla\Phi)\Big[|u_p|^2 + |v_p|^2\Big]\frac{\partial n(\varepsilon_{p0})}{\partial\varepsilon} = \\ &=\frac{e\hbar}{m}\nabla\Phi\Big[\frac{n_e}{2} + \sum_p \frac{p^2}{3m}\frac{\partial n(\varepsilon_{p0})}{\partial\varepsilon}\Big] = \\ &= \frac{e\hbar}{2m}\nabla\Phi\Big[n_e - \frac{4}{3}\sum_p \frac{p^2}{2m}\Big(-\frac{\partial n(\varepsilon_{p0})}{\partial\varepsilon}\Big)\Big] = \\ &= e(n_e - n_n)\frac{\hbar}{2m}\nabla\Phi = en_s\boldsymbol{v}_s.\end{aligned}$$

The terms with the odd powers on momentum $\boldsymbol{p}$ vanish in the sum due to symmetry.

The quantity

$$n_n = \frac{4}{3}\sum_p \frac{p^2}{2m}\Big(-\frac{\partial n(\varepsilon_p)}{\partial\varepsilon_p}\Big) = \frac{4}{3}\int\frac{d^3p}{(2\pi\hbar)^3}\frac{p^2}{2m}\Big(-\frac{\partial n(\varepsilon_p)}{\partial\varepsilon_p}\Big),$$

where $\varepsilon_p = (\xi_p^2 + \Delta^2)^{1/2}$, is called the *normal electron density* in a superconductor and quantity $n_s = n_e - n_n$ is referred to as the *superconducting electron density*. The both quantities are temperature-dependent. The superconducting electron density vanishes in the normal state at $T \geqslant T_c$, i.e. $n_s(\Delta = 0) = 0$. At zero temperature, it equals the total electron density $n_s(T = 0) = n_e$. In fact,

$$n_n(T) = \frac{1}{3\pi^2\hbar^3}\int_{-\mu}^{\infty} d\xi \frac{p^3(\xi)}{4T\cosh^2\frac{\sqrt{\xi^2+\Delta^2}}{2T}} \approx \frac{p_F^3}{3\pi^2\hbar^3}\int_{-\infty}^{\infty}\frac{d\xi}{4T\cosh^2\frac{\sqrt{\xi^2+\Delta^2}}{2T}},$$

where we have extended the lower integration limit to $-\infty$ due to presence of strong inequalities $T \ll \mu, \Delta \ll \mu$. The contribution to the integration is given by the region of the momenta near the Fermi surface $|\xi| \ll \mu$. For the same reasons we can put approximately $p(\xi) \approx p(0) = (2m\mu)^{1/2} = p_F$. Involving the relation of the Fermi momentum p_F with the total electron density n, we obtain the following expression for the fraction of normal electrons in terms of the *Yosida function* $Y(T)$:

$$\frac{n_n(T)}{n} = Y(T) \quad \text{where} \quad Y(T) = \int_{-\infty}^{\infty} d\xi \left(-\frac{\partial n}{\partial \varepsilon}\right) = \int_{-\infty}^{\infty} \frac{d\xi}{4T\cosh^2 \frac{\sqrt{\xi^2+\Delta^2(T)}}{2T}}.$$

Let us give the limiting expressions for the Yosida function

$$Y(T) = \begin{cases} \left(\frac{2\pi\Delta}{T}\right)^{1/2} e^{-\Delta/T} \sim e^{-\Delta/T}, & \Delta \gg T \to 0, \\ 1 - \frac{7\zeta(3)}{4\pi^2}\frac{\Delta^2}{T_c^2} = 1 - 2\frac{T_c - T}{T_c}, & T \sim T_c \gg \Delta. \end{cases}$$

Lastly, we have the following expression for the current in a superconductor in the magnetic field:

$$\boldsymbol{j}(\boldsymbol{r}) = \frac{e}{2m} n_s \left(\hbar \nabla\varphi - \frac{2e}{c}\boldsymbol{A}(\boldsymbol{r})\right).$$

This expression leads us to the *F. and H. London equation* on the account of $\operatorname{curl}(\nabla\varphi) = 0$ and $\boldsymbol{h} = \operatorname{curl} \boldsymbol{A}$

$$\operatorname{curl} \boldsymbol{j} = -\frac{e^2 n_s}{mc}\boldsymbol{h}.$$

This equation is quite sufficient to explain the specific physical property of a superconductor, namely *Meissner effect* or expulsion of magnetic field from the superconductor volume. In fact, the magnetic field strength $\boldsymbol{h}$ and the current density $\boldsymbol{j}$ are connected with the Maxwell equation

$$\operatorname{curl} \boldsymbol{h} = \frac{4\pi}{c}\boldsymbol{j}.$$

This equation together with another one $\operatorname{div} \boldsymbol{h} = 0$ results in

$$\nabla^2 \boldsymbol{h} = \frac{4\pi e^2 n_s}{mc^2}\boldsymbol{h} = \frac{\boldsymbol{h}}{\delta^2}$$

where

$$\delta = \left(\frac{mc^2}{4\pi e^2 n_s}\right)^{1/2}.$$

The length δ is called the *London penetration depth* of magnetic field into the superconductor.

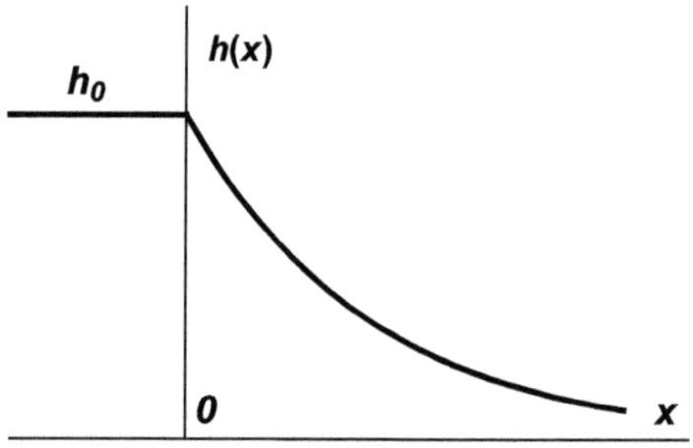

Fig. 6.5 The magnetic field penetration into a superconductor. The field at the surface is h_0

Let superconductor occupy the half-space $x > 0$. The magnetic field is parallel to the plane surface of the superconductor in the z-axis direction. Then we disclose immediately the exponential decay into the depth of the superconductor

$$\boldsymbol{h}(x) = \boldsymbol{h}_0 e^{-x/\delta}.$$

The vector $\boldsymbol{h}_0$ is the magnetic field strength at the superconductor surface and the x-axis, directed into the superconductor, is normal to the surface (Fig. 6.5). To conclude, the superconducting state of a metal results in expelling the magnetic field from the metal bulk. The magnetic induction $\boldsymbol{B}$, determined as a magnetic field average over the superconductor volume V

$$\boldsymbol{B} = \frac{1}{V}\int\limits_V \boldsymbol{h}\, dV = \langle \boldsymbol{h} \rangle_V,$$

will vanish. The screening superconducting current flows only in the near-surface superconductor layer of the thickness about penetration depth δ:

$$j_y(x) = \frac{ch_0}{4\pi\delta} e^{-x/\delta}.$$

From the expression for current $\boldsymbol{j}(\boldsymbol{r})$ it follows a remarkable property of superconducting state, namely *magnetic flux quantization*. In the superconductor bulk the current density vanishes and, therefore, we have

$$\nabla\varphi = \frac{2e}{\hbar c}\boldsymbol{A}(\boldsymbol{r}).$$

Integration around the closed contour C lying completely inside the superconductor bulk and application of the Stokes theorem for vector $\boldsymbol{h} = \text{curl}\,\boldsymbol{A}$ will deliver us the following equation for the magnetic field flux ϕ across the surface enclosed within the contour:

$$\phi = \oint \boldsymbol{A}\cdot d\boldsymbol{l} = \frac{\hbar c}{2e}\oint \nabla\varphi\cdot d\boldsymbol{l} = \frac{\hbar c}{2e}[\varphi].$$

Here $[\varphi]$ is the phase increment of superconducting order parameter $\Delta(\boldsymbol{r})$ after a full passage of contour C. Since the superconducting order parameter must be single-valued, its magnitude must be the same after a full passage of contour C. In other words, phase φ of order parameter can only be changed by an integer multiple of 2π. Then we arrive at the result

$$\phi = n\phi_0\,, \quad \phi_0 = \frac{\pi\hbar c}{e} = 2.07\cdot 10^{-7}\mathrm{G}\cdot\mathrm{cm}^2$$

where n is any integer. The quantity ϕ_0 is called the *magnetic flux quantum*. (The inverse of the flux quantum $K_J = 1/\phi_0$ is the *Josephson constant*).

As an example of magnetic flux quantization, we can indicate a massive hollow cylinder with the wall thickness much larger as compared with the magnetic field penetration depth. When the hollow cylinder is placed into the magnetic field parallel to the cylinder axis, the magnitude of the magnetic flux threading the cylinder hole will be an integer multiple of flux quantum.

6.6 The Ginzburg–Landau Functional

To describe the behavior of superconductor in the magnetic field, we can use the Bogoliubov–de Gennes equations. However, this is not a simple problem from the mathematical point of view. On the other hand, in the vicinity of the superconducting transition point it is possible to find the simplified description for the behavior of superconductor in the magnetic field. Such description is referred to as the theory of the *Ginzburg–Landau functional.*

Let us use expression for the superconducting electron density near the transition temperature

$$n_s = n\frac{7\zeta(3)}{4\pi^2}\frac{|\Delta|^2}{T_c^2} = 2n\frac{T_c - T}{T_c}$$

which allows us to rewrite the expression for the superconducting current as

$$\begin{aligned} \boldsymbol{j}(\boldsymbol{r}) = \frac{e}{2m}n\frac{7\zeta(3)|\Delta|^2}{4\pi^2 T_c^2}\left(\hbar\nabla\varphi - \frac{2e}{c}\boldsymbol{A}\right) = \frac{e}{m}\left(\hbar\nabla\varphi - \frac{2e}{c}\boldsymbol{A}\right) = \\ = -\frac{ie\hbar}{2m}\left(\psi^*\nabla\psi - \psi\nabla\psi^*\right) - \frac{2e^2}{mc}|\psi|^2\boldsymbol{A}\,. \end{aligned}$$

Here, instead of superconducting order parameter $\Delta(\boldsymbol{r})$ according to

$$\psi(\boldsymbol{r}) = \sqrt{\frac{7\zeta(3)n}{8\pi^2 T_c^2}}\Delta(\boldsymbol{r}) = \sqrt{\frac{7\zeta(3)n}{8\pi^2 T_c^2}}|\Delta(\boldsymbol{r})|e^{i\varphi(\boldsymbol{r})},$$

we have introduced the complex quantity $\psi(\boldsymbol{r})$ called the *wave function of Cooper pairs condensate* in the Ginzburg–Landau functional theory. The formula, obtained for the current density $\boldsymbol{j}(\boldsymbol{r})$, coincides formally with the quantum-mechanical expression for the current density at the motion of a particle with the wave function $\psi(\boldsymbol{r})$, mass $2m$, and charge $2e$ in the magnetic field.

The formula for current density $\boldsymbol{j}(\boldsymbol{r})$ allows us to determine the expression for the spatially inhomogeneous terms in the free energy functional $\mathcal{F}(\psi^*, \psi, \boldsymbol{A})$, depending on magnetic field $\boldsymbol{h} = \text{curl}\, \boldsymbol{A}$ in the superconducting transition region. For the variation of free energy with respect to the vector potential, we have

$$\delta\mathcal{F} = -\frac{1}{c}\int \boldsymbol{j}\,\delta\boldsymbol{A}\,d^3r + \int \frac{\boldsymbol{h}\,\delta\boldsymbol{h}}{4\pi}d^3r = -\frac{1}{c}\int \boldsymbol{j}\,\delta\boldsymbol{A}\,d^3r + \int \frac{\boldsymbol{h}\,\text{curl}(\delta\boldsymbol{A})}{4\pi}d^3r.$$

Hence we can restore the expression for the spatially inhomogeneous terms $\mathcal{F}_\nabla$ in the free energy functional of superconductor

$$\mathcal{F}_\nabla = \int\left[\frac{1}{4m}\left|\left(-i\hbar\nabla - \frac{2e}{c}\boldsymbol{A}\right)\psi\right|^2 + \frac{(\text{curl}\,\boldsymbol{A})^2}{8\pi}\right]d^3r.$$

In fact, the variation of this expression with respect to vector potential $\boldsymbol{A}$ leads us to the starting variation $\delta\mathcal{F}$. Herewith, we do not consider the surface terms that may appear since, so far, we restrict ourselves only with the bulk contribution to variation $\delta\mathcal{F}_\nabla$. The total expression for the free energy functional of superconductor $\mathcal{F}$ can be achieved by augmenting the spatially homogeneous terms to $\mathcal{F}_\nabla$. Thus, we have finally

$$\mathcal{F}[\psi(\boldsymbol{r}), \psi^*(\boldsymbol{r}), \boldsymbol{A}(\boldsymbol{r})] = \int F[\psi(\boldsymbol{r}), \psi^*(\boldsymbol{r}), \boldsymbol{A}(\boldsymbol{r})]d^3r$$

where the density of free energy F equals

$$F = \frac{\gamma}{2}\left|\left(-i\hbar\nabla - \frac{2e}{c}\boldsymbol{A}\right)\psi\right|^2 + \alpha(T)|\psi|^2 + \frac{\beta}{2}|\psi|^4 + \frac{(\text{rot}\,\boldsymbol{A})^2}{8\pi}.$$

The expansion coefficients α, β, and γ are as follows:

$$\alpha(T) = \frac{6\pi^2}{7\zeta(3)}\frac{T_c}{\mu}(T - T_c), \quad \beta = \frac{9\pi^2}{14\zeta(3)}\frac{T_c^2}{\mu^2 N(0)} \quad \text{and} \quad \gamma = \frac{1}{2m}.$$

The first three terms are associated with manifesting the superconducting state. The last term is the energy of magnetic field $\boldsymbol{h} = \text{curl}\,\boldsymbol{A}$.

The Ginzburg–Landau functional, in essence, represents the first terms in expanding the thermodynamic potential difference into a series in the smallness and slowness of spatial variation of the superconducting order parameter. We would like to mention the following aspects inherent in the phenomenological Landau theory of phase transitions based on the expansion in the powers and gradients of order parameter.

Due to interaction of a superconductor with the magnetic field, the free energy functional must be invariant with regard to the gauge transformation of vector potential $\boldsymbol{A} \to \boldsymbol{A} + \nabla f$. This can be achieved by putting the order parameter as a complex variable. The invariance of functional can be realized with gaining an additional phase of wave function

$$\psi \to \psi \exp\left(\frac{ie^*}{\hbar c} f\right).$$

Here, e^* is the parameter characterizing the magnetic field effect and having the dimension of electric charge. The real magnitude of the free energy functional and its gauge invariance can be achieved by introducing terms like $|\psi|^2$ and $|\psi|^4$ independent of the order parameter phase. As a result, the phenomenological Ginzburg–Landau functional gets a general form

$$\mathcal{F} = \int d^3r \left[\frac{\gamma}{2}\left|\left(-i\hbar\nabla - \frac{e^*}{c}\boldsymbol{A}\right)\psi\right|^2 + \alpha(T)|\psi|^2 + \frac{\beta}{2}|\psi|^4 + \frac{(\operatorname{curl}\boldsymbol{A})^2}{8\pi}\right].$$

Coefficient $\alpha(T) = \alpha_0(T - T_c)$ vanishes at $T = T_c$ and $\alpha(T > T_c) > 0$. The other two ones β and γ are positive and their possible temperature dependence can be neglected near $T = T_c$.

The normalization or units of measurement for the wave function have elements of variability and can be chosen within the framework of the Ginzburg–Landau functional from the convenience arguments. The magnitude of the proportionality coefficient in relation $\psi(\boldsymbol{r}) \sim \Delta(\boldsymbol{r})$ can vary by redefining or selecting the units of measurement for coefficients α, β, and γ. The microscopic theory of superconductivity allows us to set the magnitudes for α, β, γ, and effective charge $e^* = 2e$.

The *Ginzburg–Landau equations* are derived by minimizing the free energy functional and equating the variational derivatives to zero. So,

$$\delta\mathcal{F}_s = \int d^3r \Big[\alpha\psi\,\delta\psi^* + \beta|\psi|^2\psi\,\delta\psi^* + \\ + \frac{\gamma}{2}\left(i\hbar\nabla\delta\psi^* - \frac{2e}{c}\boldsymbol{A}\,\delta\psi^*\right)\left(-i\hbar\nabla\psi - \frac{2e}{c}\boldsymbol{A}\psi\right)\Big].$$

To put $\delta\psi^*$ beyond the brackets, we use the identity below

$$\operatorname{div}(\boldsymbol{b}\,\delta\psi^*) = \delta\psi^* \operatorname{div}\boldsymbol{b} + \boldsymbol{b}\cdot\nabla\delta\psi^*$$

where $\boldsymbol{b} = -i\hbar\nabla\psi^* - (2e/c)\boldsymbol{A}\psi$. Let us represent the integral as follows:

$$\int d^3r\,(\boldsymbol{b}\cdot\nabla\delta\psi^*) = -\int d^3r\,\delta\psi^* \operatorname{div}\boldsymbol{b} + \int d^3r \operatorname{div}(\boldsymbol{b}\delta\psi^*).$$

The last integral transforms into the surface one. Then,

$$\delta\mathcal{F}_s = \int d^3r\left[\alpha\psi + \beta|\psi|^2\psi + \frac{\gamma}{2}\left(-i\hbar\nabla\psi - \frac{2e}{c}\boldsymbol{A}\right)^2\psi\right]\delta\psi^* + \\ + \frac{\gamma}{2}\oint\limits_S \delta\psi^*\left(-i\hbar\nabla\psi - \frac{2e}{c}\boldsymbol{A}\right)d\boldsymbol{S}.$$

The requirement $\delta\mathcal{F}/\delta\psi^* = 0$ gives under condition of vanishing the surface integral

$$\frac{\gamma}{2}\left(-i\hbar\nabla - \frac{2e}{c}\boldsymbol{A}(\boldsymbol{r})\right)^2\psi(\boldsymbol{r}) + \alpha\psi(\boldsymbol{r}) + \beta|\psi(\boldsymbol{r})|^2\psi(\boldsymbol{r}) = 0.$$

Varying with respect to $\psi(\boldsymbol{r})$ results in the complex conjugate equation. As is expected, the variation by vector potential and $\delta\mathcal{F}/\delta\boldsymbol{A}(\boldsymbol{r}) = 0$ lead us to the Maxwell equation with superconducting current density $\boldsymbol{j}(\boldsymbol{r})$

$$\text{curl curl}\,\boldsymbol{A} = \text{curl}\,\boldsymbol{h} = \frac{4\pi}{c}\boldsymbol{j}(\boldsymbol{r}), \quad \boldsymbol{j} = \gamma\left(-ie\hbar\left(\psi^*\nabla\psi - \psi\nabla\psi^*\right) - \frac{4e^2}{c}|\psi|^2\boldsymbol{A}\right).$$

For the superconductor-vacuum or superconductor-isolator boundary, the so-called *natural boundary condition* can be used with the cancelation of the surface integral originating from varying the functional over ψ^*

$$\frac{i\gamma}{2}\oint dS\,\delta\psi^*\boldsymbol{n}\cdot\left(-i\hbar\nabla\psi - \frac{2e}{c}\boldsymbol{A}\psi\right).$$

Accordingly, the boundary condition reads

$$\boldsymbol{n}\cdot\left(-i\hbar\nabla - \frac{2e}{c}\boldsymbol{A}(\boldsymbol{r})\right)\psi(\boldsymbol{r})\Big|_S = 0.$$

Here $\boldsymbol{n}$ is the outer normal to the superconductor surface S. This boundary condition means the lack of superconducting current across the boundary of superconductor, i.e.

$$j_n = (\boldsymbol{n}\cdot\boldsymbol{j})|_S = 0.$$

The boundary condition for the magnetic field $\boldsymbol{h}(\boldsymbol{r})$ is the continuity of the field at the boundary.

If we require the fulfillment of condition $(\boldsymbol{n}\,\boldsymbol{j}) = 0$ at the superconductor boundary or non-penetration of superconducting current to the adjacent matter, we obtain a weaker boundary condition

$$\boldsymbol{n}\cdot\left(-i\hbar\nabla - \frac{2e}{c}\boldsymbol{A}(\boldsymbol{r})\right)\psi(\boldsymbol{r})\Big|_S = i\frac{\hbar}{b}\psi(\boldsymbol{r})\Big|_S$$

where b is a real constant having the dimension of length. Such type of boundary condition describes better the superconductor-normal metal junction. The Ginzburg–Landau equations together with the Maxwell one and boundary conditions represent a complete set of equations governing the behavior of a superconductor in the magnetic field near the superconducting transition temperature.

Let us consider homogeneous state of superconductor $\nabla\psi = 0$ in zero magnetic field $\boldsymbol{A} = 0$. Two solutions of the Ginzburg–Landau are possible. One corresponds to the normal state $\psi = 0$ and is possible at all temperatures. The second, $\psi = \psi_0$ with $\psi_0^2 = -\alpha(T)/\beta$, exists only at $T \leqslant T_c$ and corresponds to the superconducting state. The behavior of functional F_s as a function of order parameter is shown in Fig. 6.6. Nonzero value of order parameter $\psi = \psi_0$ at $T \leqslant T_c$ corresponds to the minimum of the functional per unit volume

$$F_s = \alpha\psi_0^2 + \beta\psi_0^4/2 = -\alpha^2(T)/(2\beta).$$

The transition to the superconducting state is second-order one since $\psi = 0$ and $\psi \sim (T_c - T)^{1/2}$ at the transition point.

The quantity having the sense of the *condensation energy* $\alpha^2(T)/(2\beta)$ is customary to express in the units of magnetic field energy according to

$$\frac{H_c^2}{8\pi} = \frac{\alpha^2(T)}{2\beta} \quad \text{where} \quad H_c^2(T) = 4\pi\frac{\alpha^2(T)}{\beta}.$$

The magnetic field $H_c(T)$ is called the *thermodynamic critical field* and $H_c(T) \sim (T_c - T)$ near T_c. The physical meaning of the thermodynamic critical field becomes clear with comparing the energies of superconducting and normal state in the magnetic field.

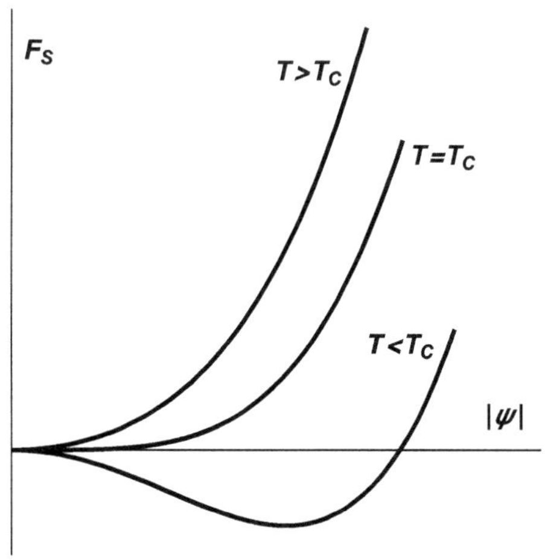

Fig. 6.6 The behavior of the energy free functional as a function of order parameter at various temperatures

In the external magnetic field induced with the constant currents in magnetic coils, it is necessary to handle with the thermodynamic potentials dependent on the magnetic field strength $\boldsymbol{H}$ rather than the potentials dependent on magnetic induction $\boldsymbol{B}$ determined as an averaged magnetic field. The relationship between the Gibbs $\tilde{F}(\boldsymbol{H})$ and Helmholtz $F(\boldsymbol{B})$ free energy potentials is governed with the Legendre transformation and we have for their specific densities

$$\tilde{F}(T, \boldsymbol{H}) = F(T, \boldsymbol{B}) - \frac{\boldsymbol{H}\boldsymbol{B}}{4\pi}.$$

In the normal state one has $\psi = 0$ and $\boldsymbol{B} = \boldsymbol{H}$. Thus,

$$\tilde{F}_n(T, \boldsymbol{H}) = -\frac{\boldsymbol{H}\boldsymbol{B}}{4\pi} = -\frac{H^2}{4\pi}.$$

For the superconducting state, we find provided that the magnetic field is completely expelled from the superconductor volume

$$\tilde{F}_s(T, \boldsymbol{H}) = -\frac{\alpha^2(T)}{2\beta} - 0 = -\frac{H_c^2(T)}{8\pi}.$$

Next, we must compare two values of the potentials and choose the smallest one. Therefore, if $H < H_c(T)$, the superconducting Meissner state with $\psi = \psi_0$ and $\boldsymbol{B} = 0$ is energetically more favorable.[14] On the contrary, if $H > H_c(T)$, the normal state with $\psi = 0$ and $\boldsymbol{B} = \boldsymbol{H}$ is energetically more favorable. The phase transition at $H = H_c(T)$ should be first-order one since the order parameter changes jump-like.

The occurrence of superconductivity, as the magnetic field decreases, can be imagined in a different way without jump-like expulsion of magnetic field when the superconducting phase nuclei appear spontaneously at some magnetic field. Since in this case the order parameter ψ should be small, we may restrict ourselves with the linearized Ginzburg–Landau equation

$$\frac{\gamma}{2}\left(-i\hbar\nabla - \frac{2e}{c}\boldsymbol{A}(\boldsymbol{r})\right)^2 \psi(\boldsymbol{r}) + \alpha\psi(\boldsymbol{r}) = 0.$$

Let magnetic field of strength H be directed along the z-axis and we take the Landau gauge $\boldsymbol{A} = (0, Hx, 0)$. Then we rewrite the Ginzburg–Landau equation as follows:

$$\frac{\gamma}{2}\left(-i\hbar\nabla - \frac{2e}{c}\boldsymbol{A}(\boldsymbol{r})\right)^2 \psi(\boldsymbol{r}) = -\alpha\psi(\boldsymbol{r}).$$

[14] In other words, the magnetic field expulsion occurs because the superconducting state has the smaller energy as compared with the normal state at the same temperature. If the superconducting state of a metal would have the same energy as the normal one, the magnetic field expulsion would be unfavorable.

This equation looks like the Schrödinger equation for the motion of charged particle with mass $m^* = 1/\gamma$ and charge $e^* = 2e$. The quantity $-\alpha$ plays a role of particle energy E. The properties of this equation are well known from the quantum mechanics. The solutions which vanish at the infinity take place at the discrete energy levels or *Landau levels* equal to

$$E = \hbar\omega\big(n + 1/2\big), \quad \omega = \frac{e^* H}{m^* c} = \frac{2e\gamma}{c} H, \quad n = 0,\ 1\,, 2, \ldots$$

The energy levels degenerate in the orbit center position of an electron. The energy minimum corresponds to the lowest level $n = 0$. Then,

$$\frac{\hbar}{2}\frac{2e\gamma}{c} H_{c2} = -\alpha(T).$$

The magnetic field

$$H_{c2}(T) = |\alpha(T)| \frac{c}{\hbar e\gamma}, \quad H_{c2}(T) \sim T_c - T,$$

at which the spontaneous nucleation of superconducting phase becomes possible, is referred to as the *upper critical field*. Due to homogeneity of a superconductor, the nucleus of superconducting phase may appear at any point of the superconductor.

It is interesting to compare the upper critical field H_{c2} with the thermodynamic critical field H_c. For this purpose, we consider a ratio of these two fields

$$\frac{H_{c2}(T)}{H_c(T)} = \frac{|\alpha| c}{\hbar e\gamma} \frac{\sqrt{\beta}}{\sqrt{4\pi}\,|\alpha|} = \frac{c\sqrt{\beta}}{\hbar e\gamma\sqrt{8\pi}} \sqrt{2} = \varkappa\sqrt{2}.$$

The dimensionless parameter $\varkappa$ introduced above is called the *Ginzburg–Landau parameter*. Since this parameter is temperature-independent, it can serve as a characteristic for superconductors with respect to its behavior in the magnetic field.

Since the Ginzburg–Landau parameter is dimensionless, it can be represented as a ratio of two lengths. The magnetic field penetration depth $\delta(T)$ into superconductor can be selected as one of the lengths. As for the second one, we will take the *coherence length* $\xi(T)$describing the typical length at which the superconducting parameter changes in the lack of magnetic field. In fact, we have

$$\frac{\gamma}{2}(-i\hbar\nabla)^2\psi + \alpha\psi = 0 \quad \text{or} \quad \xi^2\nabla^2\psi + \psi = 0$$

where the coherence length is defined according to

$$\xi(T) = \left(\frac{\hbar^2\gamma}{2|\alpha(T)|}\right)^{1/2} \sim \big(T_c - T\big)^{-1/2}.$$

One can readily see that $\varkappa = \delta/\xi$ as well

$$\frac{\delta}{\xi} = \left(\frac{c^2}{16\pi\gamma e^2\psi_0^2}\right)^{1/2} \frac{\sqrt{2|\alpha|}}{\hbar\gamma^{1/2}} = \frac{c}{e\sqrt{16\pi\gamma}} \frac{\sqrt{\beta}}{\sqrt{|\alpha|}} \frac{\sqrt{2|\alpha|}}{\hbar\gamma^{1/2}} = \frac{c\sqrt{\beta}}{\hbar e\gamma\sqrt{8\pi}} = \varkappa.$$

The upper critical field H_{c2} can be expressed in terms of coherence length $\xi(T)$ and *magnetic flux quantum* $\phi_0 = \pi c e/\hbar$

$$H_{c2} = |\alpha(T)|\frac{c}{\hbar e\gamma} = \frac{1}{2\pi}\frac{2|\alpha(T)|}{\hbar^2\gamma}\frac{\pi c\hbar}{e} = \frac{\phi_0}{2\pi\xi^2}.$$

This formula means that the magnetic flux equal to the quantum one ϕ_0 passes through the region of area $2\pi\xi^2$, the length ξ being the typical size of superconducting nucleus.

As is noted above, a superconducting nucleus can appear at any place in the superconductor. For the superconducting nucleus emerging at an arbitrary[15] point x_0, we can take the solution of the linearized Ginzburg–Landau equation in the Landau gauge $\boldsymbol{A} = (0, Hx, 0)$ as

$$\psi \sim \exp\left(\frac{ix_0}{\xi^2}y\right)\exp\left(-\frac{(x-x_0)^2}{2\xi^2}\right).$$

This solution corresponds to the lowest energy level $n = 0$ at $H = H_{c2}$. Due to spatial homogeneity of a superconductor we can assume that the general solution for the order parameter is some regular periodical structure (vortex lattice), i.e. linear combination of the solutions centered at the regular intervals

$$\psi(x, y) = \sum_{k=-\infty}^{\infty} C_k e^{ik\frac{x_0}{\xi^2}y} e^{-\frac{(x-kx_0)^2}{2\xi^2}} \quad (k \text{ is an integer}).$$

To determine the spatial structure $\psi(\boldsymbol{r})$ emerging in the magnetic fields somewhat lower than H_{c2}, it is necessary to involve nonlinear term $|\psi|^2\psi$ in the Ginzburg–Landau equation. The full analysis gives energetically most favorable solution having the symmetry of triangular vortex lattice with the spatial period of about ξ.

Depending on the magnitude of the Ginzburg–Landau parameter $\varkappa$, larger or smaller than $1/\sqrt{2}$ ($H_{c2} > H_c$ or $H_{c2} < H_c$), we should expect two various behaviors different in kind for a superconductor in the magnetic field. Accordingly, we can separate the superconductors into two types, namely *type-I superconductors* with $H_{c2} < H_c$ and *type-II ones* with $H_{c2} > H_c$.

Thus, the normal-superconducting state transition in the type-I superconductors is jump-like at field $H = H_c$ with the complete expulsion of magnetic field from the

---

[15] Usually, describing the electron motion in the magnetic field, point x_0 is referred to as the electron orbit center and the quantity $p_y = \hbar k_y = \hbar x_0/\xi^2$ is said as a y-component of electron momentum.

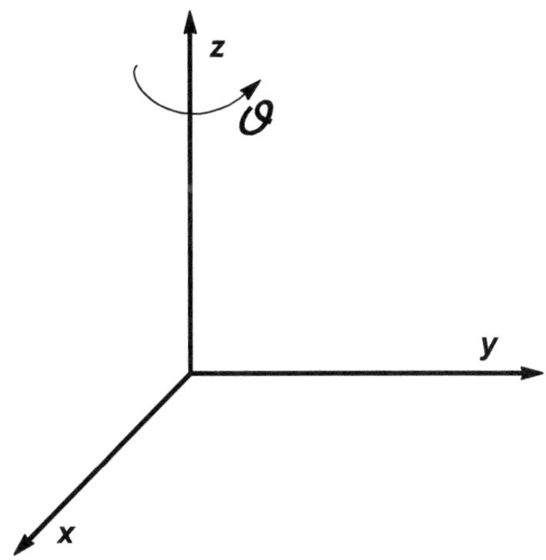

Fig. 6.7 The system of coordinates

superconductor bulk. In the type-II superconductor, the transition to the superconducting state occurs via gradual expulsion of normal phase with the superconducting nuclei appearing in the various regions of the superconductor. Therefore, we can expect that the Meissner effect will be incomplete and the magnetic field expulsion from the bulk will be smooth within some interval of magnetic field.

As we have seen above, the normal state of superconductor becomes absolutely unstable with respect to appearing the superconducting phase nuclei as the magnetic field decreases to the magnitude of upper critical field $H_{c2}(T)$. For the type-I superconductors, this means that the intermediate range of magnetic fields $H_{c2} < H < H_c$ alone can exhibit the metastable states[16] in the normal state.

The similar question can be raised here. What is the magnitude of magnetic field when the superconducting state becomes unstable against spontaneous nucleation of the normal phase in the Meissner state with the full magnetic field expulsion from the superconductor bulk? We start to clarify this question by analyzing the state of the superconductor in which the order parameter phase changes by an integer multiple of 2π along some closed path

$$\psi(\boldsymbol{r}) = |\psi|e^{i\varphi(\boldsymbol{r})}, \quad \varphi(\boldsymbol{r}) = n\ \arctan(y/x) \quad \text{where} \quad n \text{ is an integer.}$$

Here $\arctan(y/x) = \vartheta$ is the azimuthal angle in the cylindrical frame (ρ, ϑ, z) (Fig. 6.7). The z-axis, corresponding to radius $\rho = 0$, is a singular line called the *vortex line*[17] or *vortex filament*. The phase gradient, equal to

$$\nabla\varphi = \left(-n\frac{y}{\rho^2}, n\frac{x}{\rho^2}, 0\right), \quad |\nabla\varphi| = \frac{n}{\rho} \quad \text{and} \quad \rho = \sqrt{x^2 + y^2},$$

---

[16] Here we can trace the similar analogy with an existence of the metastable state region and spinodal in other first-order phase transitions, e.g. gas-liquid transition.

[17] For brevity, vortex as well.

results in emerging the circulating superconducting current around the vortex line

$$j = \frac{c}{4\pi\delta^2}\frac{\hbar c}{2e}\frac{n}{\rho}$$

where δ is the magnetic field penetration depth. Since the superconducting current vanishes on the contour infinitely far away from the vortex line, the total magnetic flux carried by a single vortex is quantized

$$\int \boldsymbol{h}\, dx\, dy = \oint \boldsymbol{A}\, d\boldsymbol{l} = \oint \frac{\hbar c}{2e}\nabla\varphi\, d\boldsymbol{l} = \frac{\hbar c}{2e}2\pi n = n\phi_0 \,.$$

Let us consider vortex line from the hydrodynamic point of view as a singular line in the velocity distribution under potential flow of liquid. Using the relation of current $\boldsymbol{j} = en_s\boldsymbol{v}_s$ with the superconducting velocity $\boldsymbol{v}_s$ and the penetration depth $\delta^{-2} = 4\pi e^2 n_s/mc^2$ with the density of superconducting electrons n_s, we have

$$v_s = \frac{\hbar}{2m}\frac{n}{\rho}, \quad \text{also} \quad \boldsymbol{v}_s = \frac{\hbar}{2m}\nabla\varphi.$$

Hence it is seen that the vortex line is characterized with the definite value of velocity circulation Γ around the closed contour L encircling the vortex filament. The circulation of velocity $\boldsymbol{v}_s$ equals

$$\Gamma = \oint_L \boldsymbol{v}_s\, d\boldsymbol{l} = 2\pi\varkappa \quad \text{where} \quad \varkappa = n\hbar/m.$$

This value is independent of choosing the integration contour. If we multiply the velocity circulation Γ by two electron masses $2m$, we get the quantity to be subjected to quantization in quantum mechanics. One may say that the vortex line in the superconductor is an analog of vortex filaments known in the hydrodynamics of fluids. The velocity $\boldsymbol{v}_s$ has a singularity at the vortex line. Since the superconducting current $\boldsymbol{j}_s$ and velocity $\boldsymbol{v}_s$ increase unlimitedly near the vortex line, the description given above for the immediate region within the vicinity of vortex line requires more detailed approach.

To satisfy the unambiguity condition directly at the vortex line $\rho = 0$, the order parameter must exactly be zero $\psi(\boldsymbol{\rho} = 0) = 0$, corresponding to the value of order parameter in the normal state. The quantized vortex is an example of *topological defect*.

In the general case, the quantitative description of vortex line is a complicated mathematical problem. However, such description simplifies essentially for the type-II superconductors with the large Ginzburg–Landau parameter $\varkappa \gg 1$. In this case the central region of vortex or *vortex core* represents approximately the normal phase region, the order parameter changing from zero value in the core center to the equilibrium value ψ_0 outside the core (Fig. 6.8). The typical scale of the core

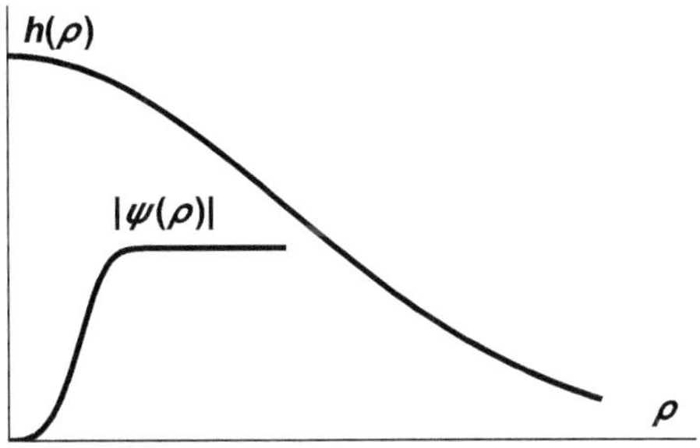

Fig. 6.8 The behavior of order parameter $\psi(\rho)$ and magnetic field $h(\rho)$. The vortex core size ξ is much smaller than the distance δ at which the magnetic field decays

radius is about the coherence length ξ. The magnetic field of vortex $\boldsymbol{h}(\rho)$ (Fig. 6.8) changes on the typical scale of about penetration depth δ much larger as compared with ξ in our case. Therefore, in first approximation we believe that the magnetic flux passes mainly across the region outside the core where $|\psi| \approx \psi_0$. In this case we can employ the following equations:

$$\operatorname{rot} \boldsymbol{h} = \frac{4\pi}{c}\boldsymbol{j} \quad \text{and} \quad \boldsymbol{j} = \frac{c}{4\pi\delta^2}\left(\frac{\phi_0}{2\pi}\nabla\varphi - \boldsymbol{A}\right).$$

Here δ is the penetration depth corresponding to the order parameter magnitude $|\psi_0|$. Applying *curl* to the both sides of the first equation, we obtain the equation for the magnetic field $\boldsymbol{h}(\rho)$ induced with the vortex

$$\boldsymbol{h} + \delta^2 \operatorname{curl}\operatorname{curl} \boldsymbol{h} = \boldsymbol{h} - \delta^2\nabla^2\boldsymbol{h} = n\phi_0\boldsymbol{e}_z\delta_2(\boldsymbol{\rho}) \quad (\rho \gtrsim \xi).$$

Here $\boldsymbol{e}_z$ is unit vector specifying the direction of vortex line and n is the number of magnetic flux quanta ϕ_0. The solution of this equation is expressed in terms of the modified Bessel function of the second kind $K_0(x)$

$$\boldsymbol{h}(\rho) = \boldsymbol{e}_z\frac{n\phi_0}{2\pi\delta^2}K_0(\rho/\delta) = \boldsymbol{e}_z\frac{n\phi_0}{2\pi\delta^2}\begin{cases}\ln\delta/\rho, & \delta \gtrsim \rho \gtrsim \xi,\\ \sqrt{\frac{\pi\delta}{2\rho}}\exp(-\rho/\delta), & \rho \gtrsim \delta.\end{cases}$$

The magnetic field $h(0)$ in the vortex center remains finite, achieving the maximum magnitude about

$$h(0) \approx n\frac{\phi_0}{2\pi\delta^2}\ln\frac{\delta}{\xi}.$$

The magnetic field decays exponentially rapid for the distances exceeding the penetration depth δ.

The nucleation of a vortex is associated with an additional energy. Calculating the energy of a vortex at $\varkappa \gg 1$, we can neglect the condensation energy of superconducting phase in the vortex core. Then the main contribution to the vortex energy comes from the region outside the vortex core $\rho \gtrsim \xi$ and represents a sum of energies of magnetic field and superconducting current. The vortex line energy per unit line, i.e. *line tension*, is approximately given by the formula

$$\epsilon_n = \int\limits_{\rho \gtrsim \xi} d^2\rho \left\{ \frac{\boldsymbol{h}^2}{8\pi} + \frac{\delta^2}{8\pi}\left(\frac{4\pi}{c}\boldsymbol{j}\right)^2 \right\} \approx n^2 \frac{\phi_0^2}{(4\pi\delta)^2} \ln\frac{\delta}{\xi}, \quad \delta \gg \xi.$$

The contribution from the core region $\rho \lesssim \xi$ to the vortex energy is about $(H_c^2/8\pi)\pi\xi^2 \sim \phi_0^2/(4\pi\delta^2)$, i.e. it is approximately by a factor $\ln \varkappa \gg 1$ as less. Accordingly, this contribution is insignificant within the logarithmic accuracy. Since $\epsilon_n \sim n^2$ and $n^2 \geqslant |n|$, the state of $|n|$ vortices each with the flux quantum ϕ_0 is energetically more favorable than the state with the single vortex having $|n|$ quanta of magnetic flux.

In the external magnetic field the formation of a vortex, i.e. normal phase nucleus, can be energetically more favorable.[18] In fact, in this case we should deal with the Gibbs free energy thermodynamic potential:

$$\tilde{\mathcal{F}}(T, \boldsymbol{h}) = \mathcal{F}(T, \boldsymbol{B}) - \int\limits_V \frac{\boldsymbol{hB}}{4\pi} dV$$

where the magnetic field strength $\boldsymbol{h}$ is an independent quantity unlike magnetic induction $\boldsymbol{B}$. Here V is the volume of the superconductor with the vortex line of length L. The presence of vortex line gives the contribution to the energy equal to $L\epsilon_n$. Thus, as compared with the Meissner state characterized by the magnetic induction $\boldsymbol{B} = 0$, we find the following variation of Gibbs free energy thermodynamic potential

$$\delta\tilde{\mathcal{F}}(T, \boldsymbol{h}) = L\epsilon_n - \int\limits_V \frac{\boldsymbol{hB}}{4\pi} dV.$$

We have according to definition of the magnetic induction as a mean quantity of magnetic field strength $\boldsymbol{h}$ averaged over the volume of the superconductor

$$\boldsymbol{B}V = \int\limits_V \boldsymbol{h}\, dV = L \int \boldsymbol{h}\, d^2\rho = n\phi_0 L.$$

[18] In the presence of magnetic field it is necessary to take into account the energy for the magnetization of the vortex in the magnetic field. The energy equals $-\boldsymbol{Mh}$ where $\boldsymbol{M}$ is the total magnetic moment of the vortex. Using relation $\boldsymbol{j} = c \operatorname{curl} \boldsymbol{m}$ where $\boldsymbol{m}$ is the magnetic moment density and employing the Maxwell equation $\operatorname{curl}(\boldsymbol{h} - 4\pi\boldsymbol{m}) = 0$, we find the following formula for the total magnetic moment per unit length of vortex

$$4\pi M = 4\pi \iint m\, d^2\rho = \iint h d^2\rho = n\phi_0.$$

The magnetic field when the vortex nucleation becomes energetically favorable should be found from the condition $\epsilon_n - Mh < 0$. The threshold field equals $h = 4\pi\epsilon_n/(n\phi_0)$ is minimal at $n = 1$.

Then

$$\delta\tilde{\mathcal{F}} = L\left(\epsilon_n - \frac{h}{4\pi}n\phi_0\right)$$

and the formation of vortex or normal phase nucleus becomes favorable if the magnetic field exceeds the threshold magnitude $H_n = 4\pi\epsilon_n/(n\phi_0)$. This magnitude is minimal for the vortex with one flux quantum $n = 1$. The corresponding critical field equal to

$$H_{c1} = \frac{4\pi\epsilon_1}{\phi_0}$$

is called the *lower critical field* and determines the upper boundary for the Meissner state in the superconductor. For $\varkappa \gg 1$, we have the following estimate within the logarithmic accuracy:

$$H_{c1} \approx \frac{\phi_0}{4\pi\delta^2}\ln\varkappa.$$

To relate the field H_{c1} with the thermodynamic critical field H_c and upper critical field H_{c2}, we have the formulas

$$H_{c1} = H_c\frac{\ln\varkappa}{\varkappa\sqrt{2}} = H_{c2}\frac{\ln\varkappa}{2\varkappa^2} \quad (\varkappa \gg 1).$$

As the temperature approaches the critical one, the lower critical field decreases according to $H_{c1}(T) \sim (T_c - T)$.

In type-II superconductors one has $H_{c1} < H_c < H_{c2}$ (Fig. 6.9). As the external magnetic field exceeds the critical magnitude H_{c1} and grows further, the number of vortices grows and the magnetic field penetration into the superconductor bulk increases as well. The Meissner effect becomes incomplete and the state of the superconductor proves to be *mixed* or *vortex* one in the intermediate $H_{c1} < H < H_{c2}$ range of magnetic fields. The thermodynamic critical field H_c turns out to be unremarkable for the physical properties in type-II superconductors. The density of vortex line number N is connected with the magnetic induction by the formula $B = N\phi_0$. The mean distance between vortex line depends on the magnetic field approximately as

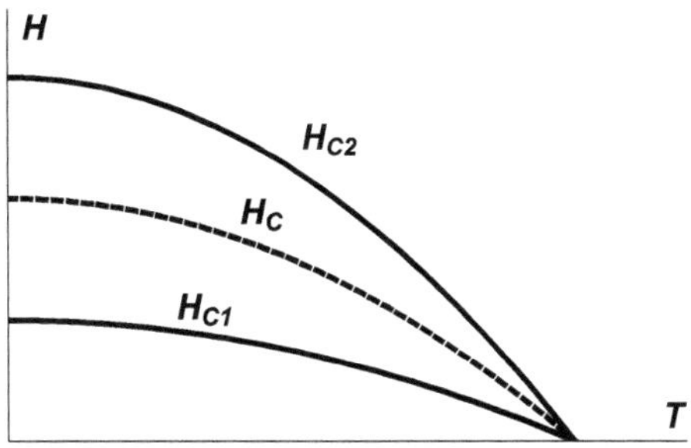

Fig. 6.9 The temperature behavior of critical fields in the type-II superconductor

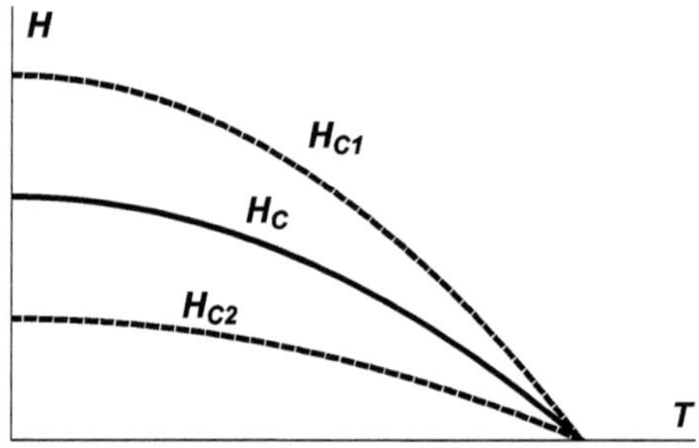

Fig. 6.10 The temperature behavior of critical fields in the type-I superconductor. The dashed lines show the boundaries of possible metastability for the Meissner and normal states

$(\phi_0/B)^{1/2}$. The vanishing of superconductivity at H_{c2} is a second-order phase transition. The mean distance between the vortex lines becomes about the core size $\xi(T)$ and the cores representing the normal phase regions merge practically with each other. In the mixed state, the vortices arrange the regular periodical structure called the *Abrikosov vortex lattice* or *flux line lattice*.This is usually a triangular lattice.

In Fig. 6.10, the diagram is given for a type-I superconductor in the magnetic field-temperature variables. Since the superconducting-normal phase transition in field H_c is a first-order one, the range between the fields H_{c2} and H_{c1} can be considered as a region of metastable phases, respectively, normal and superconducting states. In other words, it is possible to overcool the normal phase below the curve H_c at least to the field H_{c2} being the line of absolute instability for the normal line. Also, it is impossible, in principle, to overheat the superconducting phase in the Meissner state above the magnitude H_{c1}.

In conclusion, we derive the Maxwell equal area rule in its application to the normal-superconducting phase transition. Let we have two phases, normal and superconducting, in the same magnetic field of strength $\boldsymbol{H}$. The corresponding magnetic inductions $\boldsymbol{B}_n$ and $\boldsymbol{B}_s$ are determined with the derivatives of the Gibbs free energy potentials $\tilde{F}_n(\boldsymbol{H})$ and $\tilde{F}_s(\boldsymbol{H})$ with respect to the magnetic field $\boldsymbol{H}$

$$\frac{\boldsymbol{B}_n}{4\pi} = -\frac{\partial \tilde{F}_n(\boldsymbol{H})}{\partial \boldsymbol{H}} \quad \text{and} \quad \frac{\boldsymbol{B}_s}{4\pi} = -\frac{\partial \tilde{F}_s(\boldsymbol{H})}{\partial \boldsymbol{H}}.$$

Next, we note that the magnetic induction $\boldsymbol{B}_n$ in the normal phase equals the magnetic field strength $\boldsymbol{H}$ provided that the normal phase has no magnetic properties. For the superconducting phase, we have relation $\boldsymbol{B}_s = \boldsymbol{H} + 4\pi \boldsymbol{M}_s(\boldsymbol{H})$ where $\boldsymbol{M}_s$ is the magnetization of the superconductor. In other words,

$$\frac{\boldsymbol{B}_s - \boldsymbol{H}}{4\pi} = \boldsymbol{M}_s(\boldsymbol{H}) = \frac{\partial}{\partial \boldsymbol{H}}\big[\tilde{F}_n(\boldsymbol{H}) - \tilde{F}_s(\boldsymbol{H})\big].$$

Integrating this equation in the limits between zero and infinite magnetic fields, we obtain

$$\int_0^\infty \boldsymbol{M}_s(\boldsymbol{H}) \cdot d\boldsymbol{H} = \big[\tilde{F}_n(\boldsymbol{H}) - \tilde{F}_s(\boldsymbol{H})\big]_0^\infty = \tilde{F}_s(0) - \tilde{F}_n(0).$$

Here we have involved that the superconducting state breaks down in sufficiently large magnetic field and, therefore, the Gibbs free energy potentials $\tilde{F}_s$ and $\tilde{F}_n$ are the same. In zero magnetic field, according to the definition of the thermodynamic critical field H_c, we have $\tilde{F}_s(0) - \tilde{F}_n(0) = -H_c^2/8\pi$. So, we arrive at the simple integral equality for the magnetization in the superconducting phase

$$\int_0^{\infty} \boldsymbol{M}_s(\boldsymbol{H}) \cdot d\boldsymbol{H} = -\frac{H_c^2}{8\pi}. \tag{6.1}$$

For type-II and type-I superconductors, the graphic examples of the Maxwell equal area rule are given in Figs. 6.11 and 6.12. An elementary straight-line approximation of the magnetization curve within the magnetic field interval H_{c1} and H_{c2} can readily give us a simple estimate for the relation between the magnitudes H_{c1} and H_{c2} as $H_{c1}H_{c2} = H_c^2$. Obviously, this relation supposes the location of thermodynamic critical field H_c between H_{c1} and H_{c2}.

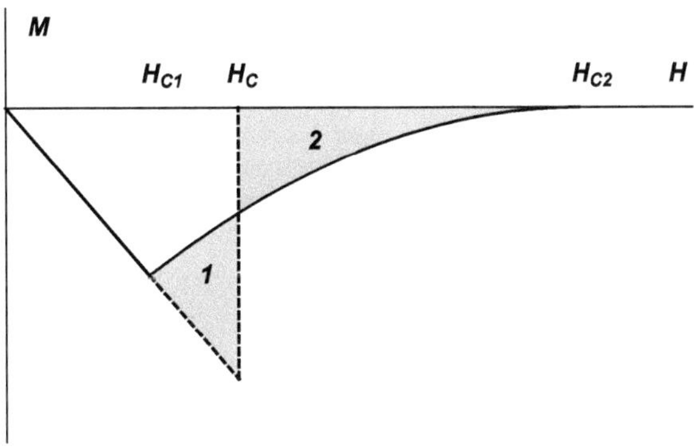

Fig. 6.11 The schematic of magnetization M versus magnetic field H for a type-II superconductor. The shaded segments 1 and 2 have the equal areas. The Meissner state is described with line $M = -H/4\pi$ and magnetic susceptibility $\chi = -1/4\pi$ or permeability $\mu = 0$

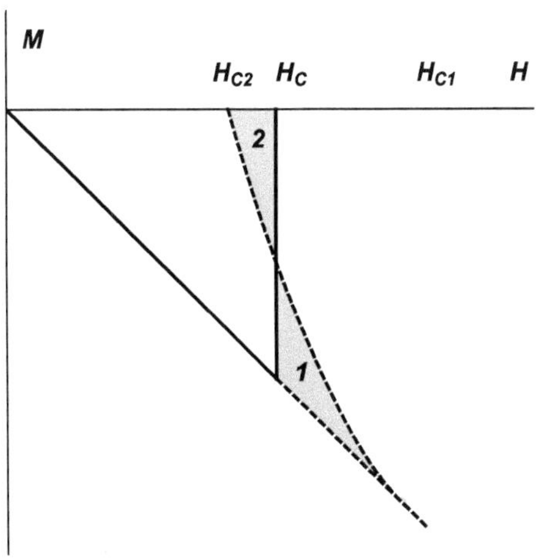

Fig. 6.12 The schematic of magnetization M versus magnetic field H for a type-I superconductor. The shaded segments 1 and 2 have the equal areas. The Meissner state is described with the line $M = -H/4\pi$. The dashed line, connecting the fields H_{c1} and H_{c2}, corresponds to the thermodynamically unstable state associated with magnetic susceptibility $\chi < -1/4\pi$ or permeability $\mu < 0$

Problems

1. Write the Ginzburg–Landau functional and corresponding equations in the dimensionless units.

Solution. Let us denote F as a density of the functional according to $\mathcal{F} = \int F\, d^3r$ and go over to new units

$$\psi' = \frac{\psi}{\psi_0}, \quad \boldsymbol{h}' = \frac{\boldsymbol{h}}{H_c\sqrt{2}}, \quad \boldsymbol{r}' = \frac{\boldsymbol{r}}{\delta}, \quad \text{and} \quad \boldsymbol{A}' = \frac{\boldsymbol{A}}{H_c\sqrt{2}\,\delta}.$$

The density of functional F and Ginzburg–Landau equations read (we omit the dashes in new variables)

$$F = \frac{H_c^2}{4\pi}\int d^3r\left[\frac{1}{2}\left|\left(\frac{-i\nabla}{\varkappa} - \boldsymbol{A}\right)\psi\right|^2 - |\psi|^2 + \frac{|\psi|^4}{2} + (\operatorname{rot}\boldsymbol{A})^2\right],$$

$$\left(-i\frac{\nabla}{\varkappa} - \boldsymbol{A}\right)^2\psi - \psi + |\psi|^2\psi = 0,$$

$$\operatorname{curl}\operatorname{curl}\boldsymbol{A} = -\frac{i}{\varkappa}(\psi^*\nabla\psi - \psi\nabla\psi^*) - \boldsymbol{A}|\psi|^2.$$

Hence it is seen that the physical properties of a superconductor in the Ginzburg–Landau theory are completely determined with the single temperature-independent parameter $\varkappa$.

2. Write the analog of the Clausius–Clapeyron equation for the superconductor-normal metal phase transition in the magnetic field in the case of type-I superconductor.

Solution. A role of transition pressure is played with the critical field $H_c(T)$ at which the first-order phase transition takes place. Instead of volume the magnetization M is a variable conjugated to the magnetic field. The magnetization changes jump-like at the transition point from magnitude $M_s = -H_c/4\pi$ in the superconducting state (since magnetic induction $B = H + 4\pi M = 0$ and $M = -H/4\pi$) to magnitude $M_n = 0$ in the normal state. Accordingly we arrive at the relation

$$\frac{dH_c}{dT} = \frac{L(T)}{T(M_s - M_n)} = -4\pi\frac{L(T)}{TH_c(T)},$$

$L(T)$ being the latent heat of transition. We find for the latent heat

$$L(T) = -\frac{T}{4\pi}H_c\frac{dH_c}{dT} = -T\frac{d}{dT}\left(\frac{H_c^2(T)}{8\pi}\right).$$

Since $L > 0$, the heat absorbs at the transition from the superconducting to the normal state. Writing $L(T)$ as $T(S_n - S_s)$, we find the difference in the entropies for the phases in equilibrium

$$S_n(T) - S_s(T) = -\frac{1}{4\pi}H_c\frac{dH_c}{dT}.$$

If the transition occurs in zero magnetic field, latent heat will be $L = 0$ since the critical field $H_c = 0$ at $T = T_c$ and derivative dH_c/dT remains finite. Therefore, the superconducting transition is of second-order one in zero magnetic field. As a result, the specific heat has a discontinuity at the transition point (*Rutger's formula*):

$$C_n - C_s = T\frac{d}{dT}(S_n - S_c)\big|_{T=T_c} = -\frac{T}{4\pi}\left(\frac{dH_c}{dT}\right)^2\bigg|_{T=T_c}.$$

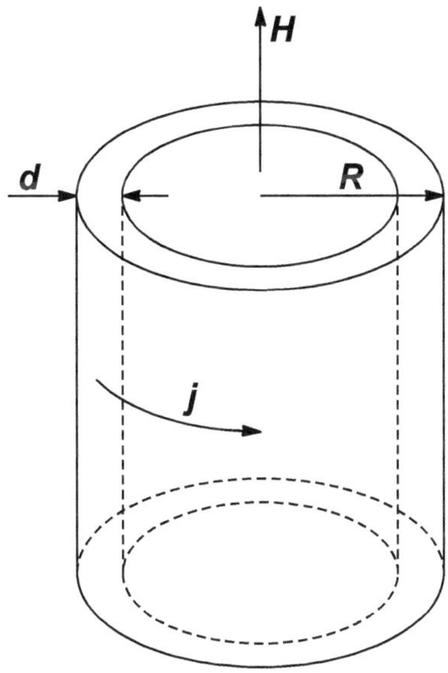

Fig. 6.13 The hollow superconducting cylinder of radius R and wall thickness d in the magnetic field H

6.7 The Little–Parks Effect

Let us consider superconducting hollow thin-wall cylinder of radius R and thickness of wall $d \ll R$ (Fig. 6.13). The homogeneous magnetic field H is directed along the cylinder axis. Below we study the behavior of the superconducting transition temperature as a function of magnetic field.

To simplify our description, we put the wall thickness d to be small as compared with the coherence length ξ and magnetic field penetration depth δ. This inequality allows us in first approximation to treat the modulus of order parameter as constant [19] inside the superconductor, i.e. $|\psi| = \text{const}$. For $\delta \ll R$ and $d \ll R$, it is also possible to neglect any variation of vector potential $\boldsymbol{A}$ between the cylinder walls and make no distinction in the strength of magnetic field inside and outside the cylinder.

Let us choose vector potential $\boldsymbol{A}$ in the gauge having only the azimuthal coordinate A_ϑ in the cylindrical coordinates (ρ, ϑ, z). Then we substitute the order parameter as

$$\psi = |\psi| e^{i\varphi(\boldsymbol{r})}$$

into the expression for the thermodynamic potential density of superconductor

$$F_s = \frac{\gamma}{2}\left|\left(-i\hbar\nabla - \frac{2e}{c}\boldsymbol{A}\right)\psi\right|^2 + \alpha|\psi|^2 + \frac{\beta}{2}|\psi|^4 + \frac{h^2}{8\pi}\,.$$

[19] This approximation satisfies the boundary condition at the superconductor surface $\boldsymbol{n}\big(-i\hbar\nabla - 2e\boldsymbol{A}/c\big)\psi = 0$ as well.

Analyzing the gradient term yields

$$F_s = \frac{\gamma\hbar^2}{2}\left(\nabla\varphi - \frac{2\pi}{\phi_0}\boldsymbol{A}\right)^2|\psi|^2 + \alpha|\psi|^2 + \frac{\beta}{2}|\psi|^4 + \frac{h^2}{8\pi}.$$

Then, using that $\nabla\varphi$ and $\boldsymbol{A}$ are constant over the cross-section of the cylinder, we can represent $\nabla\varphi - (2\pi/\phi_0)\boldsymbol{A}$ as an average for the integral along the circumference normal to the cylinder axis

$$\nabla\varphi - \frac{2\pi}{\phi_0}\boldsymbol{A} = \frac{1}{2\pi R}\oint\left(\nabla\varphi - \frac{2\pi}{\phi_0}\boldsymbol{A}\right)d\boldsymbol{l} = \\ = \frac{1}{2\pi R}\left([\Delta\varphi] - \frac{2\pi}{\phi_0}\Phi\right) = \frac{1}{R}\left(n - \frac{\Phi}{\phi_0}\right).$$

Here $[\Delta\varphi] = 2\pi n$ is a phase increment of order parameter along the circle round the cylinder axis and n is an arbitrary integer providing us an unambiguity of the order parameter at the same point in the space. The magnetic flux across the cylinder cross-section equals $\Phi = \pi R^2 H$ and ϕ_0 is the magnetic flux quantum. We find finally

$$F_s = \frac{\gamma\hbar^2}{2R^2}\left(n - \frac{\Phi}{\phi_0}\right)^2|\psi|^2 + \alpha|\psi|^2 + \frac{\beta}{2}|\psi|^4 + \frac{h^2}{8\pi}.$$

The requirement for the minimum of thermodynamic potential gives the condition for choosing the integer n from the following inequality:

$$n - \frac{1}{2} < \frac{\Phi}{\phi_0} < n + \frac{1}{2}.$$

The following equation determines the critical temperature T_c of superconducting transition as a function of magnetic flux Φ:

$$\alpha(T_c) + \frac{\gamma\hbar^2}{2R^2}\left(n - \frac{\Phi}{\phi_0}\right)^2 = 0, \quad \alpha(T_c) = \alpha_0(T_c - T_{c0}).$$

The relative variation of critical temperature equals

$$\frac{T_c - T_{c0}}{T_{c0}} = -\frac{\gamma\hbar^2}{2R^2\alpha_0 T_{c0}}\left(n - \frac{\Phi}{\phi_0}\right)^2 \sim -\frac{\xi^2(T=0)}{R^2}\left(n - \frac{\Phi}{\phi_0}\right)^2.$$

The behavior $\Delta T_c/T_{c0}$ as a function of flux Φ consists of the periodically reiterated parabolic arcs (Fig. 6.14) with period ϕ_0 and maximum ratio:

$$\frac{T_c - T_{c0}}{T_{c0}} = -\frac{\gamma\hbar^2}{8R^2\alpha_0 T_{c0}} \sim -\frac{\xi^2(T=0)}{4R^2}.$$

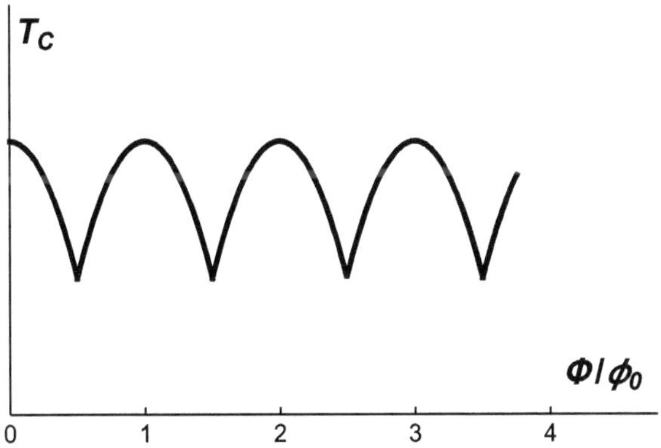

Fig. 6.14 The critical temperature T_c versus magnetic flux Φ

To conclude, the *Little–Parks effect* is a periodical variation of superconducting transition temperature as a function of the magnetic flux threading across the hollow cylinder. Note that, due to small wall thickness $d \ll \delta$ of cylinder, the superconductor cannot seize the magnetic flux, entailing no quantization of magnetic flux. The experimental observation of the Little–Parks effect has demonstrated the validness of the BCS theory conclusions on the Cooper pairing of electrons and the requirement of the gauge invariance for the vector potential.

6.8 Critical Current in a Thin Plate

Within the framework of the Ginzburg–Landau functional we study below the behavior of critical current in the thin plate of thickness $d \ll \xi$ and δ, the thickness being smaller than both coherence length ξ and magnetic field penetration depth δ. Let the current of density $\boldsymbol{j}_s$ flow along the plate and be directed along the z-axis, as is shown in Fig. 6.15. The x-axis is normal to the plate.

In a first approximation the condition $d \ll \xi$ provides us the constancy of superconducting order parameter ψ over the cross-section of the plate. This condition also agrees with the boundary condition $d\psi/dx = 0$ on the both sides of the plate. Otherwise, we have $d\psi/dx \sim \psi/d$ and contribution of gradient term $(\nabla\psi)^2$ to the Ginzburg–Landau functional becomes too large and energetically unfavorable.

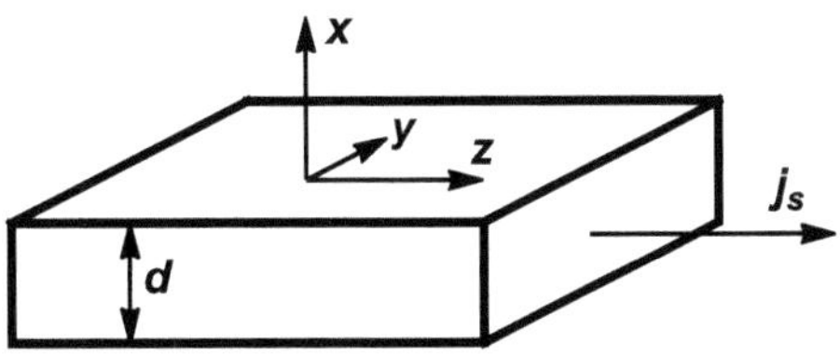

Fig. 6.15 The plate geometry

The second condition $d \ll \delta$ allows us to neglect a possible variation of magnetic field $\boldsymbol{h}$ and vector potential $\boldsymbol{A}$ over the cross-section of the plate. The relative magnitude of the corrections has a smallness of about $(d/\delta)^2 \ll 1$. In our approximation the vector-potential equals

$$\boldsymbol{A} = \frac{c\boldsymbol{j}_s}{4\gamma e^2|\psi|^2}\,.$$

The magnetic field proves to be negligibly small: $\boldsymbol{h} = \operatorname{curl}\boldsymbol{A} \approx 0$.

Correspondingly, we get for the density of the Ginzburg–Landau free energy

$$F_s[\psi] = \frac{1}{8\gamma e^2}\frac{j_s^2}{|\psi|^2} + \alpha|\psi|^2 + \frac{\beta}{2}|\psi|^4.$$

Minimizing the free energy density F_s with respect to $|\psi|^2$ results in the following equation:

$$-\frac{1}{8\gamma e^2}\frac{j_s^2}{|\psi|^4} + \alpha + \beta|\psi|^2 = 0.$$

Putting $|\psi| = \psi_0 f$ where $\psi_0^2 = -\alpha/\beta$ is the equilibrium magnitude of the superconducting order parameter, we obtain

$$j_s = j_c\frac{3\sqrt{3}}{2}f^2(1-f^2)^{1/2}, \quad j_c = \frac{4}{3\sqrt{3}}\frac{e\gamma\hbar}{\xi(T)} \sim (T_c - T)^{3/2}.$$

Here the quantity j_c means the *critical density* of superconducting current. Let us analyze behavior of current j_s as a function of parameter f^2 (Fig. 6.16). We see that, as the current density j_s increases from zero value, the parameter f^2 decreases from value $f^2 = 1$ to $f^2 = 2/3$ at which the superconducting current density reaches the maximum magnitude j_c. For $j_s > j_c$, there are no nonzero values f^2 and the plate will be in the normal and nonsuperconducting state. Therefore, the quantity j_c is the maximum magnitude of the current density compatible with the superconducting state. At the transition to the normal state when $j_s = j_c$, the superconducting order parameter jumps from $\sqrt{2/3}\,\psi_0$ to zero value.

The physical meaning of critical density of superconducting current can be comprehended from the relation below

$$j_c = \frac{1}{3\sqrt{6}}\frac{cH_c(T)}{\pi\delta_0(T)} = en_s(T)v_c(T).$$

Here $H_c(T)$ is the thermodynamic critical field, $\delta_0(T)$ is the penetration depth at the order parameter ψ_0, and $n_s(T)$ is the density of superconducting electrons. The quantity with the dimension of velocity

$$v_c(T) = \frac{1}{3\sqrt{3}}\frac{\hbar v_F}{p_F\xi(T)},$$

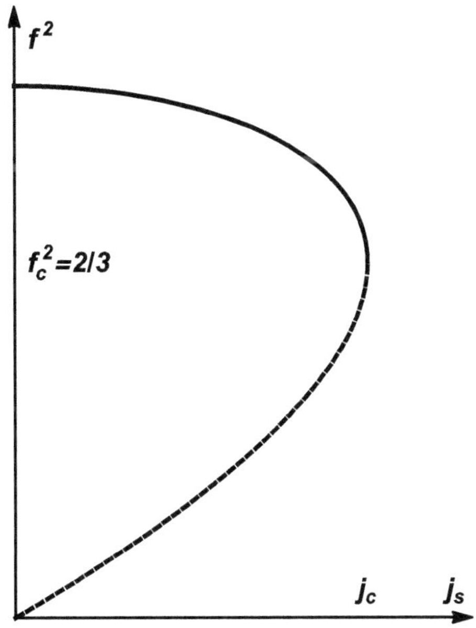

Fig. 6.16 The square of superconducting order parameter versus the current density (the variables are normalized). At $j_s > j_c$ the order parameter jumps to zero value. The dashed line shows the region of non-physical behavior

where v_F and p_F are the Fermi velocity and momentum, has a meaning of the critical velocity for elementary excitations or quasiparticles when the breakdown of Cooper pairs occurs.

In this sense, the critical velocity is proportional to the superconducting gap $\Delta(T)$ and $v_c \sim \Delta(T)/p_F$.

Chapter 7
Weakly Non-ideal Bose Gas

7.1 The Gross–Pitaevskii Equation

The laser cooling technique exploiting the Doppler effect[1] and manipulation with the alkali gases of Rb, Na, Li in magneto-optical traps after evaporative cooling stage allow one to achieve extremely low temperatures as hundreds of nanokelvin. In spite of the limited number of atoms in a trap[2] and low particle density, the realization of ultralow temperatures in the traps delivers a possibility for the experimental observation and study of the *Bose–Einstein condensation phenomenon*. From the theoretical point of view, the low particle density in the traps makes it possible to describe the properties of the Bose–Einstein condensate within the framework of weakly non-ideal Bose gas model. Below we consider the properties of weakly non-ideal condensed Bose gas of zero spin particles, i.e. the gas with small energy of interparticle interaction.

The basis for the theory of weakly non-ideal condensed Bose gas is an assumption that almost all the particles are in the condensate, i.e. in the ground state with zero momentum. The number of excited overcondensate particles is small, $N_{p\neq 0} = N - N_p \ll N$. Under conditions of dilute and condensed Bose gas, the pair collisions between particles of small momenta play a key role in the interparticle interaction. In first approximation due to low velocities of colliding particles we can restrict ourselves only with the contribution from the s-scattering[3] of particles. The diluteness of a gas and involvement of pair particle collisions alone suppose the fulfillment of inequality $na^3 \ll 1$ where n is the particle concentration. In other words, the mean

[1] The Doppler cooling process relies on absorbing a photon by an atom with the next spontaneous re-emission of a photon with the frequency larger than that of the initially absorbed photon, thereby lowering the average kinetic energy of atoms in the trap.

[2] Usually, within 10^4– 10^6 particles.

[3] If the s-scattering amplitude of slow particles is a, the p-scattering amplitude will be by a factor $(pa/\hbar)^2 \ll 1$ as smaller, magnitude p being the typical momentum of the particles. The typical scattering lengths of alkali atoms are about 100 a_B where a_B is the Bohr radius.

S. N. Burmistrov, *Statistical and Condensed Matter Physics*,
https://doi.org/10.1007/978-981-97-9072-2_7

spacing between particles is much larger as compared with the scattering length, i.e. $n^{-1/3} \gg a$. Under such approximations, the matrix element of two-particle coupling $g_{\boldsymbol{pp}'}$ can be replaced with its value at zero momenta $g_{00} = g = \text{const}$. In the coordinate representation this corresponds to the point-like interaction between particles as $U(\boldsymbol{r} - \boldsymbol{r}') = g\delta(\boldsymbol{r} - \boldsymbol{r}')$.

The coupling constant g is related with the s-scattering length a by means of $g = 4\pi\hbar^2 a/m$. The coupling between the Bose particles is assumed to be *repulsive*, i.e. $g > 0$. For the interparticle attraction $g < 0$, the state of weakly non-ideal gas proves to be thermodynamically unstable[4] with producing a denser state than the gas one. The temperature of the gas is always supposed to be much lower as compared with the Bose–Einstein condensation temperature.

So, we write down the effective Hamiltonian for the system of N spinless bosons of mass m in the external potential $V(\boldsymbol{r})$ of a trap

$$\hat{H} = \sum_{i=1}^{N}\left(\frac{\hat{\boldsymbol{p}}_i^2}{2m} + V(\boldsymbol{r}_i)\right) + \frac{1}{2}\sum_{i\neq j}^{N} U(\boldsymbol{r}_i - \boldsymbol{r}_j)$$

where $\boldsymbol{p}_i$ is the momentum operator for ith particle and $U(\boldsymbol{r}_i - \boldsymbol{r}_j) = g\delta(\boldsymbol{r}_i - \boldsymbol{r}_j)$ is the interaction operator for two particles i and j. In first approximation[5] the many-particle wave function $\Psi(\boldsymbol{r}_1, \boldsymbol{r}_2, \dots, \boldsymbol{r}_N)$ in the system of weakly interacting N spinless bosons can be represented as a symmetrized product of one-particle wave functions. In the Bose–Einstein condensed state, all the bosons are supposed in the same state with certain one-particle wave function $\phi(\boldsymbol{r})$. Therefore, the wave function $\Psi(\boldsymbol{r}_1, \boldsymbol{r}_2, \dots, \boldsymbol{r}_N)$ can approximately be written as

$$\Psi(\boldsymbol{r}_1, \boldsymbol{r}_2, \dots, \boldsymbol{r}_N) \approx \prod_{i=1}^{N} \phi(\boldsymbol{r}_i)$$

where one-particle function $\phi(\boldsymbol{r})$ is normalized as usual $\int d\boldsymbol{r}\, |\phi(\boldsymbol{r})|^2 = 1$. Let us substitute wave function $\Psi(\boldsymbol{r}_1, \boldsymbol{r}_2, \dots, \boldsymbol{r}_N)$ into the integral

$$E = \int \Psi^*(\boldsymbol{r}_1, \boldsymbol{r}_2, \dots, \boldsymbol{r}_N)\hat{H}\Psi(\boldsymbol{r}_1, \boldsymbol{r}_2, \dots, \boldsymbol{r}_N) d\boldsymbol{r}_1\, d\boldsymbol{r}_2 \dots d\boldsymbol{r}_N$$

and calculate the energy of the system with wave function Ψ

[4] The similar condensed picture in the gas with attraction can be realized only for the metastable long-living state provided that the gas is spatially confined with a trap and the number of gas particles does not exceed some critical value.

[5] This is also the Hartree–Fock approximation or self-consistent field approximation.

$$E[\Psi] = \sum_{i=1}^{N} \int d\boldsymbol{r}_i \left(\frac{\hbar^2}{2m} |\nabla\phi(\boldsymbol{r}_i)|^2 + \phi^*(\boldsymbol{r}_i)V(\boldsymbol{r}_i)\phi(\boldsymbol{r}_i) \right) +$$

$$+\frac{1}{2}\sum_{i\neq j}^{N} \int \phi^*(\boldsymbol{r}_i)\phi^*(\boldsymbol{r}_j)U(\boldsymbol{r}_i - \boldsymbol{r}_j)\phi(\boldsymbol{r}_j)\phi(\boldsymbol{r}_i)d\boldsymbol{r}_i\, d\boldsymbol{r}_j =$$

$$= \int d\boldsymbol{r} \left[N \left(\frac{\hbar^2}{2m} |\nabla\phi(\boldsymbol{r})|^2 + V(\boldsymbol{r})|\phi(\boldsymbol{r})|^2 \right) + \frac{N(N-1)}{2} g|\phi(\boldsymbol{r})|^4 \right].$$

For the physical clarity, it is convenient to change the normalization of function $\phi(\boldsymbol{r})$ and, instead, introduce function $\psi(\boldsymbol{r})$ according to $\psi(\boldsymbol{r}) = \sqrt{N}\phi(\boldsymbol{r})$. The function $\psi(\boldsymbol{r})$ will be called the *condensate wave function*. We should not ascribe the literal meaning to such terminology commonly used. The condensate wave function in its strict meaning plays a role of order parameter and it should not also be confused with the genuine many-particle wave function of the bosonic system.

So, using the inequality $N \gg 1$ and putting $N(N-1) \approx N^2$, we rewrite the expression for the energy of the system as the following functional:

$$E[\psi(\boldsymbol{r}), \psi^*(\boldsymbol{r})] = \int d\boldsymbol{r} \left[\frac{\hbar^2}{2m} |\nabla\psi(\boldsymbol{r})|^2 + V(\boldsymbol{r})|\psi(\boldsymbol{r})|^2 + \frac{g}{2}|\psi(\boldsymbol{r})|^4 \right],$$

$$\int d\boldsymbol{r}\, |\psi(\boldsymbol{r})|^2 = N.$$

As we can see from the last formula, the quantity $|\psi(\boldsymbol{r})|^2 = n(\boldsymbol{r})$ can be treated as a condensate density.

To find the optimum behavior of condensate wave function $\psi(\boldsymbol{r})$, we must minimize the energy functional[6] under conserving the total number of particles N. After introducing the Lagrange multiplier μ as a chemical potential of the gas, it is necessary to minimize the functional $E - \mu N$ with respect to ψ and ψ^*. The condition $\delta(E - \mu N)/\delta\psi^* = 0$ yields[7] the *stationary Gross–Pitaevskii equation*

$$-\frac{\hbar^2}{2m}\nabla^2\psi(\boldsymbol{r}) + V(\boldsymbol{r})\psi(\boldsymbol{r}) + g|\psi(\boldsymbol{r})|^2\psi(\boldsymbol{r}) = \mu\psi(\boldsymbol{r}),$$

governing the spatial behavior of condensate wave function. The Gross–Pitaevskii equation resembles the nonlinear Schrödinger one.

The Gross–Pitaevskii equation gives the following magnitude of chemical potential:

$$\mu = g|\psi(\boldsymbol{r})|^2 = gn$$

[6] Note that, on substituting $-i\hbar\nabla \to -i\hbar\nabla - (2e/c)\boldsymbol{A}$, the functional goes over to the Ginzburg–Landau one.

[7] The conjugate condition $(\delta E - \mu N)/\delta\psi = 0$ entails the conjugate equation.

for the homogeneous state of condensed Bose gas occupying volume V in the absence of external field $V(\boldsymbol{r}) = 0$. This corresponds completely to relation $\mu = \partial E/\partial N$ where

$$E = \frac{N(N-1)}{2} gV \approx g\frac{N^2}{2V}.$$

The pressure of condensed gas equals

$$P = -\frac{\partial E}{\partial V} = \frac{gn^2}{2} = \frac{g\rho^2}{2m^2}$$

where $\rho = mn$ is the mass density of a gas. The sound velocity, determined by derivative $c = (\partial P/\partial \rho)^{1/2}$, reads

$$c = \frac{\sqrt{g\rho}}{m} = \sqrt{\frac{gn}{m}}.$$

For the thermodynamic stability of homogeneous state, it is necessary $c^2 > 0$, i.e. repulsion $g > 0$.

The healing length[8] ξ describes the spatial scale over which the condensate wave function varies and equals

$$\xi = \frac{\hbar}{2\sqrt{mgn}} = \frac{\hbar}{2\sqrt{m\mu}}.$$

This estimate results from comparing the interaction energy gn with the kinetic energy of about $\hbar^2/(2m\xi^2)$ per one particle.

The approximation neglecting the kinetic term in the Gross–Pitaevskii equation is referred to as the *Thomas–Fermi approximation*. Then the distribution of condensate density is described with the simple equation

$$n(\boldsymbol{r}) = |\psi(\boldsymbol{r})|^2 = \frac{\mu - V(\boldsymbol{r})}{g}.$$

The spatial boundaries of the condensate are determined from condition $V(\boldsymbol{r}) = \mu$.

The condition of smallness for the typical energy gn of interparticle interaction as compared with the Bose–Einstein condensate temperature $T_c \sim \hbar^2 n^{2/3}/m$ can be interpreted in terms of small gas density by introducing the *gas parameter* na^3. The smallness of gas parameter $na^3 \ll 1$ means that the particle-particle scattering length a is significantly smaller than the mean distance between the particles $n^{-1/3}$. The volume which radius is about the healing length ξ contains a large number of particles or $n\xi^3 \gg 1$.

[8] The coherence length in the condensed matter literature.

Problems

1. Find the spatial behavior of the condensate near an impenetrable wall.

Solution. Let the wall be at the plane $x = 0$ and the condensate occupy the half-space $x > 0$. The condensate wave function $\psi(x)$ can be taken as real and it satisfies the equation

$$-\frac{\hbar^2}{2m}\psi''(x) + g\psi^3(x) = \mu\psi(x)$$

with the boundary conditions $\psi(x = 0) = 0$ and $\psi(x = \infty) = \psi_0 = (\mu/g)^{1/2}$. Multiplying the both sides of the equation by $\psi'(x)$, we find the first integral

$$-\frac{\hbar^2}{2m}\psi'^2(x) + \frac{g}{2}\psi^4(x) - \mu\psi^2(x) = -\frac{\mu}{2}\psi_0^2.$$

The solution of the equation results in the wave function

$$\psi(x) = \psi_0 \tanh(x/2\xi), \quad \xi = \hbar/(2\sqrt{m\mu})$$

which approaches its bulk value over the distance about the healing length ξ.

2. Using the Thomas–Fermi approximation, find the condensate cloud size R, chemical potential μ, and energy E for the gas of N bosons with mass m in the symmetrical harmonic trap with frequency ω.

Solution. Neglecting the gradient term in the Gross–Pitaevskii equation, we have for the condensate density $n(r) = |\psi(r)|^2$

$$n(r) = \frac{1}{g}\left(\mu - \frac{m\omega^2 r^2}{2}\right) = \frac{m\omega^2}{2g}(R^2 - r^2), \quad \mu = \frac{m\omega^2 R^2}{2}.$$

Here R is the size of condensate cloud. The total particle number N in the trap equals

$$N = \int n(r)4\pi r^2 dr = \frac{m\omega^2}{2g}4\pi\int_0^R (R^2 - r^2)r^2 dr = \frac{4\pi}{15}\frac{m\omega^2}{g}R^5.$$

Hence we find the cloud radius and chemical potential

$$R = \left(\frac{15gN}{4\pi m\omega^2}\right)^{1/5} = l\left(\frac{15aN}{l}\right)^{1/5}, \quad g = \frac{4\pi\hbar^2 a}{m};$$

$$\mu = \frac{m\omega^2}{2}\left(\frac{15gN}{4\pi m\omega^2}\right)^{2/5} = \frac{\hbar\omega}{2}\left(\frac{15aN}{l}\right)^{2/5}, \quad l = \left(\frac{\hbar}{m\omega}\right)^{1/2}.$$

Here a is the scattering length of particles and l is the oscillatory length of the trap. Using that $\mu = \partial E/\partial N$ and $\mu \sim N^{2/5}$, we obtain the total energy of the gas

$$E = \frac{5}{7}\mu N \sim N^{7/5}.$$

The validity of the Thomas–Fermi approximation implies the fulfillment of the inequality

$$\frac{\hbar^2}{2m}\frac{\psi}{R^2} \ll \mu\psi$$

leading to the following inequality for the total particle number $N \gg l/a$ in the trap. The inequality does not contradict the necessary requirement $N \gg 1$.

3. The Bose star, consisting of ultralight bosons of mass m, is considered as an array of self-gravitating Bose–Einstein condensate. Assuming the validity of the Gross–Pitaevskii equation and the description of the condensate with the aid of single wave function $\psi(\boldsymbol{r})$, estimate the radius for the star of mass M, using the following spherically symmetrical variation[9] function:

$$\psi(\boldsymbol{r}) = A\exp\left(-\frac{r^2}{2R^2}\right), \quad A = \left(\frac{M}{\pi^{3/2}R^3 m}\right)^{1/2}$$

with variational parameter R.

Solution. Let us write the total star energy with condensate wave function ψ as a sum of kinetic energy E_{kin}, interparticle interaction energy E_{int}, energy of the condensate in the gravitational field E_{gr}, and the gravitational field energy E_{f}:

$$E = E_{\text{kin}} + E_{\text{int}} + E_{\text{gr}} + E_{\text{f}} = \int d^3r\left(\frac{\hbar^2}{2m}|\nabla\psi|^2 + \frac{g}{2}|\psi|^4 + \rho(\boldsymbol{r})\Phi(\boldsymbol{r}) + \frac{(\nabla\Phi)^2}{8\pi G}\right).$$

Here g is the coupling constant between particles, $\rho(\boldsymbol{r}) = m|\psi(\boldsymbol{r})|^2$ is the mass density of condensate, $\Phi(\boldsymbol{r})$ is the gravitational potential, and G is the gravitational constant. The search of the energy minimum by varying over $\psi(\boldsymbol{r})$ and $\Phi(\boldsymbol{r})$ under constancy of total number of particles results in the Gross–Pitaevskii equation

$$\left[-\frac{\hbar^2}{2m}\nabla^2 + g|\psi(\boldsymbol{r})|^2 + m\Phi(\boldsymbol{r})\right]\psi(\boldsymbol{r}) = \mu\psi(\boldsymbol{r})$$

and in the Poisson equation determining the gravitational field

$$\nabla^2\Phi = 4\pi G\rho(\boldsymbol{r}).$$

The straightforward calculation for the kinetic and interaction energies gives the following answers:

$$E_{\text{kin}} = \frac{3}{4\pi}\frac{\hbar^2}{mR^2}\frac{M}{m} \quad \text{and} \quad E_{\text{int}} = \frac{\sqrt{2}}{8\pi^{3/2}}\frac{g}{R^3}\frac{M^2}{m^2}.$$

The calculation of gravitational energy E_{gr} is readily performed by means of the Fourier transformation. In fact,

$$\int d^3r\,\rho(\boldsymbol{r})\Phi(\boldsymbol{r}) = \int\frac{d^3k_1}{(2\pi)^3}\frac{d^3k_2}{(2\pi)^3}\rho(\boldsymbol{k}_1)\Phi(\boldsymbol{k}_2)e^{i(\boldsymbol{k}_1+\boldsymbol{k}_2)\boldsymbol{r}} = \int\frac{d^3k}{(2\pi)^3}\rho(-\boldsymbol{k})\Phi(\boldsymbol{k}),$$

and we use the simple relation between the Fourier transforms for the gravitational potential $\Phi(\boldsymbol{k})$ and condensate density $\rho(\boldsymbol{k})$

$$\Phi(\boldsymbol{k}) = -\frac{4\pi G}{k^2}\rho(\boldsymbol{k}).$$

This relation is a consequence from the Poisson equation. Then the simple calculation gives

$$E_{\text{gr}} = -4\pi G\int\frac{d^3k}{(2\pi)^3}\frac{\rho(\boldsymbol{k})\rho(-\boldsymbol{k})}{k^2} = -\sqrt{\frac{2}{\pi}}\frac{GM^2}{R} \quad \text{at} \quad \rho(\boldsymbol{k}) = Me^{-k^2R^2/4}.$$

The energy of gravitational field is calculated in the analogous manner. As is expected, we have

$$E_{\text{f}} = -\frac{1}{2}E_{\text{gr}}.$$

[9] The spherically symmetrical localized distribution is often called the axionic solution.

Finally, we find the total energy

$$E(R) = \frac{3}{4\pi}\frac{\hbar^2}{mR^2}\frac{M}{m} + \frac{\sqrt{2}}{8\pi^{3/2}}\frac{g}{R^3}\frac{M^2}{m^2} - \sqrt{\frac{1}{2\pi}}\frac{GM^2}{R}.$$

The energy minimum is delivered with the radius R_0 equal to

$$R_0 = \frac{3}{4}\sqrt{\frac{2}{\pi}}\frac{\hbar^2}{Gm^2M}\left(1+\sqrt{1+\frac{\pi}{3}g\frac{Gm^2M^2}{\hbar^4}}\right).$$

For clarity, we rewrite the answer in terms of boson Compton length $\lambda = \hbar/mc$, Planck mass $M_p = (\hbar c/G)^{1/2}$ where c is the light velocity and a is the scattering length according to $g = 4\pi\hbar^2 a/m$

$$R_0 = \frac{3}{4}\sqrt{\frac{2}{\pi}}\lambda\frac{M_p^2}{mM}\left(1+\sqrt{1+\frac{4\pi^2}{3}\frac{a}{\lambda}\frac{M^2}{M_p^2}}\right).$$

For $M \gg M_p$, we reach the limiting behavior independent of the star mass

$$R_0 \approx \sqrt{\frac{3\pi}{2}}\sqrt{a\lambda}\frac{M_p}{m}.$$

4. Find the stability condition for a dilute gas mixture of bosons and spin-polarized fermions of spin 1/2 and mass m_f at zero temperature. The constants of boson-boson and fermion-boson coupling are equal to g and λ, respectively.

Solution. Since the fermions are completely spin-polarized, the s-scattering of fermions with the same spin directions vanishes and the coupling between the fermions with small momenta can be neglected in first approximation. For the gas approximation at $T = 0$, we can put that all bosons are in the ground state with zero momentum. The total energy for the gas mixture of volume V is a sum of kinetic energy of fermions E_{kin} and energy E_{int} of boson-boson and boson-fermion interactions

$$\mathcal{E} = \mathcal{E}_{\text{kin}} + \mathcal{E}_{\text{int}} = \int d\boldsymbol{r}\, E = \int d\boldsymbol{r}\left(\frac{3}{5}\varepsilon_f n_f + \frac{1}{2}g n_b n_b + \lambda n_b n_f\right).$$

Here $\varepsilon_F = p_F^2/2m_f$ is the Fermi energy, $p_F = \hbar(6\pi^2 n_f)^{1/3}$ is the Fermi momentum, n_f and n_b are the fermion and boson densities, respectively. The chemical potentials of fermions and bosons will be equal to the derivatives of total energy $\mathcal{E}$ with respect to the particle numbers $N_f = \int d\boldsymbol{r}\, n_f$ and $N_b = \int d\boldsymbol{r}\, n_b$

$$\mu_f = \partial\mathcal{E}/\partial N_f = \varepsilon_f + \lambda n_b \quad \text{and} \quad \mu_b = \partial\mathcal{E}/\partial N_b = g n_b + \lambda n_f\,.$$

The necessary condition for the stability of mixture against fluctuations of densities $n_f \to n_f + \delta n_f$ and $n_b \to n_b + \delta n_b$ under the fixed particle numbers N_f and N_b requires the positiveness of second-order energy variation with respect to the density fluctuations

$$\delta^2\mathcal{E} = \int d\boldsymbol{r}\left(\frac{1}{2}\frac{\partial^2 E}{\partial n_b^2}(\delta n_b)^2 + \frac{\partial^2 E}{\partial n_b \partial n_f}(\delta n_b)(\delta n_f) + \frac{1}{2}\frac{\partial^2 E}{\partial n_f^2}(\delta n_f)^2\right) > 0.$$

The energy variation over the density fluctuations vanishes since the number of bosons and fermions is fixed.

In order to have the quadratic form positive definite, we must put the following conditions for its coefficients:

$$\begin{vmatrix} \partial^2 E/\partial n_b^2 & \partial^2 E/\partial n_b \partial n_f \\ \partial^2 E/\partial n_f \partial n_b & \partial^2 E/\partial n_f^2 \end{vmatrix} > 0 \quad \text{and} \quad \partial^2 E/\partial n_b^2 > 0.$$

Hence, $g > 0$ and $g(10\varepsilon_F/9n_f) - \lambda^2 > 0$. The density of fermions should not exceed the critical value

$$n_f < n_{f,\,\mathrm{cr}} = \frac{4\pi^2}{3}\left(\frac{\hbar^2 g}{\lambda^2 m_f}\right)^3.$$

If the inverse inequality is valid, the homogeneous state of the Fermi-Bose gas mixture becomes unstable and the mixture separates into two phases.

5. Find the stability condition for the binary gas mixture of spin-polarized fermions of masses m_1 and m_2 at zero temperature. The coupling constant between the different fermions equals g.

Solution. Since the fermions are completely polarized, the s-scattering between identical fermions vanishes and the coupling between the same fermions can be neglected. The total energy for the binary mixture of fermions is a sum of kinetic energies of fermions and interaction energy

$$\mathcal{E} = \int E\, d\boldsymbol{r} = \int d\boldsymbol{r}\left(\frac{3}{5}n_1\varepsilon_{F1} + \frac{3}{5}n_2\varepsilon_{F2} + g n_1 n_2\right).$$

Here n_1 and n_2 are the densities of mixture components and $\varepsilon_{F1}, \varepsilon_{F2}$ are the corresponding Fermi energies.

The necessary condition for the mixture to be stable against the density fluctuations $n_1 \to n_1 + \delta n_1$ and $n_2 \to n_2 + \delta n_2$ under fixed particle numbers N_1 and N_2 is a positive energy variation in second order with respect to density fluctuations

$$\delta^2\mathcal{E} = \int d\boldsymbol{r}\left(\frac{1}{2}\frac{\partial^2 E}{\partial n_1^2}(\delta n_1)^2 + \frac{\partial^2 E}{\partial n_1 \partial n_2}(\delta n_1)(\delta n_2) + \frac{1}{2}\frac{\partial^2 E}{\partial n_2^2}(\delta n_2)^2\right) > 0.$$

The first-order energy variation over the density fluctuations vanishes since the number of fermions of each kind is fixed. The derivative $\partial^2 E/\partial n_i^2$ is connected with the density of states at the Fermi surface

$$N(\varepsilon_{Fi}) = \frac{m_i p_{Fi}}{2\pi^2\hbar^3}, \quad \varepsilon_{Fi} = \frac{p_{Fi}^2}{2m_i} \quad (i = 1,\, 2)$$

according to $\partial^2 E/\partial n_i = 1/N(\varepsilon_{Fi})$.

The quadratic form of two variables is positive definite if its coefficients satisfy the following condition:

$$\begin{vmatrix} N^{-1}(\varepsilon_{F1}) & g \\ g & N^{-1}(\varepsilon_{F2}) \end{vmatrix} > 0 \quad \text{or} \quad g^2 N(\varepsilon_{F1}) N(\varepsilon_{F2}) < 1.$$

Note that the condition, obtained for the stability of binary fermion mixture against its separation, is independent of the sign of coupling constant g, i.e. it is the same both for attraction and repulsion between different fermions.

7.2 Dynamics of the Bose–Einstein Condensate

To study the dynamics of the Bose–Einstein condensate, it is necessary to generalize the Gross–Pitaevskii equation for the non-stationary case. For this purpose, let us write the equation for the time evolution of annihilation operator $\hat{\Psi}(\boldsymbol{r}, t)$ in the Heisenberg representation

$$i\hbar\frac{\partial}{\partial t}\hat{\Psi}(\boldsymbol{r}, t) = \left[\hat{H}, \hat{\Psi}(\boldsymbol{r}, t)\right] = \left[-\frac{\hbar^2\nabla^2}{2m} + V(\boldsymbol{r}) + g\hat{\Psi}^{+}(\boldsymbol{r}, t)\hat{\Psi}(\boldsymbol{r}, t)\right]\hat{\Psi}(\boldsymbol{r}, t)$$

where $\hat{H}$ is the effective Hamiltonian of weakly non-ideal Bose gas. First, we take the following arguments into account. The gas is sufficiently rarified, effective inter-particle interaction is small enough, and almost all the particles are in the condensate. Thus, we replace operator $\hat{\Psi}(\boldsymbol{r}, t)$ with the condensate wave function $\psi(\boldsymbol{r}, t)$ and arrive at the *non-stationary Gross–Pitaevskii equation*

$$i\hbar\frac{\partial}{\partial t}\psi(\boldsymbol{r}, t) = \left[-\frac{\hbar^2\nabla^2}{2m} + V(\boldsymbol{r}) + g\psi^*(\boldsymbol{r}, t)\psi(\boldsymbol{r}, t)\right]\psi(\boldsymbol{r}, t).$$

As an additional reason for justifying the non-stationary Gross–Pitaevskii equation, we can apply the principle of least action as

$$\delta\int_{t_i}^{t_f} L\,dt = 0,$$

considering the following Lagrangian $L[\psi^*, \psi]$:

$$L[\psi^*), \psi] = \int i\hbar\psi^*(\boldsymbol{r}, t)\frac{\partial\psi(\boldsymbol{r}, t)}{\partial t}d\boldsymbol{r} - E[\psi, \psi^*].$$

Varying over $\delta\psi^*(\boldsymbol{r}, t)$ yields the same equation

$$i\hbar\frac{\partial\psi(\boldsymbol{r}, t)}{\partial t} = \frac{\delta E}{\delta\psi^*(\boldsymbol{r}, t)} = \left[-\frac{\hbar^2\nabla^2}{2m} + V(\boldsymbol{r}) + g|\psi(\boldsymbol{r}, t)|^2\right]\psi(\boldsymbol{r}, t).$$

To have the non-stationary equation consistent with the time-independent one, the condensate wave function $\psi(\boldsymbol{r}, t)$ must evolve in time as $\exp(-i\mu t/\hbar)$ where μ is the chemical potential. The resulting phase multiplier can be explained as follows. In fact, the wave function represents the matrix element of operator $\hat{\Psi}(\boldsymbol{r}, t)$ between the ground states with N particles and $N - 1$ particles

$$\psi(\boldsymbol{r}, t) = \langle N - 1|\hat{\Psi}|N\rangle \sim e^{-i(E_N - E_{N-1})t/\hbar}$$

since the time evolution of states $|N - 1\rangle$ and $|N\rangle$ is proportional to $e^{-iE_{N-1}t/\hbar}$ and $e^{-iE_N t/\hbar}$, respectively. The difference $E_N - E_{N-1}$ is just the chemical potential μ.

Let us derive several hydrodynamic consequences from the non-stationary Gross–Pitaevskii equation. We have for ψ and ψ^* in the time-independent external potential $V(\boldsymbol{r})$

$$\begin{aligned} i\hbar\frac{\partial\psi}{\partial t} &= -\frac{\hbar^2}{2m}\nabla\psi + V(\boldsymbol{r})\psi + g|\psi|^2\psi, \\ -i\hbar\frac{\partial\psi^*}{\partial t} &= -\frac{\hbar^2}{2m}\nabla\psi^* + V(\boldsymbol{r})\psi^* + g|\psi|^2\psi^*. \end{aligned}$$

Then we multiply the both sides of equations by ψ^* and ψ, respectively, and subtract one equation from the other

$$i\hbar\frac{\partial|\psi|^2}{\partial t} = -\frac{\hbar^2}{2m}\left(\psi^*\nabla^2\psi - \psi\nabla^2\psi^*\right).$$

Using the identity

$$\operatorname{div}\left(\psi\nabla\psi^* - \psi^*\nabla\psi\right) = \psi\nabla^2\psi^* - \psi^*\nabla^2\psi,$$

we arrive at the *continuity equation* for the gas density $\rho = m|\psi|^2$

$$\frac{\partial\rho}{\partial t} + \operatorname{div}\boldsymbol{j} = 0,$$

meaning the mass conservation law. Here $\boldsymbol{j}$ is the mass density flux determined by the following formula:

$$\boldsymbol{j} = -\frac{i\hbar}{2}\left(\psi^*\nabla\psi - \psi\nabla\psi^*\right) = \hbar|\psi|^2\nabla\varphi.$$

In the last equality, the quantity φ is the phase of condensate wave function $\psi(\boldsymbol{r}, t)$ in the agreement with the usual relation

$$\psi(\boldsymbol{r}, t) = |\psi(\boldsymbol{r}, t)|e^{i\varphi(\boldsymbol{r},t)} = \sqrt{n(\boldsymbol{r}, t)}e^{i\varphi(\boldsymbol{r},t)}.$$

Writing the vector of mass density flux as $\boldsymbol{j} = \rho\boldsymbol{v}_s$ ($\rho = mn$) and putting the relation

$$\boldsymbol{v}_s(\boldsymbol{r}, t) = \frac{\hbar}{m}\nabla\varphi(\boldsymbol{r}, t),$$

we determine the vector $\boldsymbol{v}_s$ which should be called the *velocity of condensate flow*. The velocity of condensate flow has a property of *potential flow* or *irrotational flow*

$$\operatorname{curl}\boldsymbol{v}_s = \frac{\hbar}{m}\operatorname{curl}\nabla\varphi = 0.$$

This property is a distinctive attribute of superfluid liquids.

The non-stationary Gross–Pitaevskii equation allows us to obtain the momentum conservation law in the following form:

$$\frac{\partial j_i}{\partial t} + \frac{\partial\Pi_{ik}}{\partial x_k} = -n\frac{\partial V(\boldsymbol{r})}{\partial x_i}$$

where the momentum flux density tensor Π_{ik} is given by the formula

$$\Pi_{ik} = P\delta_{ik} + \left(\frac{\hbar}{2m}\right)^2 \left(\frac{\partial\psi^*}{\partial x_i}\frac{\partial\psi}{\partial x_k} - \psi^*\frac{\partial^2\psi}{\partial x_i\partial x_k} + \frac{\partial\psi}{\partial x_i}\frac{\partial\psi^*}{\partial x_k} - \psi\frac{\partial^2\psi^*}{\partial x_i\partial x_k}\right).$$

Here $P = gn^2/2$ is the pressure in the condensate and this term with pressure expresses *Pascal's law*.

The equation for phase $\varphi(\boldsymbol{r}, t)$ of condensate wave function can be derived from the non-stationary Gross–Pitaevskii equation, using $\psi = \sqrt{n}\exp(i\varphi)$ and continuity equation $\partial n/\partial t + \operatorname{div}(n\boldsymbol{v}_s) = 0$ for the particle density $n(\boldsymbol{r}, t)$. So, we find for the gas in the external potential $V(\boldsymbol{r})$

$$-\hbar\frac{\partial\varphi}{\partial t} = \left(\frac{1}{2}m\boldsymbol{v}_s^2 + V(\boldsymbol{r}) + gn - \frac{\hbar^2\nabla^2\sqrt{n}}{2m\sqrt{n}}\right), \quad \boldsymbol{v}_s = \frac{\hbar}{m}\nabla\varphi.$$

The density gradient term proportional to $\hbar^2/2m$ is referred to as the *quantum pressure* and is responsible for the quantum effects in inhomogeneous Bose condensate. Both phase and continuity equations constitute a closed set of equations equivalent to the non-stationary Gross–Pitaevskii one. Note that the neglect of the quantum pressure term corresponds to the Thomas–Fermi approximation. For the spherically symmetric trap of frequency ω, this approximation is well applicable provided that the particle number N exceeds significantly a ratio of oscillatory length $l = \sqrt{\hbar/m\omega}$ to the scattering length a, i.e. $N \gg l/a$.

7.3 Quantized Vortex in a Weakly Non-ideal Bose Gas

As we have seen above, the velocity of the Bose condensate flow is a gradient of scalar $\boldsymbol{v}_s = (\hbar/m)\nabla\varphi$ and, respectively, curl $\boldsymbol{v}_s = 0$, until the phase $\varphi(\boldsymbol{r})$ of wave function $\psi(\boldsymbol{r})$ has no singularities. Let us imagine some closed contour C and consider the phase increment $\Delta\varphi$ around the contour. Should we require the single-valued wave function, the phase increment $\Delta\varphi$ must be a multiple of 2π on the return to the same point, i.e.

$$\Delta\varphi = \oint_C (\nabla\varphi)\cdot d\boldsymbol{l} = 2\pi l$$

where $l = 0, \pm 1, \pm 2, \ldots$ is an integer. Hence we have the quantized circulation for the velocity of the condensate flow

$$\oint_C \boldsymbol{v}_s d\boldsymbol{l} = \frac{2\pi\hbar}{m}l = l\varkappa, \quad \varkappa = \frac{2\pi\hbar}{m},$$

and the quantity $\varkappa$ is called the *circulation quantum*.

A simple example of the $l \neq 0$ state is given by the following behavior of the condensate wave function

$$\psi(\boldsymbol{r}) = |\psi| e^{i\theta(\boldsymbol{r})}, \quad \theta(\boldsymbol{r}) = l \arctan(y/x).$$

Here θ is the azimuthal angle in the cylindrical frame (ρ, θ, z). The phase gradient equals

$$\nabla\theta = \left(-l\frac{y}{\rho^2}, l\frac{x}{\rho^2}, 0\right), \quad |\nabla\theta| = \frac{l}{\rho} \quad \text{and} \quad \rho = \sqrt{x^2 + y^2}.$$

The z-axis, corresponding to $\rho = 0$, is a singular line called the *vortex line*. Accordingly, we have for curl $\boldsymbol{v}_s$

$$\text{curl}\,\boldsymbol{v}_s = l\varkappa\,\delta(\boldsymbol{\rho})\boldsymbol{e}_z,$$

$\boldsymbol{e}_z$ being a unit vector in the z-axis direction.

The phase increment around any contour encircling the vortex line remains the same and equals $2\pi l$. The magnitude of wave function vanishes at the vortex line $\rho = 0$, i.e. $\psi(\rho = 0) = 0$, in order to satisfy the single-valued condition of condensate wave function. The quantized vortex is an example of a *topological*[10] *defect*.

Let us study the behavior of condensate wave function $\psi(\boldsymbol{r})$ in the vicinity of vortex line. We write the wave function in the cylindrical frame (ρ, φ, z) as

$$\psi(\boldsymbol{r}) = f(\rho)e^{il\varphi}$$

and substitute it into the stationary Gross–Pitaevskii equation. This entails the following equation:

$$-\frac{\hbar^2}{2m}\left(f''(\rho) + \frac{f'(\rho)}{\rho} - \frac{l^2}{\rho^2}f(\rho)\right) + gf^3(\rho) = \mu f(\rho).$$

Then we go over to the dimensionless variables according to

$$f = f_0 y(x) = \sqrt{n}\,y(x), \quad x = \rho/\xi \quad \text{and} \quad \xi = \hbar/\sqrt{2m\mu}, \quad \mu = gn$$

[10] From the mathematical point of view this is a continuous mapping of the spatial points of closed path C onto the unit-radius circle (1-sphere). The mapping is performed with phase $\varphi(\boldsymbol{r})$ of wave function ψ. Each homotopy class of such mapping consists of all loops that wind around the circle l times with a positive or negative sign depending on the direction of winding. The vortex line is assigned to the element of fundamental 1-sphere group $\pi_1(S^1)$ isomorphic to the additive group of integers $\mathbb{Z}$. The topological characteristic or *topological charge* of vortex line is an integer winding number l and remains unchanged until the path C with its deformation does not cross the vortex line.

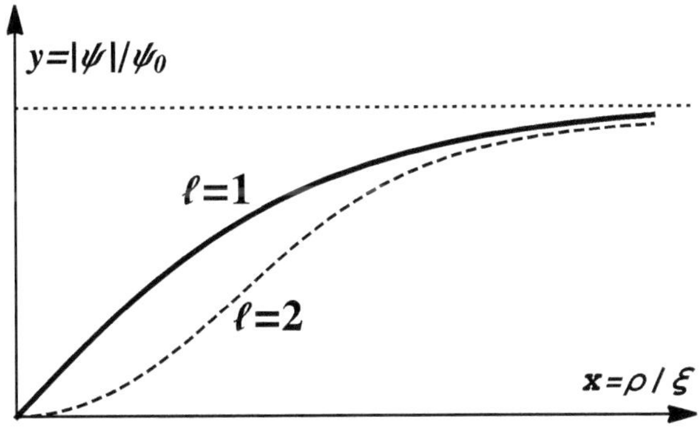

Fig. 7.1 The behavior of condensate wave function versus distance from the vortex line axis. The solid line is $l = 1$ and the dashed one is $l = 2$

where ξ is the healing length and n is the gas density far away from the vortex line. As a result, function $y(x)$ satisfies the equation

$$y'' + \frac{y'}{x} + \left(1 - \frac{l^2}{x^2}\right)y - y^3 = 0$$

with conditions $y(\infty) = 1$ and $y(0) = 0$. This equation is subject to the numerical solution. For $x \to 0$, we have $y \sim x^{|l|}$ and $y \sim 1 - l^2/(2x^2)$ at $x \to \infty$. The region beside the vortex line, where the condensate density is highly depleted, has the size of the order of the healing length ξ and is referred to as the *vortex core*. In Fig. 7.1 the qualitative behavior for the ratio $y(x) = |\psi(x)|/\psi_0$ is shown as a function of the distance from the vortex line for $l = 1$ and $l = 2$.

Let us turn to calculating the linear tension of rectilinear quantized vortex or energy per unit length. To find the energy of vortex, we should determine the difference between the total energy of the gas with the vortex and the energy of the homogeneous vortex-free gas with the same particle number N. In this case, the gas density far away from the vortex will differ from the density n_0 of homogeneous vortex-free gas. This leads to a number of mathematical inconveniences as comparing the energy E_v in the vortex state and energy E_0 in the vortex-free state.

In such situation, it is more preferably to analyze the energy of the gas as a function of chemical potential μ, i.e. the following thermodynamic potential:

$$\tilde{E}(\mu) = E(N) - \mu N = \int d\rho \left(\frac{\hbar^2}{2m}|\nabla\psi|^2 + \frac{1}{2}g|\psi|^4 - \mu|\psi|^2\right),$$

and to compare the energies if the gas with and without vortex at the fixed chemical potential. For the gas with the vortex, this trick provides us the same physical parameters far away from the vortex as in the homogeneous vortex-free state, in particular, the same density n_0 due to simple relation $\mu = gn_0$.

The energy of the vortex per unit length reads

$$\Delta E = \tilde{E}_v - \tilde{E}_0 = \int d^2\rho \left[\frac{\hbar^2}{2m}|\psi|^2 + \frac{1}{2}g|\psi|^4 - \mu|\psi|^2 - \left(\frac{1}{2}\psi_0^4 - \mu\psi_0^2 \right) \right] =$$
$$= \int d^2\rho \left(\frac{\hbar^2}{2m}|\psi|^2 + \frac{g}{2}\left(\psi_0^2 - |\psi|^2\right)^2 \right).$$

Writing the condensate wave function ψ with the vortex line at axis $\rho = 0$ in the cylindrical coordinates ρ and φ according to $\psi(\rho, \varphi) = \sqrt{n(\rho)}\exp(il\varphi)$, we arrive at the following expression:

$$\Delta E_l = \int 2\pi\rho\, d\rho \left[\frac{\hbar^2}{2m}\left(\frac{\partial\sqrt{n}}{\partial\rho} \right)^2 + \frac{\hbar^2 l^2}{2m}\frac{n}{\rho^2} + \frac{g}{2}\left(n_0 - n\right)^2 \right]$$

or in the dimensionless variables

$$\Delta E_l = \frac{\pi\hbar^2 n_0}{m} \int_0^{R/\xi} \left[\left(\frac{dy}{dx} \right)^2 + l^2\frac{y^2}{x^2} + \frac{1}{2}\left(1 - y^2\right)^2 \right] x\, dx.$$

Here R is the size of the system and we imply below that $R \gg \xi$.

The main contribution to ΔE_l is gained from the centrifugal term l^2/x^2 due to slow $\sim 1/x$ decay of the condensate flow velocity. As a result, there appears a large logarithmic integral and we have within the logarithmic accuracy

$$\Delta E_l = \frac{\pi n_0 \hbar^2 l^2}{m} \ln\left(C_l \frac{R}{\xi} \right), \quad R \gg \xi.$$

Here C_l is the number of about unity, depending on the number of circulation number. The vortex energy increases with the size of the system as ln R. The result for ΔE_l indicates that the vortex with several circulation quanta has the larger energy than several vortices of one-quantum circulation but with the same total circulation. Accordingly, the vortex with $|l| > 1$ should be unstable against its transition into the vortices of one-quantum circulation.

Problem

Estimate the linear tension of rectilinear vortex with unit circulation, using the following variational function for the condensate density:

$$n(\rho) = n_0 \frac{\rho^2}{\rho^2 + a^2\xi^2}$$

with variational parameter a.

Solution. Substituting $y(x) = x/\sqrt{a^2+x^2}$ ($x = \rho/\xi$) into the expression for vortex energy ΔE_1 and calculating several integrals, we find

$$\Delta E_1 = \frac{\pi\hbar^2 n_0}{m}\left(\frac{1}{4} + \ln\frac{R}{a\xi} + \frac{a^2}{4}\right).$$

Here we have taken the inequality $R \gg \xi$ into account and put the upper limit equal to the infinity in the convergent integrals. Minimizing $\Delta E_1(a)$ over the variational parameter a entails the optimum value $a = \sqrt{2}$. As a result, we have

$$\Delta E_1 = \frac{\pi\hbar^2 n_0}{m}\ln\left(C_1\frac{R}{\xi}\right) \quad \text{where} \quad C_1 = \frac{e^{3/4}}{\sqrt{2}} = 1.49\ldots$$

7.4 Elementary Excitations in a Weakly Non-ideal Bose Gas

Let us turn to studying the dynamical properties of homogeneous weakly non-ideal Bose gas. For this purpose, we use the non-stationary Gross–Pitaevskii equation

$$i\hbar\frac{\partial}{\partial t}\psi(\boldsymbol{r},t) = \left(-\frac{\hbar^2\nabla^2}{2m} + g|\psi(\boldsymbol{r},t)|^2\right)\psi(\boldsymbol{r},t)$$

and analyze the small perturbations for the wave function $\psi(\boldsymbol{r},t)$ of the condensate from its equilibrium state described by the following wave function:

$$\psi_0(t) = \sqrt{n}e^{-i\mu t/\hbar} \quad \text{and} \quad \psi_0^*(t) = \sqrt{n}e^{i\mu t/\hbar}.$$

Here $\mu = gn$ is the chemical potential and $n = |\psi_0|^2$ is the condensate density. Let wave function be $\psi = \psi_0 + \delta\psi$ and $\psi^* = \psi_0^* + \delta\psi^*$. Next, we linearize the Gross–Pitaevskii equation for function ψ and conjugate one ψ^* over small perturbations $\delta\psi$ and $\delta\psi^*$. Finally, we have

$$\begin{aligned}
i\hbar\frac{\partial\,\delta\psi}{\partial t} &= -\frac{\hbar^2}{2m}\nabla^2\delta\psi + 2gn\,\delta\psi + g\psi_0^2\,\delta\psi^*,\\
-i\hbar\frac{\partial\,\delta\psi^*}{\partial t} &= -\frac{\hbar^2}{2m}\nabla^2\delta\psi^* + 2gn\,\delta\psi^* + g\psi_0^{*2}\,\delta\psi.
\end{aligned}$$

The solution for a set of two linear equations is sought as a sum of independent Fourier harmonics classified by wave vector $\boldsymbol{k}$ and amplitudes $u_{\boldsymbol{k}}$ and $v_{\boldsymbol{k}}$

$$\begin{aligned}
\delta\psi(\boldsymbol{r},t) &= e^{-i\mu t/\hbar}\sum_{\boldsymbol{k}}\left[u_{\boldsymbol{k}}e^{i\boldsymbol{kr}-i\omega_{\boldsymbol{k}}t} + v_{\boldsymbol{k}}^*e^{-i\boldsymbol{kr}+i\omega_{\boldsymbol{k}}t}\right],\\
\delta\psi^*(\boldsymbol{r},t) &= e^{i\mu t/\hbar}\sum_{\boldsymbol{k}}\left[v_{\boldsymbol{k}}e^{i\boldsymbol{kr}-i\omega_{\boldsymbol{k}}t} + u_{\boldsymbol{k}}^*e^{-i\boldsymbol{kr}+i\omega_{\boldsymbol{k}}t}\right].
\end{aligned}$$

As a result, we have a pair of interrelated equations for determining $u_{\boldsymbol{k}}$ and $v_{\boldsymbol{k}}$. In our simple case the amplitudes[11] can be chosen as real quantities

$$\hbar\omega_{\boldsymbol{k}} u_{\boldsymbol{k}} = \left(\frac{\hbar^2 k^2}{2m} + 2gn - \mu\right) u_{\boldsymbol{k}} + gn v_{\boldsymbol{k}},$$
$$-\hbar\omega_{\boldsymbol{k}} v_{\boldsymbol{k}} = \left(\frac{\hbar^2 k^2}{2m} + 2gn - \mu\right) v_{\boldsymbol{k}} + gn u_{\boldsymbol{k}}.$$

The condition for existing the nontrivial solutions is zero determinant composed of the coefficients in front of unknown variables u_k and v_k. Then we arrive at the following equation relating the frequency ω_k and wave vector $\boldsymbol{k}$:

$$\hbar^2\omega_k^2 = \left(\frac{\hbar^2 k^2}{2m} + 2gn - \mu\right)^2 - (gn)^2 = \left(\frac{\hbar^2 k^2}{2m} + gn\right)^2 - (gn)^2.$$

Hence we obtain the dispersion relation

$$\varepsilon_k = \hbar\omega_k = \sqrt{\left(\frac{\hbar^2 k^2}{2m}\right)^2 + \frac{\hbar^2 k^2}{m} gn}$$

called the *Bogoliubov excitation spectrum*. Note that the spectrum is sound-like in the region of small wave vectors $k \ll 1/\xi$, ξ being the healing length

$$\varepsilon_k = \hbar\omega_k = \hbar c k \quad \text{where} \quad c = \sqrt{\frac{gn}{m}} = \sqrt{\frac{\partial P}{\partial \rho}}.$$

Emphasize also that the magnitude of velocity c coincides with the thermodynamic determination of sound velocity. In the region of large wave vectors $k \gg 1/\xi$, the excitation spectrum has a usual particle-like $\varepsilon_k = \hbar^2 k^2/2m$ behavior since the interaction between the bosons can be neglected.

A ratio of amplitudes u_k and v_k equals

$$L_k = \frac{v_k}{u_k} = \frac{1}{gn}\left(\hbar\omega_k - \frac{\hbar^2 k^2}{2m} - gn\right) = -\frac{gn}{\varepsilon_k + \sqrt{\varepsilon_k^2 + (gn)^2}}.$$

In general, the normalization of amplitudes u_k and v_k here remains still indefinite. As it concerns both the physical interpretation and the quantum mechanical applications, the choice $u_k^2 - v_k^2 = 1$ or

[11] The algebraic relations for $u_{\boldsymbol{k}}$ and $v_{\boldsymbol{k}}$ are analogous to those for amplitudes $u_{\boldsymbol{k}}$ and $v_{\boldsymbol{k}}$ in the BCS theory.

$$u_k = \frac{1}{\sqrt{1-L_k^2}} \quad \text{and} \quad v_k = \frac{L_k}{\sqrt{1-L_k^2}}.$$

is very desirable and convenient.

In fact, let us consider how the energy of a gas varies at small oscillations of condensate wave function near the equilibrium state. For this purpose, again it is more convenient to analyze the energy of the gas as a function of chemical potential μ, i.e. to study thermodynamic potential $\tilde{E}(\mu) = E - \mu N$. The choice of such thermodynamic potential $\tilde{E}$ is mathematically preferable. The point is that, for the states under fixed chemical potential μ, the stationary equilibrium state is described with the wave function $\psi_0(t) = \exp(-i\mu t/\hbar)$. Let us find variation of the energy

$$\tilde{E}(\mu) = \int d\boldsymbol{r}\left[\frac{\hbar^2}{2m}|\nabla\psi|^2 + \frac{g}{2}|\psi|^4 - \mu|\psi|^2\right]$$

at small perturbations from equilibrium $\psi = \psi_0 + \delta\psi$ and $\psi^* = \psi_0^* + \delta\psi^*$ to second order in perturbations $\delta\psi$ and $\delta\psi^*$. So,

$$\Delta\tilde{E} = \tilde{E}(\psi^*, \psi) - \tilde{E}(\psi_0^*, \psi_0) =$$

$$= \frac{\delta\tilde{E}}{\delta\psi}\delta\psi + \frac{\delta\tilde{E}}{\delta\psi^*}\delta\psi^* + \frac{1}{2}\frac{\delta^2\tilde{E}}{\delta\psi^2}\delta\psi^{*2} + \frac{\delta^2\tilde{E}}{\delta\psi^*\delta\psi}\delta\psi\delta\psi^* + \frac{1}{2}\frac{\delta^2\tilde{E}}{\delta\psi^{*2}}\delta\psi^2 + \cdots$$

As is expected, the linear terms in $\delta\psi^*$ and $\delta\psi$ vanish due to $\mu = g|\psi_0|^2$ at equilibrium. For second order in perturbation, we find

$$\Delta\tilde{E} = \int d\boldsymbol{r}\left[\frac{\hbar^2|\nabla\delta\psi|^2}{2m} + \frac{g}{2}\left(\psi_0^{*2}\delta\psi^2 + 4|\psi_0|^2\delta\psi^*\delta\psi + \psi_0^2\delta\psi^{*2}\right) - \mu\delta\psi^*\delta\psi\right].$$

The substitution $\delta\psi$ and $\delta\psi^*$ after calculation with using the formulas

$$\sum_{k,k'}\int \alpha_k\beta_{k'}e^{i(\boldsymbol{k}-\boldsymbol{k}')\boldsymbol{r}}d\boldsymbol{r} = \sum_k \alpha_k\beta_k \quad \text{and} \quad u_k v_k\frac{\hbar^2k^2}{2m} + \frac{gn}{2}\left(u_k + v_k\right) = 0$$

leads us to the following relation for the perturbation of energy:

$$\Delta\tilde{E} = \sum_k\left(\frac{\hbar^2k^2}{2m} + gn\right)\left(u_k^2 + v_k^2\right) + 2gnu_kv_k = \sum_k \hbar\omega_k\left(u_k^2 - v_k^2\right).$$

From the viewpoint of stability for the ground unexcited state of a gas the quantity $\hbar\omega_k(u_k^2 - v_k^2)$ must be strictly positive for each mode with wave vector $\boldsymbol{k}$. Provided that normalization $u_k^2 - v_k^2 = 1$ is chosen for amplitudes u_k and v_k, the quantity $\hbar\omega_k$ can be interpreted as an energy of elementary excitation from the ground state. Since the number of elementary excitations in the same state is unlimited due to spinless Bose–Einstein statistics, the energy for the excited state of a gas can be represented as

$$\Delta \tilde{E} = \sum_{k} \hbar\omega_{k} N_{k}$$

where N_k is the number of elementary excitation with the given wave vector $\boldsymbol{k}$. Such choice of normalization $u_{\boldsymbol{k}}^2 - v_{\boldsymbol{k}}^2 = 1$ allows us to interpret the elastic excitations in the Bose–Einstein condensed gas as a set of quanta of elastic strain field or phonons with the corresponding dispersion.

Problems

1. Derive the equation determining the small density oscillations of a Bose–Einstein condensed gas in the external potential $V(\boldsymbol{r})$ within the framework of the Thomas–Fermi approximation.

Solution. We linearize the corresponding equations written in the Thomas–Fermi approximation

$$\frac{\partial n}{\partial t} + \operatorname{div}(n\boldsymbol{v}_s) = 0 \quad \text{and} \quad m\frac{\partial \boldsymbol{v}_s}{\partial t} + \nabla\left(\frac{1}{2}m\boldsymbol{v}_s^2 + V(\boldsymbol{r}) + gn\right) = 0,$$

assuming the small density variation $\delta n(\boldsymbol{r}, t) = n(\boldsymbol{r}, t) - n_0(\boldsymbol{r})$. The linearized equations read

$$\frac{\partial \delta n}{\partial t} + \operatorname{div}(n_0\boldsymbol{v}_s) = 0 \quad \text{and} \quad m\frac{\partial \boldsymbol{v}_s}{\partial t} + g\nabla\delta n = 0$$

where the unperturbed stationary density profile $n_0(\boldsymbol{r})$ satisfies condition $gn_0(\boldsymbol{r}) - V(\boldsymbol{r}) = \mu$. Eliminating the condensate velocity $\boldsymbol{v}_s$, we obtain the equation desired

$$\frac{\partial^2 \delta n}{\partial t^2} = \operatorname{div}\left[c^2(\boldsymbol{r})\nabla\,\delta n\right]$$

where $mc^2(\boldsymbol{r}) = \mu - V(\boldsymbol{r}) = gn_0(\boldsymbol{r})$. The quantity $c(\boldsymbol{r})$ can be thought as a local sound velocity.

2. Using the Thomas–Fermi approximation for the gas cloud in the spherically symmetrical harmonic trap $V(r) = m\omega^2 r^2/2$, find the frequency Ω of the radial breathing mode in which the radial velocity does not change the sign.

Solution. Let the gas cloud radius be R. This radius corresponds to the chemical potential $\mu = m\omega^2 R^2/2$ and the square of local sound velocity $c^2 = \omega^2(R^2 - r^2)/2$. Taking the spherical symmetry oscillations of gas density δn into account, we need to solve the following equation:

$$\frac{\partial^2 \delta n}{\partial t^2} = \frac{1}{r^2}\frac{\partial}{\partial r}\left(r^2 c^2(r)\frac{\partial\,\delta n}{\partial r}\right).$$

Let us seek for the solution as $\delta n(r, t) = e^{-i\Omega t} f(r)$. Then, we have

$$\frac{\Omega^2}{\omega^2} f(r) = rf'(r) - \frac{R^2 - r^2}{2}\left(f''(r) + \frac{2}{r}f'(r)\right).$$

The solution, which has no oscillations in the radial direction and satisfies the particle conservation condition $\int f(r)r^2 dr = 0$, looks like $f(r) \sim (3R^2 - 5r^2)$. The substitution $f(r)$ into the last equation gives the *breathing mode frequency*

$$\Omega = \sqrt{5}\,\omega.$$

Emphasize that this frequency $\sqrt{5}\,\omega$ proves to be higher than frequency 2ω for the similar oscillating mode of non-interacting ideal condensed gas in the same trap.

7.5 Depletion of the Bose–Einstein Condensate

Below we discuss such phenomenon as a depletion of condensate in the non-ideal Bose system on the example of weakly non-ideal spinless Bose gas. For the dilute gas with the particle concentration n, it is usually assumed that the radius for the forces of interparticle interaction is small as compared with the mean interparticle spacing about $n^{-1/3}$. This allows us to be restricted with the binary interactions alone and to neglect the ternary collisions in first approximation.

Let us write the corresponding Hamiltonian of weakly non-ideal Bose gas with the particles of mass m in the field operator representation

$$\hat{H} = \int \frac{\hbar^2}{2m} \nabla\psi^+(\boldsymbol{r})\psi(\boldsymbol{r})d\boldsymbol{r} + \frac{1}{2}\int \psi^+(\boldsymbol{r}')\psi^+(\boldsymbol{r})U(\boldsymbol{r}'-\boldsymbol{r})\psi(\boldsymbol{r})\psi(\boldsymbol{r}')d\boldsymbol{r}'d\boldsymbol{r}.$$

Then we go over from the coordinate representation to the momentum one in accordance with

$$\psi(\boldsymbol{r}) = \sum_p b_p \frac{e^{i\boldsymbol{pr}/\hbar}}{\sqrt{V}} \quad \text{and} \quad \psi^+(\boldsymbol{r}) = \sum_p b_p^+ \frac{e^{-i\boldsymbol{pr}/\hbar}}{\sqrt{V}}$$

where b_p^+ and b_p are the creation and annihilation operators with momentum $\boldsymbol{p}$ in volume V. As a result, we have

$$\hat{H} = \sum_p \frac{p^2}{2m} b_p^+ b_p + \frac{1}{2V}\sum_{q,p,p'} U_q b_{p+q}^+ b_{p'-q}^+ b_p b_{p'}$$

and $U_q = \int U(\boldsymbol{r})e^{-i\boldsymbol{qr}/\hbar}d\boldsymbol{r}$ is the Fourier-transform of interaction potential $U(\boldsymbol{r})$. (The origin of factor $1/V$ is associated with summation over momentum.) Next, since in the condensed state and at low temperatures the small momenta play a key role, we neglect the dependence of Fourier-transform U_q on the momentum and take its value $g = U_0$ at $\boldsymbol{q} = 0$. The coupling constant g can be related to[12] the s-scattering length a with formula $g = 4\pi\hbar^2 a/m$.

So, we analyze the Hamiltonian below

$$\hat{H} = \sum_p \frac{p^2}{2m} b_p^+ b_p + \frac{g}{2V}\sum_{q,p,p'} b_{p+q}^+ b_{p'-q}^+ b_p b_{p'}.$$

Let us make the following approximations resulting from the sufficiently low temperatures as compared with the condensation temperature. First, the occupation number of particle states with zero $p = 0$ momentum is much larger than the occupation

[12] The cross-section is $d\sigma_s = (m/4\pi\hbar^2)^2(2|U_0|)^2 d\Omega$ for the s-scattering of two identical particles in the Born approximation. On the other hand, the s-scattering amplitude is connected with the cross-section as $d\sigma_s = (2a)^2 d\Omega$.

numbers for the particle states with nonzero $p \neq 0$ momenta, i.e. $N_{p=0} = N_0 \sim N$ and $N_{p\neq 0} = N - N_0 \ll N_0$. Second, using both $b_0^+ b_0 = N_0$, $b_0 b_0^+ = 1 + N_0$ and macroscopic condition $N_0 \gg 1$, we do not discern $b_0^+ b_0$ and $b_0 b_0^+$, treating b_0^+ and b_0 as ordinary commutative numbers equal to $b_0^+ = b_0 = \sqrt{N_0}$. Third, in the Hamiltonian we retain only the terms $a^+_{p\neq 0}$ and $a_{p\neq 0}$ not higher than quadratic. A simple calculation leads us to the Hamiltonian

$$\hat{H} = \sum_p \frac{p^2}{2m} b_p^+ b_p + \frac{g}{2V} b_0^+ b_0^+ b_0 b_0 +$$
$$+ \frac{g}{2V} \sum_{p\neq 0} \left(4 b_0^+ b_p^+ b_0 b_p + b_p^+ b_{-p}^+ b_0 b_0 + b_0^+ b_0^+ b_p b_{-p}\right).$$

The terms having only one operator b_p^+ or b_p vanish due to momentum conservation for the interaction between particles. Then, replacing the operators b_0^+ and b_0 with $\sqrt{N_0}$, we have

$$\hat{H} = \sum_p \frac{p^2}{2m} b_p^+ b_p + \frac{g}{2V} N_0^2 + \frac{g}{2V} N_0 \sum_{p\neq 0} \left(4 b_p b_p + b_p^+ b_{-p}^+ + b_p b_{-p}\right).$$

The number of particles N_0 in the condensate is unknown. On the other hand, the total number of particles N can be represented as follows:

$$N = N_0 + \sum_{p\neq 0} b_p^+ b_p = N_0 + \frac{1}{2} \sum_{p\neq 0} \left(b_p^+ b_p + b_{-p}^+ b_{-p}\right).$$

Accordingly,

$$\hat{H} = \sum_{p\neq 0} \frac{p^2}{2m} b_p^+ b_p + \frac{g}{2V} \left(N - \sum_p b_p^+ b_p\right)^2 +$$
$$+ \frac{g}{2V} N_0^2 \sum_{p\neq 0} \left(4 b_p^+ b_p + b_p^+ b_{-p}^+ + b_p b_{-p}\right).$$

Again, retaining only the quadratic terms in b_0^+ and b_0 and replacing N_0 with N in the same approximation, we finally arrive at the following Hamiltonian:

$$\hat{H} = E_0 + \frac{1}{2} \sum_{p\neq 0} \left[\left(\frac{p^2}{2m} + gn\right)\left(b_p^+ b_p + b_{-p}^+ b_{-p}\right) + gn\left(b_p^+ b_{-p}^+ + b_p b_{-p}\right)\right].$$

Here $n = N/V$ is the gas density and $E_0 = gN^2/2V$ is the energy of ground state. The corresponding chemical potential $\mu = \partial E_0/\partial N$ equals $\mu = gn$.

As we can see, the last term in the brackets is non-diagonal. The problem is to transform to the form as $\sum_p \varepsilon_p \beta_p^+ \beta_p$ where operators β_p^+ and β_p satisfy the commutation rules $[\beta_p, \beta_p^+] = 1$ for the creation and annihilation operators of bosons. In this case the quantity ε_p acquires a sense of the energy for elementary excitation with momentum $\boldsymbol{p}$.

The necessary diagonalization of Hamiltonian will be reached with the aid of the *Bogoliubov uv-transformation*

$$\begin{pmatrix} b_p & b_{-p} \\ b_{-p}^+ & b_p^+ \end{pmatrix} = \begin{pmatrix} u_p & v_p \\ v_p & u_p \end{pmatrix} \begin{pmatrix} \beta_p & \beta_{-p} \\ \beta_{-p}^+ & \beta_p^+ \end{pmatrix} \quad \text{and} \quad \begin{vmatrix} u_p & v_p \\ v_p & u_p \end{vmatrix} = 1.$$

The choice of normalization $u_p^2 - v_p^2 = 1$ or

$$u_p = \frac{1}{\sqrt{1 - L_p^2}} \quad \text{and} \quad v_p = \frac{L_p}{\sqrt{1 - L_p^2}}$$

keeps the commutation condition for the creation and annihilation operators β_p^+ and β_p of elementary excitations. Substituting the operators b_p^+ and b_p expressed via β_p^+ and β_p, we have

$$\begin{aligned} \hat{H} = E_0 &+ \sum_{p\neq 0} \frac{1}{1 - L_p^2}\left[\left(\frac{p^2}{2m} + gn\right) L_p^2 + gnL_p\right] + \\ &+ \frac{1}{2}\sum_{p\neq 0} \frac{1}{1 - L_p^2}\left[\left(\frac{p^2}{2m} + gn\right)(1 + L_p^2) + 2gnL_p\right](\beta_p^+\beta_p + \beta_{-p}^+\beta_{-p}) + \\ &+ \frac{1}{2}\sum_{p\neq 0} \frac{1}{1 - L_p^2}\left[\left(\frac{p^2}{2m} + gn\right) L_p + gn(1 + L_p^2)\right](\beta_p^+\beta_p^+ + \beta_{-p}\beta_{-p}). \end{aligned}$$

The operator-independent term means the renormalization of the ground state energy. To have the Hamiltonian diagonal, the factor in the front of non-diagonal term should be put zero by choosing L_p, i.e.

$$2L_p\left(\frac{p^2}{2m} + gn\right) + gn\left(1 + L_p^2\right) = 0.$$

Solving the quadratic equation yields

$$L_p = \frac{1}{gn}\left[-\frac{p^2}{2m} - gn + \sqrt{\left(\frac{p^2}{2m} + gn\right)^2 - (gn)^2}\,\right].$$

The sign before the square root is taken positive in order to have the positive energy of excitations as well. The excitation energy will be determined below. So, we have

$$\hat{H} = E_0 - \frac{1}{2}\sum_{p\neq 0}\left(\frac{p^2}{2m} + gn - \varepsilon_p\right) + \frac{1}{2}\sum_{p\neq 0}\varepsilon_p(\beta_p^+\beta_p + \beta_{-p}^+\beta_{-p}).$$

The energy of elementary excitation with momentum p is given by the formula of the Bogoluibov spectrum

$$\varepsilon_p = \sqrt{\left(\frac{p^2}{2m} + gn\right)^2 - (gn)^2} = \sqrt{\left(\frac{p^2}{2m}\right)^2 + \frac{p^2}{m}gn}\,.$$

The spectrum proves to be sound-like in the region of small $p \ll \sqrt{mgn}$ momenta. The sound velocity equals $c = \sqrt{gn/m}$. The excitations are the waves of compression and rarefaction in the density of a gas. For the large momentum region, the interaction between the particles plays no significant role and the spectrum acquires the character typical for free particles.

As we have seen above, the energy for the excited state of a gas can be written as $\sum_{p\neq 0}\varepsilon_p n_p$ where $n_p = \beta_p^+\beta_p$ are the occupation numbers of elementary excitations. To calculate the equilibrium distribution $N_{p\neq 0}$ for the genuine particles of a gas in the overcondensate states, it is necessary to average the particle number operator $\hat{N}_p = b_p^+ b_p$ with the Gibbs distribution. As usual, the average of some variable A means the following:

$$\langle A\rangle = \frac{\mathrm{tr}\,(Ae^{-H/T})}{\mathrm{tr}\,e^{-H/T}}.$$

Then, making the Bogoliubov transformation, we find

$$N_p(T) = \langle b_p^+ b_p\rangle =$$
$$= \frac{1}{1-L_p^2}\left[\langle\beta_p^+\beta_p\rangle + L_p\langle\beta_{-p}^+\beta_p^+\rangle + L_p\langle\beta_p\beta_{-p}\rangle + L_p^2\langle\beta_{-p}\beta_{-p}^+\rangle\right],$$

the expression being valid for momenta $\boldsymbol{p}\neq 0$. Since the creation and annihilation operators β_p^+ and β_p have the same properties as operators b_p^+ and b_p for the genuine particles, the first and last terms alone from the whole sum remain nonzero. Accordingly,

$$N_p(T) = \frac{1}{1-L_p^2}\left[\langle\beta_p^+\beta_p\rangle + L_p^2\langle\beta_{-p}\beta_{-p}^+\rangle\right] = \frac{L_p^2}{1-L_p^2} + \frac{1+L_p^2}{1-L_p^2}\langle\beta_p^+\beta_p\rangle.$$

The calculation of thermodynamic average $\langle\beta_p^+\beta_p\rangle$, which should be performed with the elementary excitation Hamiltonian $\sum_{p\neq 0}\varepsilon_p\beta_p^+\beta_p$, leads us to the ordinary Bose–Einstein distribution

$$n_p(T) = \langle\beta_p^+\beta_p\rangle = \left(e^{\varepsilon_p/T} - 1\right)^{-1}$$

and, therefore,

$$N_p(T) = \frac{L_p^2}{1-L_p^2} + \frac{1+L_p^2}{1-L_p^2} n_p(T) = v_p^2 + \frac{u_p^2 + v_p^2}{e^{\varepsilon_p/T} - 1}.$$

The number of particles in the condensate reads

$$N_0 = N - \sum_{p \neq 0} N_p(T) = N - V \int \frac{d^3 p}{(2\pi\hbar)^3} N_{p\neq 0}(T).$$

Let us start first from the case of zero temperature $T = 0$. Then the number of overcondensate particles equals

$$N - N_0(T=0) = V \int \frac{d^3 p}{(2\pi\hbar)^3} \frac{L_p^2}{1-L_p^2} =$$

$$= V \int \frac{d^3 p}{(2\pi\hbar)^3} \frac{1}{2} \frac{(gn)^2}{\varepsilon_p\left(\varepsilon_p + \sqrt{\varepsilon_p^2 + (gn)^2}\right)}.$$

The integration is straightforwardly to perform with the aid of substitution $y(x) = x + 1 + \sqrt{(x+1)^2 - 1}$ where $x = p^2/(2mgn)$. So, we have

$$N_0 = N - V\frac{(mgn)^{3/2}}{4\pi^2\hbar^3} \int_1^\infty dy \frac{y-1}{y^{5/2}} = N - V\frac{(mgn)^{3/2}}{3\pi^2\hbar^3} =$$

$$= N\left(1 - \frac{(mg)^{3/2} n^{1/2}}{3\pi^2\hbar^3}\right).$$

Non-ideality of bosonic gas and interaction between particles result in such phenomenon as a *depletion of the condensate* even at zero temperature. Let us express the answer for the relative condensate depletion in terms of the s-scattering amplitude a

$$\frac{N - N_0}{N} = \frac{8}{3\sqrt{\pi}} (na^3)^{1/2}.$$

In the experiments on alkali metal atoms in traps the condensate depletion is usually about several percents. The validity of the calculations performed implies the inequality $na^3 \ll 1$.

The finite temperature entails the *thermal depletion* of the condensate as well. The change of the condensate particle number due to temperature is given by the following formula:

$$\Delta N_0(T) = N_0(T) - N_0(T=0) = -\sum_{p\neq 0}\frac{1+L_p^2}{1-L_p^2}n_p(T) =$$

$$= -V\int\frac{d^3p}{(2\pi\hbar)^3}\frac{\sqrt{\varepsilon_p^2+(gn)^2}}{\varepsilon_p}\frac{1}{e^{\varepsilon_p/T}-1} =$$

$$= -V\frac{\sqrt{2}m^{3/2}}{2\pi^2\hbar^3}\int\limits_0^\infty d\varepsilon\frac{\left[\left(\varepsilon^2+g^2n^2\right)^{1/2}-gn\right]^{1/2}}{e^{\varepsilon/T}-1}.$$

For the sufficiently low $T \ll gn$ temperature region, the main contribution to the integral comes from the energy $\varepsilon \lesssim T$ region. We have approximately

$$\Delta N_0(T) = -V\frac{\sqrt{2}m^{3/2}}{2\pi^2\hbar^3}\frac{1}{\sqrt{2gn}}\int\limits_0^\infty\frac{\varepsilon\, d\varepsilon}{e^{\varepsilon/T}-1} = -V\frac{mT^2}{12c\hbar^3}$$

where $c = \sqrt{gn/m}$ is the sound velocity. Note that the temperature behavior T^2 differs from that $T^{3/2}$ for the ideal Bose gas. In conclusion, we emphasize that the ratio chemical potential $\mu = gn$ to the critical temperature T_c is small in units of gas parameter, i.e. $gn/T_c \sim an^{1/3} \ll 1$.

7.6 Mixture of Two Bose–Einstein Condensed Gases

The usage of magneto-optical traps for confining the alkali metal atoms makes it possible to study the mixtures of several Bose–Einstein condensates. In what follows, we consider the binary Bose-condensed gas mixture composed of two various Bose particles. The numbers of particles are N_1, N_2 and their masses are m_1, m_2, respectively. The wave function for the condensed mixture in the Hartree–Fock approximation can be written as a product of one-particle wave functions ϕ_1 and ϕ_2

$$\Psi(\boldsymbol{r}_1,\ldots,\boldsymbol{r}_{N_1};\boldsymbol{r}'_1,\ldots,\boldsymbol{r}'_{N_2}) \approx \prod_{i=1}^{N_1}\phi_1(\boldsymbol{r}_i)\prod_{k=1}^{N_2}\phi_2(\boldsymbol{r}'_k).$$

Here $\boldsymbol{r}_i$ corresponds to the type-1 particles and $\boldsymbol{r}'_k$ does to the type-2 particles. The interaction between the particles depends on their type. We denote the corresponding coupling constants[13] as g_{11}, $g_{12} = g_{21}$, and g_{22}. The following formula gives the energy for the mixture of volume V in the homogeneous state

[13] The coupling constants are $g_{ik} = 2\pi\hbar^2 a_{ik}/m_{ik}$ where a_{ik} are the corresponding scattering lengths and $m_{ik} = m_i m_k/(m_i + m_k)$ are the reduced masses $(i, k = 1, 2)$.

$$E = g_{11}\frac{N_1(N_1-1)}{V} + g_{12}\frac{N_1 N_2}{V} + g_{22}\frac{N_2(N_2-1)}{V}.$$

On the analogy with the single-component case, we introduce two wave functions $\psi_1 = \sqrt{N_1}\phi_1$ and $\psi_2 = \sqrt{N_2}\phi_2$ for each of the condensates and write the following energy functional of gas mixture:

$$E[\psi_1(\boldsymbol{r}), \psi_2(\boldsymbol{r})] = \int d\boldsymbol{r}\left[\frac{\hbar^2}{2m_1}|\nabla\psi_1|^2 + \frac{\hbar^2}{2m_2}|\nabla\psi_2|^2 + \right.$$
$$\left. + \frac{1}{2}g_{11}|\psi_1|^4 + g_{12}|\psi_1|^2|\psi_2|^2 + \frac{1}{2}g_{22}|\psi_2|^4\right].$$

Here, as above, we have neglected the corrections $1/N_1$ and $1/N_2$ as compared with unity.

The variation of energy functional over ψ_1 and ψ_2 under constancy of particle number for each component results in introducing the chemical potentials μ_1 and μ_2 and finally we get the Gross–Pitaevskii equations for each of the mixture components

$$-\frac{\hbar^2}{2m_1}\nabla^2\psi_1 + g_{11}|\psi_1|^2\psi_1 + g_{12}|\psi_2|^2\psi_1 = \mu_1\psi_1,$$
$$-\frac{\hbar^2}{2m_2}\nabla^2\psi_2 + g_{22}|\psi_2|^2\psi_1 + g_{12}|\psi_1|^2\psi_2 = \mu_1\psi_2.$$

For the spatially homogeneous mixture with the component densities $n_1 = |\psi_1|^2$ and $n_2 = |\psi_2|^2$, the Gross–Pitaevskii equations give the following relationship between the chemical potentials and the densities:

$$\mu_1 = g_{11}n_1 + g_{12}n_2 \quad \text{and} \quad \mu_2 = g_{12}n_1 + g_{22}n_2.$$

It is of fundamental interest to study the stability of homogeneous state of a mixture against its phase separation into two separate fractions or phases. For this purpose, we should analyze how the energy of a mixture changes at small density fluctuations δn_1 and δn_2 of each condensate component. Due to conserving the particle numbers, i.e. $\int d\boldsymbol{r}\, \delta n_{1,2} = 0$, the energy variation δE vanishes in first order in δn_1 and δn_2. In second order we obtain a simple formula

$$\delta^2 E = \int d\boldsymbol{r}\left(g_{11}\delta n_1^2 + 2g_{12}\delta n_1\delta n_2 + g_{22}\delta n_2^2\right).$$

For the stability of spatially homogeneous state of a mixture, it is required to have the positive quadratic form composed of coefficients g_{11}, g_{12}, and g_{22}. Accordingly, this entails the following condition:

$$g_{11} > 0 \quad \text{and} \quad g_{11}g_{22} > g_{12}^2.$$

Provided this condition breaks down, the binary mixture of Bose condensates separates into two spatially divided fractions or phases, each of them being the Bose condensate composed of particles of the same type.

The Gross–Pitaevskii equations allows us to calculate the surface tension σ for the interface between two condensates. Let the interface be plane. The condensate of type-1 particles is on the left-hand side of the interface and its density equals $n_1 = \psi_{10}^2$ $(x \to -\infty)$ far away from the interface. Accordingly, on the right-hand side from the interface there is a condensate of type-2 particle with density $n_2 = \psi_{20}^2$ $(x \to +\infty)$ at the right-hand infinity. Let us write the Gross–Pitaevskii equation for the wave functions $\psi_1(x)$ and $\psi_2(x)$ each of the two condensates

$$
\begin{aligned}
-\frac{\hbar^2}{2m_1}\frac{d^2\psi_1}{dx^2} + \left(g_{11}\psi_1^2 + g_{12}\psi_2^2\right)\psi_1 = \mu_1\psi_1 \quad (\mu_1 = g_{11}n_1),\\
-\frac{\hbar^2}{2m_2}\frac{d^2\psi_2}{dx^2} + \left(g_{12}\psi_1^2 + g_{22}\psi_2^2\right)\psi_2 = \mu_2\psi_2 \quad (\mu_2 = g_{22}n_2)
\end{aligned}
$$

where μ_1 and μ_2 denote the chemical potentials constant over the space. We can take the condensate wave functions to be real.

In the limit $g_{12} \gg g_{11}, g_{22}$ the terms with $g_{12}\psi_2^2\psi_1$ and $g_{12}\psi_1^2\psi_2$ in each equations are predominant at $g_{12} \to \infty$. Then in first approximation we can put $\psi_2^2\psi_1 = 0$ and $\psi_1^2\psi_2 = 0$ in each of equations in order to compensate the divergence at $g_{12} \to \infty$. In first approximation we can expect that $\psi_2(x) = 0$ at $x < 0$ and $\psi_1(x) = 0$ at $x > 0$. This means physically that the penetration of particles of one type to the region occupied with the other type particles is energetically unfavorable due to large values g_{12}.

So, in this approximation we have for the wave functions

$$
\begin{aligned}
-\frac{\hbar^2}{2m_1}\psi_1'' + g_{11}\psi_1^3 = \mu_1\psi_1, \quad x < 0,\\
-\frac{\hbar^2}{2m_2}\psi_2'' + g_{22}\psi_2^3 = \mu_2\psi_2, \quad x > 0,
\end{aligned}
$$

with the boundary conditions $\psi_{1,2}'(\pm\infty) = 0$ and $\psi_1^2(-\infty) = \mu_1/g_{11}$, $\psi_2^2(\infty) = \mu_2/g_{22}$. Each of the equations can be solved with multiplying the equation by $\psi_{1,2}'$, respectively, and obtaining the first integral with its next integration. As a result of first approximation of wave functions perturbed by the interface, we have

$$
\begin{aligned}
\psi_1(x) \approx \psi_{10}\tanh\frac{|x|}{2\xi_1}\,\vartheta(-x), \quad \xi_1 = \frac{\hbar}{2\sqrt{m_1\mu_1}}, \quad \mu_1 = g_{11}n_1,\\
\psi_2(x) \approx \psi_{20}\tanh\frac{x}{2\xi_2}\,\vartheta(x), \quad \xi_2 = \frac{\hbar}{2\sqrt{m_2\mu_2}}, \quad \mu_2 = g_{22}n_2
\end{aligned}
$$

where $\xi_{1,2}$ is the healing length for the corresponding condensate.

It is convenient to analyze the energy of the whole system as a function of chemical potentials μ_1 and μ_2

$$\widetilde{E}(\mu_1,\mu_2)=E(N_1,N_2)-\mu_1N_1-\mu_2N_2=\int d^3r\left(\frac{\hbar^2|\nabla\psi_1|^2}{2m_1}+\frac{\hbar^2|\nabla\psi_2|^2}{2m_2}+\right.$$
$$\left.+\frac{g_{11}}{2}|\psi_1|^4+g_{12}|\psi_1|^2|\psi_2|^2+\frac{g_{22}}{2}|\psi_2|^4-\mu_1|\psi_1|^2-\mu_2|\psi_2|^2\right).$$

Due to simple relations $\mu_1=g_{11}n_1$ and $\mu_2=g_{22}n_2$ such choice of variables will provide us directly the values of condensate densities n_1 and n_2 far away from the interface as if the interface is absent.

The transient layer between two phases results in an additional energy which magnitude per unit interface area is called the surface tension coefficient σ. The latter will be determined by the following expression

$$\sigma=\int\limits_{-\infty}^{\infty}dx\left(\frac{\hbar^2\psi_1'^2}{2m_1}+\frac{\hbar^2\psi_2'^2}{2m_2}+\frac{g_{11}}{2}\psi_1^4+g_{12}\psi_1^2\psi_2^2+\frac{g_{22}}{2}\psi_2^4-\right.$$
$$\left.-\mu_1\psi_1^2-\mu_2\psi_2^2\right)-\int\limits_{-\infty}^{0}dx\left(\frac{g_{11}}{2}\psi_{10}^4-\mu_1\psi_{10}^2\right)-\int\limits_{0}^{\infty}dx\left(\frac{g_{22}}{2}\psi_{20}^4-\mu_2\psi_{20}^2\right).$$

This expression can be simplified. Let us multiply each equation of two initial Gross–Pitaevskii ones for ψ_1 and ψ_2 by ψ_1' and ψ_2', respectively. After their summing, we arrive at the first integral

$$-\frac{\hbar^2\psi_1'^2}{2m_1}-\frac{\hbar^2\psi_2'^2}{2m_2}+\frac{g_{11}}{2}\psi_1^4+g_{12}\psi_1^2\psi_2^2+\frac{g_{22}}{2}\psi_2^4-\mu_1\psi_1^2-\mu_2\psi_2^2=\text{const}.$$

Substituting the boundary values for the wave functions at $x\to\pm\infty$, we find the following magnitudes for the constant

$$\text{const}|_{x=-\infty}=-\frac{1}{2}g_{11}n_1^2\quad\text{and}\quad\text{const}|_{x=\infty}=-\frac{1}{2}g_{22}n_2^2.$$

To match the boundary conditions, we must require the fulfilment of the equality

$$P_1=\frac{1}{2}g_{11}n_1^2=\frac{1}{2}g_{22}n_2^2=P_2.$$

From the physical point of view this means the equality of pressures P_1 and P_2 for the adjacent condensates. The equality of the pressures is one of the necessary conditions for the thermodynamic equilibrium between two condensates.

Using the first integral and condition of pressure equality, we obtain the final expression for the surface tension coefficient

$$\sigma = 2\int_{-\infty}^{\infty} dx\left(\frac{\hbar^2\psi_1'^2}{2m_1} + \frac{\hbar^2\psi_2'^2}{2m_2}\right).$$

We see that the surface tension is wholly due to spatial variation of wave functions in the near-interface region. Substituting the approximate expressions found for the wave functions, we have

$$\sigma \approx \frac{8}{3}\frac{g_{11}n_1^2}{2}\xi_1 + \frac{8}{3}\frac{g_{22}n_2^2}{2}\xi_2 = \frac{8}{3}P(\xi_1+\xi_2),$$

$P = P_1 = P_2$ being the pressure in the both condensates. The equality condition for the pressures determines a ratio of condensate volumes V_1/V_2 after separating the mixture. The ratio equals $\sqrt{g_{11}}N_1/\sqrt{g_{22}}N_2$ where $N_{1,2}$ is the number of particles of each type.

For $g_{12}^2 \to g_{11}g_{22}$, the surface tension σ vanishes, demonstrating behavior proportional to $(g_{12}^2 - g_{11}g_{22})^{1/2}$.

In conclusion, let us turn to studying the dynamical properties of binary condensate mixture. The generalization of the non-stationary Gross–Pitaevskii equation lead us to the following system of two equations:

$$i\hbar\frac{\partial\psi_1}{\partial t} = -\frac{\hbar^2}{2m_1}\nabla^2\psi_1 + g_{11}|\psi_1|^2\psi_1 + g_{12}|\psi_2|^2\psi_1 = \frac{\partial E}{\delta\psi_1^*},$$

$$i\hbar\frac{\partial\psi_2}{\partial t} = -\frac{\hbar^2}{2m_2}\nabla^2\psi_2 + g_{12}|\psi_1|^2\psi_2 + g_{22}|\psi_2|^2\psi_2 = \frac{\partial E}{\delta\psi_2^*}.$$

The stationary equilibrium solution like

$$\psi_1(t) = \sqrt{n_1}e^{-i\mu_1 t/\hbar} \quad\text{and}\quad \psi_2(t) = \sqrt{n_2}e^{-i\mu_2 t/\hbar}$$

gives us the values of the chemical potentials familiar already

$$\mu_1 = g_{11}n_1 + g_{12}n_2 \quad\text{and}\quad \mu_2 = g_{12}n_1 + g_{22}n_2.$$

To determine the elementary excitation spectrum in the binary mixture, we consider the small perturbations of the condensate wave functions as $\psi_i \to \psi_i + \delta\psi_i$ and $\psi_i^* \to \psi_i^* + \delta\psi_i^*$. Then, we find the linearized equations which the perturbation of the first component obeys

$$i\hbar\frac{\partial\psi_1}{\partial t} = -\frac{\hbar^2}{2m_1}\nabla\delta\psi_1 + (2g_{11}n_1 + g_{12}n_2)\delta\psi_1 + g_{11}\psi_1^2\delta\psi_1^* + \\ + g_{12}\psi_1\psi_2^*\delta\psi_2 + g_{12}\psi_1\psi_2\delta\psi_2^*,$$

$$i\hbar\frac{\partial\psi_1^*}{\partial t} = -\frac{\hbar^2}{2m_1}\nabla\delta\psi_1^* + \big(2g_{11}n_1 + g_{12}n_2\big)\delta\psi_1^* + g_{11}\psi_1^{*2}\delta\psi_1 + \\ + g_{12}\psi_1^*\psi_2\delta\psi_2^* + g_{12}\psi_1^*\psi_2^*\delta\psi_2,$$

and the second component perturbation does

$$i\hbar\frac{\partial\psi_2}{\partial t} = -\frac{\hbar^2}{2m_2}\nabla\delta\psi_2 + \big(2g_{22}n_2 + g_{12}n_1\big)\delta\psi_2 + g_{22}\psi_2^2\delta\psi_2^* + \\ + g_{12}\psi_2\psi_1^*\delta\psi_1 + g_{12}\psi_2\psi_1\delta\psi_1^*,$$

$$i\hbar\frac{\partial\psi_2^*}{\partial t} = -\frac{\hbar^2}{2m_2}\nabla\delta\psi_2^* + \big(2g_{22}n_2 + g_{12}n_2\big)\delta\psi_2^* + g_{22}\psi_2^{*2}\delta\psi_2 + \\ + g_{12}\psi_2^*\psi_1\delta\psi_1^* + g_{12}\psi_2^*\psi_1^*\delta\psi_1.$$

The solution of a set of linear equations is commonly sought in terms of the decomposition over the Fourier harmonics with amplitudes u_{ik} and v_{ik} $(i = 1, 2)$ depending on wave vectors $\boldsymbol{k}$

$$\delta\psi_1(\boldsymbol{r}, t) = e^{-i\mu_1 t/\hbar}\sum_{\boldsymbol{k}}\big[u_{1k}e^{i\boldsymbol{kr}-i\omega_{\boldsymbol{k}}t} + v_{1k}^* e^{-i\boldsymbol{kr}+i\omega_{\boldsymbol{k}}t}\big],$$

$$\delta\psi_2(\boldsymbol{r}, t) = e^{-i\mu_2 t/\hbar}\sum_{\boldsymbol{k}}\big[u_{2k}e^{i\boldsymbol{kr}-i\omega_{\boldsymbol{k}}t} + v_{2k}^* e^{-i\boldsymbol{kr}+i\omega_{\boldsymbol{k}}t}\big].$$

The similar relations must be written for the conjugate variables $\delta\psi_1(\boldsymbol{r}, t)^*$ and $\delta\psi_2^*(\boldsymbol{r}, t)$ as well. The substitution of these expansions into the non-stationary Gross–Pitaevskii equations results in the four equations for determining the four amplitudes u_{ik} and v_{ik} $(i = 1, 2)$

$$\hbar\omega_{\boldsymbol{k}}u_{1k} = \big(\eta_1 + g_{11}n_1 + g_{12}n_2 - \mu_1\big)u_{1k} + \Delta_{11}v_{1k} + \Delta_{12}u_{2k} + \Delta_{12}v_{2k},$$
$$-\hbar\omega_{\boldsymbol{k}}v_{1k} = \big(\eta_1 + g_{11}n_2 + g_{12}n_2 - \mu_1\big)v_{1k} + \Delta_{11}u_{1k} + \Delta_{12}v_{2k} + \Delta_{12}u_{2k},$$
$$\hbar\omega_{\boldsymbol{k}}u_{2k} = \big(\eta_2 + g_{22}n_2 + g_{12}n_1 - \mu_2\big)u_{2k} + \Delta_{22}v_{2k} + \Delta_{12}u_{1k} + \Delta_{12}v_{1k},$$
$$-\hbar\omega_{\boldsymbol{k}}v_{2k} = \big(\eta_2 + g_{22}n_2 + g_{12}n_1 - \mu_2\big)v_{2k} + \Delta_{22}u_{2k} + \Delta_{12}v_{1k} + \Delta_{12}u_{1k}.$$

In our simple case the amplitudes u_{ik} and v_{ik} $(i = 1, 2)$ prove to be real. For brevity, we have introduced the following notations:

$$\eta_1 = \frac{\hbar^2k^2}{2m_1} + \Delta_{11}, \qquad \eta_2 = \frac{\hbar^2k^2}{2m_2} + \Delta_{22},$$
$$\Delta_{11} = g_{11}n_1, \quad \Delta_{12} = g_{12}\sqrt{n_1n_2}, \quad \Delta_{22} = g_{22}n_2.$$

Nontrivial solution for a set of linear equations exists if the determinant, composed of the coefficients before unknown amplitudes, is zero

$$\begin{vmatrix} \eta_1 - \hbar\omega_k & \Delta_{11} & \Delta_{12} & \Delta_{12} \\ \Delta_{11} & \eta_1 + \hbar\omega_k & \Delta_{12} & \Delta_{12} \\ \Delta_{12} & \Delta_{12} & \eta_2 - \hbar\omega_k & \Delta_{22} \\ \Delta_{12} & \Delta_{12} & \Delta_{22} & \eta_2 + \hbar\omega_k \end{vmatrix} = 0.$$

The frequency of elementary excitations in the binary condensate mixture can be determined from the following biquadratic equation:

$$(\hbar\omega_k)^4 - (\varepsilon_1^2 + \varepsilon_2^2)(\hbar\omega_k)^2 + \varepsilon_1^2\varepsilon_2^2 - 4\Delta_{12}^2(\eta_1 - \Delta_{11})(\eta_2 - \Delta_{22}) = 0.$$

Here $\varepsilon_i = (\eta_i^2 - \Delta_{ii}^2)^{1/2}$ is the Bogoluibov spectrum of elementary excitations in each mixture component separately. As is seen from the dispersion equation, the binary mixture of condensates has already two branches ω_{1k} and ω_{2k} for elementary excitations. The square of the frequency for each branch is given by

$$\omega_{1,2}^2(k) = \frac{1}{2\hbar^2}\left(\varepsilon_1^2 + \varepsilon_2^2 \pm \sqrt{(\varepsilon_1^2 - \varepsilon_2^2)^2 + 16\Delta_{12}^2(\eta_1 - \Delta_{11})(\eta_2 - \Delta_{22})}\right).$$

The coupling constant g_{12} between the particles of different type results in hybridization for the Bogoliubov spectra of separate condensates.

Most interesting range of spectra is long wave one $k \to 0$. Here we discover two sound branches $\omega_{1,2}(k) = u_{1,2}k$ with the following sound velocities determined by the relations:

$$u_{1,2}^2 = \frac{1}{2}\left(u_1^2 + u_2^2 \pm \sqrt{(u_1^2 - u_2^2)^2 + \frac{4g_{12}^2}{g_{11}g_{22}}u_1^2u_2^2}\right)$$

where $u_1 = \sqrt{g_{11}n_1/m_1}$ and $u_2 = \sqrt{g_{22}n_2/m_2}$ are the sound velocities in each separate condensate. The requirement of positive square for the sound velocity, necessary for the phase stability of the mixture against its separation, returns us to the familiar condition $g_{11}g_{22} > g_{12}^2$ for the interparticle coupling constants.

Chapter 8
Theory of Superfluidity

8.1 Thermodynamics of Superfluid Helium

Two stable helium isotopes,[1] ^{3}He and ^{4}He, are known. The ^{3}He atoms obey the Fermi–Dirac statistics but the ^{4}He atoms do the Bose–Einstein one. In liquid[2] ^{4}He at temperature 2.17 K, called λ-*point*, there occurs a second-order phase transition from the normal He I state to the superfluid He II state. The superfluid state of a liquid, first of all, is characterized with its ability to flow through the thin capillaries without any friction, displaying no viscous effects typical for viscid normal liquid.

At absolute zero temperature the liquid is in the energetically lowest or ground state. At nonzero temperature the liquid goes over to an excited state. For sufficiently low temperatures, this will be a weakly excited state which can be described as a set of elementary excitations or quasi-particles having a definite momentum $\boldsymbol{p}$ and corresponding energy $\varepsilon(\boldsymbol{p})$.

In liquid ^{4}He the elementary excitation dispersion law $\varepsilon(\boldsymbol{p})$, obtained from the neutron scattering experiments, has a non-monotonic behavior (Fig. 8.1). After the initial linear growth at sufficiently small momenta the excitation energy $\varepsilon(\boldsymbol{p})$ reaches the maximum, then decreases and passes across the minimum $\Delta = \varepsilon(p_0)$ at certain momentum p_0 with the further growth of excitation energy as the momentum continues to grow. As the energy becomes of the order of 2Δ, the further dependence $\varepsilon(\boldsymbol{p})$ upon momentum becomes almost horizontal. The latter region of spectrum is often referred to as the *Pitaevskii plateau*.

The initial linear segment $\varepsilon(\boldsymbol{p}) = up$ is called the *sound* or phonon spectrum. The elementary excitations are *phonons* or *sound longitudinal oscillations* propagating in the liquid with the sound velocity, $u = 240$ m/s.

The energy spectrum in the vicinity of the minimum at $p = p_0$ is well approximated with the parabolic function

---

[1] The other isotopes are unstable. For example, the half-life of ^{6}He isotope is about 0.8 s.

[2] The ^{4}He solidification pressure is about 25 bar.

S. N. Burmistrov, *Statistical and Condensed Matter Physics*,
https://doi.org/10.1007/978-981-97-9072-2_8

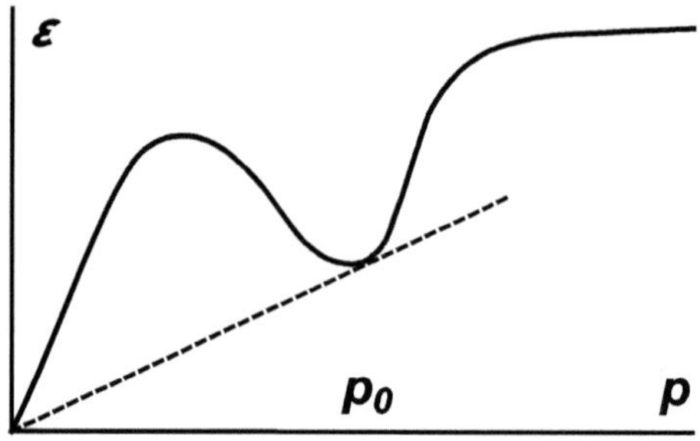

Fig. 8.1 The excitation spectrum in liquid ^{4}He. The dashed line shows the tangent which slope determines the critical velocity

$$\varepsilon_r(\boldsymbol{p}) = \Delta + \frac{(p-p_0)^2}{2m}\,.$$

The elementary excitations in the momentum range $p \sim p_0$ are commonly called *rotons* and the magnitude Δ is the *roton gap*. The roton gap Δ, momentum p_0, effective mass m as well as sound velocity are pressure-dependent and equal to

$$\Delta = 8.7\,\mathrm{K}, \quad p_0/\hbar = 1.9\cdot 10^8\,\mathrm{s}^{-1}, \quad m = 0.16M \quad (\Delta \ll p_0^2/2m),$$

at zero pressure. Here M is the mass of ^{4}He atom.

The thermodynamically equilibrium distribution of elementary excitations in ^{4}He obeys the Bose distribution function with zero chemical potential

$$n(\boldsymbol{p}) = \frac{1}{e^{\varepsilon(\boldsymbol{p})/T} - 1}\,.$$

Zero magnitude of chemical potential results from the fact that the number of elementary excitations is not fixed but determined by the thermodynamic equilibrium condition. The latter means that we must have a minimum of free energy with respect to the number of excitations, i.e.

$$\left(\frac{\partial F}{\partial N}\right)_{V,T} = \mu = 0.$$

For $\mu = 0$, the grand thermodynamic potential Ω coincides with free energy F. Thus, to find the thermodynamic functions[3], it is sufficient to calculate the following integral:

$$F = \Omega = -T\int \frac{V d^3p}{(2\pi\hbar)^3}\ln\left(1 - e^{-\varepsilon_p/T}\right)$$

[3] More precisely, we imply the difference in their magnitudes at the given temperature T and, in particular, at zero temperature $T = 0$.

or after the single integration by parts

$$F = -\frac{1}{3}\int \frac{p\,\partial\varepsilon/\partial p}{e^{\varepsilon_p/T}-1}\frac{Vd^3p}{(2\pi\hbar)^3}\,.$$

For the analytical estimate of the above integral, it is a very good approximation to represent it as a sum of two contributions, namely phonon F_{ph} and roton F_r, i.e. $F \approx F_{ph} + F_r$, with the exception of the narrow region near the λ-point of superfluid transition.

Calculating the phonon contribution with dispersion $\varepsilon_{ph} = up$ yields the following thermodynamic functions per unit volume. The free energy reads

$$F_{ph} = \Omega_{ph} = -\frac{1}{3}\int \frac{pu}{e^{up/T}-1}\frac{d^3p}{(2\pi\hbar)^3} = -\frac{4\pi^5 T}{45}\left(\frac{T}{2\pi\hbar u}\right)^3.$$

Accordingly, energy, entropy, and specific heat of phonon gas will equal

$$E_{ph} = -3F_{ph} = \frac{4\pi^5 T}{15}\left(\frac{T}{2\pi\hbar u}\right)^3,$$

$$S_{ph} = -\frac{\partial F_{ph}}{\partial T} = \frac{16\pi^5}{45}\left(\frac{T}{2\pi\hbar u}\right)^3,$$

$$C_{ph} = T\frac{\partial S_{ph}}{\partial T} = \frac{16\pi^5}{15}\left(\frac{T}{2\pi\hbar u}\right)^3.$$

Calculating the roton contribution, we take into account that the roton momenta lie within the region close to momentum p_0 and inequality $\Delta \gg T$ is valid. In this case the Bose distribution, in essence, crosses over to the Boltzmann one. The integral below determines the contribution of the roton spectrum segment into the free energy per unit volume:

$$F_r = \Omega_r \approx -T\int e^{-\varepsilon_r/T}\frac{d^3p}{2\pi\hbar)^3} = -Te^{-\Delta/T}\int e^{\frac{(p-p_0)^2}{2mT}}\frac{4\pi p^2dp}{8\pi^3\hbar} \approx$$

$$\approx -T\frac{p_0^2e^{-\Delta/T}}{2\pi^2\hbar^3}\int\limits_{-\infty}^{\infty} e^{\frac{(p-p_0)^2}{2mT}}dp = -T\frac{2p_0^2\sqrt{mT}}{(2\pi\hbar^2)^{3/2}}e^{-\Delta/T} = -TN_r,$$

$$N_r = \frac{2p_0^2\sqrt{mT}}{(2\pi\hbar^2)^{3/2}}e^{-\Delta/T}$$

where $N_r(T)$ is the number of rotons in unit volume of helium. Estimating the integral, we have kept inequality $p_0^2/2m \gg T$ in mind. In addition, the last integral is gained within the narrow region $|p - p_0| \lesssim T$.

Differentiating the free energy, we find the roton contribution to the entropy and specific heat of helium II

$$S_r = -\frac{\partial F_r}{\partial T} = N_r\left(\frac{\Delta}{T} + \frac{3}{2}\right),$$
$$C_r = T\frac{\partial S_r}{\partial T} = N_r\left(\frac{\Delta^2}{T^2} + \frac{\Delta}{T} + \frac{3}{4}\right).$$

In the low temperature region, $T \lesssim 0.8$ K the phonon contribution to the thermodynamic functions is predominant. On the contrary, for the high temperature region $T \gtrsim 0.8$ K the thermodynamic behavior of superfluid helium is mainly governed by the roton contribution.

8.2 The Euler Equation of Ideal Fluid

Before proceeding to studying the properties of the superfluid flow, we consider the thermodynamic derivation of the Euler equation for ideal (inviscid) isotropic liquid. Deriving the equation, we use only the thermodynamic identities and the laws of conserving the mass, momentum, energy, and entropy. The entropy conserves only due to our assumption on the ideality of the fluid and absence of any dissipative effects.

Consider the energy of unit liquid volume $E = E(S, \rho, \boldsymbol{j})$ as a function of the following thermodynamic variables: entropy S, liquid density ρ, and mass density flux or momentum per unit volume $\boldsymbol{j}$. The energy differential

$$dE(S, \rho, \boldsymbol{j}) = T\,dS + \mu\,d\rho + \boldsymbol{v}\cdot d\boldsymbol{j}$$

determines the conjugate thermodynamic variables as temperature T, chemical potential per unit mass μ, and fluid velocity $\boldsymbol{v}$. The conjugate thermodynamic potential $A(T, \mu, \boldsymbol{v})$ and its differential will equal

$$A(T, \mu, \boldsymbol{v}) = E - TS - \mu\rho - \boldsymbol{v}\cdot\boldsymbol{j},$$
$$dA(T, \mu, \boldsymbol{v}) = -S\,dT - \rho\,d\mu - \boldsymbol{j}\cdot d\boldsymbol{v}.$$

The quantity $P = -A = -E + TS + \mu\rho + \boldsymbol{v}\cdot\boldsymbol{j}$ should be referred to as *pressure* if we determine the pressure $P = -\partial(EV)/\partial V$ as a minus-sign derivative of total energy EV of a liquid with respect volume V under constant total mass ρV, total entropy SV, and total momentum $\boldsymbol{j}V$.

Then we write down the conservation laws for mass, momentum, energy, and entropy in the differential form

$$\frac{\partial\rho}{\partial t} + \operatorname{div}\boldsymbol{j} = 0, \quad \frac{\partial j_i}{\partial t} + \frac{\partial\Pi_{ik}}{\partial x_k} = 0,$$
$$\frac{\partial E}{\partial t} + \operatorname{div}\boldsymbol{Q} = 0, \quad \frac{\partial S}{\partial t} + \operatorname{div}\boldsymbol{F} = 0,$$

where we should specify *mass density flux* (momentum density) $\boldsymbol{j}$, *energy flux density* $\boldsymbol{Q}$, *entropy flux density* $\boldsymbol{F}$, *momentum flux density tensor* Π_{ik}.

Then, we sum all four equations, in advance multiplying each of them by the corresponding factor T, ρ or $\boldsymbol{v}$, and involving that

$$\frac{\partial E}{\partial t} = T\frac{\partial S}{\partial t} + \mu\frac{\partial \rho}{\partial t} + \boldsymbol{v}\cdot\frac{\partial \boldsymbol{j}}{\partial t}\,.$$

We find after identical transformations

$$\begin{aligned}\operatorname{div}\boldsymbol{Q} = {} & \frac{\partial}{\partial x_i}\big[TF_i + \mu j_i + v_k(\Pi_{ik} + A\delta_{ik})\big] - \\ & - (F_i - Sv_i)\frac{\partial T}{\partial x_i} - (j_i - \rho v_i)\frac{\partial \mu}{\partial x_i} - (\Pi_{ik} + A\delta_{ik} - j_k v_i)\frac{\partial v_k}{\partial x_i}\,.\end{aligned}$$

Provided that any energy dissipation in the liquid is absent, the magnitudes of fluxes $\boldsymbol{Q}$, $\boldsymbol{F}$, $\boldsymbol{j}$, and Π_{ik} must be the functions of thermodynamic variables alone, e.g. T, ρ, and $\boldsymbol{j}$, and be independent of their derivatives with respect to coordinates x_i and time t. In addition, the right-hand side of the above equation should be an expression in the form of total derivative or divergence. Thus, it is necessary to put the following for the dissipationless fluxes:

$$\begin{aligned} j_i = \rho v_i\,, \qquad F_i = Sv_i\,, \\ \Pi_{ik} = -A\delta_{ik} + j_k v_i = P\delta_{ik} + \rho v_i v_k\,, \\ Q_i = TF_i + \mu j_i + v_k(\Pi_{ik} + A\delta_{ik}) = TF_i + \mu j_i + (\boldsymbol{v}\cdot\boldsymbol{j})v_i\,. \end{aligned}$$

For clearness, we rewrite the first two and the last equalities in the vector form

$$\boldsymbol{j} = \rho\boldsymbol{v}, \quad \boldsymbol{F} = S\boldsymbol{v} \quad \text{and} \quad \boldsymbol{Q} = T\boldsymbol{F} + \mu\boldsymbol{j} + (\boldsymbol{v}\cdot\boldsymbol{j})\boldsymbol{v}.$$

The term $P\delta_{ik}$ in the momentum flux density tensor means that the pressure in a fluid at rest is isotropic, i.e. pressure is transmitted equally in all directions (*Pascal's law*).

The substitution of tensor Π_{ik} into the momentum conservation equation and the subsequent use of mass conservation law lead us to the *Euler equation* of ideal fluid

$$\rho\left(\frac{\partial \boldsymbol{v}}{\partial t} + (\boldsymbol{v}\cdot\nabla)\boldsymbol{v}\right) = -\nabla P.$$

This equation is one of the main equations describing the hydrodynamics of fluids. Underline that, deriving the equation, we have completely neglected any possible energy dissipation processes due to viscosity and heat conduction.

Problem

1. A self-gravitational gaseous cloud can be described with the aid of continuity equation, Euler equation for an ideal fluid (gas), and Poisson equation for the gravitational potential ϕ. Assuming the self-gravitational gaseous cloud to be homogeneous and equilibrium, find its density oscillation spectrum and estimate its radius when the cloud becomes unstable against the collapse under the influence of self-gravitational forces (*the Jeans collapse criterion*).

Solution. Let us write down the corresponding equations

$$\frac{\partial \rho}{\partial t} + \operatorname{div}(\rho \boldsymbol{v}) = 0, \quad \frac{\partial \boldsymbol{v}}{\partial t} + (\boldsymbol{v}\nabla)\boldsymbol{v} = -\frac{\nabla P}{\rho} - \nabla\phi \text{ and } \nabla^2\phi = 4\pi G\rho.$$

Here ρ is the density, $\boldsymbol{v}$ is the velocity of a gas, P is the pressure, and G is the gravitational constant. The equilibrium state satisfies two equations

$$\nabla P = -\rho\nabla\phi \text{ and } \nabla^2\phi = 4\pi G\rho.$$

Let us introduce small perturbations of density, pressure, and gravitational potential

$$\rho'(\boldsymbol{r},t) = \rho + \delta\rho(\boldsymbol{r},t), \quad P'(\boldsymbol{r},t) = P + \delta P(\boldsymbol{r},t) \text{ and } \phi'(\boldsymbol{r},t) = \phi + \delta\phi(\boldsymbol{r},t).$$

Then we linearize the equations above in the perturbations from the equilibrium values ρ, P, and ϕ. So, we find

$$\frac{\partial\,\delta\rho}{\partial t} + \rho \operatorname{div}\boldsymbol{v} = 0, \quad \frac{\partial \boldsymbol{v}}{\partial t} = -\frac{\nabla\delta P}{\rho} - \nabla\delta\phi = -\frac{u^2}{\rho}\delta\rho - \nabla\delta\phi \text{ and } \nabla^2\delta\phi = 4\pi G\,\delta\rho,$$

$u^2 = (\partial P/\partial\rho)_s$ being the square of adiabatic sound velocity. Next, we eliminate velocity $\boldsymbol{v}$ with the aid of differentiating the continuity equation with respect to time and find

$$\frac{\partial^2\delta\rho}{\partial t^2} - u^2\nabla^2\delta\rho - 4\pi\rho G\,\delta\rho = 0.$$

Hence we obtain the density oscillation spectrum of a gaseous cloud

$$\omega^2(k) = u^2k^2 - 4\pi\rho G.$$

The homogeneous state of a gaseous cloud is possible if the wave vector k for the cloud density perturbations satisfies the condition $\omega^2(k) > 0$. Accordingly,

$$k > k_J = \frac{\sqrt{4\pi\rho G}}{u}.$$

For the cloud of radius R, the possible minimum wave vector will be of the order of $k_m \sim \pi/R$. Accordingly, from the condition $k_m > k_J$, we arrive at the following estimate for the cloud radius when the internal gas pressure in the cloud can still counteract the gravitational forces of attraction

$$R \lesssim \lambda_J = \frac{u}{\sqrt{\rho G}}.$$

The length scale λ_J is known as the *Jeans length*. The Jeans instability is considered as one of the reasons for the formation of inhomogeneities in the Universe.

8.3 Superfluid and Normal Flows. The Hydrodynamics of Superfluid Liquid

Most fascinating property of helium II is its ability to flow without friction and viscous effects through the thin capillaries. However, such ability of superfluid flow in He II will be restricted with some critical velocity v_c of the fluid flow.

Let liquid ^{4}He be in the ground state at zero temperature and flow at uniform constant velocity $\boldsymbol{v}$. If the liquid is subjected to friction against the capillary walls, a fraction of its kinetic energy converts into heat energy. This results in heating the liquid and its transition into an excited state. The excited state of the liquid is characterized by the presence of elementary excitations and, therefore, friction or viscosity of the liquid should appear as a generation of elementary excitations. The finite lifetime or damping of the generated elementary excitations will result in relaxing the liquid into the ground state and slowing down the fluid flow with corresponding energy dissipation. Thus, the necessary condition for the dissipationless and superfluid flow of a liquid is an exclusion of spontaneous generation of elementary excitations.

Let us consider the reference frame in which the liquid is at rest and let $\varepsilon(\boldsymbol{p})$ be the energy of the elementary excitation generated with momentum $\boldsymbol{p}$. According to the Galilean principle of relativity for the inertial reference frames, the energy of the liquid in the laboratory reference frame, in which the liquid flows at constant and uniform velocity $\boldsymbol{v}$, proves to be equal to

$$\varepsilon(\boldsymbol{p}) + \boldsymbol{p}\boldsymbol{v} + M\boldsymbol{v}^2/2.$$

Here M is the total liquid mass and $M\boldsymbol{v}^2/2$ is the total kinetic energy of the liquid as a whole. Thus, the emergence of elementary excitation in the liquid flowing at velocity $\boldsymbol{v}$ changes the energy of the liquid[4] by quantity $\varepsilon(\boldsymbol{p}) + \boldsymbol{p}\boldsymbol{v}$. The spontaneous generation of elementary excitation becomes energetically favorable provided $\varepsilon(\boldsymbol{p}) + \boldsymbol{p}\boldsymbol{v} < 0$. The generation of excitations is most favorable if momentum $\boldsymbol{p}$ and flow velocity $\boldsymbol{v}$ are antiparallel. So, the necessary condition for the dissipationless fluid flow

$$v < v_c = \min \frac{\varepsilon(p)}{p}$$

will determine the maximum allowable or *critical velocity* of the liquid flow and is called the *Landau criterion of superfluidity*.

The minimum of ratio $\varepsilon(p)/p$ is given by the extremum condition $d(\varepsilon/p)/dp = 0$ resulting in

$$\frac{d\varepsilon}{dp} = \frac{\varepsilon}{p}.$$

Geometrically, this condition corresponds to the point at which the curve $\varepsilon(p)$ is tangent to a straight line starting from the coordinate origin (Fig. 8.1). For the excitation

---

[4] One may say the elementary excitation energy is subjected to the Doppler shift.

spectrum[5] in liquid ^{4}He, this point lies somewhat on the right-hand side of the roton minimum point p_0 and the critical velocity

$$v_c = \frac{\sqrt{p_0^2 + 2m\Delta} - p_0}{m} \approx \frac{\Delta}{p_0}$$

equals approximately 58 m/s. In a weakly non-ideal Bose gas with the Bogoliubov excitation spectrum the critical velocity equals the sound velocity. For the ideal condensed Bose gas with the particle-like excitation spectrum $\varepsilon = p^2/2M$, the critical velocity v_c proves to be zero and the phenomenon of superfluidity is absent in spite of the presence of condensate.

The experimental study of the liquid He II flow through the ultrathin capillaries and ultra-narrow slits (superleaks)[6] shows that not all the helium mass[7] but only its fraction does flow through the capillaries. This fraction is naturally called the *superfluid component* and its flow is specified with some *superfluid density* ρ_s and *superfluid velocity* $\boldsymbol{v}_s$. The remaining helium fraction, stopped with the capillary, is called the *normal component* on the analogy with the behavior of viscid normal fluid. The normal component is characterized with some *normal density* ρ_n and *normal velocity* $\boldsymbol{v}_n$.

Such behavior of liquid He II indicates that two independent flows, superfluid and normal, realize simultaneously in the liquid at the finite temperature. Since the total mass of the liquid remains unchangeable, a sum of the superfluid and normal component masses should be equal to the total mass of the liquid. Accordingly, a sum of superfluid and normal densities equals the total density of the liquid ρ

$$\rho_n + \rho_s = \rho.$$

The fraction of the flowing superfluid component ρ_s/ρ reduces as the temperature increases and vanishes completely at the λ-point of the superfluid transition.

An existence of normal component in liquid He II is self-consistently to associate with the thermal elementary excitations. The latter ones are always present in the thermally equilibrium concentration at $T > 0$ and increase their number as the temperature grows. However, a possibility to change the number of elementary excitations due to additional spontaneous generation is absent for small flow velocities $v < v_c$. The fluid will stay in the same state with the same energy and, therefore, can maintain the dissipationless flow of the liquid fraction decoupled with the transport of elementary excitations.

[5] Along with the above-considered type of elementary excitations, the quantized vortex-like excitations are possible and observed in a rotating superfluid He II. The critical velocity for the vortex generation is determined by $v_{cr}R \sim \hbar/M_{^4\mathrm{He}}$ where R is the typical linear size of a vessel or magnitude of roughness at the vessel walls.

[6] E.g., nano-dispersed silver powder or nano-porous glass (vycor) is usually applied as an obstacle for the helium-II flow.

[7] This property is used to purify the superfluid ^{4}He from the ^{3}He isotope impurities.

In what follows, we will employ the above-described thermodynamic approach for determining the equations for the two-fluid hydrodynamics of isotropic superfluid liquid on the neglect of possible energy dissipation effects. To derive, we use the conservation laws of mass, momentum, energy, entropy, and the Galilean transformation in addition. Similarly to an ideal fluid the superfluid component flow is potential or *irrotational*, i.e.

$$\operatorname{curl} \boldsymbol{v}_s = 0,$$

in the whole space occupied with the superfluid component. The vector curl $\boldsymbol{v}_s$ is called *vorticity*. The irrotational condition should not be disturbed while the flow velocities do not reach the critical magnitudes and no coupling between the normal and superfluid components appears.

From the physical point of view an existence of two independent flows in the fluid instead of one flow will require to augment the number of variables to describe the state of the fluid. Thus, the thermodynamic variables inherent in usual normal fluid will be augmented with the superfluid velocity $\boldsymbol{v}_s$ and conjugate variable called the mass flux $\boldsymbol{j}_s$ of superfluid component. So, the energy density of superfluid liquid $E = E(S, \rho, \boldsymbol{j}, \boldsymbol{v}_s)$ should depend on the four variables such as: entropy S, density ρ, mass flux density or momentum density $\boldsymbol{j}$, and superfluid velocity $\boldsymbol{v}_s$.

As a first step, let us write down the conservation laws for mass, momentum, energy, and entropy in the differential form

$$\begin{aligned} &\tfrac{\partial \rho}{\partial t} + \operatorname{div} \boldsymbol{j} = 0, \quad \tfrac{\partial j_i}{\partial t} + \tfrac{\partial \Pi_{ik}}{\partial x_k} = 0, \\ &\tfrac{\partial E}{\partial t} + \operatorname{div} \boldsymbol{Q} = 0, \quad \tfrac{\partial S}{\partial t} + \operatorname{div} \boldsymbol{F} = 0. \end{aligned}$$

The elimination of the energy dissipation effects from our consideration means the conservation of the total entropy.

In the two-fluid hydrodynamics these four equations should be augmented with one more equation describing the dynamics of superfluid velocity $\boldsymbol{v}_s$. On the analogy with an ordinary fluid this should be an equation for the time derivative with respect to superfluid velocity $\boldsymbol{v}_s$. To satisfy the condition of irrotational flow curl $\boldsymbol{v}_s = 0$, we can expect the equation analogous to the Euler equation for an ideal liquid

$$\frac{\partial \boldsymbol{v}_s}{\partial t} + \nabla \psi = 0.$$

Here $\psi = \psi(S, \rho, \boldsymbol{j}, \boldsymbol{v}_s)$ is a scalar function to be determined together with the fluxes of energy $\boldsymbol{Q}$, entropy $\boldsymbol{F}$, mass $\boldsymbol{j}$, and the momentum flux density tensor Π_{ik}.

To find the expressions for the flows, as a first step it is convenient to choose the reference frame K_0 in which the superfluid component is at rest, i.e. $\boldsymbol{v}_s = 0$, but the normal component alone streams with the counterflow velocity $\boldsymbol{w} = \boldsymbol{v}_n - \boldsymbol{v}_s$. Since the superfluid component is at rest in this reference frame and does not participate in the liquid flow, we can expect that the properties of the liquid in the reference frame K_0 will be the same as those of ordinary normal fluid. This conjecture allows us to determine the corresponding fluxes $\boldsymbol{j}_0$, $\boldsymbol{Q}_0$, $\boldsymbol{F}_0$, and Π_{0ik}.

To find the fluxes in the laboratory reference frame K in which the superfluid component flows at velocity $\boldsymbol{v}_s$, we use the Galilean transformation connecting the corresponding physical quantities in the different reference frames.

For the energy density $E_0 = E_0(S, \rho, \boldsymbol{j}_0) = E(S, \rho, \boldsymbol{j}, \boldsymbol{v}_s = 0)$ in the reference frame K_0, we write down a usual differential

$$dE_0 = T\,dS + \mu\,d\rho + \boldsymbol{w}\cdot d\boldsymbol{j}_0.$$

Here, the temperature T, chemical potential μ, and counterflow velocity $\boldsymbol{w}$ are defined as thermodynamic variables conjugated to variables S, ρ, and $\boldsymbol{j}_0$. The flux vector $\boldsymbol{j}_0$ as a vector in isotropic liquid can only be directed along the counterflow one $\boldsymbol{w}$, i.e.

$$\boldsymbol{j}_0 = \rho_n \boldsymbol{w}.$$

The scalar $\rho_n = \rho_n(S, \rho, \boldsymbol{w}^2)$ determines the magnitude of normal component density and, in general, depends on entropy (temperature), total density of a liquid ρ, and the square of counterflow velocity $\boldsymbol{w}$.

Accordingly, we can write down the following expressions for the other fluxes in the reference frame K_0:

$$\Pi_{0ik} = P\delta_{ik} + \rho_n w_i w_k,$$
$$\boldsymbol{Q}_0 = T\boldsymbol{F}_0 + \mu\boldsymbol{j}_0 + (\boldsymbol{w}\cdot\boldsymbol{j}_0)\boldsymbol{w} \quad \text{and} \quad \boldsymbol{F}_0 = S\boldsymbol{w}.$$

The pressure P and its differential dP are related to the conjugate thermodynamic potential $A = E_0 - TS - \mu\rho - \boldsymbol{w}\cdot\boldsymbol{j}$ in accordance with

$$P = -A = -E_0 + TS + \mu\rho + \boldsymbol{w}\cdot\boldsymbol{j}_0,$$
$$dP = -dA = S\,dT + \rho\,d\mu + \boldsymbol{j}_0\cdot d\boldsymbol{w}.$$

In the reference frames K and K_0, the quantities of our interest can be related with the aid of the Galilean transformations

$$E = E_0 + \boldsymbol{j}_0\cdot\boldsymbol{v}_s + \frac{\rho\boldsymbol{v}_s^2}{2},$$
$$\boldsymbol{j} = \boldsymbol{j}_0 + \rho\boldsymbol{v}_s, \qquad \boldsymbol{F} = \boldsymbol{F}_0 + S\boldsymbol{v}_s,$$
$$\Pi_{ik} = \Pi_{0ik} + \rho v_{si} v_{sk} + v_{si} j_{0k} + j_{0i} v_{sk},$$
$$Q_i = Q_{0i} + \left(E_0 + \boldsymbol{j}_0\cdot\boldsymbol{v}_s + \frac{\rho\boldsymbol{v}_s^2}{2}\right)v_{si} + \frac{\boldsymbol{v}_s^2}{2} j_{0i} + \Pi_{0ik} v_{sk}.$$

Substituting the values of fluxes in reference frame K_0, after the algebraic calculations we find the values of the fluxes expressed via normal $\boldsymbol{v}_n$ and superfluid $\boldsymbol{v}_s$ velocities in the laboratory reference frame K

$$j = \rho_n \boldsymbol{w} + \rho \boldsymbol{v}_s = \rho_n(\boldsymbol{v}_n - \boldsymbol{v}_s) + (\rho_n + \rho_s)\boldsymbol{v}_s = \rho_n \boldsymbol{v}_n + \rho_s \boldsymbol{v}_s,$$
$$\boldsymbol{F} = S\boldsymbol{w} + S\boldsymbol{v}_s = S(\boldsymbol{v}_n - \boldsymbol{v}_s) + S\boldsymbol{v}_s = S\boldsymbol{v}_n,$$
$$\Pi_{ik} = P\delta_{ik} + \rho_n v_{ni} v_{nk} + (\rho - \rho_n) v_{si} v_{sk} = P\delta_{ik} + \rho_n v_{ni} v_{nk} + \rho_s v_{si} v_{sk},$$
$$\boldsymbol{Q} = \Big(\mu + \frac{\boldsymbol{v}_s^2}{2}\Big)\boldsymbol{j} + TS\boldsymbol{v}_n + \rho_n \boldsymbol{v}_n(\boldsymbol{v}_n - \boldsymbol{v}_s)\cdot\boldsymbol{v}_n.$$

As we see, the mass flux density $\boldsymbol{j}$ and momentum flux density tensor Π_{ik} include the terms associated with the both normal and superfluid component flows. Emphasize that this property does not concern the entropy transfer. In fact, according to the formula for the entropy flux $\boldsymbol{F} = S\boldsymbol{v}_n$, we see that the entropy flux is connected with the normal component flow alone. The superfluid component does not contribute to the entropy transfer. *The transfer of entropy is convective-like*. Note also that, for an equality between normal and superfluid velocities $\boldsymbol{v}_n = \boldsymbol{v}_s = \boldsymbol{v}$, the expressions of the fluxes resemble those for the corresponding ones in a normal non-superfluid liquid.

Now it remains to determine the scalar function ψ which gradient governs the time derivative of the superfluid velocity. This can be done, e.g. using the energy conservation law and above-derived expressions for the energy density and fluxes in the laboratory reference frame K. Calculating yields a simple answer after considerable reductions

$$\frac{\partial E}{\partial t} + \operatorname{div} \boldsymbol{Q} = -\rho_s(\boldsymbol{v}_n - \boldsymbol{v}_s)\cdot\left(\frac{\partial \boldsymbol{v}_s}{\partial t} + \nabla\Big(\mu + \frac{\boldsymbol{v}_s^2}{2}\Big)\right) + \rho \boldsymbol{v}_n \cdot (\boldsymbol{v}_s \times \operatorname{curl} \boldsymbol{v}_s) = 0.$$

The requirement of zero right-hand side of equation together with the irrotational flow condition $\operatorname{curl} \boldsymbol{v}_s = 0$ for the superfluid component will give us the scalar function $\psi = \mu + \boldsymbol{v}_s^2/2$ and equation of motion for the superfluid component velocity

$$\frac{\partial \boldsymbol{v}_s}{\partial t} + \nabla\Big(\mu + \frac{\boldsymbol{v}_s^2}{2}\Big) = 0.$$

Let us write down the other hydrodynamic equations of superfluid liquid

$$\frac{\partial \rho}{\partial t} + \operatorname{div} \boldsymbol{j} = 0, \qquad \boldsymbol{j} = \rho_n \boldsymbol{v}_n + \rho_s \boldsymbol{v}_s,$$
$$\frac{\partial \boldsymbol{j}}{\partial t} + \boldsymbol{v}_s \operatorname{div} \boldsymbol{j} + (\boldsymbol{j}\cdot\nabla)\boldsymbol{v}_s + \boldsymbol{j}_0 \operatorname{div} \boldsymbol{v}_n + (\boldsymbol{v}_n\cdot\nabla)\boldsymbol{j}_0 = -\nabla P,$$
$$\frac{\partial S}{\partial t} + \operatorname{div}(S\boldsymbol{v}_n) = 0, \qquad \boldsymbol{j}_0 = \rho_n(\boldsymbol{v}_n - \boldsymbol{v}_s).$$

The complete equations of two-fluid hydrodynamics are rather complicated since the scalar variables μ, ρ, S, and P in the above equations are also the functions of counterflow velocity $\boldsymbol{v}_n - \boldsymbol{v}_s$ and the separate calculation and determination for them are required as well.

8.4 Normal and Superfluid Densities

In what follows, we consider the question about calculating and determining the normal component density ρ_n. Any elementary excitation has some momentum $\boldsymbol{p}$ and thus contributes to the total momentum of a liquid. The following integral determines the momentum per unit volume or momentum density corresponding to the equilibrium distribution function $n(\boldsymbol{p})$

$$\boldsymbol{j} = \int \boldsymbol{p} n(\boldsymbol{p}) \frac{d^3 p}{(2\pi\hbar)^3}.$$

Obviously, momentum density vector vanishes due to symmetry $n(\boldsymbol{p}) = n(-\boldsymbol{p})$. If there is some relative velocity between the liquid and elementary excitations, the momentum per unit volume is no longer to be zero. Nonzero momentum means that the motion of excitations relative to the liquid will be accompanied with the transfer of some liquid mass.

Let all elementary excitations move as a whole at certain velocity $\boldsymbol{v}$ relative to the liquid at rest. In the reference frame in which the excitations are immobile, the liquid flows at velocity $(-\boldsymbol{v})$. Due to the Doppler shift the energy of elementary excitations in such reference frame becomes equal to $\varepsilon(\boldsymbol{p}) + \boldsymbol{p}(-\boldsymbol{v})$. The distribution function of excitations takes the form $n(\varepsilon_{\boldsymbol{p}} - \boldsymbol{p}\boldsymbol{v})$ and the following integral gives the momentum density resulted from the motion of elementary excitations:

$$\boldsymbol{j} = \int \boldsymbol{p} n(\varepsilon_{\boldsymbol{p}} - \boldsymbol{p}\boldsymbol{v}) \frac{d^3 p}{(2\pi\hbar)^3}.$$

Decomposing in small velocities $\boldsymbol{v}$, we have the answer

$$\boldsymbol{j} \approx -\int \boldsymbol{p}(\boldsymbol{p}\boldsymbol{v}) \frac{\partial n}{\partial \varepsilon_p} \frac{d^3 p}{(2\pi\hbar)^3} = -\boldsymbol{v}\frac{1}{3}\int p^2 \frac{\partial n}{\partial \varepsilon_p} \frac{d^3 p}{(2\pi\hbar)^3}.$$

Deriving the last equality, we have averaged over the whole directions of vector $\boldsymbol{p}$. Rewriting this relation as $\boldsymbol{j} = \rho_{\text{ex}}\boldsymbol{v}$, we see that nonzero flux of elementary excitations is inevitably accompanied with transferring some fraction of the liquid and characterized by some effective density ρ_{ex}. The latter, in general, depends on velocity $\boldsymbol{v}$ and distribution function $n(\varepsilon)$. The fraction of the liquid mass transferred equals ρ_{ex}/ρ, ρ being the total mass of the liquid.

Let us suppose that in He II there are two homogeneous flows, namely normal at velocity $\boldsymbol{v}_n$ and superfluid at velocity $\boldsymbol{v}_s$. In the reference frame comoving together with the superfluid component the momentum density or momentum per unit volume will be equal to the following flux of normal component:

$$\boldsymbol{j}_n = \boldsymbol{j} - \rho\boldsymbol{v}_s = \rho_n(\boldsymbol{v}_n - \boldsymbol{v}_s) = \rho_n\boldsymbol{w}.$$

On the other hand, we can express the momentum of the normal component flowing at the counterflow velocity $\boldsymbol{w} = \boldsymbol{v}_n - \boldsymbol{v}_s$ by means of the integral

$$\boldsymbol{j}_n = \int \boldsymbol{p} n(\varepsilon_p - \boldsymbol{p}\boldsymbol{w}) \frac{d^3p}{(2\pi\hbar)^3}\,.$$

The integral sums the momenta of all elementary excitations. Comparing both expressions above, we arrive at the equation determining the magnitude of normal density $\rho_n = \rho_n(w^2)$

$$\rho_n \boldsymbol{w} = \int \boldsymbol{p} n(\varepsilon_p - \boldsymbol{p}\boldsymbol{w}) \frac{d^3p}{(2\pi\hbar)^3}\,.$$

For the small values of counterflow velocity $\boldsymbol{w}$, the expansion of the integrand results in the following magnitude of normal density in superfluid He II:

$$\rho_n = \frac{1}{3}\int p^2\left(-\frac{\partial n(\varepsilon_p)}{\partial p}\right)\frac{d^3p}{(2\pi\hbar)^3}\,.$$

Let us calculate first the contribution to the normal density from the phonon segment of excitation spectrum with dispersion $\epsilon_p = up$

$$\rho_{n,ph}(w)\boldsymbol{w} = \int \boldsymbol{p}\left[\exp\left(\frac{up - \boldsymbol{p}\boldsymbol{w}}{T}\right) - 1\right]^{-1}\frac{d^3p}{(2\pi\hbar)^3}.$$

Choosing the z axis in the direction of the counterflow vector $\boldsymbol{w}$, we have

$$\rho_{n,ph}(w)w = \frac{2\pi}{8\pi^3\hbar^3}\int\limits_0^\infty p^2dp\int\limits_0^\pi \sin\theta d\theta\frac{p\cos\theta}{e^{(up-pw\cos\theta)/T} - 1} =$$

$$= \frac{T^4}{4\pi^2\hbar^3u^4}\int\limits_{-1}^{1}\frac{x\,dx}{(1 - xw/u)^4}\int\limits_0^\infty\frac{\varepsilon^3d\varepsilon}{e^\varepsilon - 1} = \frac{T^4}{4\pi^2\hbar^3u^4}\frac{8}{3}\frac{w/u}{(1 - w^2/u^2)^3}\frac{\pi^4}{15}.$$

Hence the phonon contribution to the normal component density equals

$$\rho_{n,ph}(w) = \frac{\rho_{n,ph}(0)}{(1 - w^2/u^2)^3}\,, \quad \rho_{n,ph}(0) = \frac{2\pi^2T^4}{45u^5\hbar^3} = \frac{4}{3}\frac{E_{ph}(T)}{u^2}.$$

As the counterflow velocity increases, the normal component density grows and diverges at $w = u$. This indicates a principal impossibility of superfluid motion at velocities $w > u$ and limitations of the theory at the flow velocities comparable with the sound one.

As concerns the similar calculation for the contribution of the roton dispersion segment, we can use the Boltzmann distribution instead of the Planck one due to strong inequality $\Delta \gg T$

$$\rho_{n,r}(w)\boldsymbol{w} = \int \boldsymbol{p} \exp\left(-\frac{\varepsilon_p - \boldsymbol{p}\boldsymbol{w}}{T}\right)\frac{d^3p}{(2\pi\hbar)^3}, \quad \varepsilon_p = \Delta + \frac{(p-p_0)^2}{2m},$$

$$\rho_{n,r}(w) = 3\rho_{n,r}(0)\frac{\tilde{w}\cosh\tilde{w} - \sinh\tilde{w}}{\tilde{w}^3}, \quad \tilde{w} = \frac{wp_0}{T}$$

$$\text{and} \quad \rho_{n,r}(0)\rho_{n,r}(w=0) = \frac{p_0^2}{3T}N_r(T).$$

Here $N_r(T)$ is the number of rotons in unit volume of helium. The superfluid component density ρ_n in He II as well as the thermodynamic functions is adequately described by a sum of phonon and roton contributions, i.e. $\rho_n = \rho_{n,ph} + \rho_{n,r}$. As a rule, for most of problems it is possible to ignore the dependence of thermodynamic variables on the counterflow velocity $\boldsymbol{w}$. The point is that the ratios w/u and wT/p_0 are noticeably small in the flow velocity region where superfluid He properties are observed and used.

Problem

1. Find the flux density $\boldsymbol{i}$ in the gas of thermodynamically equilibrium phonons for its flow as a whole at velocity $\boldsymbol{w}$ relative to the liquid. The phonon dispersion is $\varepsilon = up$.

Solution. The phonon flux density is determined with the integral

$$\boldsymbol{i} = \int \boldsymbol{v}n(\boldsymbol{p})\frac{d^3p}{(2\pi\hbar)^3},$$

$\boldsymbol{v} = \partial\varepsilon/\partial\boldsymbol{p}$ being the velocity of excitations. For the flow of elementary excitations as a whole in the liquid at velocity $\boldsymbol{w}$, the distribution function changes according to $n(\varepsilon_{\boldsymbol{p}}) \to n(\varepsilon_{\boldsymbol{p}} - \boldsymbol{p}\boldsymbol{w})$ as a result of the Doppler shift of excitation energy. So, we have

$$\boldsymbol{i} = \int \boldsymbol{v}n(\varepsilon_p - \boldsymbol{p}\boldsymbol{w})\frac{d^3p}{(2\pi\hbar)^3}.$$

Let z axis be taken in the direction of vector $\boldsymbol{w}$. Then we have for the density of the flux directed along the z axis

$$i = \frac{2\pi}{8\pi^3\hbar^3}\int_0^\infty p^2dp\int_0^\pi \sin\theta d\theta\frac{u\cos\theta}{e^{(up-pw\cos\theta)/T}-1} =$$

$$= \frac{u}{2\pi^2\hbar^3}\left(\frac{T}{u}\right)^3\int_0^\infty\frac{\epsilon^2 d\epsilon}{e^\epsilon - 1}\int_{-1}^{1}\frac{x\,dx}{(1-xw/u)^3} = \frac{u}{2\pi^2\hbar^3}\left(\frac{T}{u}\right)^3 2\zeta(3)\frac{2w/u}{(1-w^2/u^2)^2} =$$

$$= \frac{2w}{(1-w^2/u^2)^2}N_{ph}(T) \quad \text{where} \quad N_{ph}(T) = \frac{T^3}{\pi^2\hbar^3u^3}\zeta(3).$$

Here $N_{ph}(T)$ is the number of phonons in the unit He II volume. We can relate the normal component flux density $\boldsymbol{j}_n$ to the phonon number flux density of the phonon number $\boldsymbol{i}$ as

$$\boldsymbol{j}_n = \frac{\pi^4}{45\zeta(3)}\frac{T}{u^2}\frac{\boldsymbol{i}}{1-w^2/u^2}.$$

2. The same for the gas of rotons.

Solution. Let us write down the expression of the roton flux density, involving the roton velocity $\boldsymbol{v} = \partial\varepsilon_p/\partial\boldsymbol{p}$

$$\boldsymbol{i} = \int \boldsymbol{v}\exp\left(-\frac{\Delta + \frac{(p-p_0)^2}{2m} - \boldsymbol{pw}}{T}\right)\frac{d^3p}{(2\pi\hbar)^3}, \quad \boldsymbol{v} = \frac{p-p_0}{mp}\boldsymbol{p}.$$

The z axis is taken in the direction of vector $\boldsymbol{w}$. We have for the density of the flux directed along the z axis

$$i = \frac{2\pi}{8\pi^3\hbar^3}e^{-\Delta/T}\int_0^\infty p^2dp\int_0^\pi \sin\theta d\theta\left(\frac{p-p_0}{m}\cos\theta\right)e^{-\frac{(p-p_0)^2}{2mT}}e^{\frac{pw\cos\theta}{T}} =$$

$$= \frac{e^{-\Delta/T}}{4\pi^2\hbar^3}\int_0^\infty dp\frac{p^2(p-p_0)}{m}e^{-\frac{(p-p_0)^2}{2mT}}\int_{-1}^1 xe^{\frac{pwx}{T}}dx \approx$$

$$\approx w\frac{e^{-\Delta/T}}{4\pi^2\hbar^3}\int_0^\infty dp\frac{2}{3}\frac{p^3(p-p_0)}{mT}e^{-\frac{(p-p_0)^2}{2mT}}.$$

Here we have restricted ourselves with the approximation of small counterflow velocity $w \ll T/p_0$. The estimate of the last integral yields a simple relation in the $p_0^2 \ll 2mT$ approximation

$$\boldsymbol{i} = N_r\boldsymbol{w} \quad \text{and} \quad N_r = \frac{2p_0^2(mT)^{1/2}}{(2\pi\hbar^2)^{3/2}}e^{-\Delta/T},$$

N_r being the number of rotons per unit volume. Relating the normal component flux density $\boldsymbol{j}_n$ and the density of the roton number flux, we get the formula

$$\boldsymbol{j}_n = \frac{p_0^2}{3T}\boldsymbol{i}.$$

3. The ^{3}He atoms diluted in small concentration in superfluid ^{4}He can be described as an ideal gas of Fermi excitations (*impuritons*) with momentum $\boldsymbol{p}$ and energy $\varepsilon_p = -\varepsilon_0 + p^2/2m^*$ where ε_0 is the dilution energy of a ^{3}He atom and m^* is the effective ^{3}He mass. At low velocities the ^{3}He impurities do not interact with the superfluid component of liquid ^{3}He-^{4}He mixture and are completely dragged along with the normal component flow.

Find the impurity contribution to the normal component density in a dilute liquid ^{3}He-^{4}He mixture with concentration N_3 of ^{3}He atoms.

Solution. At low velocities of flow the normal component density is determined by the integral

$$\rho_{n,i} = \frac{1}{3}\int p^2\left(-\frac{\partial n_i}{\partial\varepsilon}\right)\frac{2d^3p}{(2\pi\hbar)^3}.$$

Here $n_i(\varepsilon)$ is the Fermi–Dirac distribution function of impurities. The factor 2 takes the nuclear spin 1/2 of ^{3}He atoms into account. While calculating, it is convenient to include the dilution energy ε_0 into the chemical potential. Then, we have

$$\rho_{n,i} = \frac{2m^*}{3}\int \varepsilon\left(-\frac{\partial n_i}{\partial\varepsilon}\right)\frac{2d^3p}{(2\pi\hbar)^3}.$$

Integrating by parts yields

$$\rho_{n,i} = m^*\int n_i(\varepsilon)\frac{2d^3p}{(2\pi\hbar)^3} = m^*N_3.$$

The superfluid d-phase of liquid ^{3}He-^{4}He mixture is an example of superfluid system in which the normal component density remains finite[8] as $T \to 0$.

8.5 The First and Second Sounds in Superfluid Liquids

Let us apply the above-derived hydrodynamical equations of superfluid liquid for studying the sound propagation in a superfluid liquid. As usual, the liquid flow velocity in sound oscillations is assumed to be sufficiently low and such thermodynamic variables as pressure and temperature are close to the equilibrium magnitudes. Then, we linearize a set of hydrodynamical equations, neglecting all possible second and higher order terms. A set of the linearized equations reduces to

$$\frac{\partial \rho}{\partial t} + \operatorname{div} \boldsymbol{j} = 0, \quad \frac{\partial \boldsymbol{j}}{\partial t} = -\nabla P; \quad \boldsymbol{j} = \rho_n \boldsymbol{v}_n + \rho_s \boldsymbol{v}_s \quad \text{and} \quad \rho = \rho_n + \rho_s,$$
$$\frac{\partial (\sigma\rho)}{\partial t} + \sigma\rho \operatorname{div} \boldsymbol{v}_n = 0, \quad \frac{\partial \boldsymbol{v}_s}{\partial t} + \nabla\mu = 0;$$
$$\sigma = \frac{S}{\rho} \quad \text{and} \quad \frac{\nabla P}{\rho} = \sigma\nabla T + \nabla\mu.$$

For further convenience, we have introduced the specific entropy σ instead of entropy density S according to $\sigma = S/\rho$.

Eliminating the momentum density $\boldsymbol{j}$ from the first two equations, we have the equation for the relation between the oscillations of pressure P and density ρ. This equation is similar to that known for the normal fluid

$$\frac{\partial^2 \rho}{\partial t^2} = \nabla^2 P.$$

Using the thermodynamic relation $d\mu = -\sigma\, dT + dP/\rho$, we find from the second and fourth equations

$$\rho_n \frac{\partial}{\partial t}(\boldsymbol{v}_n - \boldsymbol{v}_s) = -\sigma\rho\nabla T.$$

Using the first and third equation yields

$$\rho_s \operatorname{div} (\boldsymbol{v}_n - \boldsymbol{v}_s) = -\frac{\rho}{\sigma}\frac{\partial \sigma}{\partial t}.$$

Eliminating the counterflow $(\boldsymbol{v}_n - \boldsymbol{v}_s)$ from the last two equations, we arrive at the equations relating the oscillations of specific entropy σ and temperature T

[8] It is expected that the impurity ^{3}He component transmits to the superfluid state at ultralow temperatures. Such phase transition is not observed down to 100 μK.

$$\frac{\partial^2 \sigma}{\partial t^2} = \frac{\sigma^2 \rho_s}{\rho_n} \nabla^2 T.$$

Let us introduce the deviations of pressure δP and temperature δT from their equilibrium values. The deviations of density $\delta\rho$ and entropy $\delta\sigma$ from their equilibrium values can be represented as

$$\delta\rho = \frac{\partial \rho}{\partial P}\delta P + \frac{\partial \rho}{\partial T}\delta T \quad \text{and} \quad \delta\sigma = \frac{\partial \sigma}{\partial P}\delta P + \frac{\partial \sigma}{\partial T}\delta T.$$

Substituting them into the equations for the time derivatives of density and entropy gives a set of two equations

$$\frac{\partial \rho}{\partial P}\frac{\partial^2 \delta P}{\partial t^2} - \nabla^2 \delta P = -\frac{\partial \rho}{\partial T}\frac{\partial^2 \delta T}{\partial t^2},$$
$$\frac{\partial \sigma}{\partial T}\frac{\partial^2 \delta T}{\partial t^2} - \frac{\sigma^2 \rho_s}{\rho_n}\nabla^2 \delta T = \frac{\partial \sigma}{\partial P}\frac{\partial^2 \delta P}{\partial t^2}.$$

As usual, we seek the solution of equations as a plane wave propagating in the x axis direction at frequency ω and velocity u. In other words, the oscillations of pressure $\delta P(x,t)$ and temperature $\delta T(x,t)$ are proportional to the general factor $\exp[-i\omega(t - x/u)]$. Then,

$$\left(u^2\frac{\partial \rho}{\partial P} - 1\right)\delta P + u^2\frac{\partial \rho}{\partial T}\delta T = 0,$$
$$u^2\frac{\partial \sigma}{\partial P}\delta P + \left(u^2\frac{\partial \sigma}{\partial T} - \frac{\sigma^2 \rho_s}{\rho_n}\right)\delta T = 0.$$

Equating the determinant of this system to zero yields the equation which determines the sound velocity u

$$u^4\left(\frac{\partial \rho}{\partial P}\frac{\partial \sigma}{\partial T} - \frac{\partial \rho}{\partial T}\frac{\partial \sigma}{\partial P}\right) - u^2\left(\frac{\partial \sigma}{\partial T} + \frac{\sigma^2 \rho_s}{\rho_n}\frac{\partial \rho}{\partial P}\right) + \frac{\sigma^2 \rho_s}{\rho_n} = 0.$$

The equation can be rewritten in more obvious form if we use the thermodynamic relation between the specific heat at constant volume (constant density) $c_V = T(\partial\sigma/\partial T)_V$ and constant pressure $c_P = T(\partial\sigma/\partial T)_P$

$$c_V = T\left(\frac{\partial \sigma}{\partial T}\right)_P - T\frac{\left(\frac{\partial \sigma}{\partial P}\right)_T\left(\frac{\partial \rho}{\partial T}\right)_P}{\left(\frac{\partial \rho}{\partial P}\right)_T}.$$

As a result, we get

$$\left(\frac{u^2}{u_1^2} - 1\right)\left(\frac{u^2}{u_2^2} - 1\right) = 1 - \frac{c_V}{c_P}$$

where we denote

$$\frac{1}{u_1^2} = \frac{c_V}{c_P}\left(\frac{\partial \rho}{\partial P}\right)_T = \left(\frac{\partial \rho}{\partial P}\right)_S,$$
$$\frac{1}{u_2^2} = \frac{c_V}{c_P}\frac{\rho_n(\partial\sigma/\partial T)_P}{\sigma^2\rho_s} = \frac{\rho_n(\partial\sigma/\partial T)_V}{\sigma^2\rho_s}.$$

The biquadratic equation derived determines two possible sound velocities. The difference $c_P - c_V$ in the specific heats is proportional to the square of thermal expansion coefficient $\alpha_V = -\rho^{-1}(\partial\rho/\partial T)_P$. At low temperatures in superfluid ^{4}He the magnitude α_V is negligible and we can make no difference between the specific heats c_V and c_P, putting them practically equal to each other with the exception of close vicinity of λ-point of superfluid transition. This circumstance simplifies the solution of the equation and we have for the roots

$$u_1 = \sqrt{\left(\frac{\partial P}{\partial \rho}\right)_\sigma} \quad \text{and} \quad u_2 = \sqrt{\frac{\sigma^2\rho_s}{\rho_n\left(\partial\sigma/\partial T\right)_\rho}}.$$

The first root u_1 determines the velocity of *first sound* or ordinary sound existing above the superfluid transition temperature. The oscillations of pressure or density propagate at such velocity. In the first sound wave the normal and superfluid component velocities coincide $\boldsymbol{v}_n = \boldsymbol{v}_s$ if the thermal expansion coefficient is neglected, i.e. $\alpha_V = 0$. This means that the normal and superfluid components oscillate in-phase. In this sense the first sound is similar to the usual sound in normal fluid.

The second root u_2 determines the velocity of *second sound*. The undamped oscillations of temperature propagate at this velocity. The wave of second sound[9] is a specific attribute of superfluids. At the λ-point of superfluid transition the second sound velocity vanishes together with the superfluid density. For the low temperature $T \to 0$ limit when elementary excitations in ^{4}He are phonons, the second sound velocity is $u_2 = u_1/\sqrt{3}$. In the $\alpha_V = 0$ approximation, the density and pressure in the second sound wave do not oscillate and the total mass flow is absent $\boldsymbol{j} = \rho_n\boldsymbol{v}_n + \rho_s\boldsymbol{v}_s = 0$. This means that the superfluid and normal components oscillate in antiphase and compensate mutually the mass flows of the both components. The second sound can be interpreted as undamped waves of compression and rarefaction in the gas of elementary excitations (phonons) since the temperature oscillations induce the oscillations in the number and density of elementary oscillations (phonons). Unlike superfluid liquid where the heat transfer has a convective nature, in normal fluid the heat transfer and temperature oscillations are diffusive and thus damping due to finite heat conduction.

[9] The *third sound* refers usually to the wave processes in the thin helium films absorbed on the substrate. The *fourth sound* is a propagation of sound in the thin capillaries when the normal component motion is completely retarded, i.e. $\boldsymbol{v}_n = 0$ but $\boldsymbol{v}_s \neq 0$.

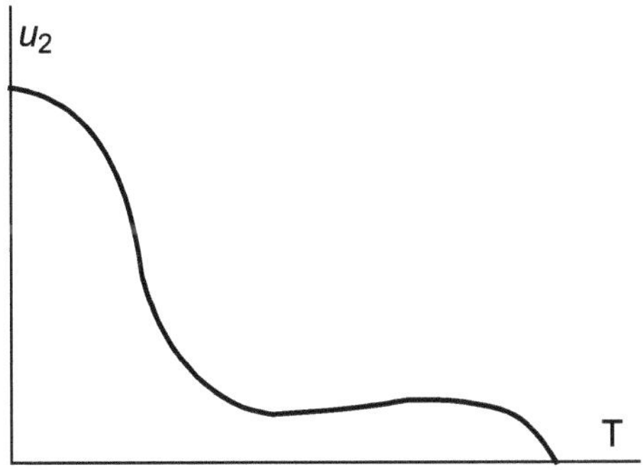

Fig. 8.2 The temperature behavior of second sound velocity u_2 in He II. Velocity $u_2 = 0$ at λ-point of superfluid transition ($\rho_s = 0$). In the low temperature region below about 0.5 K the velocity u_2 is governed by the phonon excitations and equal to $u_1/\sqrt{3}$

Nonzero thermal expansion coefficient results in the effect of coupling[10] between first and second sounds. To the extent $\alpha_V \neq 0$ in first sound there also arise temperature oscillations along with the pressure and density oscillations and, strictly speaking $\boldsymbol{v}_n \neq \boldsymbol{v}_s$. In its turn, in the second sound there appear the pressure and density oscillations in addition to the temperature oscillations. The total mass flow in the second sound wave does not vanish rigorously, i.e. $\boldsymbol{j} \neq 0$. The temperature behavior of second sound velocity $u_2(T)$ is shown in Fig. 8.2.

8.6 Quantized Vortices in a Rotating Superfluid Liquid

If we rotate a bucket with He II and assume the irrotational curl $\boldsymbol{v}_s = 0$ motion of the superfluid component, the normal component alone should start to rotate with the bucket but the superfluid component will stay immobile. The normal excitations, colliding and interacting with the bucket walls, are dragged with the rotation like an ordinary viscid fluid. The superfluid component, which does not interact with the walls, remains at rest and immobile. Such situation conserves while the velocity of the bucket walls is sufficiently small and the walls create no elementary excitations which could involve the superfluid component into rotation. The *quantized vortices* or singular vortex lines in the superfluid velocity distribution have been found as elementary excitations violating the irrotational motion and dragging the superfluid component into rotation.

Let us write down the condition of irrotational motion curl $\boldsymbol{v}_s = 0$ with the aid of the Stokes theorem in the equivalent integral form

$$\varkappa = \oint_C \boldsymbol{v}_s d\boldsymbol{l} = 0$$

[10] The effect in degree is characterized with the Landau–Placzek ratio $(c_P - c_V)/c_V$.

as zero circulation $\varkappa$ of superfluid velocity round some closed contour C. Next, we examine the quantity $M\varkappa/2\pi$, where M is the ^{4}He atom mass, from the viewpoint of an adiabatic invariant. According to the principles of quantum mechanics this quantity should be subjected to quantization as

$$\frac{1}{2\pi}M\oint_C \boldsymbol{v}_s d\boldsymbol{l} = n\hbar \quad \text{where } n \text{ is an integer}$$

and, correspondingly, we have for the circulation

$$\oint_C \boldsymbol{v}_s d\boldsymbol{l} = \varkappa_n = n\varkappa, \quad \varkappa = \frac{2\pi\hbar}{M}.$$

The quantity $\varkappa$ is the *circulation quantum*.

It follows immediately from this relation that two essentially different situations arise in a superfluid at $n = 0$ and $n \neq 0$. If $n = 0$, the vorticity is curl $\boldsymbol{v}_s = 0$ and in any simply connected region the superfluid component remains at rest ($\boldsymbol{v}_s \equiv 0$) regardless of rotation of the liquid. For nonzero circulation, there appears a motion of the superfluid component around some singular vortex lines.

Let us consider the simplest example of rectilinear vortex line parallel to the z axis with nonzero $\varkappa_n$ circulation around any closed contour encircling the vortex line. The superfluid velocity has the azimuthal component alone

$$v_s(r) = \frac{\varkappa_n}{2\pi r} = n\frac{\varkappa}{2\pi r}, \quad \varkappa = \frac{2\pi\hbar}{M}$$

and r is the distance from the vortex line. The flow lines are the circles whose planes are normal to the vortex line and the centers are at the vortex line. As we see, the superfluid velocity magnitude grows unlimitedly on approaching to the vortex line center. At the same time, the superfluid component density decreases but the normal component density increases. The superfluid component density strictly vanishes at the vortex line. The *vortex core* means the region around the vortex line where the superfluid density is negligibly small and no superfluid motion. The core size is about correlation length ξ being a few interatomic distance with the exception of the close vicinity of λ-point.

The kinetic energy of rectilinear quantized vortex per its unit length (or linear tension) is given by

$$E_n = \int \frac{\rho_s \boldsymbol{v}_s^2}{2} d^2r = \frac{\rho_s}{2}\int_{\xi}^{R}\left(\frac{n\varkappa}{2\pi r}\right)^2 2\pi r\, dr = n^2\frac{\rho_s\varkappa^2}{4\pi}\ln\frac{R}{\xi}.$$

Estimating the large integral within the logarithmic accuracy, we have chosen the length ξ as a vortex core radius of about interatomic spacing and length $R \gg \xi$ as some external radius of the vortex. The latter can be the bucket radius or the mean distance between vortices. The creation of vortex-like excitation is associated with an additional energy. Since $E_n \sim n^2$ and $n^2 \geqslant |n|$, the state of $|n|$ vortices with the one-quantum circulation $\varkappa$ is energetically more favorable as compared with the state of one vortex but carrying the $|n|$ circulation quanta at once. The total excitation energy for a *vortex ring* of radius R can be estimated as $2\pi R E_1$.

To find the critical angular velocity Ω_{cr}, at which the first quantized vortex will appear in a rotating bucket, it is necessary to deal with the thermodynamic potential $\tilde{E}(\boldsymbol{\Omega}) = E - \boldsymbol{L}\boldsymbol{\Omega}$ as a function of angular velocity, $\boldsymbol{L}$ being the angular momentum of the total liquid. The spontaneous creation of a vortex becomes energetically favorable if the inequality $\tilde{E}(\boldsymbol{\Omega}) < 0$ is achieved with the growth of angular velocity Ω.

We can clarify the necessity to analyze the thermodynamic potential $\tilde{E}(\boldsymbol{\Omega})$ as follows. In the reference frame in which the liquid flows at velocity $\boldsymbol{v}$, the elementary excitation energy experiences the Doppler shift $\varepsilon \to \varepsilon + \boldsymbol{pv}$ and the condition $\varepsilon + \boldsymbol{pv} < 0$ permits a spontaneous creation of an excitation. In our case, if we consider an excitation at distance r from the bucket axis z, the superfluid component flows at velocity $\boldsymbol{v} = -\boldsymbol{\Omega} \times \boldsymbol{r}$ and

$$\varepsilon + \boldsymbol{pv} = \varepsilon - \boldsymbol{p}\cdot(\boldsymbol{\Omega} \times \boldsymbol{r}) = \varepsilon - \boldsymbol{l}(\boldsymbol{r}) \cdot \Omega$$

where $\boldsymbol{l}(\boldsymbol{r}) = \boldsymbol{r} \times \boldsymbol{p}$ is the angular momentum. Integrating the last relation over the whole liquid volume, we arrive at $E(\boldsymbol{\Omega}) - \boldsymbol{L}\boldsymbol{\Omega}$ where $E = \int \varepsilon\, dV$ and $\boldsymbol{L} = \int \boldsymbol{l}\, dV$.

Prompting the symmetrical hints, we expect that the minimum magnitude of the thermodynamic potential $\tilde{E}(\boldsymbol{\Omega})$ realizes for location of a vortex line at the center of the bucket and parallel to its symmetry axis. The angular momentum of vortex line is parallel to the rotation z axis. Its magnitude per unit length of rectilinear line is

$$L = \int\limits_{r \leqslant R} d^2r [\boldsymbol{r} \times \rho_s \boldsymbol{v}_s]_z = \rho_s \int d^2r\, r v_s = \frac{n}{2} \varkappa \rho_s R^2,$$

ρ_s being the superfluid component density. We have taken here into account that the superfluid velocity vector has no z component of the velocity in the direction of the bucket axis. From condition $\tilde{E}(\boldsymbol{\Omega}) = E - \boldsymbol{L}\boldsymbol{\Omega} = 0$ we find the minimum magnitude of the *critical angular velocity* of rotation when the creation of the first vortex becomes favorable

$$\Omega_{cr} = \frac{\varkappa}{2\pi R^2} \ln \frac{R}{\xi}.$$

Of course, this angular velocity corresponds to one quantum circulation $\varkappa$. The critical angular velocity Ω_{cr} matches the critical linear velocity

$$v_{cr} = \Omega_{cr} R = \frac{\hbar}{2MR} \ln \frac{R}{\xi}.$$

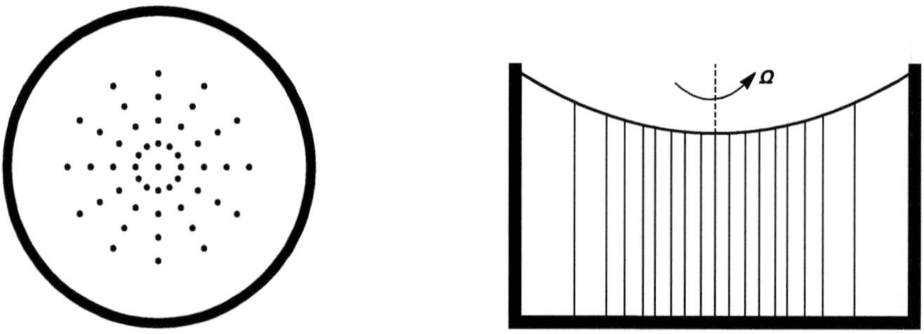

Fig. 8.3 Vortices in the rotating bucket with helium. *At the lhs*: top view. *At the rhs*: side view

By the order of the magnitude, this estimate remains correct for the critical flow velocity at which quantized vortices are generated in the capillaries and orifices with the typical diameter $2R$.

The presence of quantized vortices will imitate the rotation of the superfluid component as a whole. The density of vortices N, i.e. the number of vortices per unit area, can be related with vorticity curl $\boldsymbol{v}_s$, using the Stokes theorem as $N = \frac{|\text{curl}\, \boldsymbol{v}_s|}{\varkappa}$. Since $|\text{curl}\,(\boldsymbol{\Omega} \times \boldsymbol{r})| = 2\Omega$, the angular velocity-dependent density of vortices in the bucket can be estimated as $N = 2\Omega/\varkappa$. Thus, when the bucket with superfluid helium rotates, the emergence of quantized vortices produces a vortex-like motion with $|\text{curl}\, \boldsymbol{v}_s| = 2\Omega$ as it takes place for rotation of normal viscid fluid. Herewith, as before, the magnitude of vorticity $|\text{curl}\, \boldsymbol{v}_s| = 0$ vanishes in the vortex-free regions. As the rotation velocity enhances, the vortices start to occupy practically the whole bucket volume. However, a small region beside the bucket walls still remains completely vortex-free. The resulting picture is schematically shown in Fig. 8.3.

Chapter 9
Magnetism

9.1 Types of Magnetic Structures

The carriers of magnetism in a condensed matter are usually the atoms of transition elements with non-zero magnetic moments or spins. The magnetic moment in an atom appears due to existence of unfilled electronic d- or f-shells. These are $3d$ elements of the iron group, rare earth $4f$ elements, and $5f$ actinides. As a rule, the atoms of these elements are responsible for the manifestation of magnetism.

The spontaneous magnetic ordering at sufficiently low temperatures is observed in the huge number of wide variety of substances. For the magnetic ordering, it is unnecessary to have the spatially ordered crystalline structure in a substance. The magnetic ordering exists in amorphous disordered systems as well.

Let us mention the most common types of magnetic ordering as a few examples. The simplest structure is *ferromagnetic* when all magnetic moments or spins are directed parallel to each other, realizing non-zero value of magnetization $\langle \boldsymbol{M} \rangle \neq 0$ (Fig. 9.1a). As simple examples of ferromagnets, we note the following chemical elements: Fe, Co, Ni, Gd or compounds EuO, MnS, USe.

The *antiferromagnetic* ordering is primarily specified with the lack of the mean magnetization $\langle \boldsymbol{M} \rangle = 0$. The simplest antiferromagnetic structure is collinear, composed of two equivalent ferromagnetic sublattices $\boldsymbol{M}_1$ and $\boldsymbol{M}_2$ with the oppositely directed magnetic moments or spins, resulting in $\langle \boldsymbol{M}_1 - \boldsymbol{M}_2 \rangle \neq 0$ (Fig. 9.1b). The typical examples of antiferromagnets are as follows: Mn, Cr, Sm, Eu, CoO, MnO, UO_2.

The *ferrimagnetic* ordering has also the collinear structure. The magnetic moments in two magnetic sublattices $\boldsymbol{M}_1$ and $\boldsymbol{M}_2$ are antiparallel but the magnetic moments in the various sublattices differ in their magnitudes. Accordingly, there appears nonzero resultant magnetization $\langle \boldsymbol{M}_1 + \boldsymbol{M}_2 \rangle \neq 0$. The various sublattices here consist of different atoms or ions (Fig. 9.2c). Often these are double oxides, e.g. $NiOFe_2O_3$, $Gd_3Fe_5O_{12}$, $FeOFe_2O_3$, $MnGe_2$.

The non-collinear magnetic ordering can be imagined as two magnetic sublattices ordered initially in the opposite directions and then canted in the same direction

S. N. Burmistrov, *Statistical and Condensed Matter Physics*,
https://doi.org/10.1007/978-981-97-9072-2_9

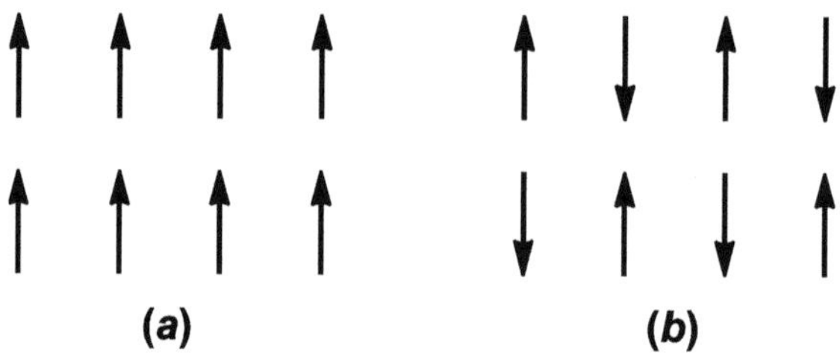

Fig. 9.1 **a** The ferromagnetic ordering; **b** The antiferromagnetic ordering

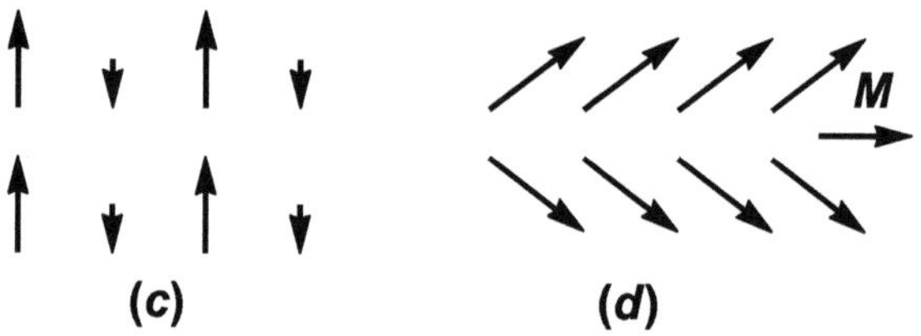

Fig. 9.2 **c** Ferrimagnetism; **d** Canted antiferromagnetism (***M*** is the resultant weak magnetization)

(Fig. 9.2d). As a rule, at small angle of inclination this type of magnetic ordering is referred to as *canted antiferromagnetism* or, infrequently, weak ferromagnetism since $\langle|\boldsymbol{M}_1 + \boldsymbol{M}_2|\rangle \ll \langle|\boldsymbol{M}_1 - \boldsymbol{M}_2|\rangle$. The examples of such compounds are hematite α-Fe_2O_3, $SmFeO_3$, $CoCO_3$.

There exist *spiral* or *helical* magnetic structures as well. For instance, we can indicate the structure of *antiferromagnetic helicoid* (Fig. 9.3) in rare earth terbium

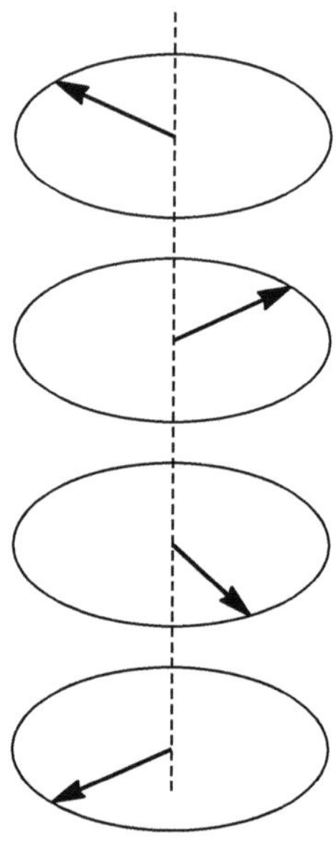

Fig. 9.3 The magnetization vector in the antiferromagnetic helicoid

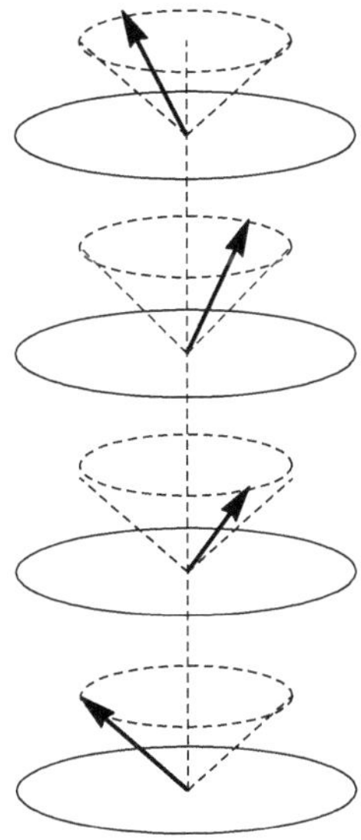

Fig. 9.4 The ferromagnetic helicoid

between 219 and 230 K or dysprosium between 85 and 174 K. At lower temperatures these substances experience the magnetic transition to the state of collinear ferromagnetic.

Erbium below 19 K has a structure of *ferromagnetic helicoid* (Fig. 9.4) and in the temperature region 19–53 K the structure is more complicated, namely *cycloidal antiferromagnetic* (Fig. 9.5).

In the temperature region close to the magnetic ordering temperature there can arise *sinusoidal* ordering (Fig. 9.6) when it still remains energetically favorable to change the magnitude of the magnetic moment with maintaining its direction along some selected axis z according to $\langle M_z \rangle \sim \sin qz$.

Unlike the ferromagnetic ordering, the antiferromagnetic-type interaction between the spins can lead to the effects of multiple degeneration of the ground state with the minimum energy. A typical example of such situation can be illustrated with putting three spins at the corners of an equilateral triangle with antiferromagnetic coupling between them (Fig. 9.7a). The total number of possible states is $2^3 = 8$. Two of them when all three spins are oriented in the same direction do not meet the ground state. For the other six states, two spins are directed to reduce the coupling energy but the third one has to be oriented in the unfavaroble direction from the energetic viewpoint. The energy of all these six states proves to be the same, i.e. the ground state is sixfold degenerate. This simple example illustrates the phenomenon of geometrical *frustration* when the spatial position of particles conflicts with the interparticle interaction and results in the macroscopic number of the ground states in the system.

The frustration effects are typical for the triangular, hexagonal, or *kagome* lattices. The latter ones consist of equilateral triangles and regular hexagons, arranged so that each hexagon is surrounded with the triangles and *vice versa*. The examples of such

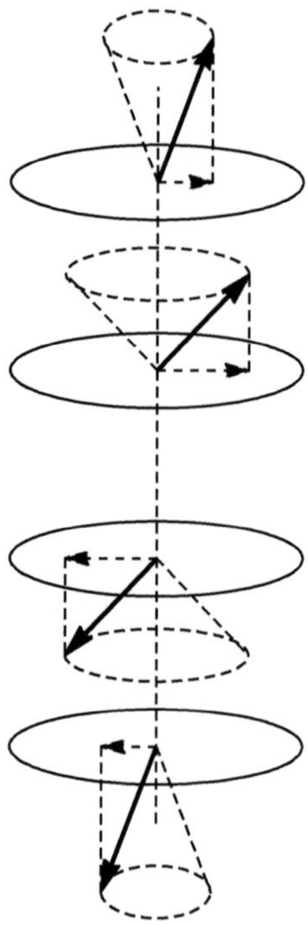

Fig. 9.5 The cycloidal structure of magnetization

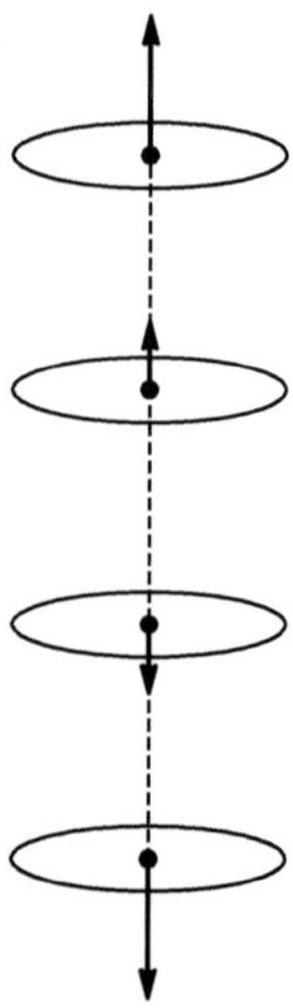

Fig. 9.6 The sinusoidal ordering

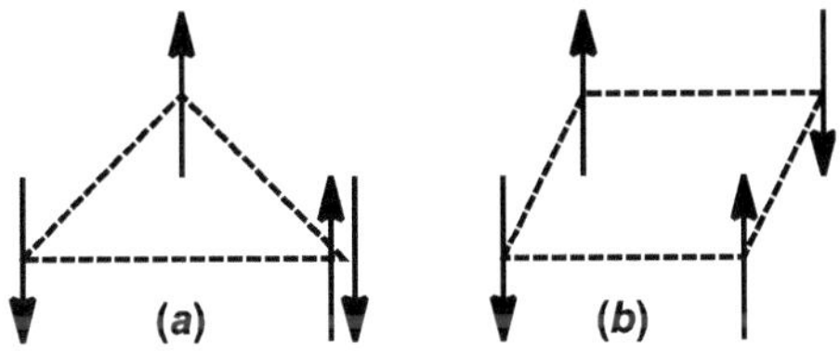

Fig. 9.7 **a** The triangular lattice. The both directions of the spin at the right-hand spin are unfavorable from the viewpoint of minimizing the nearest neighbor interaction. **b** The square lattice

spatial structures can be found in the jarosite $KFe_3(SO_4)_2\,(OH)_6$ and herbertsmithite $ZnCu_3(OH)_6Cl_2$ minerals. In the square[1] lattice, on the contrary, the ground state is not degenerate if we disregard the direction of the spin ordering axis. The ground state will be a simple collinear antiferromagnetic ordering consisting of two ferromagnetic sublattices and proves to be the most energetically favorable (Fig. 9.7b).

The magnetic structure can experimentally be determined by observing the diffraction patterns of neutron scattering.

9.2 The Ferromagnetic Ordering

An existence of magnetic structure in the condensed media leads to emerging additional elementary excitations and specific magnetic branches in the energy spectrum of excitations. In its turn, this results in changing the thermodynamic medium as compared with its nonmagnetic state. As a rule, the main type of magnetic interaction is an exchange interaction which can be represented as the *Heisenberg Hamiltonian*

$$\hat{H} = -\frac{1}{2}\sum_{a\neq b} J(\boldsymbol{R}_{ab})\hat{\boldsymbol{S}}_a\hat{\boldsymbol{S}}_b\,, \quad \boldsymbol{R}_{ab} = \boldsymbol{R}_a - \boldsymbol{R}_b\,.$$

The sum is taken over all atom spins $\boldsymbol{S}_a$ located at points $\boldsymbol{R}_a$. It is convenient to augment the *exchange integral* J with $J(\boldsymbol{R}_{ab} = 0) = 0$ and then write the exchange Hamiltonian, as follows:

$$\hat{H} = -\frac{1}{2}\sum_{ab} J(\boldsymbol{R}_{ab})\hat{\boldsymbol{S}}_a\cdot\hat{\boldsymbol{S}}_b\,, \quad \boldsymbol{R}_{ab} = \boldsymbol{R}_a - \boldsymbol{R}_b\,.$$

Along with the exchange interaction, there exist both dipole interaction between the magnetic moments of atoms and spin–orbit interaction between the magnetic moment and the electric field induced by the particles in the medium.

[1] The frustration effects are also possible if there is an interaction not only between the nearest neighboring spins but the interaction exists between the next neighboring spins and has a long-range character.

Let us introduce operator of magnetic moment density

$$\hat{\boldsymbol{M}}(\boldsymbol{r}) = \mu \sum_a \hat{\mathbf{S}}_a \delta(\boldsymbol{r} - \boldsymbol{R}_a)$$

where μ is the *effective Bohr magneton*. Then we represent the Heisenberg Hamiltonian in the following form:

$$\hat{H} = -\frac{1}{2\mu^2} \sum_{ab} \int d^3r\, d^3r'\, J(\boldsymbol{r} - \boldsymbol{r}')\delta(\boldsymbol{r} - \boldsymbol{R}_a)\delta(\boldsymbol{r} - \boldsymbol{R}_b) =$$

$$= -\frac{1}{2\mu^2} \int d^3r\, d^3r'\, J(\boldsymbol{r} - \boldsymbol{r}') \sum_a \hat{\mathbf{S}}_a \delta(\boldsymbol{r} - \boldsymbol{R}_a) \sum_b \hat{\mathbf{S}}_b \delta(\boldsymbol{r} - \boldsymbol{R}_b) =$$

$$= -\frac{1}{2\mu^2} \int d^3r\, d^3r'\, J(\boldsymbol{r} - \boldsymbol{r}')\hat{\boldsymbol{M}}(\boldsymbol{r})\hat{\boldsymbol{M}}(\boldsymbol{r}').$$

This Hamiltonian conforms to the free energy F_{ex} which should be calculated with the aid of the equilibrium density matrix $\hat{\rho} = \exp[(F - \hat{H})/T]$, as follows:

$$F_{\text{ex}} = \operatorname{tr}(\hat{\rho}\hat{H}) = \langle \hat{H} \rangle = -\frac{1}{2\mu^2} \int d^3r\, d^3r'\, J(\boldsymbol{r} - \boldsymbol{r}')\langle \hat{\boldsymbol{M}}(\boldsymbol{r})\hat{\boldsymbol{M}}(\boldsymbol{r}')\rangle \approx$$

$$\approx -\frac{1}{2\mu^2} \int d^3r\, d^3r'\, J(\boldsymbol{r} - \boldsymbol{r}')\langle \hat{\boldsymbol{M}}(\boldsymbol{r})\rangle \langle \hat{\boldsymbol{M}}(\boldsymbol{r}')\rangle.$$

Here we have used the *mean-field approximation* or *self-consistent approximation*, replacing the average for the product of operators with the product of the averages of operators. This approximation is usually justified well below the Curie temperature. Taking into account that the average for operator $\hat{\boldsymbol{M}}(\boldsymbol{r})$ equals the magnetization vector $\boldsymbol{M}(\boldsymbol{r})$, we get

$$F_{\text{ex}} = -\frac{1}{2\mu^2} \int d^3r\, d^3r'\, J(\boldsymbol{r} - \boldsymbol{r}')\boldsymbol{M}(\boldsymbol{r})\boldsymbol{M}(\boldsymbol{r}').$$

If the magnetization is constant in the space $\boldsymbol{M}(\boldsymbol{r}) = \text{const}$, we have

$$F_{\text{ex}} = -\frac{1}{2\mu^2} \int d^3r\, J(0)\boldsymbol{M}^2, \quad J(0) = \int d^3r\, J(\boldsymbol{r})$$

and $J(0)$ is zero Fourier-transform of the exchange integral. When the magnetization varies slowly in the space, we can take into account that exchange integral $J(\boldsymbol{r})$ decays rapidly with the distance. Let us expand $\boldsymbol{M}(\boldsymbol{r})$ to second-order terms

$$M_i(\boldsymbol{r}') = M_i(\boldsymbol{r}) + (x_k' - x_k)\frac{\partial M_i(\boldsymbol{r})}{\partial x_k} + \frac{1}{2}(x_k' - x_k)(x_l' - x_l)\frac{\partial^2 M_i(\boldsymbol{r})}{\partial x_k \partial x_l} + \ldots$$

and substitute this expansion into the expression for the free energy. Then,

$$F_{\text{ex}} = -\frac{1}{2\mu^2}\int d^3r\, M_i^2(\boldsymbol{r})\int d^3r'\, J(\boldsymbol{r}')-$$
$$-\frac{1}{2\mu^2}\int d^3r\, M_i\frac{\partial M_i}{\partial x_k}\int d^3r'\, x_k' J(\boldsymbol{r}')-$$
$$-\frac{1}{2\mu^2}\int d^3r\, M_i\frac{\partial^2 M_i}{\partial x_k \partial x_l}\cdot\frac{1}{2}\int d^3r'\, J(\boldsymbol{r}')x_k x_l + \cdots$$

For simplicity, we restrict ourselves with the rather general case of centrally symmetrical lattice of the spin sites, i.e.

$$\int d^3r\, \boldsymbol{r} J(\boldsymbol{r}) = 0,$$

and introduce the *coefficient of magnetic stiffness* according to

$$\alpha_{ik} = \frac{1}{2\mu^2}\int d^3r\, J(\boldsymbol{r})x_i x_k\,.$$

Thus, the second term in the expansion vanishes and the third one plays a role of *energy of inhomogeneity*

$$F_{\text{inh}} = -\frac{1}{2}\int d^3r\, \alpha_{kl} M_l(\boldsymbol{r})\frac{\partial^2 M_i}{\partial x_k \partial x_l} =$$
$$= -\frac{\alpha_{kl}}{2}\int d^3r\, \frac{\partial}{\partial x_k}\left(M_i\frac{\partial M_i}{\partial x_l}\right) + \frac{\alpha_{kl}}{2}\int d^3r\, \frac{\partial M_i}{\partial x_k}\frac{\partial M_i}{\partial x_l}\,.$$

The integrand with the brackets corresponds to the surface integral which can be disregarded if we are interested only in the bulk contribution, i.e. free energy density. The second possibility is to choose the natural boundary condition at the surface of a ferromagnet in order to nullify this surface contribution. Accordingly, we can put the following condition[2] at the ferromagnet surface S:

$$(\boldsymbol{\nu}\cdot\nabla \boldsymbol{M})\big|_S = 0,$$

$\boldsymbol{\nu}$ being the normal to the surface.

So, we have for free energy F_{ex} and its density w_{ex}

$$F_{\text{ex}} = \int d^3r\, w_{\text{ex}} \quad \text{where} \quad w_{\text{ex}} = -\frac{1}{2\mu^2}J(0)\boldsymbol{M}^2 + \frac{1}{2}\alpha_{kl}\frac{\partial M_i}{\partial x_k}\frac{\partial M_i}{\partial x_l}.$$

[2] The continuity condition is also used for the normal component of the energy flux density.

For the spin lattice of cubic symmetry, the coefficient is $\alpha_{kl} = \alpha\delta_{kl}$ and the inhomogeneity energy density can be written as follows:

$$w_{\text{inh}} = \frac{\alpha}{2}\frac{\partial M_i}{\partial x_k}\frac{\partial M_i}{\partial x_k} \equiv \frac{\alpha}{2}(\nabla \boldsymbol{M})^2.$$

Let us estimate the magnitude of the *magnetic stiffness coefficient* α, supposing that exchange integral J decays rapidly at the interatomic distance a about the spacing between the neighboring spins. Then, α is of the order of

$$\alpha \sim \frac{J}{\mu^2}a^3 \cdot a^2 \sim \frac{\Theta a^3}{\mu^2}a^2.$$

Here we have introduced the Curie temperature Θ equal approximately to the magnitude of exchange integral J between the neighboring spins. Taking into account that the saturation magnetization in a ferromagnet is about $M_s \sim \mu/a^3$, we obtain

$$\alpha \sim \frac{\Theta}{\mu M_s}a^2.$$

Usually, for the ferromagnets with the Curie temperatures large as compared with several kelvins, the inequality $\Theta/(\mu M_s) \gg 1$ is valid and, therefore, $\sqrt{\alpha} \gg a$.

The exchange interaction does not specify the magnetization direction in the ordinary space. The alignment of magnetization in the space is connected with the relativistic interactions whose energy depends on the orientation of the magnetic moments with respect to the crystallographic axes. The magnitude of the relativistic interactions is usually much less than the exchange one to degree $v/c \ll 1$ and reaches approximately $(v/c)^2 w_{\text{ex}}$, where v is the velocity of electrons in an atom and c is the light velocity. Here, first of all, we mention the *dipole-dipole interaction* between two magnetic moments. Another known type of relativistic interaction is the *spin-orbit coupling* which can be described as an interaction of an atom spin with the electric field of ion lattice.

The spin-orbit and dipole-dipole interactions contain both the terms having the same symmetry like the exchange interaction and the terms depending on the direction of the spins relative to the spatial axes. The sum of first terms can be included into the exchange interaction Hamiltonian but the second ones can be represented in the following form:

$$\hat{H}_{an} = \frac{1}{2}\mu^2 \sum_{a\neq b}\sum_{i,k} \beta_{ik}(\boldsymbol{R}_{ab})\hat{S}_{a,i}\hat{S}_{b,k} \quad (i, k = x, y, z).$$

Introducing the magnetization operator and calculating the free energy with the anisotropic Hamiltonian $\hat{H}_{an}$ like in the case of the exchange interaction, we obtain the anisotropic part of the free energy expressed in terms of the magnetization vector $\boldsymbol{M}(\boldsymbol{r})$ in the general form

$$W_{an} = \frac{1}{2}\int d^3r\, d^3r'\, \beta_{ik}(\boldsymbol{r}-\boldsymbol{r}')M_i(\boldsymbol{r})M_k(\boldsymbol{r}').$$

Since function $\beta_{ik}(\boldsymbol{r}-\boldsymbol{r}')$ decays rather rapidly with the distance, we can take the local approximation $\beta_{ik}(\boldsymbol{r}-\boldsymbol{r}') \approx \beta_{ik}\delta(\boldsymbol{r}-\boldsymbol{r}')$ as a first one. Then,

$$W_{an} = \int d^3r\, w_{an} \quad \text{and} \quad w_{an} = \frac{1}{2}\beta_{ik}M_i(\boldsymbol{r})M_k(\boldsymbol{r})$$

where w_{an} is the density of anisotropy energy. It is natural that its expression should be time-reversal invariant $\boldsymbol{M} \to -\boldsymbol{M}$ and represent a scalar. The symmetry of anisotropy coefficient β_{ik} reflects the spatial symmetry of a ferromagnet.

For a uniaxial ferromagnet, we write down

$$w_{an} = \frac{\beta_1}{2}(M_x^2 + M_y^2) + \frac{\beta_2}{2}M_z^2\,.$$

Since $M_x^2 + M_y^2 + M_z^2 = M^2$, the part of the anisotropy energy can be included into the exchange one depending on M^2 alone. The following representation is more convenient:

$$w_{an} = \frac{\beta}{2}M_z^2 = \frac{\beta}{2}(\boldsymbol{n}\boldsymbol{M})^2,$$

$\boldsymbol{n}$ being unit vector in the direction of the anisotropy axis. In a uniaxial ferromagnet with, the *easy axis* of magnetization $\beta < 0$ the spontaneous magnetization direction will coincide with that of anisotropy axis z. For the case of *easy plane* $\beta < 0$, the magnetic moment will be aligned in the plane normal to the anisotropy axis.

In a crystal of cubic symmetry $\beta_{ik} = \beta\delta_{ik}$ and in the energy of anisotropy it is necessary to involve the terms of higher order in magnetization

$$w_{an} = \beta(M_x^2M_y^2 + M_x^2M_z^2 + M_y^2M_z^2) \Leftrightarrow -\frac{\beta}{2}(M_x^4 + M_y^4 + M_z^4).$$

For $\beta > 0$, the magnetic moment aligns in the direction of spatial axes x, y, z. For $\beta < 0$, the alignment is along the diagonals of a cube. On the whole, the free energy must be scalar invariant to replacing as $\boldsymbol{M} \to -\boldsymbol{M}$ and correspond to the spatial symmetry of a magnet.

9.3 Total Energy of a Ferromagnet

In zero magnetic field $H = 0$ the free energy $F[\boldsymbol{M}(\boldsymbol{r}), \nabla\boldsymbol{M}(\boldsymbol{r})]$ of a ferromagnet can be represented as

$$F = \int d^3r\, w(\boldsymbol{M}, \nabla\boldsymbol{M}), \quad w(\boldsymbol{M}, \nabla\boldsymbol{M}) = \frac{\alpha_{ik}}{2}\frac{\partial\boldsymbol{M}}{\partial x_i}\frac{\partial\boldsymbol{M}}{\partial x_k} + w_{an} + f(\boldsymbol{M}^2)$$

where $f(\boldsymbol{M}^2)$ has mainly the exchange origin. To find the behavior of free energy as a function of magnetic field strength $\boldsymbol{H}$, we employ the thermodynamic definition for the derivative of the Gibbs free energy $\widetilde{F} = \widetilde{F}(\boldsymbol{M}, \boldsymbol{H})$ with respect to magnetic field

$$\left(\frac{\partial \widetilde{F}}{\partial \boldsymbol{H}}\right)_M = -\frac{\boldsymbol{B}}{4\pi} = -\frac{\boldsymbol{H} + 4\pi \boldsymbol{M}}{4\pi} = -\frac{\boldsymbol{H}}{4\pi} - \boldsymbol{M}.$$

Hence, taking $\widetilde{F}(\boldsymbol{M}, 0) = F(\boldsymbol{M}, 0)$ into account, we have

$$\widetilde{F}(\boldsymbol{M}, \boldsymbol{H}) = F(\boldsymbol{M}, 0) - \boldsymbol{M}\boldsymbol{H} - \frac{\boldsymbol{H}^2}{8\pi}.$$

It is straightforwardly to find the expression for the Helmholtz free energy F dependent on the magnetic induction $\boldsymbol{B}$

$$F(\boldsymbol{M}, \boldsymbol{B}) = \widetilde{F} + \frac{\boldsymbol{H}\boldsymbol{B}}{4\pi} = F(\boldsymbol{M}, 0) - \boldsymbol{M}\boldsymbol{H} - \frac{\boldsymbol{H}^2}{4\pi} - \frac{\boldsymbol{H}^2}{8\pi} = F(\boldsymbol{M}, 0) + \frac{\boldsymbol{H}^2}{8\pi}.$$

Expressing $\boldsymbol{H}$ in terms of the magnetic induction vector $\boldsymbol{B}$, we arrive at the final expression for the Helmholtz free energy of a ferromagnet

$$F(\boldsymbol{M}, \boldsymbol{B}) = F(\boldsymbol{M}, 0) + \frac{(\boldsymbol{B} - 4\pi \boldsymbol{M})^2}{8\pi}.$$

In conclusion, we emphasize that, depending on the experimental and external conditions, it is necessary to pay attention[3] which of two thermodynamic potentials $\widetilde{F}(\boldsymbol{H})$ or $F(\boldsymbol{B})$ should be chosen for determining the thermodynamic state of a condensed matter in the magnetic field.

Problem

1. There is a ferromagnet with the easy-axis anisotropy. Find the direction of magnetization $\boldsymbol{M}$ in magnetic field $\boldsymbol{H}$ normal to the easy axis. At the temperatures noticeably below the Curie temperature, the magnitude of magnetization can be put to be constant.

Solution. Let us take anisotropy axis $\boldsymbol{n}$ in the z-axis direction. The angle ϑ is that between magnetization $\boldsymbol{M}$ and z-axis. We write down the free Gibbs energy of a ferromagnet in the magnetic field $\boldsymbol{H}$

$$\widetilde{F}(\boldsymbol{M}, \boldsymbol{H}) = F_0(M^2) + \frac{\beta}{2} M^2 \sin^2\vartheta - MH \sin\vartheta - \frac{H^2}{8\pi}.$$

Here $\beta > 0$ in accordance with the easy axis anisotropy. The equilibrium state means the minimum condition $\partial \widetilde{F}/\partial\vartheta = 0$ for the free energy, i.e.

$$\beta M \sin\vartheta \cos\theta - H \cos\vartheta = 0.$$

[3] For condensed media in the electric field, the same problem takes place for the choice between thermodynamic potential $F(\boldsymbol{D})$ dependent on electric induction $\boldsymbol{D}$ and thermodynamic potential $\widetilde{F}(\boldsymbol{E})$ dependent on electric field strength $\boldsymbol{E}$.

Hence we find the following two solutions:

$$\cos\vartheta = 0, \quad \widetilde{F}_1 = F_0 + \frac{\beta M^2}{2} - MH,$$
$$\sin\vartheta = \frac{H}{\beta M}, \quad \widetilde{F}_2 = F_0 + \frac{H^2}{2\beta} - \frac{H^2}{\beta} = F_0 - \frac{H^2}{2\beta}.$$

Comparing the free energies, we see that

$$\widetilde{F}_1 - \widetilde{F}_2 = \frac{\beta M^2}{2} - MH + \frac{H^2}{2\beta} = \frac{\beta}{2}\left(M - \frac{H}{\beta}\right)^2 \geqslant 0.$$

Therefore, the state with the magnetization direction rotated at angle $\sin\vartheta = H/\beta M$ will be energetically favorable. In the field $H > \beta M$ the magnetization is aligned in the magnetic field direction and lies in the plane normal to the easy axis.

9.4 Ferromagnet Near the Curie Point

In accordance with the Landau phenomenological theory of second-order phase transitions, we should decompose the free energy of a ferromagnet into a series in powers of order parameter. For the particular case of a ferromagnet, this is a decomposition in powers of spontaneous magnetization

$$F(\boldsymbol{M}) = F_0 + A(T)\boldsymbol{M}^2 + \frac{b}{2}\boldsymbol{M}^4 + \cdots$$

The Curie temperature Θ is determined with condition $A(T = \Theta) = 0$. Near the Curie temperature we put $A(T) = a_0(T - \Theta)$ with $a_0 > 0$. The latter inequality assumes that the ferromagnetic state realizes at low $T < \Theta$ temperatures.

We have in the magnetic field H

$$\Delta\widetilde{F}(\boldsymbol{H}) = \widetilde{F}(\boldsymbol{H}) - F_0 = A(T)\boldsymbol{M}^2 + \frac{b}{2}\boldsymbol{M}^4 - \boldsymbol{M}\boldsymbol{H} - \frac{\boldsymbol{H}^2}{8\pi}.$$

Then we find from the extremum $\partial\widetilde{F}/\partial\boldsymbol{M} = 0$

$$2\boldsymbol{M}(A + b\boldsymbol{M}^2) = \boldsymbol{H}.$$

In zero field $\boldsymbol{H} = 0$ two states are possible. The first is nonmagnetic $M = 0$ and possible at all temperatures. The second is magnetic with $M^2 = M_0^2 = -A/b$ and possible only if $T < \Theta$. For the nonmagnetic state, $\Delta F = 0$. For the magnetic state, $\Delta F = -A^2/b < 0$. For the low $T < \Theta$ temperatures, there appears a magnetic state more favorable than the nonmagnetic one.

Near the Curie temperature ($T < \Theta$) we have the behavior of magnetization $M(T) \sim (\Theta - T)^{\beta}$ with the critical exponent $\beta = 1/2$ typical in the Landau theory. At the temperatures well below the Curie temperature the magnetization tends to the saturation $M \to M_s$. Differentiating the equation for the magnetization with respect to magnetic field, we find the magnetic susceptibility $\chi = \partial M/\partial H$

$$2\chi(A + bM^2) + 4bM^2\chi = 1 \quad \text{or} \quad \chi = \frac{1}{2(A + 3bM^2)}.$$

In the low field the magnetization is $M \approx M_0(T)$ and we disclose the critical behavior $\chi \sim (|T - \Theta|)^{-\gamma}$ with critical exponent $\gamma = 1$

$$\chi(T) = \begin{cases} \frac{1}{2A(T)} = \frac{1}{2a_0(T-\Theta)}, & T > \Theta, \\ -\frac{1}{4A(T)} = \frac{1}{4a_0(\Theta-T)}, & T < \Theta. \end{cases}$$

The behavior with critical exponent $\gamma = 1$ corresponds to the *Curie law*. Such behavior takes place under the low field condition $H < H_t$. The magnitude H_t can be estimated with condition $|A(T)| \gg 3bM^2 = 3b\chi^2H^2$. Hence, we have the estimate $H_t \sim |A(T)|/(b\chi^2) \sim |A^3(T)|/b \sim a_0^3|T - \Theta|^3/b$.

In the high magnetic field limit, the critical behavior of magnetization is governed by the critical exponent δ according to $M \sim H^{1/\delta}$. For the fields $H \gg H_t$, we find approximately $2bM^3 \approx H$, correspondingly, $M = (H/2b)^{1/3}$ and susceptibility $\chi \sim H^{-2/3}$. Thus, the critical exponent is $\delta = 3$ in the region of high magnetic field.

To conclude, we concern the applicability of the Landau phenomenological theory within the framework of self-consistent field for describing the critical behavior near the phase transition temperature. As we have seen above, we should require that the closeness to the phase transition temperature would be larger than the Ginzburg–Levanyuk number, i.e. $|\tau| \gg \mathrm{Gi}$. For the typical ferromagnets, this number is about a few hundredths.

9.5 Dynamics of Magnetization. The Landau–Lifshitz Equation

The motion of an isolated magnetic moment $\boldsymbol{m}$ in the external magnetic field $\boldsymbol{H}$ is a precession around the magnetic field direction and is described with the equation

$$\frac{\partial \boldsymbol{m}}{\partial t} = \gamma \boldsymbol{m} \times \boldsymbol{H}.$$

Here γ is the *gyromagnetic ratio*, i.e. ratio of magnetic moment to the mechanical angular moment of a particle, $\gamma = g\mu_B/\hbar$ where g is the *g-factor* or *Landé g-factor*, and $\mu_B = e\hbar/2mc$ is the *Bohr magneton*.

In a ferromagnet, if its temperature is well below the Curie temperature, due to strong exchange spin-spin interaction we can approximately put that the magnetization vector conserves its magnitude, changing the magnetization direction alone.

Thus, we may expect the time evolution magnetization $\boldsymbol{M}(\boldsymbol{r},t)$ in a ferromagnet looks like a precession in some effective magnetic field

$$\frac{\partial \boldsymbol{M}(\boldsymbol{r},t)}{\partial t} = \gamma \boldsymbol{M}(\boldsymbol{r},t) \times \boldsymbol{H}_{\text{eff}}.$$

Let us try to estimate $\boldsymbol{H}_{\text{eff}}$. We start from the equation for the time derivative of magnetic moment density operator $\hat{\boldsymbol{M}}(\boldsymbol{r},t)$

$$-i\hbar\frac{\partial \hat{\boldsymbol{M}}}{\partial t} = \left[\hat{\mathcal{H}}, \hat{\boldsymbol{M}}\right].$$

The right-hand side of equation is a commutator of the Hamiltonian $\hat{\mathcal{H}}$ with the magnetic moment density operator. The Hamiltonian can be represented as an expansion of general form in powers of operator $\hat{\boldsymbol{M}}$

$$\hat{\mathcal{H}}\left[\hat{\boldsymbol{M}}\right] = \sum_n \int d\boldsymbol{r}_1 d\boldsymbol{r}_2 \dots d\boldsymbol{r}_n \, f_{i_1 i_2 \dots i_n}(\boldsymbol{r}_1, \boldsymbol{r}_2, \dots \boldsymbol{r}_n) \times$$
$$\times \hat{\boldsymbol{M}}_{i_1}(\boldsymbol{r}_1,t)\hat{\boldsymbol{M}}_{i_2}(\boldsymbol{r}_2,t)\dots\hat{\boldsymbol{M}}_{i_n}(\boldsymbol{r}_n,t)$$

where $f_{i_1 i_2 \dots i_n}(\boldsymbol{r}_1, \boldsymbol{r}_2, \dots \boldsymbol{r}_n)$ are the coordinate functions symmetrical with respect to any permutation of a pair of variables $(i_k\, \boldsymbol{r}_k)$ and $(i_l\, \boldsymbol{r}_l)$.

Then we calculate the commutator

$$\left[\hat{\mathcal{H}}, \hat{\boldsymbol{M}}_\alpha(\boldsymbol{r},t)\right] = \sum_n \int d\boldsymbol{r}_1 d\boldsymbol{r}_2 \dots d\boldsymbol{r}_n \, f_{i_1 i_2 \dots i_n}(\boldsymbol{r}_1, \boldsymbol{r}_2, \dots \boldsymbol{r}_n) \times$$
$$\times \sum_{k=1}^{n} \hat{\boldsymbol{M}}_{i_1} \dots \hat{\boldsymbol{M}}_{i_{k-1}} \left[\hat{\boldsymbol{M}}_{i_k}(\boldsymbol{r}_k,t),\ \hat{\boldsymbol{M}}_{i_\alpha}(\boldsymbol{r},t)\right] \hat{\boldsymbol{M}}_{i_{k+1}} \dots \hat{\boldsymbol{M}}_{i_n},$$

using the following value of commutator:

$$\left[\hat{\boldsymbol{M}}_\beta(\boldsymbol{r},t), \hat{\boldsymbol{M}}_\alpha(\boldsymbol{r}',t)\right] = i\hbar\gamma\, e_{\beta\alpha\gamma} \hat{\boldsymbol{M}}_\gamma(\boldsymbol{r},t)\delta(\boldsymbol{r}-\boldsymbol{r}')$$

where $e_{\beta\alpha\gamma}$ is the unit antisymmetric tensor. As a result of integrating over coordinate $\boldsymbol{r}_k$, we get

$$\frac{\partial \hat{\boldsymbol{M}}(\boldsymbol{r},t)}{\partial t} = -\gamma \sum_n \sum_{k=1}^{n} \int d\boldsymbol{r}_1 \dots d\boldsymbol{r}_{k-1} d\boldsymbol{r}_{k+1} \dots d\boldsymbol{r}_n \times$$
$$\times f_{i_1 i_2 \dots i_n}(\boldsymbol{r}_1, \dots \boldsymbol{r}_{k-1}, \boldsymbol{r}, \boldsymbol{r}_{k+1}, \dots \boldsymbol{r}_n) \times$$
$$\times \hat{\boldsymbol{M}}_{i_1}(\boldsymbol{r}_1,t)\dots\hat{\boldsymbol{M}}_{i_{k-1}}(\boldsymbol{r}_{k-1},t)\hat{\boldsymbol{M}}_{i_{k+1}}(\boldsymbol{r}_{k+1},t)\dots\hat{\boldsymbol{M}}_{i_n}(\boldsymbol{r}_n,t) e_{i_k\alpha\gamma}\hat{\boldsymbol{M}}_\gamma(\boldsymbol{r},t).$$

The macroscopic description implies the transfer from the operators to its mean values. The latter ones are calculated with the aid of density matrix $\hat{\rho}$, in general, non-equilibrium. Accordingly, the magnetization is determined as an average of magnetic moment density operator

$$\boldsymbol{M}(\boldsymbol{r},t) = \mathrm{tr}\left(\hat{\rho}\hat{\boldsymbol{M}}(\boldsymbol{r},t)\right) = \langle \hat{\boldsymbol{M}}(\boldsymbol{r},t)\rangle.$$

In what follows, we assume sufficiently slow variations of magnetization in time so that the quasi-local equilibrium has time to be established and the self-consistent field approximation is valid as well. In other words, the average for the product of operators equals the product of the averages of the operator values

$$\langle \hat{\boldsymbol{M}}_1(\boldsymbol{r}_1,t)\ldots\hat{\boldsymbol{M}}_n(\boldsymbol{r}_n,t)\rangle = \langle \hat{\boldsymbol{M}}_1(\boldsymbol{r}_1,t)\rangle\ldots\langle\hat{\boldsymbol{M}}_n(\boldsymbol{r}_n,t)\rangle = \\ = \boldsymbol{M}_1(\boldsymbol{r}_1,t)\ldots\boldsymbol{M}_n(\boldsymbol{r}_n,t).$$

Averaging the equation of evolution for the magnetic moment density under such assumptions, we arrive at the following equation for the magnetization $\boldsymbol{M}(\boldsymbol{r},t)$ dynamics:

$$\frac{\partial M_\alpha(\boldsymbol{r},t)}{\partial t} = -\gamma\sum_n\sum_{k=1}^{n}\int d\boldsymbol{r}_1\ldots d\boldsymbol{r}_{k-1}d\boldsymbol{r}_{k+1}\ldots d\boldsymbol{r}_n\times \\ \times f_{i_1i_2\ldots i_n}(\boldsymbol{r}_1,\ldots\boldsymbol{r}_{k-1},\boldsymbol{r},\boldsymbol{r}_{k+1},\ldots\boldsymbol{r}_n)\times \\ \times\boldsymbol{M}_{i_1}(\boldsymbol{r}_1,t)\ldots\boldsymbol{M}_{i_{k-1}}(\boldsymbol{r}_{k-1},t)\boldsymbol{M}_{i_{k+1}}(\boldsymbol{r}_{k+1},t)\ldots\boldsymbol{M}_{i_n}(\boldsymbol{r}_n,t)\epsilon_{i_k\alpha\gamma}\boldsymbol{M}_\gamma(\boldsymbol{r},t).$$

Here we will take into account that the quantities obtained are already the ordinary variables, but the operators, and their order is insignificant for us. This gives a factor n instead of a sum over k. Then, changing the order of the indexes in antisymmetric tensor $e_{i_k\alpha\gamma}$, we get

$$\frac{\partial M_\alpha(\boldsymbol{r},t)}{\partial t} = -\gamma\sum_n n\int d\boldsymbol{r}_2\ldots d\boldsymbol{r}_n f_{i_1\ldots i_n}(\boldsymbol{r},\boldsymbol{r}_2\ldots\boldsymbol{r}_n)\times \\ \times M_{i_2}(\boldsymbol{r}_2,t)\ldots M_{i_n}(\boldsymbol{r}_n,t)e_{\alpha\gamma i_1}M_\gamma(\boldsymbol{r},t) = -\gamma\frac{\delta W}{\delta M_{i_1}(\boldsymbol{r},t)}e_{\alpha\gamma i_1}M_\gamma(\boldsymbol{r},t)$$

where we have introduced the variational derivative $\delta W/\delta M_{i_1}$ for the quantity

$$W[\boldsymbol{M}] = \sum_n\int d\boldsymbol{r}_1\ldots d\boldsymbol{r}_n\, f_{i_1\ldots i_n}(\boldsymbol{r}_1,\ldots\boldsymbol{r}_n)M_{i_1}(\boldsymbol{r}_1,t)\ldots M_{i_n}(\boldsymbol{r}_n,t).$$

It is readily to see that under our assumptions, the quantity $W[\boldsymbol{M}]$ is the free energy functional, i.e. the mean value of Hamiltonian $W[\boldsymbol{M}] = \langle\mathcal{H}[\hat{\boldsymbol{M}}]\rangle$. This can be checked from the relation

$$\langle\hat{\boldsymbol{M}}(\boldsymbol{r}_1,t)\ldots\hat{\boldsymbol{M}}(\boldsymbol{r}_n,t)\rangle = \langle\boldsymbol{M}(\boldsymbol{r}_1,t)\rangle\ldots\langle\boldsymbol{M}(\boldsymbol{r}_n,t)\rangle.$$

Finally, we arrive at the phenomenological expression called the *Landau–Lifshitz equation* describing the dynamics of magnetization in a ferromagnet. This equation describes the precession of magnetization vector in the effective magnetic field $\boldsymbol{H}_{\text{eff}}$

$$\frac{\partial \boldsymbol{M}(\boldsymbol{r},t)}{\partial t} = \gamma \boldsymbol{M}(\boldsymbol{r},t) \times \boldsymbol{H}_{\text{eff}}(\boldsymbol{r},t).$$

The effective magnetic field is the minus-signed variational derivative of free energy W taken with respect to magnetization

$$\boldsymbol{H}_{\text{eff}}(\boldsymbol{r},t) = -\frac{\delta W}{\delta \boldsymbol{M}(\boldsymbol{r},t)} \quad \text{or} \quad \delta W = -\int \boldsymbol{H}_{\text{eff}}(\boldsymbol{r},t)\delta \boldsymbol{M}(\boldsymbol{r},t)\, d^3r.$$

Emphasize that the Landau–Lifshitz equation does not involve the possible processes of the longitudinal and transverse relaxation of magnetization and the energy dissipation processes as well.

Zero effective magnetic field $\boldsymbol{H}_{\text{eff}} = 0$ is the condition that determines the equilibrium magnitudes of magnetization and corresponds to the free energy minimum. As an example, we consider a uniaxial ferromagnet, namely easy axis. Its free energy reads

$$W = \int d^3r \left(\frac{\alpha_{ij}}{2} \frac{\partial \boldsymbol{M}}{\partial x_i} \frac{\partial \boldsymbol{M}}{\partial x_j} + f(\boldsymbol{M}^2) - \frac{\beta}{2}(\boldsymbol{M}\boldsymbol{n})^2 - \frac{H^2}{8\pi} - \boldsymbol{M}\boldsymbol{H} \right)$$

where $\boldsymbol{n}$ is unit vector directed along the anisotropy axis and the anisotropy constant is positive $\beta > 0$. Then the effective magnetic field $\boldsymbol{H}_{\text{eff}}$ equals

$$\boldsymbol{H}_{\text{eff}} = \boldsymbol{H} + \alpha_{ij} \frac{\partial^2 \boldsymbol{M}}{\partial x_i \partial x_j} - 2\boldsymbol{M} f'(\boldsymbol{M}^2) + \beta \boldsymbol{n}(\boldsymbol{M}\boldsymbol{n}).$$

For the homogeneous external field and constant magnetization $\boldsymbol{M}(\boldsymbol{r}) = \text{const} = \boldsymbol{M}_0$, we get the following equation for equilibrium:

$$\boldsymbol{H} - 2\boldsymbol{M}_0 f'(\boldsymbol{M}_0^2) + \beta \boldsymbol{n}(\boldsymbol{M}_0 \boldsymbol{n}) = 0.$$

In zero external field $\boldsymbol{H} = 0$ the magnetization vector $\boldsymbol{M}_0$ will be parallel to the easy axis, i.e. anisotropy axis $\boldsymbol{n}$. In the high magnetic field limit the magnetization is oriented in the direction of the external magnetic field.

9.6 Spin Waves in a Ferromagnet

Let us turn to studying the small oscillations of magnetization $\boldsymbol{M}(\boldsymbol{r},t)$ in a ferromagnet near the equilibrium value $\boldsymbol{M}_0$. Along with the magnetization oscillations, there appear magnetic field oscillations around the equilibrium value $\boldsymbol{H}_0$. Below, for definiteness, we consider the easy-axis ferromagnet in which magnetization $\boldsymbol{M}_0$ and magnetic field $\boldsymbol{H}_0$ are parallel to the anisotropy axis $\boldsymbol{n}$, i.e. $\boldsymbol{M}_0 \parallel \boldsymbol{n} \parallel \boldsymbol{H}_0$.

We put

$$\boldsymbol{H}(\boldsymbol{r},t) = \boldsymbol{H}_0 + \boldsymbol{h}(\boldsymbol{r},t) \quad \text{and} \quad \boldsymbol{M}(\boldsymbol{r},t) = \boldsymbol{M}_0 + \boldsymbol{m}(\boldsymbol{r},t)$$

where $\boldsymbol{m}(\boldsymbol{r},t)$ and $\boldsymbol{h}(\boldsymbol{r},t)$ are small quantities. Then we linearize the Landau–Lifshitz equation

$$\frac{\partial \boldsymbol{M}(\boldsymbol{r},t)}{\partial t} = \gamma \boldsymbol{M}(\boldsymbol{r},t) \times \boldsymbol{H}_{\text{eff}}(\boldsymbol{r},t)$$

with respect small perturbation $\boldsymbol{h}$ and small response $\boldsymbol{m}$, involving that $\boldsymbol{H}_{\text{eff}}^{(0)} = 0$, $\partial \boldsymbol{M}_0/\partial t = 0$ and $\nabla_i \boldsymbol{M}_0 = 0$ in equilibrium. So, we have

$$\boldsymbol{H}_{\text{eff}} = \boldsymbol{H}_{\text{eff}}^{(0)} + \boldsymbol{h}_{\text{eff}} = \boldsymbol{h} + \alpha_{ij}\frac{\partial^2 \boldsymbol{m}}{\partial x_i \partial x_j} - 2\boldsymbol{m} f'(\boldsymbol{M}_0^2) -$$
$$-4\boldsymbol{M}_0(\boldsymbol{m}\boldsymbol{M}_0) f''(\boldsymbol{M}_0^2) - 2\boldsymbol{m} f'(\boldsymbol{M}_0^2) + \beta \boldsymbol{n}(\boldsymbol{m}\boldsymbol{n}) = \boldsymbol{h} + \alpha_{ij}\frac{\partial^2 \boldsymbol{m}}{\partial x_i \partial x_j} -$$
$$-\frac{\boldsymbol{M}_0\boldsymbol{H}_0 + \beta(\boldsymbol{n}\boldsymbol{M}_0)^2}{M_0^2}\boldsymbol{m} + \beta \boldsymbol{n}(\boldsymbol{m}\boldsymbol{n}) - 4\boldsymbol{M}_0(\boldsymbol{m}\boldsymbol{M}_0) f''(\boldsymbol{M}_0^2).$$

Here we have used the relation $2M_0^2 f'(M_0^2) = \boldsymbol{M}_0\boldsymbol{H}_0 + \beta(\boldsymbol{n}\boldsymbol{M}_0)^2$, resulting from the obvious equality $\boldsymbol{M}_0 \cdot \boldsymbol{H}_{\text{eff}} = 0$. As a result, we get the following linearized equation:

$$\frac{\partial \boldsymbol{m}}{\partial t} = \gamma \boldsymbol{M}_0 \times \boldsymbol{h}_{\text{eff}} =$$
$$= \gamma \boldsymbol{M}_0 \times \left(\boldsymbol{h} + \alpha_{ij}\frac{\partial^2 \boldsymbol{m}}{\partial x_i \partial x_j} - \frac{\boldsymbol{M}_0\boldsymbol{H}_0 + \beta(\boldsymbol{n}\boldsymbol{M}_0)^2}{M_0^2}\boldsymbol{m} + \beta \boldsymbol{n}(\boldsymbol{m}\boldsymbol{n}) \right).$$

Since this equation is linear, the response of magnetization $\boldsymbol{m}(\boldsymbol{r},t)$ to the disturbing alternating magnetic field $\boldsymbol{h}(\boldsymbol{r},t)$ can be written as a convolution

$$m_i(\boldsymbol{r},t) = \int \chi_{ij}(\boldsymbol{r}-\boldsymbol{r}', t-t') h_j(\boldsymbol{r}',t')\, d^3r'\, dt'.$$

Hence, for harmonic perturbation $\boldsymbol{h}(\boldsymbol{r},t) = \boldsymbol{h}(\boldsymbol{k},\omega)e^{i\boldsymbol{k}\boldsymbol{r}-i\omega t}$ with wave vector $\boldsymbol{k}$ and frequency ω, we get for the Fourier transform

$$m_i(\boldsymbol{k},\omega) = \chi_{ij}(\boldsymbol{k},\omega) h_j(\boldsymbol{k},\omega).$$

The tensor $\chi_{ij}(\boldsymbol{k},\omega)$, representing the linear response function, is referred to as the *tensor of high-frequency magnetic susceptibility*.

Using the following relations

$$\boldsymbol{m}(\boldsymbol{r}, t) = \boldsymbol{m}(\boldsymbol{k}, \omega)e^{i\boldsymbol{k}\boldsymbol{r}-i\omega t} \quad \text{and} \quad \boldsymbol{h}(\boldsymbol{r}, t) = \boldsymbol{h}(\boldsymbol{k}, \omega)e^{i\boldsymbol{k}\boldsymbol{r}-i\omega t},$$

we arrive at the algebraic equation for the Fourier transforms $\boldsymbol{m}(\boldsymbol{k}, \omega)$ and $\boldsymbol{h}(\boldsymbol{k}, \omega)$

$$-i\omega\boldsymbol{m} = \gamma \boldsymbol{M}_0 \times \left(\boldsymbol{h} - \alpha_{ij}k_ik_j\boldsymbol{m} - \frac{\boldsymbol{M}_0\boldsymbol{H}_0 + \beta(\boldsymbol{n}\boldsymbol{M}_0)^2}{M_0^2}\boldsymbol{m} + \beta\boldsymbol{n}(\boldsymbol{m}\boldsymbol{n})\right).$$

Let us direct the z-axis parallel to both the anisotropy axis $\boldsymbol{n}$ and the magnetization vector $\boldsymbol{M}_0$. Then, the x and y axes are in the perpendicular plane. We rewrite the last equation from the vector form to the coordinate one

$$-i\omega m_x = \Omega_k m_y - \gamma M_0 h_y\,, \; -i\omega m_z = 0,$$
$$-i\omega m_y = \Omega_k m_x + \gamma M_0 h_x\,, \; \Omega_k = \gamma M_0\left(\alpha_{ij}k_ik_j + \beta + \tfrac{H_0}{M_0}\right).$$

Solving this system of equations gives us the *high-frequency susceptibility tensor*

$$\chi_{ij}(\boldsymbol{k}, \omega) = \begin{pmatrix} \chi_{xx} & \chi_{xy} & 0 \\ \chi_{yx} & \chi_{yy} & 0 \\ 0 & 0 & 0 \end{pmatrix}.$$

The tensor has the both spatial and temporal dispersion with the following components:

$$\chi_{xx} = \frac{\gamma M_0 \Omega_k}{\Omega_k^2 - \omega^2}\,, \quad \chi_{yy} = \frac{\gamma M_0 \Omega_k}{\Omega_k^2 - \omega^2}\,, \quad \chi_{xy} = -\chi_{yx} = \frac{i\omega\gamma M_0}{\Omega_k^2 - \omega^2}\,.$$

Note that frequency $\omega = \Omega_k$ is a resonance one for the response to the magnetic field.

The *spin waves* are the low frequency oscillations of magnetization. Therefore, for determining their oscillation spectrum, we can use the magnetostatic approximation and disregard the electric field. Accordingly, magnetic induction $\boldsymbol{b}$, magnetic field strength $\boldsymbol{h} = \boldsymbol{b} - 4\pi\boldsymbol{m}$, and magnetization $\boldsymbol{m}$ satisfy the equations

$$\operatorname{div} \boldsymbol{b} = \operatorname{div}(\boldsymbol{h} + 4\pi\boldsymbol{m}) = 0, \quad \operatorname{curl} \boldsymbol{h} = 0.$$

Turning to the Fourier transforms, we have

$$\boldsymbol{k} \times \boldsymbol{h}(\boldsymbol{k}, \omega) = 0, \quad \boldsymbol{k}\cdot\big(\boldsymbol{h}(\boldsymbol{k}, \omega) + 4\pi\boldsymbol{m}(\boldsymbol{k}, \omega)\big) = 0.$$

The first equation assumes that $\boldsymbol{h} = \boldsymbol{k}\phi$. Substituting it into the second equation yields with $m_i = \chi_{ij}h_j$

$$\boldsymbol{k}^2\phi + 4\pi k_i\chi_{ij}k_j\phi = k_ik_j(\delta_{ij} + 4\pi\chi_{ij})\phi = k_ik_j\mu_{ij}\phi = 0.$$

Here $\mu_{ij} = \delta_{ij} + 4\pi\chi_{ij}$ is the *magnetic permeability*.

For existing non-trivial solution $\phi \neq 0$, it is necessary to vanish the following convolution with susceptibility μ_{ij}:

$$k_i \mu_{ij}(\boldsymbol{k}, \omega) k_j = 0.$$

This relation, in essence, is the *dispersion equation* which will determine the *spin wave spectrum* $\omega = \omega(\boldsymbol{k})$ or magnetization oscillation frequencies. So, we get

$$\omega(\boldsymbol{k}) = \sqrt{\Omega_k^2 + 4\pi\gamma M_0 \Omega_k \sin^2 \vartheta_k}$$

where ϑ_k is the angle between wave vector $\boldsymbol{k}$ and anisotropy axis $\boldsymbol{n}$. For $k = 0$, in the spin wave spectrum there appears an energy gap equal to

$$\omega(0) = \gamma M_0(\beta + H_0/M_0)$$

and resulting from the magnetic anisotropy and external magnetic field. In the region of sufficiently large wave vectors $\alpha_{ij}k_ik_j \gg max\{4\pi,\ \beta,\ H_0/M_0\}$, the dispersion law simplifies remarkably

$$\omega(\boldsymbol{k}) = \gamma M_0 \alpha_{ij} k_i k_j.$$

The spin wave frequency in a ferromagnet becomes directly proportional to the square of wave vector. In the isotropic case $\alpha_{ij} \sim \alpha\delta_{ij}$ and $\omega(k) = \gamma M_0 \alpha k^2$. We recall the order-of-magnitude estimate of magnetic stiffness coefficient α as $\alpha \sim a^2(\Theta/\mu M_0)$, where $\mu = g\mu_B$ is the effective magneton and $\gamma = \mu/\hbar$ is the gyromagnetic ratio. To conclude, we obtain

$$\hbar\omega(\boldsymbol{k}) \sim \Theta(ka)^2.$$

Here Θ is the Curie temperature and a is the interatomic distance.

Problem

An excitation of uniform oscillations of magnetization is used for studying the spin waves in magnets.

Determine the frequency of *uniform ferromagnetic resonance* for the easy axis-type ferromagnet in the shape of a ball. The uniform constant external magnetic field $\boldsymbol{H}_0$ is parallel to the easy magnetization axis.

Solution. On the account of the demagnetization coefficients for the ball, the magnetic field $\boldsymbol{H}_{\text{in}}$ inside the ball is related to the external uniform field $\boldsymbol{H}_0$ and magnetization $\boldsymbol{M}_0$, as follows:

$$\boldsymbol{H}_{\text{in}} + \frac{4\pi}{3}\boldsymbol{M}_0 = \boldsymbol{H}_0 \quad \text{or} \quad \boldsymbol{H}_{\text{in}} = \boldsymbol{H}_0 - \frac{4\pi}{3}\boldsymbol{M}_0.$$

The alternating components of magnetization $m(t)$, internal magnetic field $\boldsymbol{h}_{\text{in}}(t)$, and external field $\boldsymbol{h}_{\text{out}}(t)$, varying in time as $\sim e^{-i\omega t}$, are connected with the same relation

$$\boldsymbol{h}_{\text{in}} + \frac{4\pi}{3}\boldsymbol{m} = \boldsymbol{h}_{\text{out}}.$$

Since the magnetic field varies according to $\boldsymbol{h}_{\text{in}}(t) \sim e^{-i\omega t}$, we have $\boldsymbol{m} = \hat{\chi}(\omega)\boldsymbol{h}_{\text{in}}$, $\hat{\chi}(\omega) = \hat{\chi}(k = 0, \omega)$ being the high-frequency magnetic susceptibility tensor. Correspondingly, we get

$$\left(1+\frac{4\pi}{3}\hat{\chi}(\omega)\right)\boldsymbol{h}_{\text{in}}=\boldsymbol{h}_{\text{out}}.$$

The frequency of uniform ferromagnetic resonance in a ball is determined by the roots of the equation below

$$\det\left(1+\frac{4\pi}{3}\hat{\chi}(\omega)\right)=\det\left(\delta_{ik}+\frac{4\pi}{3}\chi_{ik}(\omega)\right)=0\quad(i,k=x,y,z).$$

The high-frequency susceptibility equals

$$\hat{\chi}(\omega)=\begin{pmatrix}\chi_{xx} & \chi_{xy} & 0\\ \chi_{yx} & \chi_{yy} & 0\\ 0 & 0 & 0\end{pmatrix},\quad \chi_{xx}=\chi_{yy}=\frac{\gamma M_0\Omega_0}{\Omega_0^2-\omega^2},\quad \chi_{xy}=-\chi_{yx}=\frac{i\omega\gamma M_0}{\Omega_0^2-\omega^2}.$$

The frequency Ω_0 is given by

$$\Omega_0=\gamma M_0\big(\beta+H_{\text{in}}/M_0\big)=\gamma M_0\big(\beta+H_0/M_0\ -4\pi/3\big)$$

where we have involved that the magnetic field inside the magnetized ball equals $\boldsymbol{H}_{\text{in}}=\boldsymbol{H}_0-(4\pi/3)\boldsymbol{M}_0$. The solution of equation $(1+4\pi\chi_{xx}/3)^2-(4\pi\chi_{xy}/3)^2=0$ delivers us the resonance frequency

$$\omega_{\text{res}}=\Omega_0+4\pi\gamma M_0/3=\gamma M_0\big(\beta+H_0/M_0\big).$$

The magnitude of resonance frequency depends both on the external field strength and on the shape of a ferromagnet by means of demagnetization coefficients. The inhomogeneous resonance frequencies refer to the proper oscillations of magnetization when the magnetic field and magnetization are coordinate-dependent.

9.7 Thermodynamics of Ferromagnets

The excitations in a magnetic material represent oscillations of magnetic moment (spin) around its equilibrium direction. As long as the energy of the state in the magnetic material is sufficiently small, the oscillations of magnetization are expected to be small as well. In first approximation the oscillations can be represented as a superposition of independent harmonic ones, each of them being a monochromatic plane spin wave propagating in the magnetic medium. The spin wave is specified with wave vector $\boldsymbol{k}$ and corresponding frequency $\omega(\boldsymbol{k})$.

Using the analogy with quantization of sound waves and quantum mechanical correspondence, we can interpret the small oscillations of magnetization representing the spin waves as a set quanta of elementary magnetic excitations or *magnons*. Correspondingly, the energy $\varepsilon(\boldsymbol{p})$ of a magnon with momentum[4] $\boldsymbol{p}=\hbar\boldsymbol{k}$ is determined with frequency $\omega(\boldsymbol{k})$ and wave vector $\boldsymbol{k}$ of spin wave according to $\varepsilon(\boldsymbol{p})=\hbar\omega(\boldsymbol{k})$. On the analogy with quantum mechanics the energy for a set of such elementary excitations or magnons can be given by the sum

---

[4] In the crystalline periodic spin lattice the vector $\boldsymbol{p}$ is determined within the accuracy of the reciprocal lattice vector and is a quasi momentum.

$$E = \sum_k \hbar\omega_k \big(N_k + 1/2\big)$$

where N_k is either zero or any positive integer. The numbers N_k can be interpreted as a number of magnons in the state with wave vector $\boldsymbol{k}$. Since numbers N_k can get any values, we can draw a conclusion that magnons obey the Bose statistics even if the particles composing the magnetic matter have a half-integer[5] spin.

Specifying the statistics of magnons and the law of their dispersion, we can proceed to studying the thermodynamic properties of a ferromagnet. We employ the relation

$$\Omega = -T \ln \operatorname{tr} e^{-\hat{\mathcal{H}}/T}$$

connecting the thermodynamic potential Ω with Hamiltonian $\hat{\mathcal{H}}$ of the magnetic subsystem. The Hamiltonian can be written as $\hat{\mathcal{H}} = \hat{\mathcal{H}}_0 + \hat{\mathcal{H}}_{\text{int}}$ where $\hat{\mathcal{H}}_0$ is the Hamiltonian of ideal magnon gas and $\hat{\mathcal{H}}_{\text{int}}$ is the Hamiltonian of interaction between magnons.

In the region of sufficiently low temperatures we can assume in first approximation that the magnons compose an ideal gas of Bose particles and the effects of magnon-magnon interaction are negligible. To meet these conditions, the temperature should satisfy the inequality $T \ll \Theta$, Θ being the Curie temperature. The low energy magnons with $\hbar\omega_k \lesssim T$ and, correspondingly, small wave vectors $ka \ll 1$ play the most essential role in the thermodynamic properties of a ferromagnet at the low temperatures. The wave vector $ka \ll 1$ region agrees with the estimate $\hbar\omega_k \sim \Theta(ka)^2$, a being the interatomic distance.

In what follows, we disregard the interaction between magnons and write down the magnetic contribution to the thermodynamic potential for a ferromagnet of volume V in the external magnetic field $\boldsymbol{H}$

$$\Omega_m(T, \boldsymbol{H}) = E_0(\boldsymbol{H}) + \Delta\Omega_m(T, \boldsymbol{H}) = \\ = \int \frac{V d^3k}{(2\pi)^3} \frac{\hbar\omega_k}{2} + T \int \frac{V d^3k}{(2\pi)^3} \ln\big(1 - e^{\hbar\omega_k/T}\big).$$

Here the first term $E_0(\boldsymbol{H}) = \Omega_m(0, \boldsymbol{H})$ is zero energy of magnons, giving no contribution to the temperature behavior of the thermodynamic variables. Within the framework of the macroscopic $\omega_k \sim (ka)^2$ description of magnon dispersion, this integral diverges at large values of wave vector $ka \gtrsim 1$, i.e. it is determined by the short wave asymptotic and, in general, requires of involving the microscopic theory. The second term, in essence, is the thermodynamic potential of an ideal Bose-gas with zero chemical potential.

The equilibrium magnetization density $\boldsymbol{M}(T, \boldsymbol{H})$ and magnetic contribution to the specific heat $C_m(T, \boldsymbol{H})$ are given by expressions

[5] Under oscillations of spin direction the change of the spin projection on some axis is always an integer regardless of whether a half-integer or integer spin. The strict consideration requires the determination of the commutation relations for the magnon creation and annihilation operators.

$$\boldsymbol{M} = -\frac{1}{V}\frac{\partial \Omega_m}{\partial \boldsymbol{H}} \quad \text{and} \quad C_m = -\frac{T}{V}\frac{\partial^2 \Omega_m}{\partial T^2}.$$

For an ideal magnon gas, these derivatives reduce to

$$\boldsymbol{M} = -\frac{1}{V}\frac{\partial E_0}{\partial \boldsymbol{H}} - \int \frac{d^3k}{(2\pi)^3}\frac{\hbar \partial\omega_{\boldsymbol{k}}/\partial \boldsymbol{H}}{e^{\hbar\omega_{\boldsymbol{k}}/T} - 1},$$

$$C_m = \frac{1}{V}\frac{\partial E}{\partial T} \quad \text{where} \quad E(T, \boldsymbol{H}) = \int \frac{V d^3k}{(2\pi)^3}\frac{\hbar\omega_{\boldsymbol{k}}}{e^{\hbar\omega_{\boldsymbol{k}}/T} - 1}.$$

The quantity $E(T, \boldsymbol{H})$ can be called the temperature contributions of magnons to the internal energy of a ferromagnet.

As an example, we consider below a uniaxial ferromagnet of the easy axis type. For definiteness, we suppose that the external field and magnetization are parallel to the easy axis. Accordingly, the spin wave or magnon dispersion reads

$$\omega_{\boldsymbol{k}} = \sqrt{\Omega_{\boldsymbol{k}}^2 + 4\pi\gamma M_0 \Omega_{\boldsymbol{k}} \sin^2\theta_{\boldsymbol{k}}}, \quad \Omega_{\boldsymbol{k}} = \gamma M_0(\alpha k^2 + \beta + H/M_0)$$

where $\theta_{\boldsymbol{k}}$ is the angle between the anisotropy axis and wave vector $\boldsymbol{k}$. To simplify the final formulas, we neglect also a possible anisotropy of magnetic stiffness coefficient, putting $\alpha_{ij} = \alpha\delta_{ij}$, $(i, j = x, y, z)$. Using this formula and replacing the integration variables $\boldsymbol{k}$ with $x = \hbar\omega/T$ and angle θ, we express the difference $\Delta M(T, \boldsymbol{H}) = M(T, \boldsymbol{H}) - M(0, \boldsymbol{H})$ in terms of the following integral:

$$\Delta M(T, H) = -\frac{\gamma}{(2\pi)^2}\left(\frac{T}{\hbar\gamma M_0 \alpha}\right)^{3/2} \int_0^{\pi/2} \sin\theta\, d\theta \times$$

$$\times \int_{\sqrt{\eta(\eta+2\xi\sin^2\theta)}}^{\infty} \frac{\left[\sqrt{x^2 + \xi^2\sin^4\theta} - (\eta + \xi\sin^2\theta)\right]^{1/2}}{e^x - 1} dx$$

where

$$\xi = \frac{2\pi\hbar\gamma M_0}{T} \quad \text{and} \quad \eta = \frac{\hbar\gamma(\beta M_0 + H)}{T}.$$

In the same notations as ξ and η the spin wave contribution to the specific heat is given by the integral

$$C_m(T, H) = \frac{1}{(2\pi)^2}\left(\frac{T}{\hbar\gamma M_0 \alpha}\right)^{3/2} \int_0^{\pi/2} \sin\theta\, d\theta \times$$

$$\times \int_{\sqrt{\eta(\eta+2\xi\sin^2\theta)}}^{\infty} x\left[\frac{\sqrt{x^2+\xi^2\sin^4\theta}-\xi\sin^2\theta}{x^2+\xi^2\sin^4\theta}\right]^{1/2}\left(\frac{x}{2\sinh x/2}\right)^2 dx.$$

For transparency and simplification, we demonstrate the expressions for $\Delta M(T)$ and $C_m(T)$ in the most interesting case of sufficiently high temperature when we can approximately put $\xi = 0$ and $\eta = 0$ in the integrand:

$$\Delta M(T,H) = -\frac{\zeta(3/2)}{8\pi^{3/2}}\frac{\gamma}{\hbar}\left(\frac{T}{\hbar\gamma M_0\alpha}\right)^{3/2} \sim M_0\left(\frac{T}{\Theta}\right)^{3/2}$$

$$\text{at}\quad 2\pi\hbar\gamma M_0,\ \hbar\gamma(\beta M_0+H) \ll T.$$

Here Θ is the Curie temperature. In the case of extremely low temperatures, we have

$$\Delta M(T,H) \sim -\exp\left(-\frac{\hbar\gamma(\beta M_0+H)}{T}\right)$$

$$\text{at}\quad 2\pi\hbar\gamma M_0,\ \hbar\gamma(\beta M_0+H) \gg T.$$

The temperature behavior $\Delta M/M_0 \sim -T^{3/2}$ at $T \gg 2\pi\gamma M_0$ and $\gamma(\beta M_0+H)$ means the *Bloch* $\mathrm{T}^{3/2}$ *law* typical for an isotropic ferromagnet in zero magnetic field. In the low temperature limit, the finite magnitude of energy gap in the spin wave spectrum results in the appearance of thermally activated and exponentially small behavior. The magnitude in the exponent is the minimum magnitude of energy gap at $\theta = 0$ and $\theta = \pi$.

In the same limiting cases, we have for the spin wave contribution to the specific heat of a ferromagnet at high temperatures

$$C_m(T,H) = \frac{15\zeta(3/2)}{32\pi^{3/2}}\left(\frac{T}{\hbar\gamma M_0\alpha}\right)^{3/2} \sim \left(\frac{T}{\Theta}\right)^{3/2}$$

$$\text{when}\quad 2\pi\hbar\gamma M_0,\ \hbar\gamma(\beta M_0+H) \ll T,$$

Θ being the Curie temperature. Correspondingly, we get the exponentially small correction in the low temperature limit

$$C_m(T,H) \sim \exp\left(-\frac{\hbar\gamma(\beta M_0+H)}{T}\right)$$

$$\text{when}\quad T \ll 2\pi\hbar\gamma M_0,\ \hbar\gamma(\beta M_0+H).$$

The spin wave contribution to the specific heat of a ferromagnet demonstrates the temperature behavior similar to that of magnetization. For sufficiently high temperatures, we get the Bloch $\mathrm{T}^{3/2}$ law and there is an exponentially small correction in the low temperature limit.

Calculating the spin wave effect on the thermodynamic functions of a ferromagnet, we have omitted the temperature-independent contribution to the energy of ferromagnet due to zero oscillations of magnons.

$$E_0(H) = \frac{1}{2}\sum_{\boldsymbol{k}} \hbar\omega_{\boldsymbol{k}} = \frac{1}{2}\int \hbar\omega(\boldsymbol{k})\frac{Vd^3k}{(2\pi)^3}.$$

Accordingly, we have for the zero temperature magnetization

$$M(0, H) = -\frac{1}{V}\frac{\partial E_0(H)}{\partial H} = -\frac{1}{2}\int \frac{\hbar\,\partial\omega(\boldsymbol{k})}{\partial H}\frac{d^3k}{(2\pi)^3}.$$

This integral diverges at the large values of wave vector k and is governed with the behavior of the short wave magnon spectrum $ka \gtrsim 1$ which cannot be determined within the framework of macroscopic description $ka \ll 1$. Note only that the contribution of magnon zero oscillations decreases the magnetization M_0 by the relatively small magnitude of about $(\mu M_0/\Theta)^{3/2} \ll 1$. Unlike the three-dimensional case, the effect of long wave magnon fluctuations in a two-dimensional isotropic ferromagnet proves to be so strong that results in breaking the state of spontaneous magnetization down.

Problems

1. Find the spin wave contribution to the thermodynamic variables in the low temperature limit $T \ll \hbar\omega(0)$ and $2\pi\hbar\gamma M_0$.

Solution. The main contribution is associated with the spin waves propagating in the direction of the minimum energy gap in the spectrum, i.e. $\vartheta_{\boldsymbol{k}} = 0$ and $\vartheta_{\boldsymbol{k}} = \pi$. These both values entail the same contribution. For small $\vartheta_{\boldsymbol{k}} \ll 1$, we get

$$\omega_k \approx \Omega_k + 2\pi\gamma M_0\vartheta_k^2, \quad \Omega_k = \gamma M_0(\alpha k^2 + \beta + H/M_0).$$

For such low temperatures and nonzero energy gap, we can take the magnon distribution as the Boltzmann one, neglecting the unity in the denominator of the Bose distribution and replacing $\omega(\boldsymbol{k})$ with $\omega(0)$ in the preexponential factors. Integration over k and over ϑ_k can be extended to the infinity. Then we find the low temperature correction to the magnetization

$$\Delta M(T, H) \approx -\int \frac{d^3k}{(2\pi)^3}\frac{\hbar\partial\omega(0)}{\partial H}e^{-\hbar\omega_k/T} = -2\frac{\hbar\gamma}{(2\pi)^2}e^{-\hbar\gamma(\beta M_0+H)/T}$$

$$\times\int_0^\infty k^2dk\int_0^\infty \theta d\theta e^{-\frac{\hbar\gamma M_0\alpha k^2}{T}}e^{-\frac{2\pi\hbar\gamma M_0\theta^2}{T}} = \hbar\gamma\alpha\left(\frac{T}{4\pi\hbar\gamma M_0\alpha}\right)^{5/2}e^{-\frac{\hbar\gamma(\beta M_0+H)}{T}}$$

and to the specific heat in accordance with $C_m(T, H) = V^{-1}\partial E(T, H)/\partial T$ where

$$E(T, H) \approx \int \frac{Vd^3k}{(2\pi)^3}\hbar\omega(0)e^{-\hbar\omega_k/T} = -2\frac{\hbar\gamma(\beta M_0 + H)}{(2\pi)^2}e^{-\hbar\gamma(\beta M_0+H)/T}$$

$$\times\int_0^\infty k^2dk\int_0^\infty \theta d\theta e^{-\frac{\hbar\gamma M_0\alpha k^2}{T}}e^{-\frac{2\pi\hbar\gamma M_0\theta^2}{T}} =$$

$$= \frac{V}{32\pi^{5/2}} \frac{\beta M_0 + H}{M_0} \frac{T^{5/2}}{(\hbar\gamma M_0\alpha)^{3/2}} e^{-\frac{\hbar\gamma(\beta M_0+H)}{T}}.$$

Calculating the specific heat, it is sufficient to be restricted with differentiating the exponential function

$$C_m(T, H) = \frac{(\beta + H/M_0)^2}{32\pi^{5/2}\alpha} \left(\frac{T}{\hbar\gamma M_0\alpha}\right)^{1/2} e^{-\frac{\hbar\gamma(\beta M_0+H)}{T}}.$$

2. The *magnetocaloric effect* is a change in the temperature of a magnetic material under its adiabatic magnetization (demagnetization) in the external magnetic field. Find the magnitude of the effect as a ratio of temperature change to the magnetic field variation.

Solution. The adiabatic process means that the magnetic material is under heat-insulated conditions and its entropy remains constant. Considering entropy $S(T, H)$ as a function of temperature and magnetic field, we can derive the following equation for the entropy change:

$$dS = \left(\frac{\partial S}{\partial T}\right)_H dT + \left(\frac{\partial S}{\partial H}\right)_T dH = 0.$$

The magnitude of the magnetocaloric effect equals

$$\frac{dT}{dH} = -\frac{(\partial S/\partial H)_T}{(\partial S/\partial T)_H}.$$

The derivative in the denominator is $(\partial S/\partial T)_H = C_H/T$ where C_H is the specific heat of the magnetic material in the constant magnetic field. The derivative in the nominator transforms as

$$(\partial S/\partial H)_T = \partial^2 F/\partial H \partial T = \partial^2 F/\partial T \partial H = (\partial M/\partial T)_H,$$

M being the magnetization of a magnetic material. Finally, we get

$$\frac{dT}{dH} = -\frac{T}{C_H}\left(\frac{\partial M}{\partial T}\right)_H.$$

3. There is an isotropic ferromagnet. Determine the second sound velocity in a magnon gas, assuming it ideal and neglecting completely the effects of magnon-magnon and magnon-phonon scattering.

Solution. First, we write the equations for the energy and momentum conservation in the magnon gas

$$\frac{\partial E}{\partial t} + \operatorname{div} \boldsymbol{Q} = 0 \quad \text{and} \quad \frac{\partial P_i}{\partial t} + \frac{\partial \Pi_{ij}}{\partial x_j} = 0.$$

Here E is the energy density, $\boldsymbol{Q}$ is the energy flux density, P_i is ith component of momentum density, and Π_{ij} are the components of momentum flux density tensor. These variables are determined with the following relations in terms of distribution function $n_{\boldsymbol{p}} = n_{\boldsymbol{p}}(\boldsymbol{r}, t)$:

$$E = \sum_{\boldsymbol{p}} \varepsilon_{\boldsymbol{p}} n_{\boldsymbol{p}}, \quad \boldsymbol{Q} = \sum_{\boldsymbol{p}} \boldsymbol{v}_{\boldsymbol{p}} \varepsilon_{\boldsymbol{p}} n_{\boldsymbol{p}} \quad \text{and} \quad P_i = \sum_{\boldsymbol{p}} p_i n_{\boldsymbol{p}}, \quad \Pi_{ij} = \sum_{\boldsymbol{p}} v_i p_j n_{\boldsymbol{p}}$$

where $\varepsilon_{\boldsymbol{p}}$ is the energy and $\boldsymbol{v}_{\boldsymbol{p}} = \partial\varepsilon_{\boldsymbol{p}}/\partial\boldsymbol{p}$ is the velocity of magnons with momentum $\boldsymbol{p}$.

We seek for the general solution of these equations as

$$n_p(\boldsymbol{r}, t) = n\left(\frac{\varepsilon_p - \boldsymbol{u}\boldsymbol{p}}{T}\right) = \left[\exp\left(\frac{\varepsilon_p - \boldsymbol{u}\boldsymbol{p}}{T}\right) - 1\right]^{-1}.$$

The temperature $T = T(\boldsymbol{r}, t)$ and $\boldsymbol{u} = \boldsymbol{u}(\boldsymbol{r}, t)$ are some slow functions of coordinate and time and we can restrict ourselves with a linear approximation in velocity $\boldsymbol{u}$ and temperature derivatives, assuming them small. After calculating, we find

$$C\frac{\partial T}{\partial t} + TS \operatorname{div} \boldsymbol{u} = 0,$$

$$\rho_n \frac{\partial \boldsymbol{u}}{\partial t} + S\nabla T = 0.$$

On the analogy with the case of superfluid helium the coefficients found

$$C = -\sum_p \frac{\varepsilon_p^2}{T^2} n'_p\left(\frac{\varepsilon_p}{T}\right), \quad S = -\frac{1}{3}\sum_p \frac{\varepsilon_p p v_p}{T^2} n'_p\left(\frac{\varepsilon_p}{T}\right), \quad \rho_n = -\frac{1}{3}\sum_p \frac{p^2}{T} n'_p\left(\frac{\varepsilon_p}{T}\right)$$

represent specific heat $C(T)$, entropy $S(T)$, and normal density $\rho_n(T)$ in the gas of magnon excitations.

Then, from the equations above we eliminate velocity $\boldsymbol{u}$ with the aid of divergence operator and time differentiation and, finally, obtain the equation determining the temperature oscillations in a second sound wave

$$\rho_n C \frac{\partial^2 T}{\partial t^2} - TS^2 \nabla^2 T = 0.$$

Let temperature oscillate in this wave according to the law $\exp[-i(\omega t - kx)]$. (The x-axis is directed along the wave propagation, ω is the frequency, and k is wave vector.) The velocity of oscillations is given by

$$u_2 = \frac{\omega}{k} = \sqrt{\frac{TS^2}{\rho_n C}}.$$

Next, we have after calculating the integrals for the magnon spectrum $\varepsilon_p = \Theta(pa/\hbar)^2$ where Θ is the Curie temperature and a is the interatomic spacing

$$u_2 = \sqrt{\frac{10\zeta(5/2)}{3\zeta(3/2)}} \frac{a}{\hbar}\sqrt{T\Theta} \sim \frac{a}{\hbar}\sqrt{T\Theta}.$$

For the phonon spectrum $\varepsilon = cp$, we see the familiar answer $u_2 = c/\sqrt{3}$.

The second sound in a ferromagnet has a curious specific feature. The magnon density oscillations are accompanied by the magnetic moment oscillations and, therefore, result in emerging the magnetic field of small strength. Usually, the observations of second sound in a ferromagnet are hindered by the dissipative magnon scattering processes entailing a noticeable attenuation of temperature wave oscillations and diffusive nature of sound propagation.

9.8 Antiferromagnetic Ordering

The types of antiferromagnetic ordering found in nature are very numerous and diverse in their magnetic structure. The simplest antiferromagnetic structure is a collinear two-lattice antiferromagnet with the opposite magnetizations $\boldsymbol{M}_1$ and $\boldsymbol{M}_2$

each of two *equivalent sublattices*. In the lack of magnetic field the net magnetization is $\boldsymbol{M} = \boldsymbol{M}_1 + \boldsymbol{M}_2 = 0$. Let us define additional order parameter, called the *antiferromagnetic vector* $\boldsymbol{L} = \boldsymbol{M}_1 - \boldsymbol{M}_2$ which becomes nonzero below the *Néel temperature* Θ_N with the simultaneous appearance of antiferromagnetic ordering.

For the case of two magnetic sublattices, the simplest expression for the exchange interaction has the following structure:

$$w_{ex} = f(\boldsymbol{M}_1^2, \boldsymbol{M}_2^2, \boldsymbol{M}_1\boldsymbol{M}_2) = f(\boldsymbol{M}_1^2) + f(\boldsymbol{M}_2^2) + I\boldsymbol{M}_1\boldsymbol{M}_2 .$$

For the antiferromagnetic ordering, it is necessary to have a positive exchange coupling constant $I > 0$ to provide us the energetically favorable relation $\boldsymbol{M}_1 = -\boldsymbol{M}_2$. In the opposite $I < 0$ case there appears a trivial ferromagnetic ordering.

The density of inhomogeneity energy can be written as follows:

$$w_{\text{inh}} = \frac{1}{2}\alpha_{ik}\left(\frac{\partial \boldsymbol{M}_1}{\partial x_i}\frac{\partial \boldsymbol{M}_1}{\partial x_k} + \frac{\partial \boldsymbol{M}_2}{\partial x_i}\frac{\partial \boldsymbol{M}_2}{\partial x_k}\right) + \alpha'_{ik}\frac{\partial \boldsymbol{M}_1}{\partial x_i}\frac{\partial \boldsymbol{M}_2}{\partial x_k} .$$

Then, we introduce the density of magnetic anisotropy energy in an uniaxial antiferromagnet with the anisotropy axis $\boldsymbol{n}$ according to

$$w_{an} = -\frac{\beta}{2}\left[(\boldsymbol{M}_1\boldsymbol{n})^2 + (\boldsymbol{M}_2\boldsymbol{n})^2\right] - \beta'(\boldsymbol{M}_1\boldsymbol{n})(\boldsymbol{M}_2\boldsymbol{n}),$$

β and β' being the anisotropy constants.

If $\beta > \beta'$, the anisotropy energy minimum of $w_{an} = -\beta + \beta' < 0$ realizes when the magnetizations of sublattices are oriented in the direction of anisotropy axis $\boldsymbol{n}$ and $\boldsymbol{M}_{10} = -\boldsymbol{M}_{20}$. This case of anisotropy will be called the *easy axis type*. In the opposite case $\beta < \beta'$ the anisotropy energy minimum $w_{an} = 0$ will be achieved when the sublattice magnetizations are normal to the anisotropy axis $\boldsymbol{n}$ and $\boldsymbol{M}_{10} = -\boldsymbol{M}_{20}$ as well. This type of anisotropy will be referred to as *easy plane*.

Within the framework of self-consistent field theory, we can represent the free energy density of an antiferromagnet near the Néel temperature Θ_N as an expansion in powers of magnetization vectors since antiferromagnetic vector $\boldsymbol{L}$ is sufficiently small

$$F = a\boldsymbol{M}_1^2 + \frac{b}{2}\boldsymbol{M}_1^4 + a\boldsymbol{M}_2^2 + \frac{b}{2}\boldsymbol{M}_2^4 + 2I\boldsymbol{M}_1\boldsymbol{M}_2 \quad (I > 0).$$

On the use of the following relations for magnetization $\boldsymbol{M}$ and antiferromagnetic vector $\boldsymbol{L}$

$$\boldsymbol{M}_1 = (\boldsymbol{M} + \boldsymbol{L})/2 \quad \text{and} \quad \boldsymbol{M}_2 = (\boldsymbol{M} - \boldsymbol{L})/2,$$

it is convenient to rewrite the free energy density as

$$F = \frac{a - I}{2}\boldsymbol{L}^2 + \frac{b}{16}\boldsymbol{L}^4 + \frac{a + I}{2}\boldsymbol{M}^2 + \frac{b}{16}\boldsymbol{M}^4 + \frac{b}{8}\boldsymbol{M}^2\boldsymbol{L}^2 + \frac{b}{4}(\boldsymbol{M}\boldsymbol{L})^2.$$

One should put $a(T) = I + a_0(T - \Theta_N)$ near the Néel temperature.

In zero external magnetic field the net magnetization is absent, i.e. $\boldsymbol{M} = 0$. The free energy of an antiferromagnet reduces to the expression similar to that of a ferromagnet. Correspondingly, the behavior of antiferromagnetic vector near the Néel temperature at $T \leqslant \Theta_N$ will be qualified with the critical exponent $\beta = 1/2$ typical for the self-consistent field

$$L(T) \sim (\Theta_N - T)^{\beta}, \quad \beta = 1/2.$$

In the external field, the Gibbs free energy $\widetilde{F}(\boldsymbol{M}, \boldsymbol{L}, \boldsymbol{H})$ takes the form

$$\widetilde{F}(\boldsymbol{M}, \boldsymbol{L}, \boldsymbol{H}) = \frac{a - I}{2}\boldsymbol{L}^2 + \frac{b}{16}\boldsymbol{L}^4 + \frac{a + I}{2}\boldsymbol{M}^2 + \frac{b}{16}\boldsymbol{M}^4 + \\ + \frac{b}{8}\boldsymbol{M}^2\boldsymbol{L}^2 + \frac{b}{4}(\boldsymbol{M}\boldsymbol{L})^2 - \boldsymbol{M}\boldsymbol{H} - \frac{\boldsymbol{H}^2}{8\pi}.$$

Let us study behavior of magnetic susceptibility χ in the low magnetic field. The equilibrium values of the antiferromagnetic and magnetization vectors will be determined from a set of equations

$$\frac{\partial \widetilde{F}}{\partial \boldsymbol{M}} = (a + I)\boldsymbol{M} + \frac{b\boldsymbol{M}^2}{4}\boldsymbol{M} + \frac{b\boldsymbol{L}^2}{4}\boldsymbol{M} + \frac{b}{2}(\boldsymbol{M}\boldsymbol{L})\boldsymbol{L} - \boldsymbol{H} = 0,$$

$$\frac{\partial \widetilde{F}}{\partial \boldsymbol{L}} = (a - I)\boldsymbol{L} + \frac{b\boldsymbol{L}^2}{4}\boldsymbol{L} + \frac{b\boldsymbol{M}^2}{4}\boldsymbol{L} + \frac{b}{2}(\boldsymbol{M}\boldsymbol{L})\boldsymbol{M} = 0.$$

In the low magnetic field, the magnetization is $\boldsymbol{M} \sim \boldsymbol{H}$ and the corrections to the antiferromagnetic vector $\boldsymbol{L}$ are of the order of $\boldsymbol{H}^2$. These corrections are neglected below. So, we can approximately put $\boldsymbol{L} = \boldsymbol{L}_0(T)$, i.e. $\big((a - I) + b\boldsymbol{L}_0^2/4\big)\boldsymbol{L}_0 = 0$, and get

$$\boldsymbol{M}\big(a + I + b\boldsymbol{L}_0^2/4\big) + b\boldsymbol{L}_0(\boldsymbol{M}\boldsymbol{L}_0)/2 = \boldsymbol{H}.$$

Above the Néel temperature $T > \Theta_N$ the antiferromagnetic vector vanishes $\boldsymbol{L}_0 = 0$ and magnetization is $\boldsymbol{M} = \chi\boldsymbol{H}$ where magnetic susceptibility χ is given by the formula

$$\chi = \frac{1}{a + I} = \frac{1}{2I + a_0(T - \Theta_N)}.$$

Emphasize that, unlike ferromagnetic phase transition when the magnetic susceptibility diverges at the transition point, the susceptibility of an antiferromagnet remains finite within the approximation of the self-consistent theory of phase transitions.

Below the Néel temperature there appears a selected direction in the space, resulting in the anisotropy of magnetic susceptibility. If the external magnetic field is parallel to the antiferromagnetic vector, we have

$$\chi_{\parallel} = \frac{1}{a + I + 3bL_0^2/4} = \frac{1}{4I - 2a} =$$
$$= \frac{1}{2I + 2a_0(\Theta_N - T)} \qquad (\boldsymbol{H} \parallel \boldsymbol{L}_0).$$

When the external field is normal to the antiferromagnetic vector $\boldsymbol{L}_0$, the magnetic susceptibility, equal to

$$\chi_{\perp} = \frac{1}{a + I + bL_0^2/4} = \frac{1}{a + I + I - a} = \frac{1}{2I} \quad (\boldsymbol{H} \perp \boldsymbol{L}_0),$$

is temperature-independent. For polycrystals, since all the three spatial directions of vector $\boldsymbol{L}_0$ are equiprobable, the magnetic susceptibility has the averaged magnitude

$$\chi_{\text{av}} = \frac{\chi_{\parallel}}{3} + \frac{2\chi_{\perp}}{3}.$$

The temperature behavior of magnetic susceptibility is illustrated in Fig. 9.8. In the limit of high magnetic field, the magnetic moments of the both sublattices will align in the magnetic field direction so that $\boldsymbol{L} = 0$ and $\boldsymbol{M}_1 = \boldsymbol{M}_2$.

Problem

There is a collinear antiferromagnet with the easy axis anisotropy. Find magnetization $\boldsymbol{M}$ of the antiferromagnet in magnetic field $\boldsymbol{H}$ in the direction of easy axis $\boldsymbol{n}$. Treat the magnitudes of sublattice magnetizations $\boldsymbol{M}_1$ and $\boldsymbol{M}_2$ as constant, assuming the temperature much lower than the Néel one.

Solution. In the free energy expression we select only the terms depending on the orientation of magnetic sublattices with respect to anisotropy axis $\boldsymbol{n}$ and magnetic field $\boldsymbol{H}$

$$w = I\boldsymbol{M}_1\boldsymbol{M}_2 - \frac{\beta}{2}\left[(\boldsymbol{M}_1\boldsymbol{n})^2 + (\boldsymbol{M}_2\boldsymbol{n})^2\right] - \beta'(\boldsymbol{M}_1\boldsymbol{n})(\boldsymbol{M}_2\boldsymbol{n}) - \boldsymbol{H}(\boldsymbol{M}_1 + \boldsymbol{M}_2).$$

Let magnetizations $\boldsymbol{M}_1$ and $\boldsymbol{M}_2$ have angle ϑ relative to the anisotropy axis $\boldsymbol{n}$ and magnetic field $\boldsymbol{H}$. Then the necessary part of free energy equals

$$w(\vartheta) = \left(I\cos 2\vartheta - \beta\cos^2\vartheta - \beta'\cos^2\vartheta\right)M_0^2 - 2HM_0\cos\vartheta$$

where M_0 is the magnetization magnitude for each of the sublattices.

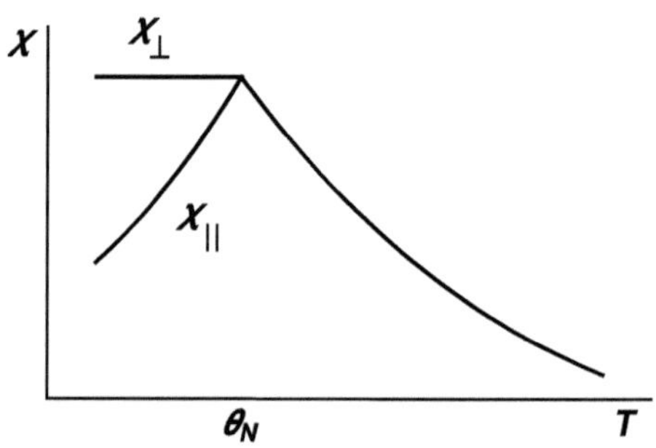

Fig. 9.8 The temperature behavior of magnetic susceptibility in an antiferromagnet. The susceptibility demonstrates an anisotropy in the behavior below the Néel temperature Θ_N

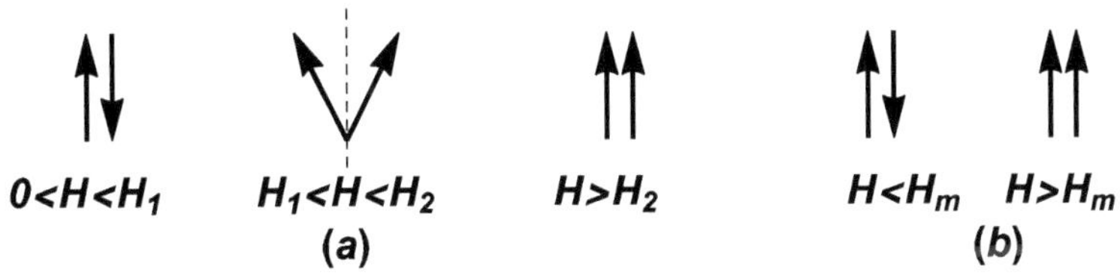

Fig. 9.9 The magnetization of the easy axis antiferromagnet as a function of external magnetic field parallel to the anisotropy axis. **a** The exchange interaction is large. **b** The exchange interaction is small

The qualitative behavior of magnetization is governed by the proportion between the exchange constant I and the anisotropy constants β and β'.
(a) For the predominant magnitude of exchange interaction $2I > \beta + \beta'$, we find two critical fields H_1 and H_2

$$H_1 = M_0\sqrt{(2I - \beta - \beta')(\beta - \beta')} \quad \text{and} \quad H_2 = 2I - \beta - \beta'.$$

Between these critical fields, the magnetization increases smoothly to the maximum magnitude (Fig. 9.9a).
(b) For the small magnitude of exchange interaction $2I < \beta + \beta'$, we disclose the *metamagnetic transition* at the critical field $H_m = I - \beta'$. The metamagnetic transition is specified with a drastic transition from the antiferromagnetic to ferromagnetic state with the maximum possible magnetization $2M_0$ (Fig. 9.9b).

9.9 Spin Waves in an Antiferromagnet

Let us consider small oscillations of magnetizations $\boldsymbol{M}_1(\boldsymbol{r}, t)$ and $\boldsymbol{M}_2(\boldsymbol{r}, t)$ for the both magnetic sublattices in an antiferromagnet. If the possible energy dissipation effects due to the magnetization oscillations are neglected, the equation of motion for each sublattice magnetization $\boldsymbol{M}_1$ and $\boldsymbol{M}_2$ will have the form of the Landau–Lifshitz equation for the magnetization precession in a ferromagnet

$$\frac{\partial \boldsymbol{M}_1(\boldsymbol{r}, t)}{\partial t} = \gamma \boldsymbol{M}_1(\boldsymbol{r}, t) \times \boldsymbol{H}_{1,\text{eff}}(\boldsymbol{r}, t),$$
$$\frac{\partial \boldsymbol{M}_2(\boldsymbol{r}, t)}{\partial t} = \gamma \boldsymbol{M}_2(\boldsymbol{r}, t) \times \boldsymbol{H}_{2,\text{eff}}(\boldsymbol{r}, t).$$

Here $\gamma = \mu/\hbar$ is the gyromagnetic ratio, $\boldsymbol{H}_{1,\text{eff}}$ and $\boldsymbol{H}_{2,\text{eff}}$ are the effective magnetic fields acting on $\boldsymbol{M}_1$ and $\boldsymbol{M}_2$. These fields are the variational derivatives of energy $\widetilde{W}$ of antiferromagnet with respect to magnetization

$$\boldsymbol{H}_{i,\text{eff}}(\boldsymbol{r}, t) = -\frac{\delta \widetilde{W}}{\delta \boldsymbol{M}_i(\boldsymbol{r}, t)} \quad (i = 1,\ 2) \quad \text{and} \quad \widetilde{W} = \int \widetilde{F}\, d^3r.$$

Involving the expression for energy $\widetilde{W}$ of antiferromagnet, we find

$$\boldsymbol{H}_{i,\mathrm{eff}} = \boldsymbol{H} - \frac{\partial \widetilde{F}}{\partial \boldsymbol{M}_i} + \nabla_k \left(\frac{\partial \widetilde{F}}{\partial (\nabla_k \boldsymbol{M}_i)} \right) \quad (i = 1,\ 2)$$

where $\boldsymbol{H}$ is the magnetic field inside the antiferromagnet and a sum over index $k = x, y, z$ is kept in mind. Next, we can determine the high-frequency susceptibility tensor of the antiferromagnetic, using the equations for the precession of magnetizations and knowing their equilibrium magnitudes. For this purpose, we put the magnetizations and magnetic field in both the equations of precession and the equations determining the magnetizations of sublattice ($i = 1,\ 2$), as follows:

$$\boldsymbol{M}_i(\boldsymbol{r}, t) = \boldsymbol{M}_{i0} + \boldsymbol{m}_i(\boldsymbol{r}, t), \quad \boldsymbol{H}(\boldsymbol{r}, t) = \boldsymbol{H}_0 + \boldsymbol{h}(\boldsymbol{r}, t).$$

Here $\boldsymbol{m}_1, \boldsymbol{m}_2$ and $\boldsymbol{h}$ are the small deviations from the equilibrium values and proportional to $\exp(i\boldsymbol{kr} - i\omega t)$. Then, it is necessary to linearize the equations and obtain a set of two linear differential equations for determining the small deviations $\boldsymbol{m}_1$ and $\boldsymbol{m}_2$. As usual, the linear system of equations is solved by means of the Fourier transformation. Expressing the Fourier transforms for deviations $\boldsymbol{m}_1$ and $\boldsymbol{m}_2$ via the Fourier transform of the alternating and perturbing magnetic field $\boldsymbol{h}$, we determine the total induced magnetization

$$\boldsymbol{m}(\boldsymbol{k}, \omega) = \boldsymbol{m}_1(\boldsymbol{k}, \omega) + \boldsymbol{m}_2(\boldsymbol{k}, \omega) = \hat{\chi}(\boldsymbol{k}, \omega)\boldsymbol{h}(\boldsymbol{k}, \omega),$$

$\hat{\chi}(\boldsymbol{k}, \omega)$ being the high-frequency susceptibility tensor of the antiferromagnet.

To simplify mathematics and clarify physics, we analyze below the easy axis-type antiferromagnet in the external magnetic field H_0 parallel to the anisotropy axis. The linearized equations on $\boldsymbol{m}_1(\boldsymbol{k}, \omega)$ and $\boldsymbol{m}_2(\boldsymbol{k}, \omega)$ are as follows:

$$-i\omega \boldsymbol{m}_1(\boldsymbol{k}, \omega) = \gamma\big(\boldsymbol{M}_{10} \times \big[\boldsymbol{h}(\boldsymbol{k}, \omega) - A_+ \boldsymbol{m}_1(\boldsymbol{k}, \omega) - B\boldsymbol{m}_2(\boldsymbol{k}, \omega)\big],$$
$$-i\omega \boldsymbol{m}_2(\boldsymbol{k}, \omega) = \gamma\big(\boldsymbol{M}_{20} \times \big[\boldsymbol{h}(\boldsymbol{k}, \omega) - A_- \boldsymbol{m}_2(\boldsymbol{k}, \omega) - B\boldsymbol{m}_1(\boldsymbol{k}, \omega)\big],$$
$$A_+ = I + \frac{H_0}{M_0} + \beta - \beta' + \alpha_{ij} k_i k_j, \quad A_- = I - \frac{H_0}{M_0} + \beta - \beta' + \alpha_{ij} k_i k_j,$$
$$\text{and} \quad B = I + \alpha'_{ij} k_i k_j.$$

Here $i, j = x, y, z$ and a sum over these indexes is implied. The alternating component of external field is $\boldsymbol{h}(\boldsymbol{k}, \omega)$. Taking the z-axis parallel to the anisotropy axis, we represent the solution of the above system as

$$\hat{\chi}(\boldsymbol{k}, \omega) = \begin{pmatrix} \chi_{xx} & \chi_{xy} & 0 \\ \chi_{yx} & \chi_{yy} & 0 \\ 0 & 0 & 0 \end{pmatrix}.$$

In zero external magnetic field $H_0 = 0$, the expressions for the tensor components simplify significantly

$$\chi_{xx} = \chi_{yy} = \frac{2}{A+B}\frac{\Omega^2}{\Omega^2-\omega^2}, \quad \chi_{xy} = -\chi_{yx} = 0,$$
$$A = A_+ = A_- = I + \beta - \beta' + \alpha_{ij}k_i k_j, \quad \Omega = \gamma M_0\sqrt{A^2 - B^2}.$$

In the finite external magnetic field $H_0 \neq 0$, the χ_{xy} and χ_{yx} components of the susceptibility tensor will be nonzero.

The functional dependence of resonant frequency Ω as a function of wave vector k is noticeably simplified since the exchange interaction, as a rule, exceeds significantly the relativistic ones, i.e. $I \gg |\beta|, |\beta'|$. Then keeping the long wave limit $\alpha_{ij}k_i k_j,\ \alpha'_{ij}k_i k_j \ll 1$ in mind and putting $A + B \approx 2I$, we get

$$\chi_{xx} = \chi_{yy} \approx \chi_0\frac{\Omega^2}{\Omega^2-\omega^2}, \quad \Omega = \gamma M_0\sqrt{2I}\sqrt{\beta - \beta' + (\alpha_{ij} - \alpha'_{ij})k_i k_j}$$

where $\chi_0 = I^{-1}$.

Often, for clarity and experimental description, it is common to represent the resonant frequency in a simplified form, neglecting a possible anisotropy in the magnetic stiffness coefficients α_{ij} and α'_{ij} (for simplicity, $\alpha_{ji} = \alpha\delta_{ij}$ and $\alpha'_{ij} = \alpha'\delta_{ij}$):

$$\Omega = \sqrt{\gamma^2 H_a^2 + (\Theta/\hbar)^2(ak)^2},$$
$$H_a = M_0\sqrt{2I(\beta - \beta')}, \quad \Theta = \mu M_0\sqrt{2I(\alpha - \alpha')}\,/a$$

where a is the interatomic distance and $\mu = \hbar\gamma$ is the effective Bohr magneton. In essence, this is the definition for two quantities H_a and Θ. The field H_a has the meaning of external magnetic field, which if applied along the anisotropy axis, will result in an instability of the antiferromagnetic ordering. The temperature Θ is of the order of the Néel one Θ_N since $\alpha, \alpha' \sim a^2\Theta_N/\mu M_0$. The difference $\alpha - \alpha'$ of the magnetic stiffness coefficients is supposed to be positive, otherwise the initial ground state of the antiferromagnet is qualified with a finite value of wave vector and would be inhomogeneous.

For greater clarity, in the approximation of large $A + B \approx 2I$ exchange interaction we give nonzero components of high-frequency susceptibility in the external magnetic field H_0 parallel to the anisotropy axis

$$\chi_{xx} = \chi_{yy} = \frac{\chi_0}{2}\Omega\left(\frac{\Omega_+}{\Omega_+^2-\omega^2} + \frac{\Omega_-}{\Omega_-^2-\omega^2}\right),$$
$$\chi_{xy} = -\chi_{yx} = i\omega\frac{\chi_0}{2}\left(\frac{\Omega}{\Omega_+^2-\omega^2} - \frac{\Omega}{\Omega_-^2-\omega^2}\right),$$
$$\Omega_\pm = \Omega \pm \gamma H_0 \quad \text{where} \quad \chi_0 = \frac{2}{A+B} \approx I^{-1}.$$

The knowledge of high-frequency susceptibility allows us to find the spin wave spectrum. For this purpose, we employ the general dispersion equation used already for determining the spin wave spectrum in ferromagnets

$$k^2 + k_i k_j \chi_{ij}(\boldsymbol{k}, \omega) = 0 \quad (i, j = x, y, z).$$

Taking into account that the components of tensor $\hat{\chi}(\boldsymbol{k}, \omega)$ are proportional to small parameter $\chi_0 \sim I^{-1} \ll 1$, we see that the spin wave frequencies coincide with the poles of tensor $\hat{\chi}(\boldsymbol{k}, \omega)$ within accuracy to the terms of the order of $\gamma M_0 \chi_0$. Hence we find the spin wave frequencies in the easy axis-type antiferromagnet

$$\omega_{1,2}(k) = \Omega_{\pm} = \sqrt{\gamma^2 H_a^2 + (\Theta/\hbar)^2 (ak)^2} \pm \gamma H_0 \quad (H_0 < H_a).$$

In the external magnetic field the spin wave spectrum splits into two independent branches. The appearance of two branches is a direct consequence of the larger number of degrees of freedom, namely two magnetic sublattices or two order parameters[6] instead of single one as in a ferromagnet. In the high external field $H_0 > H_a$ one of spin wave frequencies becomes negative, evidencing for the instability of the initial antiferromagnetic state with zero magnitization $\boldsymbol{M} = \boldsymbol{M}_{10} + \boldsymbol{M}_{20} = 0$.

As it concerns the uniform oscillations of magnetic moments in antiferromagnets of the limited sizes, we have seen that the components of tensor $\hat{\chi}(\boldsymbol{k}, \omega)$ are proportional to the small parameter $\chi_0 = I^{-1}$. Thus, the frequencies of *uniform antiferromagnetic resonance* at zero wave vector coincide approximately with the poles of susceptibility $\hat{\chi}(\boldsymbol{k}, \omega)$, i.e. with the spin wave frequencies at $k = 0$, and are practically independent of the shape of a magnet. This is in contrast to the frequencies of uniform ferromagnetic resonance dependent significantly on the shape of a ferromagnet. To conclude, the resonance frequencies will equal

$$\omega_{1,res} = \gamma(H_a + H_0) \quad \text{and} \quad \omega_{2,res} = \gamma(H_a - H_0)$$

for the easy axis antiferromagnet in the magnetic field parallel to the anisotropy axis. Underline that the measurement of resonance frequencies allows us to determine the magnitude of the anisotropy field H_a.

Let us turn now to studying the spin excitation effect on the thermodynamic functions of an antiferromagnet and calculating the spin wave contribution to the specific heat on the example of the easy axis antiferromagnet. So, we have for the specific heat

$$C_m(T) = \frac{\partial}{\partial T} \sum_{j=1,2} \int \frac{\hbar\omega_j(\boldsymbol{k})}{e^{\hbar\omega_j(\boldsymbol{k})/T} - 1} \frac{d^3k}{(2\pi)^3}.$$

[6] The total number of spin wave branches or modes is determined with the number of magnetic sublattices, i.e. with the number of magnetic atoms in the magnetic unit cell. Here we mention the same analogy between the number of atoms in the unit cell of a crystal and the number of branches in the phonon spectrum.

To simplify the calculation, we analyze the case of zero external magnetic field $H_0 = 0$. The spin wave energy $\varepsilon_k = \hbar\omega_k$ has a gap equal to μH_a. Unlike the case of ferromagnets, this energy is *linear* as a function of large wave vectors

$$\varepsilon_k = \sqrt{(\mu H_a)^2 + \Theta^2 (ka)^2} = \begin{cases} \mu H_a + (\Theta^2/2\mu H_a)(ka)^2, & ka \ll \mu H_a/\Theta, \\ \Theta(ka), & ka \gg \mu H_a/\Theta. \end{cases}$$

Substituting $\hbar\omega_k$ into the formula for the specific heat and taking an existence of two branches into account, we find the magnetic contribution to the specific heat of an antiferromagnetic

$$C_m(T) = \frac{1}{\pi^2 a^3}\left(\frac{T}{\Theta}\right)^3 \int\limits_{\xi}^{\infty} \frac{dx}{e^x - 1}\, \frac{x^2(4x^2 - 3\xi^2)}{\sqrt{x^2 - \xi^2}}, \quad \xi = \frac{\mu H_a}{T}.$$

We specify here two limiting cases

$$C_m(T) = \begin{cases} \frac{2}{(2\pi)^{3/2} a^3}\left(\frac{\mu H_a}{\Theta}\right)^3 \left(\frac{\mu H_a}{T}\right)^{1/2} \exp(-\frac{\mu H_a}{T}), & T \ll \mu H_a \ll \Theta, \\ \frac{4\pi^2}{15 a^3}\left(\frac{T}{\Theta}\right)^3, & \mu H_a \ll T \ll \Theta. \end{cases}$$

So, in the temperature $\Theta \sim \Theta_N \gg T \gg \mu H_a$ region the magnetic contribution to the specific heat of an antiferromagnet is proportional to T^3 and similar to that resulting from the phonon excitations. The condition $\Theta_N \gg \mu H_a$ is only satisfied in the antiferromagnets with the sufficiently high Néel temperature. For antiferromagnets with the Néel temperature of several tens of kelvins, the magnitudes Θ_N and μH_a become comparable. On the whole, this makes it difficult to observe the strict $C_m \sim T^3$ behavior.

As well as in ferromagnets, the spin wave excitations are responsible for the temperature-dependent contributions to the longitudinal $\chi_{\parallel}$ and transverse $\chi_{\perp}$ magnetic susceptibilities of antiferromagnets.

Problem

Determine the low temperature behavior of longitudinal static magnetic susceptibility $\chi_{\parallel}$ in the easy axis antiferromagnet due to spin wave effect.

Solution. Let the z-axis as well as magnetic field be parallel to the anisotropy axis. The corresponding component χ_{zz} or $\chi_{\parallel}$ of susceptibility tensor is determined with the derivative $\chi_{zz} = \partial M_z/\partial H_z$. Using the relation of magnetization M_z with the thermodynamic potential $\Omega(T, H)$, we determine the temperature behavior of magnetic susceptibility with the aid of the following formula:

$$\Delta\chi_{zz}(T) = \chi_{zz}(T) - \chi_{zz}(0) = \frac{\partial M_z}{\partial H_z}\Big|_{H_z=0} = -\frac{\partial^2\Omega}{\partial H_z \partial H_z}\Big|_{H_z=0}$$
$$= \frac{\partial^2}{\partial H_z \partial H_z} T \sum_{j=1,2} \int \frac{d^3k}{(2\pi)^3} \ln\big(1 - e^{-\hbar\omega_j(k)/T}\big)\Big|_{H_z=0} =$$
$$= -\frac{\partial}{\partial H_z} \sum_{j=1,2} \int \frac{d^3k}{(2\pi)^3} \frac{\hbar\partial\omega_j(k)/\partial H_z}{e^{\hbar\omega_j(k)/T} - 1}\Big|_{H_z=0}.$$

Substituting the spectrum of spin waves in the magnetic field parallel to the anisotropy z-axis

$$\hbar\omega_{1,2}(k) = \sqrt{(\mu H_a)^2 + \Theta^2(ka)^2} \pm \mu H_z = \hbar\omega_0(k) \pm \mu H_z,$$

we get

$$\Delta\chi_\parallel = \Delta\chi_{zz}(T) = 2\mu^2 \int \frac{d^3k}{(2\pi)^3} \frac{1}{4T\sinh^2(\hbar\omega_0(k)/2T)} =$$
$$= \begin{cases} \frac{2}{(2\pi)^{3/2}} \frac{\mu^2}{a^3\Theta} \left(\frac{\mu H_a}{\Theta}\right)^{3/2} \left(\frac{T}{\Theta}\right)^{1/2} \exp\left(-\frac{\mu H_a}{T}\right), & T \ll \mu H_a, \\ \frac{1}{3}\frac{\mu^2}{a^3}\frac{T^2}{\Theta^3}, & \mu H_a \ll T \ll \Theta. \end{cases}$$

The similar temperature behavior takes place for the transverse susceptibility $\Delta\chi_\perp(T)$.

9.10 Weak Ferromagnetism. The Dzyaloshinskii-Moriya Interaction

There is an antiferromagnet composed of two magnetic sublattices whose magnetic moments lie in the plane normal to the spatial symmetry axis $\boldsymbol{n}$ of the antiferromagnet. Let us imagine that one magnetic sublattice does not cross over into the other under rotations around the symmetry axis $\boldsymbol{n}$. In such situation, the free energy of the antiferromagnet admits the presence of the following term:

$$w_d = d[\boldsymbol{M}_1 \times \boldsymbol{M}_2] \cdot \boldsymbol{n} = d\left[\frac{\boldsymbol{L}+\boldsymbol{M}}{2} \times \frac{\boldsymbol{M}-\boldsymbol{L}}{2}\right] \cdot \boldsymbol{n} = 2d\boldsymbol{n} \cdot [\boldsymbol{M} \times \boldsymbol{L}].$$

This term is invariant with respect to the transformation below. Along with swapping the sublattices $\boldsymbol{M}_1 \to \boldsymbol{M}_2$ the inversion $\boldsymbol{n} \to -\boldsymbol{n}$ is also performed relative to the center between two magnetic moments. Usually, this symmetry property takes place if the unit cell in a crystal has at least two different magnetic ions. The term of such symmetry in the free energy of a magnet is known as the *Dzyaloshinskii-Moriya interaction* or *antisymmetric exchange interaction* as well. The Dzyaloshinskii–Moriya interaction has a relativistic origin and, as a rule, its magnitude is small as compared with that of ordinary exchange interaction. The vector $d\boldsymbol{n}$ is called the *Dzyaloshinskii vector.*

Let us write down the free energy density of such antiferromagnet

$$F = 2d\boldsymbol{n}\cdot[\boldsymbol{M}\times\boldsymbol{L}] + \frac{a-I}{2}\boldsymbol{L}^2 + \frac{b}{16}\boldsymbol{L}^4 + \\ + \frac{a+I}{2}\boldsymbol{M}^2 + \frac{b}{4}\boldsymbol{M}^4 + \frac{b}{8}\boldsymbol{M}^2\boldsymbol{L}^2 + \frac{b}{4}(\boldsymbol{ML})^2.$$

Here the z-axis is parallel to the symmetry axis $\boldsymbol{n}$ and the x-axis is directed along the antiferromagnetic vector $\boldsymbol{L}$ so that $L_x = L_0$. Due to smallness of the Dzyaloshinskii–Moriya interaction as compared with the exchange one, i.e. $d \ll I$, we retain only the quadratic terms with magnetization $\boldsymbol{M}$ in the free energy density. As a result, the free energy density reads

$$F = -2dM_yL_x + \frac{a-I}{2}L_x^2 + \frac{b}{16}L_x^4 + \frac{a+I}{2}M^2 + \frac{b}{8}M^2L_x^2 + \frac{b}{4}M_x^2L_x^2.$$

Minimizing the free energy with respect to magnetization $\boldsymbol{M}$ yields

$$\frac{\partial F}{\partial M_y} = -2dL_x + (a+I+bL_x^2/4)M_y = 0,$$

$$\frac{\partial F}{\partial M_x} = (a+I+3bL_x^2/4)M_x = 0, \quad \frac{\partial F}{\partial M_z} = (a+I+bL_x^2/4)M_z = 0.$$

Hence we find the magnetization

$$M_y = (d/I)L_x = (d/I)L_0(T) \quad \text{and} \quad M_x = M_z = 0.$$

Thus, in a two-sublattice antiferromagnet the Dzyaloshinskii-Moriya interaction results in appearing the magnetization $\boldsymbol{M} \neq 0$ small to the extent of $d/I \ll 1$. The magnetization $\boldsymbol{M}$ is normal both to the symmetry axis and to the magnetization direction of the magnetic sublattices, i.e. perpendicular to the antiferromagnetic vector $\boldsymbol{L}_0$. Such type of magnetic ordering is referred to as *weak ferromagnetism* or *canted antiferromagnet*. For example, it is observed in carbonates $CoCO_3$, $MnCO_3$, and in hematite α-Fe_2O_3.

Due to inequality $d/I \ll 1$ we neglect the small quadratic terms in magnetization. Then the free energy can be approximated as

$$F \approx \left(-\frac{2d^2}{I} + (a-I)\right)\frac{L_x^2}{2} + \frac{b}{16}L_x^4.$$

Accordingly, the expressions for the temperature behavior L_x and M_y below the Néel temperature take the usual form typical for the self-consistent Landau theory of phase transitions

$$L_x(T) = 2\left(\frac{-a+I+2d^2/I}{b}\right)^{1/2} = \left(4a_0/b\right)^{1/2}\left(\Theta_N - T\right)^{1/2},$$

$$M_y(T) = (d/I)L_x(T) \sim \left(\Theta_N - T\right)^{1/2} \quad \text{and} \quad a = a(T) = I + a_0(T - \Theta_0).$$

Note that the involvement of the Dzyaloshinskii–Moriya interaction shifts somewhat the initial Néel temperature Θ_0 as

$$\Theta_N = \Theta_0 + \frac{2d^2}{Ia_0}.$$

The presence of the Dzyaloshinskii–Moriya interaction in the thermodynamic potential results in a number of specific features for the behavior of canted antiferromagnet in the external magnetic field near the Néel temperature Θ_N. In the external magnetic field H we must deal with the Gibbs free energy potential

$$\widetilde{F}(\boldsymbol{H}) = F - \boldsymbol{M}\boldsymbol{H} - \boldsymbol{H}^2/8\pi.$$

Varying over both magnetization and antiferromagnetic vectors yields two equations

$$\frac{\partial \widetilde{F}}{\partial \boldsymbol{M}} = -2d(\boldsymbol{n} \times \boldsymbol{L}) + (a+I)\boldsymbol{M} + \frac{b\boldsymbol{M}^2}{4}\boldsymbol{M} + \frac{b\boldsymbol{L}^2}{4}\boldsymbol{M} + \frac{b}{2}(\boldsymbol{M}\boldsymbol{L})\boldsymbol{L} - \boldsymbol{H} = 0,$$
$$\frac{\partial \widetilde{F}}{\partial \boldsymbol{L}} = 2d(\boldsymbol{n} \times \boldsymbol{M}) + (a-I)\boldsymbol{L} + \frac{b\boldsymbol{L}^2}{4}\boldsymbol{L} + \frac{b\boldsymbol{M}^2}{4}\boldsymbol{L} + \frac{b}{2}(\boldsymbol{M}\boldsymbol{L})\boldsymbol{M} = 0.$$

Assuming the weakness of external field H, smallness $d \ll I$ of the Dzyaloshinskii–Moriya interaction and, accordingly, smallness of magnetization, we retain only the first-order terms over $d\boldsymbol{L}$, $\boldsymbol{M}$, and $\boldsymbol{H}$ in the equations above. Then we have

$$-2d(\boldsymbol{n} \times \boldsymbol{L}) + (a + I + bL^2/4)\boldsymbol{M} + (b/2)(\boldsymbol{M}\boldsymbol{L})\boldsymbol{L} = \boldsymbol{H}.$$
$$2d(\boldsymbol{n} \times \boldsymbol{M}) + (a - I + bL^2/4)\boldsymbol{L} = 0.$$

From the second equation it follows that $\boldsymbol{n}\boldsymbol{L} = 0$ and $\boldsymbol{M}\boldsymbol{L} = 0$. Hence, antiferromagnetic vector $\boldsymbol{L}$ is in the plane normal to the symmetry axis $\boldsymbol{n}$ and, in its turn, the magnetization vector $\boldsymbol{M}$ is perpendicular to the antiferromagnetic vector. As a result, we obtain the following solution of equations:

$$\boldsymbol{M} = \frac{2d(\boldsymbol{n} \times \boldsymbol{L}) + \boldsymbol{H}}{a + I + bL^2/4}.$$

The antiferromagnetic vector is determined by

$$\left(a - I - \frac{4d^2}{a + I + bL^2/4} + \frac{bL^2}{4}\right)\boldsymbol{L} = -\frac{2d(\boldsymbol{n} \times \boldsymbol{H})}{a + I + bL^2/4}.$$

Taking the large magnitude of exchange constant I into account, we finally have

$$\boldsymbol{M} = \frac{2d(\boldsymbol{n} \times \boldsymbol{L}) + \boldsymbol{H}}{2I} \quad \text{and} \quad \left(a - I - \frac{4d^2}{2I} + \frac{bL^2}{4}\right)\boldsymbol{L} = -(d/I)(\boldsymbol{n} \times \boldsymbol{H}).$$

Recalling the relation between the thermodynamic potential parameters and the Néel temperature, we represent the last equation in the apparent form

$$\left[a_0(T - \Theta_N) + bL^2/4\right]\boldsymbol{L} = -(d/I)(\boldsymbol{n} \times \boldsymbol{H}).$$

First of all, we see that, unlike a usual antiferromagnet, the magnetic field $\boldsymbol{H}$ normal to the symmetry axis $\boldsymbol{n}$ is able to induce the antiferromagnetism in the paramagnetic state above the Néel temperature at $T > \Theta_N$. The antiferromagnetic vector $\boldsymbol{L}$ of finite magnitude proves to be normal to the both anisotropy axis and magnetic field. Thus, strictly speaking, in the magnetic field the antiferromagnet with the Dzyaloshinskii–Moriya effect has no distinction in the magnetic symmetry above and below the Néel temperature Θ_N.

Let us write the approximate solution of the cubic equation for the magnitude $L(T)$ in the various limits

$$L_H(T) \approx \begin{cases} \frac{(d/I)H}{a_0(T-\Theta_N)}, & T > \Theta_N, \\ \left(\frac{4(d/I)H}{b}\right)^{1/3}, & T = \Theta_N, \\ L_0(T) + \frac{(d/I)H}{2a_0(\Theta_N - T)}, & T < \Theta_N. \end{cases}$$

Indeed, the relatively low external magnetic field induces a noticeable antiferromagnetic ordering (Fig. 9.10). The substitution of the last formulas into the expression for the susceptibility shows that the magnetic susceptibility $\chi_\perp = \partial \boldsymbol{M}/\partial \boldsymbol{H}$ exhibits a drastic maximum near the Néel temperature (Fig. 9.11). In fact,

$$\frac{\chi_\perp}{\chi_0} = \begin{cases} 1 + \frac{2d^2}{a_0 I}\frac{1}{T-\Theta_N}, & T > \Theta_N, \\ 1 + \frac{2d^2}{a_0 I}\frac{1}{2(\Theta_N - T)}, & T < \Theta_N, \end{cases}$$

where $\chi_0 = 1/(2I)$ is the susceptibility in the lack of the Dzyaloshinskii–Moriya interaction. As it concerns the magnetic response $\chi_\parallel$ to the magnetic field in the direction of the anisotropy axis, the effect of the Dzyaloshinskii–Moriya interaction

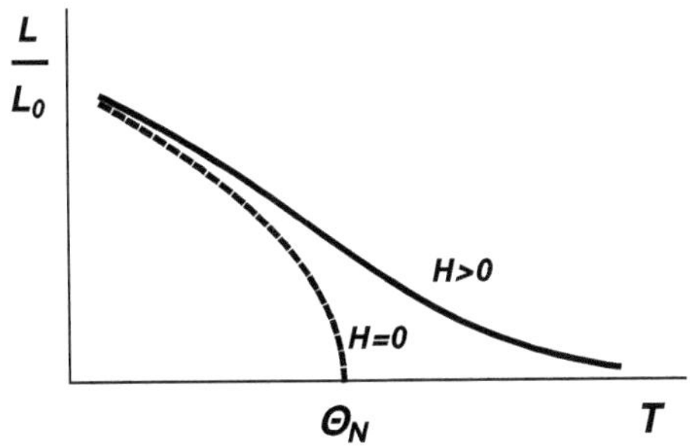

Fig. 9.10 The behavior of antiferromagnetic vector L in the magnetic field $\boldsymbol{H}$ normal to the symmetry axis $\boldsymbol{n}$. Vector is $\boldsymbol{L} \perp (\boldsymbol{n}, \boldsymbol{H})$. The dashed line corresponds to zero field $H = 0$

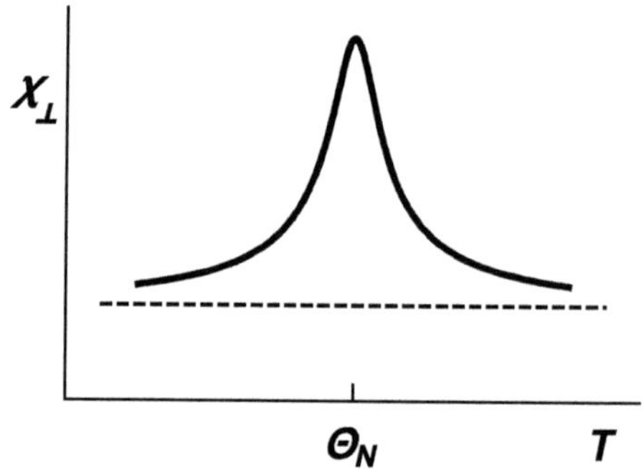

Fig. 9.11 The temperature behavior for the transverse magnetic susceptibility $\chi_\perp$ of weak ferromagnet in the close vicinity of the Néel temperature Θ_N. The dashed line shows the behavior $\chi_\perp$ if there is no Dzyloshinskii–Moriya interaction

proves to be insignificant and the magnetic susceptibility remains unchanged, i.e. $\chi_\parallel = \chi_0 = 1/(2I)$.

In conclusion, we note that, in addition to the combinations of ferromagnetic structures with the antiferromagnetic ones, the similar aggregations of the antiferromagnetic structures are also possible.

9.11 Helical Structures

The emergence of helical structures in a magnet is usually associated with the long-range nature of the exchange interaction between the spins. Here we examine some conditions for appearing such structures, using a simple example of a magnet with the uniaxial anisotropy which compels the magnetic moments to lie entirely in the basic planes normal to the anisotropy axis (Fig. 9.12). Let us imagine strong exchange interaction between the magnetic moments belonging to the same plane so that this

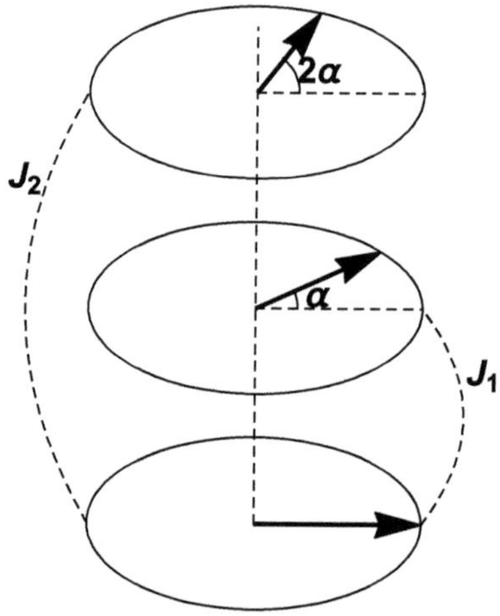

Fig. 9.12 On the emergence of antiferromagnetic helical structure. The coupling constants between the planes are J_1 and J_2

exchange interaction is responsible for the ferromagnetic ordering of the magnetic moments in each plane. The interaction of the magnetic moments from the different planes has also an exchange character but oscillates from one plane to another. So, we put the exchange coupling constant equal to $J_1 > 0$ between the magnetic moments from the nearest-neighbor basic planes and the exchange constant equal to $J_2 < 0$ between the magnetic moments from the next-nearest-neighbor planes. As a result, there may appear a helical antiferromagnetic structure due to competition between the coupling constants of ferromagnetic and antiferromagnetic signs.

Let us suppose that the direction of magnetic moments in each basic plane rotates by some angle α with the changeover to the neighbor plane. Then we write the exchange interaction energy per an atom between the magnetic moment layers as a function of the magnetic moment rotation angle α

$$W_{ex}(\alpha) = -(J_1 \cos\alpha + J_2 \cos 2\alpha).$$

Here α is the rotation angle of magnetic moment with the changeover to the neighbor plane and 2α is the rotation angle with the changeover to the next-nearest-neighbor plane.

The equilibrium value of angle is determined with condition $\partial W_{ex}/\partial\alpha = 0$ yielding the equation

$$J_1 \sin\alpha + 2J_2 \sin 2\alpha = 0.$$

This equation has two solutions

$$\sin\alpha = 0 \quad \text{and} \quad \cos\alpha = -\frac{J_1}{4J_2}.$$

The first solution exists for any relation between the exchange constants. For $J_1 > 0$, the energy minimum realizes at angle $\alpha = 0$ corresponding to the ordinary ferromagnetic ordering.

The second solution at $\alpha_0 = \arccos(-J_1/4J_2)$ corresponds to the helical antiferromagnetic ordering. It is possible only for the sufficiently slow decay $|J_2| \geqslant J_1/4$ of the interaction across the basic planes. Comparing the energies of the ferromagnetic and helical orderings

$$W_{ex}(\alpha_0) - W_{ex}(0) = \frac{(J_1 + 4J_2)^2}{8J_2} < 0,$$

we see that the helical structure will be energetically more favorable as compared with the ferromagnetic ordering. The helical period across the planes is $L = 2\pi d/\alpha_0$, d being the interplane spacing, proves to be *incommensurate* with the distance between the neighbor planes.

The appearance of the helical structure can be described with the aid of the following phenomenological expression for the inhomogeneity density term:

$$W_{\text{inh}} = \gamma \boldsymbol{M} \cdot \operatorname{curl} \boldsymbol{M} + \frac{\alpha}{2}(\nabla \boldsymbol{M})^2.$$

If $\gamma \neq 0$, the condition $\alpha > 0$ cannot provide us any stability of the homogeneous state with $\nabla \boldsymbol{M} = 0$. In fact, let we have simple spiral structure $\boldsymbol{M} = \{M \cos qz, M \sin qz, 0\}$ and its curl $\boldsymbol{M} = \{-qM \cos qz, -qM \sin qz, 0\}$. Then, the inhomogeneity energy depends on the spiral vector q and equals

$$W_{\text{inh}} = -\gamma q M^2 + \frac{\alpha}{2} q^2 M^2.$$

Hence the inhomogeneous state with nonzero spiral vector $q = \gamma/\alpha$ is energetically more favorable.

Problems

1. Find the magnetic susceptibility χ of antiferromagnetic helix in the low magnetic field H lying in the plane normal to the helix axis. The exchange constants between closest ferromagnetic layers are J_1 and J_2.

Solution. If the external magnetic field is applied in the direction normal to the anisotropy axis, the angle α_0 between the directions of magnetic moment in the nearest neighboring layers changes and depends now on the magnetic field strength. The energy of the helix in the magnetic field reads

$$W(H) = -\sum_n \big[J_1 \cos(\alpha_n - \alpha_{n-1}) + J_2 \cos(\alpha_n - \alpha_{n-2})\big] - \mu H \sum_n \cos \alpha_n$$

where μ is the magnitude of magnetic moment and α_n is the angle between the directions of the magnetic field and magnetic moment in nth layer. In the low magnetic field the helical structure will be distorted and the magnetic moments deviate from the ideal spiral position, rotating additionally by some angle in the magnetic field direction. Let us write the rotation angle α_n in nth layer as

$$\alpha_n = n\alpha_0 + \beta_n$$

where α_0 is the rotation angle in zero field $H = 0$ and $|\beta_n| \ll 1$.

We start from minimizing the energy $W(H)$ with respect to angle α_n. This yields the equation

$$J_1\big[\sin(\alpha_n - \alpha_{n-1}) - \sin(\alpha_{n+1} - \alpha_n)\big] + \\ + J_2\big[\sin(\alpha_n - \alpha_{n-2}) - \sin(\alpha_{n+2} - \alpha_n)\big] = -\mu H \sin \alpha_n\,.$$

Expanding the left-hand side of equation over small correction β_n and taking the smallness of magnetic field H on the right side of the equation, we obtain

$$J_1(2\beta_n - \beta_{n-1} - \beta_{n+1}) \cos \alpha_0 + J_2(2\beta_n - \beta_{n+2} - \beta_{n-2}) \cos 2\alpha_0 \approx -\mu H \sin(n\alpha_0).$$

We seek for the solution as

$$\beta_n = A \sin(n\alpha_0) \quad \text{where} \quad A \sim \mu H,$$

and arrive at the equation

$$A = -\frac{1}{4} \frac{\mu H}{J_1 \sin^2(\frac{\alpha_0}{2}) \cos \alpha_0 + J_2 \sin^2 \alpha_0 \cos 2\alpha_0}.$$

Next, we calculate the mean value of the magnetic moment projection onto the direction of magnetic field H

$$\langle\mu\rangle = \mu\langle\cos(n\alpha_0+\beta_n)\rangle \approx -\mu\langle\beta_n\sin(n\alpha_0)\rangle = -\mu A\langle\sin^2(n\alpha_0)\rangle = -\frac{1}{2}\mu A.$$

The magnetic susceptibility χ per one magnetic particle is given by

$$\chi = \frac{\partial\langle\mu\rangle}{\partial H} = \frac{\mu^2}{8}\frac{1}{J_1\sin^2(\frac{\alpha_0}{2})\cos\alpha_0 + J_2\sin^2\alpha_0\cos 2\alpha_0} \quad \text{and} \quad \beta_n = -\frac{2\chi}{\mu}H\sin(n\alpha_0).$$

2. Under conditions of the previous problem, find the critical magnetic field H_c when all magnetic moments in the layers are aligned in the direction of the magnetic field.

Solution. Near the critical magnetic field the magnetic moments will slightly be deviated from the magnetic field direction. Let α_n be angle of the magnetic moment in nth layer with respect to the magnetic field direction. All the angles α_n are extremely small near H_c, i.e. $|\alpha_n| \ll 1$. Linearizing the equation

$$J_1\big[\sin(\alpha_n-\alpha_{n-1}) - \sin(\alpha_{n+1}-\alpha_n)\big] + \\ + J_2\big[\sin(\alpha_n-\alpha_{n-2}) - \sin(\alpha_{n+2}-\alpha_n)\big] = -\mu H\sin\alpha_n$$

obtained in the previous problem over small $\alpha_n \to 0$, we have the following linear equation:

$$J_1(2\alpha_n-\alpha_{n-1}-\alpha_{n+1}) + J_2(\alpha_n-\alpha_{n+2}-\alpha_{n-2} = -\mu H\alpha_n.$$

Then, we seek for the solution as

$$\alpha_n = \zeta\sin(n\varphi), \quad \text{implying} \quad |\zeta| \ll 1.$$

And finally we have the relation

$$\mu H = \mu H(\varphi) = -2(J_1+J_2) + 2J_1\cos\varphi + 2J_2\cos 2\varphi = -2(J_1+J_2) - W_0(\varphi)$$

where $W_0(\varphi) = -2J_1\cos\varphi - 2J_2\cos 2\varphi$ is the energy of the system when φ is the angle between the magnetic moments of two nearest neighbor layers. The value $\mu H(\varphi)$ is maximum if energy $W_0(\varphi)$ is minimum, i.e. at the optimum angle $\varphi = \varphi_0$ determined by condition $\cos\varphi_0 = -J_1/4J_2$. So, in the magnetic fields exceeding the critical magnitude

$$H_c = -\frac{16J_2}{\mu}\sin^4\frac{\varphi_0}{2} = \frac{(J_1+4J_2)^2}{4\mu|J_2|},$$

any deviation of magnetic moments from the magnetic field direction is suppressed and the homogeneous ferromagntic ordering is established.

9.12 Quantum Theory of Spin Waves

To calculate the magnon dispersion law within the entire range of wave vectors beyond the long wave limit and to study both the thermodynamic and the kinetic properties of magnets, it is necessary to develop a microscopic theory for the physical processes in magnets. This requires more complicated and sophisticated mathematical methods.

If the magnetic moment of any particle in a magnet is disturbed from the equilibrium direction corresponding to the minimum energy, the relaxation process will represent the spin wave propagation as a rule. The energy of the wave equals that of magnet excitation. The energy of such elementary excitation can be connected with the frequency and wave vector of the wave as usual $\varepsilon_s(\boldsymbol{p}) = \hbar\omega_s(\boldsymbol{k})$ where $\boldsymbol{p} = \hbar\boldsymbol{k}$ is the momentum of elementary excitation. The elementary excitations in a magnet are called *magnons*. The types and properties of magnons depend essentially on the magnetic interactions and dimensionality of the magnetic system.

To study the magnon effect on the physical properties of magnets and describe the magnon-magnon coupling, we employ the method based on introducing the *magnon creation and annihilation operators* denoted as $b_{\boldsymbol{p}}^{+}$ and $b_{\boldsymbol{p}}$, respectively. Since the magnons obey the Bose statistics, the operators $b_{\boldsymbol{p}}^{+}$ and $b_{\boldsymbol{p}}$ should be determined on the analogy with the phonon ones. In other words, the operators should obey the Bose commutation relations

$$[b_{\boldsymbol{p}}, b_{\boldsymbol{p}'}^{+}] = \delta_{\boldsymbol{p},\boldsymbol{p}'}, \quad [b_{\boldsymbol{p}}, b_{\boldsymbol{p}'}] = 0 \quad \text{and} \quad [b_{\boldsymbol{p}}^{+}, b_{\boldsymbol{p}'}^{+}] = 0.$$

As we have seen above, it follows from these relations that operator $n_{\boldsymbol{p}} = b_{\boldsymbol{p}}^{+}b_{\boldsymbol{p}}$ has non-negative integer eigenvalues

$$b_{\boldsymbol{p}}^{+}b_{\boldsymbol{p}}\psi|..., n_{\boldsymbol{p}}, ...\rangle = n_{\boldsymbol{p}}\psi|..., n_{\boldsymbol{p}}, ...\rangle.$$

Here the eigenvector $\psi|..., n_{\boldsymbol{p}}, ...\rangle$ corresponds to the state having $n_{\boldsymbol{p}}$ magnons with momentum $\boldsymbol{p}$. The creation operator $b_{\boldsymbol{p}}^{+}$ acting on the state vector $\psi|..., n_{\boldsymbol{p}}, ...\rangle$ increases the number of magnons with momentum $\boldsymbol{p}$ by unity. The annihilation operator $b_{\boldsymbol{p}}$ decreases their number by unity according to

$$\begin{aligned} b_{\boldsymbol{p}}^{+}\psi|..., n_{\boldsymbol{p}}, ...\rangle &= \sqrt{n_{\boldsymbol{p}}+1}\,\psi|..., (n_{\boldsymbol{p}}+1), ...\rangle \\ b_{\boldsymbol{p}}\psi|..., n_{\boldsymbol{p}}, ...\rangle &= \sqrt{n_{\boldsymbol{p}}}\,\psi|..., (n_{\boldsymbol{p}}-1), ...\rangle. \end{aligned}$$

The next step is to express the magnetic Hamiltonian given in terms of spin or magnetic moment operators via the creation and annihilation magnon ones. As usual, the Hamiltonian of the magnetic system is written by means of the spin operators for the particles composing the magnetic material. Therefore, it is necessary to connect the spin operators $\hat{\boldsymbol{S}} = (\hat{S}^x, \hat{S}^y, \hat{S}^z)$ with the creation and annihilation magnon operators. Below we keep in mind a simplest Hamiltonian of exchange type to describe the isotropic ferromagnet[7] in the external magnetic field H parallel to the z-axis

$$\hat{\mathcal{H}} = -\frac{1}{2}\sum_{l\neq l'}[J_{ll'}(\hat{S}_l^x\hat{S}_{l'}^x + \hat{S}_l^y\hat{S}_{l'}^y + \hat{S}_l^z\hat{S}_{l'}^z] - \mu H\sum_l \hat{S}_l^z.$$

Here $J_{ll'} = J(\boldsymbol{R}_l - \boldsymbol{R}_{l'})$ is the exchange integral between the spins at sites $\boldsymbol{R}_l$, $\boldsymbol{R}_{l'}$, and μ is the magnetic moment of the spin S particle. For independent summation

[7] In general, the exchange constants as well as their signs may be different for the various spin components: $\hat{\mathcal{H}} = -\frac{1}{2}\sum(J^x\hat{S}^x\hat{S}^x + J^y\hat{S}^y\hat{S}^y + J^z\hat{S}^z\hat{S}^z)$, i.e. $J^x \neq J^y \neq J^z$.

over l and l', each pair of the spins occurs twice and, of course, $J_{ll'} = J_{l'l}$. The ferromagnetic state implies the inequality $J_{ll'} > 0$.

Before solving this problem, we recall the following properties of the circular $\hat{S}^{\pm} = \hat{S}^x \pm i\hat{S}^y$, and z-projection $\hat{S}^z$ operators of spin operator $\hat{\mathbf{S}}$:

$$\hat{S}^{\pm}|S_z>= \sqrt{S(S+1) - S_z(S_z \pm 1)}\,|S_z \pm 1> \quad \text{and} \quad \hat{S}^z|S_z>= S_z|S_z> .$$

Let $n = S - S_z$ be spin projection deviation from the maximum value S. Then we have for the above circular and z-projection operators

$$\hat{S}^{+}|n>= \sqrt{2S}\sqrt{1-\frac{n-1}{2S}}\,\sqrt{n}\,|n-1>,$$

$$\hat{S}^{-}|n>= \sqrt{2S}\sqrt{n+1}\sqrt{1-\frac{n}{2S}}\,|n+1>, \quad \hat{S}^z|n>= (S-n)|n> .$$

Introducing the Bose creation b^+ and annihilation b operators with the ordinary properties

$$b^+|n>= \sqrt{n+1}\,|n+1>, \quad b|n>= \sqrt{n}\,|n-1> \quad \text{and} \quad b^+b|n>= n\,|n>,$$

we finally arrive at the following realization[8] of the *Holstein–Primakoff transformation* for operators $\hat{S}^{\pm}$ and $\hat{S}^z$:

$$\hat{S}^{+} = \sqrt{2S}\sqrt{1-\frac{b^+b}{2S}}\,b, \quad \hat{S} = \sqrt{2S}b^+\sqrt{1-\frac{b^+b}{2S}}, \quad \hat{S}^z = S - b^+b.$$

Let us return to the initial Hamiltonian and perform the Holstein–Primakoff transformation for each spin at the corresponding site of the spin lattice. As a result, we arrive at the complicated form of the Hamiltonian

$$\hat{\mathcal{H}} = -\frac{1}{2}\sum_{l\neq l'} J_{ll'}\big[2Sb_l^+\left(1-\frac{b_l^+b_l}{2S}\right)^{1/2}\left(1-\frac{b_{l'}^+b_{l'}}{2S}\right)^{1/2} b_{l'} +$$

$$+S^2 - Sb_l^+b_l - Sb_{l'}^+b_{l'} + b_l^+b_lb_{l'}^+b_{l'}\big] - \mu H\sum_l (S - b_l^+b_l).$$

If we start to expand the square roots in powers $b_l^+b_l$, the Hamiltonian will represent the expansion as

$$\hat{\mathcal{H}} = \hat{\mathcal{H}}^{(0)} + \hat{\mathcal{H}}^{(2)} + \hat{\mathcal{H}}^{(3)} + \hat{\mathcal{H}}^{(4)} + \ldots$$

[8] From the formal viewpoint the eigenvalues $n = b^+b$ can run the values from 0 to ∞ and not only between 0 and $2S$. The difference is illusory since the transitions from states $n \leqslant 2S$ to those with $n > 2S$ have no place.

where

$$\hat{\mathcal{H}}^{(0)} = -\frac{1}{2}\sum_{l\neq l'} J_{ll'}S^2 - \mu H\sum_l S$$

has no operators b_l^+, b_l and corresponds to the energy of a magnet in the ground state, the spins being treated as classical variables. Then, $\hat{\mathcal{H}}^{(2)}$ is some quadratic form of these operators

$$\hat{\mathcal{H}}^{(2)} = -\frac{1}{2}\sum_{l\neq l'} J_{ll'}S\big(2b_l^+ b_{l'} - b_l^+ b_l - b_{l'}^+ b_{l'}\big) + \mu H\sum_l b_l^+ b_l,$$

and represents the Hamiltonian of non-interacting magnons. In our simplest case the third-order form $\hat{\mathcal{H}}^{(3)}$ vanishes identically. The fourth-order form $\hat{\mathcal{H}}^{(4)}$, equal to

$$\hat{\mathcal{H}}^{(4)} = -\frac{1}{2}\sum_{l\neq l'} J_{ll'}\left(b_l^+ b_l b_{l'}^+ b_{l'} - \frac{1}{2}b_l^+ b_l^+ b_l b_{l'} - \frac{1}{2}b_l^+ b_{l'}^+ b_{l'} b_{l'}\right) =$$

$$= -\frac{1}{2}\sum_{l\neq l'} J_{ll'}\left(b_l^+ b_l b_{l'}^+ b_{l'} - b_l^+ b_l^+ b_l b_{l'}\right),$$

will be responsible for the magnon-magnon interaction. For validity of our expansion, it is necessary to assume that the mean values of operator $b_l^+ b_l$ at the site are small as compared with $2S$, i.e. occupation numbers are $n_l = \langle b_l^+ b_l\rangle \ll 2S$. As a rule, this assumes the sufficiently low $T \ll J$ temperatures or large $S \gg 1$ spin limit.

In the expression for $\hat{\mathcal{H}}^{(2)}$ we go over from operators b_l^+ and b_l to their Fourier transforms in accordance with

$$b_l = \frac{1}{\sqrt{N}}\sum_{\boldsymbol{k}} b_{\boldsymbol{k}} e^{-i\boldsymbol{k}\boldsymbol{R}_l} \quad \text{and} \quad b_l^+ = \frac{1}{\sqrt{N}}\sum_{\boldsymbol{k}} b_{\boldsymbol{k}}^+ e^{i\boldsymbol{k}\boldsymbol{R}_l}$$

where $\boldsymbol{R}_l$ is the coordinate of the spin site and N is the number of sites. Then, we obtain

$$\hat{\mathcal{H}}^{(2)} = \sum_{\boldsymbol{k}}\big[\mu H + S\big(J(0) - J(\boldsymbol{k})\big)\big]b_{\boldsymbol{k}}^+ b_{\boldsymbol{k}} = \sum_{\boldsymbol{k}} \varepsilon_{\boldsymbol{k}} b_{\boldsymbol{k}}^+ b_{\boldsymbol{k}}.$$

We have used the following relations for the transformation:

$$\sum_l e^{i(\boldsymbol{k}-\boldsymbol{q})\boldsymbol{R}_l} = N\delta_{\boldsymbol{k},\boldsymbol{q}} \quad \text{and} \quad J_{ll'} = J(\boldsymbol{R}_l - \boldsymbol{R}_{l'}) = \sum_{\boldsymbol{q}} J(\boldsymbol{q})e^{i\boldsymbol{q}(\boldsymbol{R}_l - \boldsymbol{R}_{l'})}.$$

Let us discuss the result obtained. For the quantity

$$\varepsilon_{\boldsymbol{k}} = \mu H + S\big(J(0) - J(\boldsymbol{k})\big),$$

we must attribute the meaning of excitation energy for a magnon with wave vector $\boldsymbol{k}$ and frequency $\omega_k = \varepsilon_k/\hbar$. The operators b_k^+ and b_k will have a sense of the magnon creation and annihilation ones. The following excitation energy of a ferromagnet

$$E = \sum_k \varepsilon_k n_k \quad \text{where} \quad n_k = 0,\ 1,\ 2, \ldots$$

corresponds to Hamiltonian $\hat{\mathcal{H}}^{(2)}$.

Putting for simplicity that the spin positions correspond to the simple cubic lattice, we can use the following long wave expansion of the exchange integral:

$$J(\boldsymbol{k}) = \int J(\boldsymbol{r}) e^{i\boldsymbol{kr}} d^3r = \int J(\boldsymbol{r}) d^3r - \int (\boldsymbol{kr})^2 J(\boldsymbol{r}) d^3r + \cdots =$$
$$= J(0) - J_0(ka)^2 + \cdots \quad \text{where} \quad J_0 a^2 = \frac{1}{3}\int r^2 J(r) d^3r.$$

Here J_0 is the magnitude of the order of the exchange integral between the nearest neighbor spins separated with the distance equal to a. Accordingly, we have the quadratic dispersion for the magnon energy in the long wave approximation

$$\varepsilon_k \approx \mu H + S J_0 (ka)^2.$$

The presence of third-order expansion terms in Hamiltonian $\hat{\mathcal{H}}^{(3)}$ would describe the processes of merging two magnons and decaying a magnon into two other magnons. These processes would mean an existence of the magnon damping. The magnon-magnon interaction, resulted from the fourth-order terms in $\hat{\mathcal{H}}^{(4)}$, gives rise to the two-magnon scattering and, depending on a number of conditions, can result in emerging a two-magnon bound state. These effects can contribute to the thermodynamic variables of a magnet.

So far, studying the thermodynamic properties of magnets, we have believed that spin waves or magnons compose an ideal gas of non-interacting elementary excitations. To clarify the magnon-magnon interaction effect on the thermodynamics of a ferromagnet, we determine the correction to the thermodynamic potential, resulting from interaction energy $\hat{\mathcal{H}}^{(4)}$ in its first approximation. This correction is given by

$$\delta F = -T \ln(Z_0 + \delta Z) - (-T \ln Z_0) \approx -T\frac{\delta Z}{Z_0} = \frac{\operatorname{tr}(\hat{H}_1 e^{-\frac{\hat{H}_0}{T}})}{\operatorname{tr} e^{-\frac{\hat{H}_0}{T}}} = \langle \hat{H}_1 \rangle_0 .$$

Here the trace is taken over all the states of unperturbed Hamiltonian $\hat{H}_0$ and, therefore, the correction to the thermodynamic potential is simply the mean value of the perturbing Hamiltonian $\hat{H}_1$. So, we obtain the correction to the thermodynamic potential due to magnon-magnon interaction

$$\delta F_{\text{int}} = \frac{\operatorname{tr}\left(\hat{\mathcal{H}}^{(4)} e^{-\frac{\hat{\mathcal{H}}^{(0)}+\hat{\mathcal{H}}^{(2)}}{T}}\right)}{\operatorname{tr} e^{-\frac{\hat{\mathcal{H}}^{(0)}+\hat{\mathcal{H}}^{(2)}}{T}}} = \frac{\operatorname{tr}\left(\hat{\mathcal{H}}^{(4)} e^{-\frac{\hat{\mathcal{H}}^{(2)}}{T}}\right)}{\operatorname{tr} e^{-\frac{\hat{\mathcal{H}}^{(2)}}{T}}} = \langle \hat{\mathcal{H}}^{(4)} \rangle.$$

Since $\hat{\mathcal{H}}^{(0)}$ is a constant and thus it cancels. In essence, the thermodynamic averaging of Hamiltonian $\hat{\mathcal{H}}^{(4)}$ is performed with the Hamiltonian of magnons $\hat{\mathcal{H}}^{(2)}$. For the convenience of calculating the mean value $\langle \hat{\mathcal{H}}^{(4)} \rangle$, we rewrite it in terms of the Fourier transforms of creation and annihilation operators

$$\langle \hat{\mathcal{H}}^{(4)} \rangle = -\frac{1}{2} \sum_{\boldsymbol{q},\boldsymbol{k},\boldsymbol{k}'} J(\boldsymbol{q}) \big[\langle b^+_{\boldsymbol{k}-\boldsymbol{q}} b_{\boldsymbol{k}} b^+_{\boldsymbol{k}'+\boldsymbol{q}} b_{\boldsymbol{k}'} \rangle - \langle b^+_{\boldsymbol{k}} b^+_{\boldsymbol{k}'} b_{\boldsymbol{k}+\boldsymbol{k}'+\boldsymbol{q}} b_{-\boldsymbol{q}} \rangle \big].$$

The determination of the mean value $\langle \hat{\mathcal{H}}^{(4)} \rangle$ is the calculation of the trace. The necessary contribution is provided only by the diagonal matrix elements of operator $\hat{\mathcal{H}}^{(4)} \exp(-\hat{\mathcal{H}}^{(2)}/T)$. Nonzero contribution can only be delivered with the terms having the same number of creation and annihilation operators so that the magnon occupation numbers would remain the same after action of these operators. Thus, we have $\langle b^+_{\boldsymbol{k}} b^+_{\boldsymbol{k}'} \rangle = \langle b_{\boldsymbol{k}} b_{\boldsymbol{k}'} \rangle = 0$. The mean value $\langle b^+_{\boldsymbol{k}} b_{\boldsymbol{k}'} \rangle$ is nonzero only if the wave vectors $\boldsymbol{k}$ and $\boldsymbol{k}'$ are equal to each other. It is obvious that the mean value

$$\langle b^+_{\boldsymbol{k}} b_{\boldsymbol{k}} \rangle = n_{\boldsymbol{k}} = \frac{1}{e^{\varepsilon_{\boldsymbol{k}}/T} - 1},$$

is the equilibrium number of magnons in the state with wave vector $\boldsymbol{k}$ and energy $\varepsilon_{\boldsymbol{k}}$ at temperature T. Thus, the mean value is the Bose distribution $n(\varepsilon_{\boldsymbol{k}})$.

For the first term in $\hat{\mathcal{H}}^{(4)}$, such conditions can be fulfilled in two cases, namely at $\boldsymbol{q} = 0$ and at $\boldsymbol{q} = \boldsymbol{k} - \boldsymbol{k}'$. For the second term, these conditions are satisfied in two cases as well, namely at $\boldsymbol{q} = -\boldsymbol{k}$ and at $\boldsymbol{q} = -\boldsymbol{k}'$. As a result, we arrive at the mean value desired

$$\langle \hat{\mathcal{H}}^{(4)} \rangle = -\frac{1}{2} \sum_{\boldsymbol{k},\boldsymbol{k}'} \big[J(0) n_{\boldsymbol{k}} n_{\boldsymbol{k}'} + J(\boldsymbol{k} - \boldsymbol{k}') n_{\boldsymbol{k}} n_{\boldsymbol{k}'} - $$
$$ - J(-\boldsymbol{k}) n_{\boldsymbol{k}} n_{\boldsymbol{k}'} - J(-\boldsymbol{k}') n_{\boldsymbol{k}} n_{\boldsymbol{k}'} \big].$$

Thus, we obtain the correction to the thermodynamic potential due to magnon-magnon interaction

$$\delta F_{\text{int}} = \frac{1}{2} \int \frac{d^3k}{(2\pi)^3} \frac{d^3k'}{(2\pi)^3} n_{\boldsymbol{k}} U_{\text{int}}(\boldsymbol{k}, \boldsymbol{k}') n_{\boldsymbol{k}'}.$$

The sign of sum is replaced with the equivalent integration over wave vectors. Taking this symmetry $J(\boldsymbol{k}) = J(-\boldsymbol{k})$ of exchange integral into account, we represent the potential of magnon-magnon interaction $U_{\text{int}}(\boldsymbol{k}, \boldsymbol{k}')$ in the final form

$$U_{\text{int}}(\boldsymbol{k},\boldsymbol{k}') = J(\boldsymbol{k}) + J(\boldsymbol{k}') - J(\boldsymbol{k}-\boldsymbol{k}') - J(0).$$

Let us turn to estimating the integral in the low temperature region where, in essence, our correction is valid. At low temperatures the integral is gained by the region of small magnitudes of vectors $\boldsymbol{k}$ and $\boldsymbol{k}'$. Correspondingly, we can limit ourselves with expanding the energy $\varepsilon_{\boldsymbol{k}}$ and $J(\boldsymbol{k})$ over the powers of wave vector $\boldsymbol{k}$. Then the energy $\varepsilon_{\boldsymbol{k}}$ is approximated by the quadratic expression

$$\varepsilon_{\boldsymbol{k}} \approx \mu H + SJ_0(ka)^2.$$

The exchange integral $J(\boldsymbol{k})$ is an even function of vector $\boldsymbol{k}$. As we will see below, in expanding the exchange integral, it is necessary to involve fourth-order terms since the quadratic ones give zero contribution. We also, as above, assume the cubic symmetry in the spin arrangement. Then we have

$$J(\boldsymbol{k}) = J(0) - J_0a^2(k_x^2+k_y^2+k_z^2) + J_0\frac{b^4}{12}(k_x^4+k_y^4+k_z^4) + \cdots$$

(For the cube lattice symmetry of spin arrangement, we have $b = a$ where a is the size of the edge in the unit cell.) Accordingly, we arrive at the following expansion for the interaction energy:

$$U_{\text{int}}(\boldsymbol{k},\boldsymbol{k}') = 2J_0a^2\sum_{i=x,y,z} k_ik_i' + \frac{J_0b^4}{6}\sum_{i=x,y,z}\left(2k_ik_i'^3 - 3k_i^2k_i'^2 + 2k_ik_i'^3\right) + \cdots$$

On the substitution of this expression for the correction to thermodynamic potential δF_{int}, the odd terms in $\boldsymbol{k}$ or in $\boldsymbol{k}'$ give zero contribution due to averaging over the directions of vectors $\boldsymbol{k}$ and $\boldsymbol{k}'$. After averaging the last remaining term with $k_i^2k_i'^2$ over the directions, we should calculate the following integral expression:

$$\delta F_{\text{int}}(T,H) = -\frac{J_0b^4}{6}\int\frac{d^3k}{(2\pi)^3}\frac{d^3k'}{(2\pi)^3}n_{\boldsymbol{k}}k^2k'^2n_{\boldsymbol{k}'} =$$

$$= -\frac{J_0b^4}{6}\left(\int\frac{d^3k}{(2\pi)^3}k^2n_{\boldsymbol{k}}\right)^2 = -\frac{J_0b^4}{6}\left(\int_0^\infty\frac{k^2dk}{2\pi^2}\frac{k^2}{\exp\left(\frac{\mu H+SJ_0(ka)^2}{T}\right)-1}\right)^2.$$

For zero external magnetic field $H = 0$, the integration results in the following correction:

$$\delta F_{\text{int}}(T) = -\frac{J_0b^4}{6}\frac{9\zeta^2(5/2)}{256\pi^3}\left(\frac{T}{SJ_0a^2}\right)^5.$$

Using relation $C = -T\partial^2F/\partial T^2$ yields the magnon interaction contribution to the specific heat of a ferromagnet

$$\delta C_{\text{int}} = \frac{2b^4}{Sa^2} 15\pi\zeta^2(5/2)\left(\frac{T}{4\pi S J_0 a^2}\right)^4 \sim \left(\frac{T}{\Theta}\right)^4,$$

Θ being the Curie temperature.

The correction to the magnetization in the low $\mu H \ll T$ magnetic field is given by the formula $\delta M_{\text{int}} = -\partial F_{\text{int}}/\partial H$ where

$$\delta F_{\text{int}}(T, H) = -\frac{J_0 b^4}{6}\left[\frac{3}{16\pi^3}\left(\frac{T}{SJ_0 a^2}\right)^{5/2}\left(\zeta(5/2) - \zeta(3/2)\frac{\mu H}{T} + \cdots\right)\right]^2.$$

We see that the magnon-magnon interaction leads to some additional decrease of magnetization with the temperature growth

$$\delta M_{\text{int}} = -\mu\frac{b^4}{Sa^2} 3\pi\zeta(5/2)\zeta(3/2)\left(\frac{T}{4\pi S J_0 a^2}\right)^4 \sim -\mu\left(\frac{T}{\Theta}\right)^4,$$

following the same T^4 law as for the specific heat. The relations are valid while $T \ll \Theta$.

In the high $\mu H \gg T$ magnetic field the corrections have an exponentially small character

$$\delta F_{\text{int}}(T, H) = -\frac{J_0 b^4}{6}\frac{9}{256\pi^3}\left(\frac{T}{SJ_0 a^2}\right)^5 e^{-2\mu H/T},$$

$$\delta C_{\text{int}} = \frac{b^4}{Sa^2} 6\pi\frac{(\mu H T)^2}{(4\pi S J_0 a^2)^4} e^{-2\mu H/T},$$

$$\delta M_{\text{int}} = -\mu\frac{3\pi b^4}{Sa^2}\left(\frac{T}{4\pi S J_0 a^2}\right)^4 e^{-2\mu H/T}.$$

In conclusion, the interaction of magnons delivers the corrections to the thermodynamic variables in higher T/Θ approximation. The main terms in the magnon contribution to the specific heat and magnetization follow the Bloch $T^{3/2}$ behavior law.

Problems

1. Express the circular operators $\hat{S}^+$ and $\hat{S}^-$ via the Bose creation b^+ and annihilation b operators if the z-projection operator $\hat{S}^z$ is defined as an increment to the minimum value of projection, i.e. $\hat{S}^z = -S + b^+ b$.

Solution. Solution is analogous to that discussed above

$$\hat{S}^+ = \sqrt{2S}\, b^+\sqrt{1 - \frac{b^+ b}{2S}}, \quad \hat{S}^- = \sqrt{2S}\sqrt{1 - \frac{b^+ b}{2S}}\, b.$$

2. Using the anticommutation relation $\{\sigma^+, \sigma^-\} = 1$ and property $(\sigma^+)^2 = 0$, $(\sigma^-)^2 = 0$ for the raising $\sigma^+ = (\sigma^x + i\sigma^y)/2$ and lowering $\sigma^- = (\sigma^x - i\sigma^y)/2$ operators of the Pauli spin-1/2 matrices, find the transformation converting the spin operators into the fermionic creation c^+ and annihilation c operators in the case of one-dimensional chain of spin-1/2 particles.

Solution. We can think about two spin projections as if it is *empty and occupied* fermionic states. For the spin at the same site l, we can engage the operators σ_l^+ and σ_l^- ($\sigma_l^z = 2\sigma_l^+\sigma_l^- - 1$) as initial Fermi operators since they obey the fermionic anticommutation relations

$$\{\sigma_l^+, \sigma_l^-\} = 1, \quad \{\sigma_l^+, \sigma_l^+\} = \{\sigma_l^-, \sigma_l^-\} = 0.$$

However, this is not the case for the spins at the different $l \neq l'$ sites because they commute

$$[\sigma_l^+, \sigma_{l'}^-] = [\sigma_l^+, \sigma_{l'}^+] = [\sigma_l^-, \sigma_{l'}^-] = 0.$$

To satisfy the necessary condition of anticommutating the operators at the different sites of the chain, we follow the *Jordan–Wigner transformation* and introduce the following operators for the site with the number l:

$$c_l^+ = \sigma_l^+ e^{i\varphi_l} \quad \text{and} \quad c_l = e^{-i\varphi_l}\sigma_l^- \quad \text{where phase } \varphi_l \text{ equals} \quad \varphi_l = \pi\sum_{k=1}^{l-1}\sigma_k^+\sigma_k^-,$$

i.e. this sum is taken over all the sites preceding the site l. Since $\sigma^+\sigma^- = (\sigma^z + 1)/2$, i.e. zero or unity, the exponential $e^{i\varphi_l}$ can also be represented as[9]

$$e^{\pm i\varphi_l} = e^{\pm i\pi\sum_{k=1}^{l-1}\sigma_k^+\sigma_k^-} = \prod_{k=1}^{l-1} e^{\pm i\pi\sigma_k^+\sigma_k^-} = \prod_{k=1}^{l-1}(1 - 2\sigma_k^+\sigma_k^-) = \prod_{k=1}^{l-1}(-\sigma_k^z).$$

It remains to check the necessary anticommutation relations $\{c_l^+, c_{l'}\} = \delta_{ll'}$ and $\{c_l^+, c_{l'}^+\} = \{c_l, c_{l'}\} = 0$.

These relations can readily be checked if we point out that operator $\exp(\pm i\pi\sigma_l^+\sigma_l^-)$, equal to $-\sigma_l^z$, anticommutes with operators σ_l^+ and σ_l^- at the same site of the chain but any two operators referred to different sites commute always with each other, i.e. $[\sigma_l^\pm, \sigma_{l'}^\pm] = 0$ at $l \neq l'$.

3. Find the magnon or spin wave spectrum in the one-dimensional Heisenberg XY model of spin-1/2 chain, using the Jordan–Wigner transformation. The chain is described with the Hamiltonian

$$\mathcal{H} = J\sum_l (S_l^x S_{l+1}^x + S_l^y S_{l+1}^y),$$

J being the exchange integral for the neighboring spins.

Solution. Let us rewrite the Hamiltonian in terms of operators σ_l^+ and σ_l^-

$$\mathcal{H} = \frac{J}{2}\sum_l (\sigma_l^+\sigma_{l+1}^- + \sigma_l^-\sigma_{l+1}^+).$$

Next, we transform the Hamiltionan, using the identity $c_l^+c_l = \sigma_l^+\sigma_l^-$ and the Jordan–Wigner transformation

$$\sigma_l^+ = e^{-i\pi\sum_{k=1}^{l-1}c_k^+c_k}\, c_l^+, \quad \sigma_l^- = e^{i\pi\sum_{k=1}^{l-1}c_k^+c_k}\, c_l$$

for which the following identities are valid:

$$\sigma_l^+\sigma_{l+1}^- = c_l^+ e^{i\pi c_l^+c_l} c_{l+1} = c_l^+c_{l+1},$$

$$\sigma_l^-\sigma_{l+1}^+ = c_l e^{-i\pi c_l^+c_l} c_{l+1}^+ = c_l e^{-i\pi + c_l c_l^+} = -c_l c_{l+1}^+ = c_{l+1}^+ c_l,$$

$$\sigma_l^+\sigma_{l+1}^+ = c_l^+c_{l+1}^+, \qquad \sigma_l^-\sigma_{l+1}^- = -c_l c_{l+1}.$$

[9] Use the expansion $e^{(i\pi\sigma^+\sigma^-)} = 1 + i\pi\sigma^+\sigma^- + \frac{1}{2!}(i\pi)^2(\sigma^+\sigma^-)^2 + \cdots$ and $(\sigma^+\sigma^-)^n = \sigma^+\sigma^-$.

Finally, we have the equivalent Hamiltonian

$$\widehat{H} = \frac{J}{2}\sum_l (c_l^+ c_{l+1} + c_{l+1}^+ c_l)$$

which reduces the initial spin problem to that for the spinless non-interacting fermions with hopping between the nearest sites.

For the further, we go over to the Fourier representation with introducing the creation c_q^+ and annihilation c_q operators for the fermions with wave vector q

$$c_l = \frac{1}{\sqrt{N}}\sum_q c_q e^{iqx_l} \quad \text{and} \quad c_l^+ = \frac{1}{\sqrt{N}}\sum_q c_q^+ e^{-iqx_l}$$

where N is the number of sites and $x_l = la$ is the coordinate for the site of the chain with period a. As a result, we arrive at the Hamiltonian of spinless non-interacting fermions

$$\widehat{H} = \sum_q \varepsilon_q c_q^+ c_q, \quad \epsilon_q = J\cos qa.$$

Thus, relation $\epsilon_q = J\cos qa$ determines the fermionic states[10] and, correspondingly, the magnon energy spectrum is a periodic function of wave vector q with the reciprocal lattice period $2\pi/a$. Due to interaction the discrete levels of isolated spins broaden into the band of width $|J|$.

4. Estimate the temperature behavior of specific heat for the one-dimensional spin-1/2 chain.

Solution. Since the creation and annihilation operators of elementary excitations obey the fermionic commutation relations, the Fermi distribution function must be chosen as a distribution function of excitations. In addition, since the number of spin excitations is not fixed and may change, the chemical potential of excitations must be put equal to zero. Accordingly, at zero temperature in the ground state the fermionic states of negative energy $\varepsilon_q < 0$ are occupied but the positive energy states $\varepsilon_q > 0$ are free. In the ground state the magnon band proves to be exactly half-filled.

So, the specific heat of a free fermion gas is given by the formula

$$C(T) = \frac{1}{T^2}\sum_q \varepsilon_q^2 n_q(1-n_q), \quad n_q = \left[e^{\varepsilon_q/T}+1\right]^{-1}$$

where n_q is the occupation number for the state with energy ε_q. Next, we have

$$C(T) = \int_{-\pi/a}^{\pi/a} \frac{a\,dq}{2\pi}\frac{(J/T)^2\cos^2 qa}{4\cosh^2\left(\frac{J\cos qa}{2T}\right)} = \frac{K^2}{\pi}\int_0^{\pi} dx\frac{\cos^2 x}{\cosh^2(K\cos x)}, \quad K = \frac{J}{2T}.$$

For low $T \ll |J|$ temperatures, the specific heat $C(T) \approx \pi T/(3|J|)$ is a linear function of temperature. Such low temperature behavior is typical for fermions. As the temperature grows, the specific heat crosses the maximum at $T \sim |J|$ and then decays as $J^2/(8T^2)$ in the high $T \gg |J|$ temperature region.

In conclusion, we remind that the systems of one-dimensional acoustic phonons (bosons) and one-dimensional fermions with a fixed spin projection have also the same linear temperature behavior of specific heat in the low temperature region. In this sense these systems are isomorphic to each other.

---

[10] In the case of antiferromagnetic sign $J > 0$ the excited state are often called *spinons*.

Bibliography

1. L.D. Landau, E.M. Lifshitz, *Statistical Physics, Part I*, 3rd edn. (Pergamon, Oxford, 1980)
2. E.M. Lifshitz, L.P. Pitaevskii, *Statistical Physics, Part II* (Pergamon, Oxford, 1980)
3. E.M. Lifshitz, L.P. Pitaevskii, *Physical Kinetics* (Pergamon, Oxford, 1981)
4. L.D. Landau, E.M. Lifshitz, *Fluid Mechanics*, 2nd edn. (Pergamon, Oxford, 1987)
5. Yu.M. Belousov, S.N. Burmistrov, A.I. Ternov, *Problem Solving in Theoretical Physics* (Wiley-VCH, Weinheim, 2020)
6. A.Z. Patashinskii, V.L. Pokrovskii, *Fluctuation Theory of Phase Transitions* (Pergamon, Oxford, 1979)
7. J.M. Ziman, *Principles of the Theory of Solids* (Cambridge University Press, Cambridge, 1964)
8. A.A. Abrikosov, *Fundamentals of the Theory of Metals* (Groningen, North-Holland, Elsevier, 1980)
9. P.G. de Gennes, *Superconductivity of Metals and Alloys*, 2nd edn. (CRC Press, 1999) (Benjamin, New York, 1966)
10. C.J. Pethick, H. Smith, *Bose Einstein Condensation in Dilute Gases* (Cambridge University Press, Cambridge, 2002)
11. L.P. Pitaevskii, S. Stringari, *Bose-Einstein Condensation* (Oxford University Press, Oxford, 2003)
12. I.M. Khalatnikov, *An Introduction to the Theory of Superfluidity* (Westview Press, Colorado, 2000)
13. S.J. Putterman, *Superfluid Hydrodynamics* (North-Holland, Amsterdam, 1974)
14. A.I. Akhiezer, V.G. Bar'yakhtar, S.V. Peletminskii, *Spin Waves* (North-Holland, Amsterdam, 1968)
15. S.N. Burmistrov, *Physical Kinetics: Classical and Quantum Problems and Solutions* (Springer, Singapore, 2022)

S. N. Burmistrov, *Statistical and Condensed Matter Physics*,
https://doi.org/10.1007/978-981-97-9072-2

Index

S. N. Burmistrov, *Statistical and Condensed Matter Physics*,
https://doi.org/10.1007/978-981-97-9072-2

MIX
Papier aus verantwortungsvollen Quellen
Paper from responsible sources
FSC® C105338

If you have any concerns about our products,
you can contact us on
ProductSafety@springernature.com

In case Publisher is established outside the EU,
the EU authorized representative is:
Springer Nature Customer Service Center GmbH
Europaplatz 3, 69115 Heidelberg, Germany

Printed by Libri Plureos GmbH
in Hamburg, Germany